ESIPT Photochromism

ESIPT Photochromism
The Development of the Modern Views

$$OH \text{-----------} h\nu_1 \text{------------>} NH$$
$$OH \text{<-----------} h\nu_2 \text{ or } \Delta \text{------------} NH$$

Mikhail Knyazhanskiy

Copyright © 2018 by Mikhail Knyazhanskiy.

Library of Congress Control Number: 2018902392
ISBN: Hardcover   978-1-5434-7751-1
              Softcover   978-1-5434-7750-4
              eBook   978-1-5434-7749-8

All rights reserved. No part of this book may be reproduced or transmitted in any form or by any means, electronic or mechanical, including photocopying, recording, or by any information storage and retrieval system, without permission in writing from the copyright owner.

Any people depicted in stock imagery provided by Getty Images are models, and such images are being used for illustrative purposes only.
Certain stock imagery © Getty Images.

Print information available on the last page.

Rev. date: 10/10/2018

To order additional copies of this book, contact:
Xlibris
1-888-795-4274
www. Xlibris. com
Orders@Xlibris. com

Contents

Preface xv

Chapter 1
The Development Of The Conceptions On The Photochromism Of Schiff Bases(The Brief Historical Review) 1

    Conclusion 13
    References 15

Chapter 2
The Ground State Structural Transformations Of The Photochromic Anils... Solvato And Thermochromism 22

    Introduction 22
    2. 1. The topography of the Ground state PES for Salicylidene molecule. The main structures, and transition states. 23
    2. 2. The investigation of the OH and NH structures. 26
    2. 3. The OH <=>NH equilibrium, and the mechanism of the Ground State Intramolecular Proton Transfer (GSIPT) 31
    2. 4. The structure effect on the OH⇔NH equilibrium 33
        2. 4. 1. Effect of the substitutions in the phenyl rings. 33
        2. 4. 2. Effect of the ring π-electron system extension 41
        2. 4. 3. The role of the substituents in the imine bridge 47
    2. 5. The solvent effects. Solvatochromism. 52
    2. 6. The temperature effect on the OH⇔NH equilibrium. The problems of the thermochromism in the crystals and the solutions. 56
        2. 6. 1. The positive thermochromism (crystals). 57

  2. 6. 1. 1. The crucial role of the aldehyde moiety and the classification of the crystal structures. 57
  2. 6. 1. 2. Semiquantitative conception of the thermochromic process development. 67
 2. 6. 2. The Negative thermocromism –"cryochromism"*) (solution)..72
  2. 6. 2. 1. Review and discussion of the experimental findings. 72
  2. 6. 2. 2. The comparison of the Negative and Positive thermochromism natures, and the possible mechanisms of the Negative thermochromism. 80
Conclusions..85
References 87

Chapter 3
The Modern Ideas Of Photochromism And Accompanying Photoinduced Processes In Anil Molecules. 94

Introduction..94
3. 1. The characteristics and deactivation of the electronic excited states of imines of o-hydroxyaldehyde Enol structure. 95
 3. 1. 1. The excited states' deactivation and the structural transformations in the Enol form of Salicylideneanilines. 95
 3. 1. 2. Effect of the molecular structure and the medium on characteristics and deactivation of the excited state Enol form. 109
  3. 1. 2. 1. Influence of the substituents in the phenyl ring..109
  3. 1. 2. 2. The effect of the imine moiety structure. 115
  3. 1. 2. 3. Effect of the rings' conjugation. 131

- 3. 1. 2. 4. The influence of the substituents in the azomethine bridge. 134
- 3. 2. Kinetics and Mechanism of the ESIPT with generation of the fluorescent (NH)* structure... 137
  - 3. 2. 1. The general description of the phenomenon and methods of its study.. 137
  - 3. 2. 2. The ESIPT of the related structures.................... 144
  - 3. 2. 3. The study of the ESIPT of Salycilideneanilines..... 147
  - 3. 2. 4. The effect of the molecular structure on the ESIPT mechanism and dynamics. 172
    - 3. 2. 4. 1. The influence of the ring substituents (solvents and thermochromic crystals).... 172
    - 3. 2. 4. 2. The effect of the imine moiety structure on the ESIPT mechanism......................... 175
    - 3. 2. 4. 3. The effect of the ring conjugation......... 177
    - 3. 2. 4. 4. The effect of the substituents in azomethine bridge. 179
- 3. 3. The fluorescence and the excited state structural transformations initiated by the ESIPT. 188
  - 3. 3. 1. The nature of the fluorescent NH* state and the fluorescence with the Anomalous Stokes Shift. 188
  - 3. 3. 2. The molecular structure influence the ASS fluorescence 193
    - 3. 3. 2. 1. The effect of the ring substituents......... 193
    - 3. 3. 2. 2. The imine moiety structure influence the ASS fluorescence. 200
    - 3. 3. 2. 3. The effect of the ring conjugation on the ASS fluorescence.................................... 202
    - 3. 3. 2. 4. The influence of the substituents in the imine bridge. 205
  - 3. 3. 3. The development of the ideas about the post-ESIPT relaxation including the problems of the ASS fluorescence and precursor state 207

3. 4. The nature of the Photocolored product. 217
    3. 4. 1. The present ideas about structure and characteristics of photocolored form. 217
    3. 4. 2. The hypothesis of the twisted PCF structures generated via the "TICT-like" state. 228
3. 5. The generalized scheme of photochromism and side photoinduced processes of anils. 239
3. 6. Photochromism of Anils in the crystals. 244
    3. 6. 1. General information. 244
    3. 6. 2. The methods of preparation of the photochromic crystals. 246
        3. 6. 2. 1. Bulky group substitution method. 247
        3. 6. 2. 2. Clathrate crystal method. 247
        3. 6. 2. 3. Neighboring alkyl-group substitution method. 248
    3. 6. 3. The generalized view to the photo and thermochromism in the crystals. 249
Conclusion 257
References 260

Chapter 4
The Bichromophoric Systems On The Base Of The Anil Structures. 268

Introduction and classification of the structures. 268
4. 1. Homobichromophoric molecular systems 269
    4. 1. 1. The early studies of bieanils. 269
    4. 1. 2. The molecular systems with Aryl ring between imine moieties. 274
        4. 1. 2. 1 The structure and the Interfragmental ground state interaction of BSP molecule in solvents. 275

- 4. 1. 2. 2. The peculiarities of the photoinduced processes in the solvents. 277
- 4. 1. 2. 3. The photoinduced processes in the crystals. 289
- 4. 1. 3. The molecular systems with the saturated bridge between the imine moieties. 294
  - 4. 1. 3. 1. The structural and spectral peculiarities of the double enol form...............294
  - 4. 1. 3. 2. The nature of the ASS fluorescence and the PCF structures in the solvents...............297
  - 4. 1. 3. 3. The role of the interfragmental interactions, and the general schemes of the photoinduced processes in the solvents..299
  - 4. 1. 3. 4. The peculiarities of the photoinduced processes in the crystals. 306
- 4. 1. 4. Anil bichromophoric systems formed through the hydroxyl rings. 308
  - 4. 1. 4. 1. The structures with the separated anil moieties. 309
  - 4. 1. 4. 2. The double hydroxy structures with the common hydroxyphenyl ring...............309
  - 4. 1. 4. 3. The single hydroxy-structures with the common hydroxy-phenyl ring. 317
- 4. 2. Heterochromophoric molecular systems with an anil moiety. 318
  - 4. 2. 1. The bichromophors including the nonphotochromic structures. 318
    - 4. 2. 1. 1. Pyridinium cation with salicylideneaniline. 318
    - 4. 2. 1. 2. Pyrrole-salicylideneaniline bichromophoric system...............325
    - 4. 2. 1. 3. Schiff bases with the 4-aminoantipyrine structure. 329

4. 2. 2. The bichromophoric systems on the base on anil and spirocyclic structures. 329
Conclusion 332
References 335

Chapter 5
The Esipt Phpotochromism Of The Molecular Systems With The Nonanil Structures. 339

Introduction and classification of the structures. 339
5. 1. The ESIPT between the different heteroatoms (OH⇔NH). 341
    5. 1. 1. Oximes of orthohydroxyaldehydes. 341
    5. 1. 2. Photochromism of orthohydroxybenzylidenehydrazones. 343
    5. 1. 3. Orthohydroxyaryltriazines (HTrs). 346
        5. 1. 3. 1. Photoinduced structural transformations, generation, and decay of the final metastable photoproduct of HTrs. 346
        5. 1. 3. 2. The crucial role of the primary step character (ESIPT or ESIHT) in the photochromic reaction (Htrz vs SA). 356
    5. 1. 4. Photochromism of orthonitrobenzylidene derivative. 359
    5. 1. 5. Novel photochromic Dye based on the formation of the Hydrogen bond. 362
    5. 1. 6. 2-(2-Hydroxyphenyl)benzazoles, and photochromism of 2-(2-Hydroxyphenyl) benzthiazoles as a simulation of generation and structure of anil photocolored form. 364
5. 2. The structures with the ESIPT (ESIHT) between identical heteroatom. 367
    5. 2. 1. Oxygen-Oxygen (OH⇔OH) proton transfer. 367

5. 2. 1. 1. Nitrones of o-Hydroxyaldehydes, and their vinilogs. 367
5. 2. 1. 2. 2-hydroxychalcones as a "latent" (L) and prospective "true"(T) photochromic molecules. 371
5. 2. 2. The Nitrogen-Nitrogen (NH ⇔ NH)transfer. 376
5. 2. 2. 1. Phenoxazine derivatives. 376
5. 2. 2. 2. Spirans of perymidine series. 385
5. 2. 2. 3. Photochromic Metal-Dithizonate complexes. 391
5. 3. The Proton or Hydrogen transfer from Carbon atom to Heteroatoms...395
5. 3. 1. The H-transfer to Oxygen of Carbonyl group. 395
5. 3. 2. The H-transfer to Oxygen of Nitro-group. 397
5. 3. 3 The ESIHT from Carbon atom to Nitrogen one (O-Alkyl Aromatic Imines). 398
5. 4. The double-step ESIPT(ESIHT) from C atom to both O and N ones. The derivatives of 2-(2', 4'-Dinitrobenzyl) pyridine (DNBP). 399
5. 4. 1. Structure and electronic spectra of the initial form in solvents and crystals. 404
5. 4. 2. The absorption and the structures of the photocolored form(PCF) in solvents. 406
5. 4. 2. 1. The long-lived transient (I). 407
5. 4. 2. 2. The very short-lived transient (II). 411
5. 4. 2. 3. The short-lived transient (III). 412
5. 4. 3. The kinetics of the generation, and the interplay of the colored forms in the solvents........................412
5. 4. 4. The kinetics and structures' interplay of the discoloration reaction in solvents. 416
5. 4. 5. Photochromism of the crystals. 419
5. 4. 5. 1. The structure and the absorption spectra of the colored forms. 419
5. 4. 5. 2. The dynamics of photocoloration in the crystals. 421

>   5. 4. 5. 3. The date of the dark and photo-
>   discoloration in the crystals....................427
>   5. 4. 6. The mechanism and the general structural-energetic
>   scheme of the photoinduced processes. 430
>   5. 4. 6. 1. About the nature and the intrinsic
>   mechanism of the ESIPT and the GSIPT
>   in the two-step PT reaction. 430
>   5. 4. 6. 2. The generation of the (OH) and
>   accompanying colored forms. 434
>   5. 4. 6. 3. The generation of the (NH) colored form.
> 437
>   5. 4. 6. 4. The reverse ground state bleaching
>   reactions 443
>   5. 4. 6. 5. The differences between the photochromic
>   processes in the solvents and the crystals.
> 451
> Conclusion and the general discussion..454
> References 459

Chapter 6
The General Information About Possible Application Of The Esipt Photochromism. 467

> Introduction and the principle requirements for the photochromic
>   systems. 467
> 6. 1. The materials for switching and information storage. 471
> 6. 2. Liquid crystalline properties, and photochromism............480
> 6. 3. Photochromic Langmuir-Blodgett (LB) films. 481
> 6. 4. The possible applications in the analytical, physical, and
>   biological chemistry..485
> 6. 5. Materials for photoprotection. 487
> 6. 6. The prospective Photochromic materials. 491
>   6. 6. 1. Photochromic polymers (intrinsic
>   photochromism)..491

6. 6. 2 Possible amorphous molecular materials. 493
6. 6. 3. Possible perspective nanoparticles. 497
      6. 6. 3. 1. Information of the probable semi-manufactures for anil nanomaterials. 497
      6. 6. 3. 2 Photoswitchers on the base of the fluorescent photochromic nanoparticle. 502
References 515

I
Preface

The Photochromism is the reversible change of the substance' color under irradiation of light and caused by visible absorption spectra change as a result of the change of the molecular structure.

The organic compounds which Photochromism is caused by the Excited State Proton Transfer (ESIPT) take the particular place among the considerable diversity of the organic photochromic compounds owing both to the special mechanisms(of the ESIPT, and of the back, Ground State Intramolecular Proton Transfer, the GSIPT) and to the peculiarities of the photochromic properties displayed also in the rigid media (including crystals) unlike the most of other organic photochromes.

The historically caused prevalent importance among the ESIPT photochromes belongs by right to Schiff Bases, imines of ortho-hydroxyaldehydes (anils).

Their photochromic properties have been discovered in the crystals for the first time among the organic compounds in the early twenty century, and the ESIPT is practically the first known mechanism of Photochromism for the organic compounds.

These molecules are the very advantageous objects of the study. Their peculiarities of the electronic structure combine well with the high mobility of the molecular structure.

It is manifested distinctly in the formation of the absorption, luminescent, and photochemical (especially photochromic) properties caused by the shift of the tautomeric equilibrium OH⇔NH towards NH structure in the ground and the excited states.

The peculiar interest is provoked by photochromism of the crystalline anils. It is a typical example of the topochemical control in the crystalline structures and provides the most important applications.

At the same time the high opportunities of the synthesis of the novel molecular structures from this class of compounds provide a comparatively easy variation of the structural factors responsible for the photochromic properties and widely utilized in the up-to-date investigations of the novel photochromic compounds especially in the crystals.

The above peculiarities of the structure, properties and their variability ensure the heightened interest to the study and applications of photochromic anils in comparison with another ESIPT photochromes.

Historically the circumstances are so formed that Salicilideneanilines (anils) have been studied the most carefully. That tendency is reflected by the contents of the book where the greatest part of the discussion is concerned photochromic anils and their derivatives.

Nevertheless the investigations of the ESIPT photochromism of the various molecular structures different from anils are also carried out for many years, and their results in some cases can compete well with the findings of anils.

Therefore the detailed review, discussion, and generalization of the results of such studies have been conducted in the monograph for the first time.

II

Thus the present book is mainly scientifically historical character and devoted to the versatile discussion of the practically all problems concerned the experimental and theoretic studies, development, and forming of the modern ideas about the mechanisms of the ESIPT photochromism and of its applications for all various compounds of such a type on the base of the results (including also author's ones) mainly of the two decades (1990-2010), and partially of the earlier fundamental data, with taking into account the author's experience of the researches in this field.

The six chapters are included in the book.

The chapter one is a brief historic review of the three periods of the development of the ideas about anil photochromism from the discovery of the phenomenon to the present time. The short discussion of the earliest scientific works having only historical meaning has been conducted.

The discussion with the detailed list of the classical works (Weizmann research institute, Israel) and the brief consideration of the following studies of the same period with the Russian works having both a historical and the principal meaning have been shortly presented.

The chapter two is devoted to the analysis of the anil molecule structures and of the OH⇔NH equilibrium in the Ground state both in the solvents and in the crystals and its dependence on temperature (thermochromism) and on the solvent nature (solvatochromism).

In the third chapter the detailed discussion of the anil photochromism nature is conducted. It is based on the results of the numerous investigations of the nature and kinetics of the ESIPT followed by the structural transformations in the excited state involving the transient and fluorescent structures.

The particular attention has been given to various notions of the generation and the structure of the Photocolored colored form.

The chapter four is devoted to the consideration and the discussion of the bichromophoric systems containing the anil structures.

The discussion is based on the classification of such systems and also on the conception of the difference in the interaction between their structural fragments.

In the chapter five for the first time the detailed review of the real and potential ESIPT photochromes different from anils has been carried out on the base of the empiric classification of about twenty structural types.

The particular attention has been given to the photochromic mechanism of the Dinitrobenzylpyridine (DNBP) derivatives which photochromism has been discovered in the early twenties of the last century and has been studied only recently.

At last the sixth chapter is devoted to the brief description and discussion of the probable applications of the ESIPT photochromism for the creation of the new photochromic materials for various purposes including the photoswitchers in the Nonlinear Optical (NLO) materials.

The opportunity of the constructing of the nanomaterials on the base of the ESIPT photochromes is also considered.

III

So long as the notions about the mechanisms of the photochromic processes are still not sufficiently determined at present, the alternative views have been exhibited in the discussions of the various problems, and the readers have opportunity to make own choice with help of the tabular data and the schemes of the photoinduced processes reflecting different views.

In the structures with the ESIPT photochromism the basic reactions (ESIPT, GSIPT) are usually accompanied by the various photoinduced processes including electron transfer, trans-cis isomerization, energy transfer, internal rotation and so forth.

Thus the study of the ESIPT photochromism is connected with the discussion of the many fundamental photochemical and photophysical processes that can be useful for illustration of the basic processes occurring under excitation of the organic molecules considered usually in the educational courses of Molecular Photochemistry*).

Thus the monograph can be recommended not only for the researchers in the field of Photochromy of the Organic compounds but also for the broad circle of Photochemists, including both beginners and also the post graduate students, specialized in the Organic Photochemistry.

In conclusion author would like express his thanks to former collaborators-Prof. Dr. M. Strjukov(now the director of Institute of Connection), Dr. A. Ljubarska, |Dr. Ja. Tymjanskiy|, Dr. A. Metelitsa(now the director of Sc. Research. Inst. of Phys. Org. Chem.), Dr. N. Makarova, and Dr. V. Bezugliy for their direct participation in the investigations and in the active discussion of results of experiments, and also Dr. S. Aldoshin (now the Academician of Russian Academy of Sciences) for carrying out of the X-ray structural studies and joint discussion of the crystalline Photochromic anils.

---

* See e. g. N. Turro, V. Ramamurthy, J. Scaiano "Modern Molecular Photochemistry of Organic Molecules "University Science Books" Sausalito, California (2009)

Author would like to express the particular gratitude to Academician of Russian Academy of Sciences, Professor V. Minkin for the immediate participation in the statements of the problems, discussion of the results, the support of the investigations, and for the initiative in the publication of this monograph.

For moral support and big help in publication of the monograph I am obliged very much to my dear women-the daughter Natalia Rachford and the granddaughter Alina Rachford.

For the creation of the conditions for my work, long-term support, and invaluable help in our life I am Grateful so much to my nearest friend and wife, Alina Shopen.

I infinitely grateful to the wonderful Doctor and the person, Surgeon Dr. Richard. Birhle (Indiana University), who gives to me my life, health and possibility for publication of the book.

Chapter 1

The Development Of The Conceptions On The Photochromism Of Schiff Bases (The Brief Historical Review)

As a matter of fact the history of the investigations of the photochromism caused by the Excited State Intramolecular Proton Transfer (ESIPT) comes to the discovery, the very initial and the following (including up-to-day) investigations of photochromic Schiff bases-imines of o-Hydroxybenzaldehydes, well-known as "anils".

Photochromism of anils has been discovered in the beginning of the last century that is long before the correct views at their molecular structure and its transformations could been formed. At the same time unlike anils almost the all another organic ESIPT-based photochromes were obtained and studied much more later when the principle notions of the ESIPT reactions and the following structural transformations have been already elaborated and developed sufficiently (see ch. 5) with an only exception of 2-(2', 4'-Dinitrobenzyl)pyridine derivatives (DNBP's) which photochromic properties have been studied and expslained correctly only in almost half of century after their discovery in 1925 (see sec. 5. 5).

Therefore the history of the anil photochromism in the making of the up-to-day notions of its nature is a very instructive example of the close connection between the development of the ideas and the progress of the theoretical and the experimental methods of the physico-chemical investigations. The principal ideas of the anil photochromism in the solvents and the crystals and their development based on the results obtained in the various periods are presented in several reviews of the last two decades /1-9/. One can imagine the three main typical outlined historical periods in the progress of the conceptions of the anil photochromism.

From the best of our knowledge the first, the least known period, begins from the publication of O. Anselmino(1907)/10/who discovered a color change of the crystalline powder of the two different crystalline modifications of salicylideneaniline (SA) initiated by light and reverted in dark. Some later the information of photochromism of the two related compounds, salicylic acid and ether has been reported /11/.

In the first systematic investigations (A. Senier at al /12-18/), the crystalline powders of the more thirty substituted SA compounds were studied with a naked eye. The photochromism has been observed only for some of them, and the various polymorphic modifications of the same anils were discovered which differed by ability for photochromism and thermochromism (color change with temperature increase). The upper temperature limits of the observed photocoloration and the relative rates of the discoloration for different powder samples have been determined with a naked eye.

The results of the above described observations and also a lack of the visible photocoloration in the liquid solvents and an impossibility of isolation of the photocolored form from the photochromic crystalline modification are the findings of principle which keep their significance now and bound up with the modern ideas about the photochromic reaction mechanism, photocolo-red product structure and the topochemical control of photochromic and thermochromic properties. On the base of the data obtained, the photochromism is explained by formation of the molecular aggregates within the crystalline packing without any changes of the molecular and crystal structures.

In the series of the studies, M. Padoa et al/19a-e/ with use of the comparative visual observation of the photochromic transformations in the crystals, the first and the second orders of photocoloration and dark bleaching reactions respectively have been determinated for the several photochromic anils. In some cases the photobleaching reactions has also been observed. The photochromic transformations were explained by the shift of the equilibrium monomer<= >polymer towards a hypothetic polymer structure in the crystal under irradiation.

Some later/20, 21/the series of the qualitatively studied compounds has been expanded, and an attempt to understand a connection between the photochromism and the photoelectric effect in the crystalline anils has been undertaken/22/.

In the same time the important but not explained data have been obtained by H. Stobbe/23/.

These findings point out the intramolecular nature of the anil photochromism that manifested by the anil molecules adsorbed on the surfaces and dissolved in the rigid rosins.

On the othe hand in more later studies S. Bhatnagar et al /24/ suggested that color change is not a result of any structural transformations or polymerization but that of formation of the molecular aggregates responsible for photocoloration according to Senier et al /16-18/.

Such conclusions have been made on the basis of the findings about the magnetic susceptibility that doesn't change with coloration and bleaching, and has the same values for the different colored crystalline modifications of salicylidene-b-naphthylimine.

V. De Gaouk and R. Le Fevre /25/ have failed to find out the crystal structure change with photocoloration, and also the change of the absorption spectra and dipole moments of the solutions under steady –state irradiation and after dissolution of the preliminary photocolorated crystalline powders.

The erroneous assumption about parallel arrangement of the planar molecules in the photochromic crystals has been made by the authors. In such a crystalline packing the intermolecular H atom transfer OH->NH between adjacent molecules with for-mation of the colored structure can be possible without modification of the crystal structure. In authors' view the lack of the photocoloration of the o-methoxy substituted molecules testifies also to just that mechanism. As it will be clear later the incorrectness of these conclusions about the photochromic mechanism were apparently a consequents of the lack of the necessary experimental equipment and the wrong notions of the peculiarities of the o-hydroxyazo-methine molecular structure.

Nevertheless just in this work for the first time the NH (keto) form have been suggested as the structure responsible for the coloration of the crystalline anils, and in our view the work /25/ is the significant stage in development of the ideas about mechanism of anils' photochromism. In the detailed work of G. Lindemann/26/ on the investigation of the kinetics of photochromic reaction in the crystalline powder samples of salicylidene –m-toluidine the quantum efficiency of the photocoloration (P>1), the first order of the back bleaching dark reaction and its activation energy (~25kcal/mol) have been determined. The qualitative interpretation of the photoreaction's mechanism in that work differs a little from that suggested in /25/.

In the same period the various assumptions were put forward about mechanisms of the anil's photochromic reaction on the base of the conception about the electronic tautomerism/27/ or the conformational transformations/28/without a consideration of the peculiarities of the anil molecule structure.

Thus, up to the first half of the sixties the mechanism of the photochromic reactions of anils was interpreted mainly on the base of the qualitative observations without taking into account the substantial factors connected with the molecular structure and with the specific intermolecular interactions in the crystalline phase exclusively. Nevertheless in this period the photochromic crystalline anils not only have been discovered, but it was also shown that photochromism is not their common property.

The exceptional role of the rigid medium (including the crystalline state) in the formation of the visually observed photocolored form and uselessness of the steady-state physicochemical methods for the study of the photocolored structure in the unrigged media including liquid solvents have been shown.

At last the reasonable hypothesis about the photocolored form nature had been proposed which anticipated at some degree the up-to-date notions.

The need of the taking into account of the structural peculiarities of SA's(scheme 1. 1) for interpretation of the photochromic reaction has become evident after the detection of the strong Intramolecular Hydrogen Bond (IHB)OH... N, stabilizing trans(about C=N)Enol (E)structure with the coplanar "aldehyde" ring (A), C=N group and six –member cycle with IHB /29/, and also after the classical works of A. Weller/30/in which the fluorescence with the Anomalous Stokes Shift (ASS flu) in the molecules with the strong IHB(OH...N)(salicylic acid and relative structures)has been explained as a result of the Excited State Intramolecular Proton Transfer(ESIPT).

The results of these studies have become the basis of the modern ideas of the connection between the structure and the fluorescent –photochromic properties of anils.

In a general these ideas have been expressed and based experimentally as a result of the ten year series of the classical studies of M. D. Cohen, Y. Hirshberg, and J. M. Schmidt(1957-1968) /31-36/ by which the second period of the investigation of the anils' photochromism has been started. From the present viewpoint the grounded experimentally principal ideas can be described in the following way(Scheme 1. 1).

1. With So—S^*_1 excitation (irradiation by 330-360nm) of the anil molecules in the Enol trans (around C=N bond)structure in various rigid media(low temperature glass-like solutions, paraffine or polymer matrices)at low concentrations excluding the marked interactions of the dissolved molecules, the ASSflu(v~10000cm⁻¹)arises and the metastable photocolored form (λ^{max}_{abs}= 480-500nm)is generated of the azomethine molecules only with the ortho-hydroxy aldehyde group forming the strong OH.. N IHB. Thus the intramolecular (but not intermolecular!)adiabatic process in the S*1 state-ESIPT (OH)*-->(NH)*, is the basis of both the ASS flu and the photochromism.

According to the supposition of the authors the photocolored form(PC) is also generated but not observed in the liquid solvents visually(like ASS flu) due to the very low stability of the PC ground state structure under such conditions. However it can be observed and registered by a naked eye and by steady –state spectral methods as a result of its stabilization in the rigid media especially at low temperature.

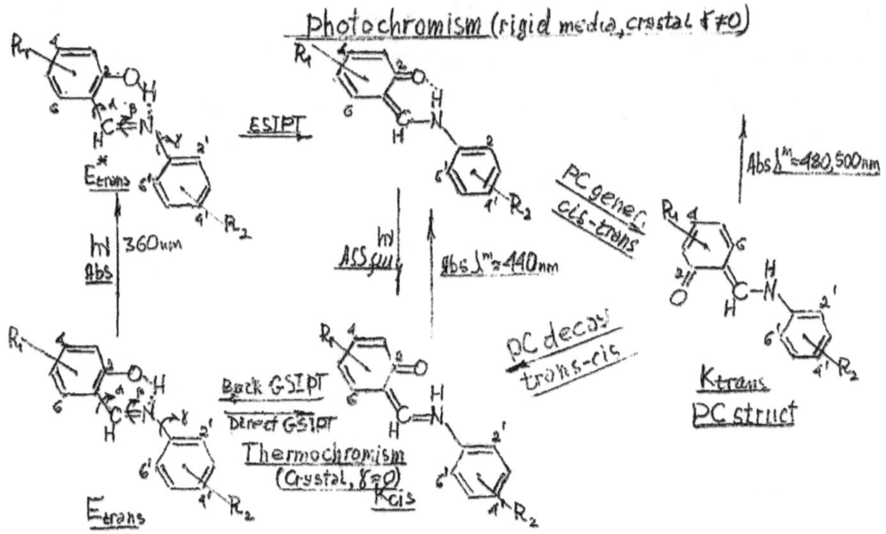

Scheme 1. 1
The general qualitative notions of the photo and thermochromism of Anils in the solvents and the crystals on the base of the findings /31-36/(see text).

The trans –keto(K-trans)structure with a respect to the C1=C7 bond with the broken H-bond NH…O is supposed for the photocolored form(Sch. 1. 1).

2. Anils which are photochromic in the rigid noncrystallic media have been divided on the two types in the crystalline state.

The crystal of the first type are characterized by a lack of photochromism and show the ASSflu. With the temperature increase the new absorption band arises(λ max~440nm)responsible for the coloration("the positive thermochromism"). Such a crystal has the specific crystalline packing that cannot provide a free volume sufficient for the considerable variations of the molecular conformation(chapter 2).

The photochromic nonluminescent at the room temperature crystalline systems belong to the second, less spread type of the crystalline structures. The absorption bands of the photocolored forms in the rigid solvents and the crystal are identical (λmax=480, 550nm). The molecules in the crystals have the planar structural fragment with the six-member cycle including IHB (OH.. N) but with the twisted aniline ring (~40-60 deg)(Scheme1. 1).

Latter, the irregular, more loose crystalline packing is created that has a free volume sufficient for the considerable variations of the molecular conformation with generation under irradiation of the metastable PC similar to that of the rigid solvents. This PC decays owing to the back dark reaction including the reverse Ground State Intramolecular Proton Transfer(GSIPT) like in the solutions.

The competition of the direct photo and the reverse dark reactions leads to the existence of the temperature limits for the observed photocoloration in the crystals.

Thus, according to above the thermo and photochromism in the crystals are caused by the intramolecular reactions including ESIPT and GSIPT. At the same time in spite of the realization of the ESIPT in the thermochromic crystals also the generation of the PC is inhibited, and on the contrary, the fluorescence and thermocoloration are promoted by the corresponding crystalline structures. Photochromism and thermochromism are mutually exclusive phenomenon and so called topochemical control of such reactions is manifested by such a way.

The numerous studies have been conducted by the various research groups on the base of the above ideas in the second period.

The interest for the new physical phenomena in the crystalline nails was shown right away after the basic investigations, and the several specifically studies of the piroelectric effect in the photochromic crystalline anils have been conducted in the sixties and seventies/38, 39/. This phenomenon can be useful not only by its direct utilization but it also clearly indicates the noncentro-symmetrical structure of the anil crystals and therefore their possible application as a nonlinear optical(NLO) media(/see 40/ and Capter6).

However the investigations of the photochromic and thermochromic properties of anils have conducted mainly by the steady-state spectroscopic and structural methods in the liquid, rigid amorphous and crystalline media at the various temperature and with use of the quantumchemical semi-empiric methods. The main results of many of these works have kept their importance up to present and have been discussed in detail in the reviews /1-4/.

Unfortunately the results of the series of the studies in the published in the Russian journals of this period and discussed in /2/, have not mentioned practically in /1, 3, 4/and therefore have remained as difficult of access for the broad groups of the scientists. However there are English full text translations of almost all articles for that period which are available in any scientific library.

At the same time one can think these results have not only historic interest but keep also a scientific significance and therefore they are worthy of a brief review in this chapter and will be discussed below (Ch 2-5) in case of need.

The anils' absorption and fluorescence have been studied as long ago as the early sixties/41/ and later/42/ by R. Nurmukhametov et al. In those studies the absorption and the fluorescence properties of the anil molecules have been investigated in the rigid glass-like solvents, and the ideas have been expressed which are similar to those of the studies /31-35/. In /41, 42/ the Zwitterionic (Zw) structure responsible for the ASSflu has been qualitatively based, and possibility of the analogous structure of the colored photoproduct was discussed also/42a, 43/.

For the first time in 1964/44a/and later/44b, c/the absorption and fluorescence of the bisazomethine compounds with the anil moieties linked directly or by the various bridge groups have been investigated in the liquid solvents(see also Ch. 4).

In the middle sixties the investigations of the photochromic and thermochromic anils have also been started in the Rostov State University. The influence of the structural factors and medium on the photochromism and fluorescence of N-aryl/45a, 45b/and N-alkyl /46-48/substituted anils has been studied for the large series of the structures by the steady-state absorption and fluorescent methods(Ch3).

The problems of the localization of the excitation and the photochromic transformations in connection with the structure of the photocolored form has been examinated on the base of the experimental and theoretic findings for the first time /45, 48, 52/(seeCh3).

The theoretical and experimental study of the keto⇔enol equilibrium in So state /49/, the physico-chemical characteristics and deactivation/50/and also the ionic structure role /51/in the excited states have been conducted in the seventies.

In the same period the photochromic and fluorescence properties of anils of o-Hydroxynaphthalidene /54/, o-mercapto/55/aldehydes, o-hydroxybenzylidene derivatives of oximes/2/ and hydrazones/2, 56/ anils of heterocyclic o-hydroxyaldehydes/2, 57/and also some model rigid structures of anils' analogues/58/ have been studied (see Ch. 2, 3).

In the early eighties the experimental and theoretical findings about photocolored product structures have been obtained/52/, and X-ray, and theoretical investigations of the crystalline structure and the intermolecular interactions in the crystal have been conducted for the known previously and the novel photochromic crystalline anils/53/. Slightly earlier for the first time the study of the photochromism and fluorescence of the polynuclear o-hydroxyazomethines and especially of "symmetrical" dianils have been carried out /59/(see Ch4).

The photoinduced processes including the intramolecular radiationless energy transfer in complex azomethine derivatives of pyridinium cations have also been investigated/60/(seeCh4). At last in the ivestigations of the metaloorganic and the various metal intracomplex compounds with the anil ligands the keto, ⇔ enol equilibrium in the ground state and photochromism have been discovered like the corresponding anil molecules /61a-d/that could shed light on the mechanism of the proton transfer in the latter(see Ch3).

As a result of the studies following the classical investigations next two decades, the extended range of anils and the simulated structures has been obtained to elucidate a mechanism of the photochromic transformations /1-3/.

As early as 1968 on the base of the spectral data the supposition about the twisted PC structure has been suggested by R. Becker and W. Richey/62/ unlike /31/, that conforms with the up-to-date notions(see Ch3).

The evolution of the topochemical principles for the control of the crystal properties has allowed synthesizing of the novel photochromic and thermochromic crystalline systems on the base of anils (see Ch3).

In the seventies-eighties the new photochromic crystalline anils have been synthesized and studied by E. Hadjoudis et al /63/, T. Kawato et al /64/ and M. Knyazhansky et al /53/. In the series of the studies the new crystalline thermochromes /65-68/, the novel thermophotochromic crystals /67/, and also the solutions with the unusual, "negative(cryo)" thermo-chromism, have been obtained by E. Hadjoudis et al/67, 68/.

In the middle eighties the successful attempts of the modification of the photochromic properties with study of their mechanism on the base of the findings of the novel structural analogs and simulated structures have been undertaken/69/(see Ch3).

The extraneous crystalline matrices (D. Higelin and H. Sixl, 1983 /70/), the Langmuir-Blodjett films (S. Kawamura et al. (1988)/71/ and the inclusion complexes with cyclodextrine (E. Hadjoudis et al. (1988)/72/) have been studied for the elucidation of the photochromic reaction mechanism and the perspectives for the applications. Latter, anils have been studied also as the potential components for the solar collectors /73/.

At the same time a lot of the key problems concerning the kinetics, mechanism and structures of the transient and the final forms of the photoreactions in the crystals and the solvent could not be solved by the experimental and theoretic means available in that period.

The investigations in such directions were conducted immediately after elaboration and utilization of the improved transient absorption, fluorescent and other spectral impulse excitation methods in the short time scales, and the development of the novel dynamic methods of the ab-initio quantumchemical calculations. The utilization of such methods signified the gradual transition toward the third, modern period in the development of the ideas about the nature and the mechanisms of the thermo-photochromism caused by the Proton transfer in the Ground and the Excited states.

Probably the first utilization of the low and middle (milli-microsecond time scales)time resolution transient absorption spectra in the solvents with variation of the solvent nature has been carry out in the middle sixties –early seventies by G. Wetermark et al/74/ and later by M. Ottolenghi et al/75/ for the study of the Photocolored structure decay.

These works(1967 and 1973) have started the combined experimental and quantum-chemical studies of the kinetics and mechanism of the photochromic reaction including the PC structure. The findings of the twisted PC structure are corroborated by the data/52, 63/.

The large series of the experiments with use of the time resolved absorption and fluorescence spectroscopy (nano-subpicosecond scales) in the solvents and the crystals have been started in the later seventies by R. Nakagaki etal/76/, R. Becker et al/77/ and E. Hadjoudis et al/65/ and continued in the middle eighties by P. Rentzepis et al/78/, J. Lewis, C. Sandorfi/79/and U. Grummit/80/ for the kinetic study of the transient structures.

The attempt of the direct spectra-structural study has been carried out in that period by the method of the Resonance Raman Spectroscopy for the elucidation of the PC structure by J. Ledbeter(1982)/81/, and H. Lee and T. Katagawa(1986)/82/(see also Ch3).

The high time resolved methods (till the picoseconds time scale) could be used with success for the study of the kinetics of the secondary post ESIPT process only involving generation and decay of the transient and the PC structures. However these methods have no a sufficiently high time resolution for the study of the very fast, primary, photoinduced structural transformation, caused by the ESIPT directly. Such ultrahigh time resolved methods of the <u>transient absorption spectroscopy (subfemto-femtosecond scale) have been elaborated and</u> used recently (the later 90th—10^{th}XXI) for the study of the ultrafast photoreactions including ESIPT with t(ESIPT))=several tens fs of many organic molecules with the IHB including photochromic anils (see Ch3).

At the same time one can believe that the development of the theoretical studies in the field of the molecular structure and reaction kinetic have led later to the highly constructive quantum-chemical ab-initio dynamic methods of the Density Functional Theory (DFT) and Time Dependent DFT(TDDFT)which results corroborate very good the recent kinetic experimental fin-dings for the ultrafast processes including the ESIPT reaction /83, 85/ unlike another quantumchemical ab-initio results/84/(see Ch3).

Thus the combined experimental and theoretical studies of the Anil photochromic structures can be prospective for the elucidation of the ESIPT photochromism nature.

Lately(through 2010)such studies are stimulated by the applications of anils as the photochromic switchers (modulators)for the Nonlinear Optical (NLO)devices /85/(see Ch6).

In the following chapters (Ch2-6) the results and the ideas of the last two decades (1990-2010) are discussed in detail in connection with the conceptions elaborated on the base of the classical studies of the sixties.

Conclusion

As it appears from above the history of the development of conceptions about anil photochromism can be divided on the three periods.

The first period has started from the discovery and qualitative discussion of the photochromism of the crystalline anils in the beginning of the last century.

The second one has begun in the sixties by the ascertainment of the photochromism nature connected with the ESIPT followed by the structural transformations in the solvents and the crystals. These studies have been a base for the numerous following, including also the recent, investigations.

At last the third period which is continued now is connected with the studies of the kinetics and the mechanisms of the photoinduced transformations including the very fast and ultrafast processes.

The development of the ideas about the nature, mechanism and kinetics of the ESIPT photochromism is closely connected with the progress of the physical and structural organic chemistry, spectral, kinetic, and the quantum-chemistry methods of organic chemistry.

Therefore the beginnig and the investigations of the second period became possible owing to the two-three decads' development of ideas about the intramolecular H-bond and the structural transformations, and also the methods of the molecular spectroscopy, the crystallochemistry and the theoretical methods.

The gradual development and the realization of the investigations of the third period are connected with the progress of the kinetic spectral and the structural methods of the high and ultra-high time resolution and the new quantum –chemical dynamic methods.

At present the investigation of the ESIPT photochromism, including photochromism of Anils is stimulated strongly by their possible applications especially as the Photochromic switches in NLO and other photoelectronic materials and also as nanomaterials. The search and the study of the novel photochromic systems with the new photoactive compounds including into the proper media by the improved experimental and theortical methods can be the next stage in the investigations of the ESIPT photochromism.

References

1. E. Hadjoudis, in Photocromism, Molecules and Systems, H. Durr and H. Bouas-Laurent(Eds), Elsevier, Amsterdam, p. 685 (1990).
2. M. Knyazhansky, and A. Metelitsa, Photoinduced Processes of Azomethine molecules and their structural analogs. Publish House of Rostov State University, Rostov-on-Don(1992)(in Russian).
3. E. Hadjoudis, Mol. Eng. 5, 301(1995).
4. Organic Photochromic and Thermochromic Comp. J. Crano, R. Guglielmetti Eds. V1. Main Photochromic Families, Plenum N. Y. (1999).
5. H. Bouas-Laurent, and H. Durr, Pure Appl. Chem. 73(4)639(2004).
6. E. Hadjoudis, and E. Mavridis, Chem. Soc. Rew. 33(9)579(2004).
7. M. Youn, in CRS Handbook of Org. Photochem. 2nd Ed. W. Harspool, F. Lenci(Ed.), CRS Press Boca Ranton, London, N. Y., Washington D. C., p 68(2004).
8. K. Animoto, and T. Kawato, Photochem. Photobiol. C. Photochem. Rev. 6., 207(2005).
9. E. Hadjoudis, S. Chatziefthimiou, and I. Mavridis, Carrent Org. Chem. 13, 269(2009).
10. O. Anselmino, Berichte, Bd. 40, 3465(1907).

11. W. Manchot, and J. Furlong, Berichte, Bd 42, 3030, 4383(1909).
12. A. Senier, and F Shepheard, J. Chem. Soc. 95, 441(1909).
13. A. Senier, and R. Clarke, J. Chem. Soc. 99, 2081(1911).
14. A. Senier, F. Shepheard, and R. Clarke, J. Chem. Soc. 101, 1950 (1912).
15. A. Senier, and R. Clarke, J. Chem. Soc. 105, 1917(1914).
16. A. Senier, and R. Forster, J. Chem. Soc. 105, 2462(1914).
17. A. Senier, and R. Forster, J. Chem. Soc. 1168(1915).
18. A. Senier, and A. Gallagher, J. Chem. Soc. 113, 28(1918).
19. M. Padoa et al., Atti R. Accad. dei Lincei(Roma), a)22 II, 500(1913), b)22 II, 576(1913), c)23 I, 95(1914), d)24 I, 828(1915), e)25 I, 808(1916).
20. P. Gallagher, Bull. Soc. Chim. France, (4)683(1921).
21. P. Gallagher, Bull. Soc. Chim. France, (4)961(1921).
22. C. Brewster, and L. Millan, J. Am. Chem. Soc. 55, 763(1963).
23. H. Stobbe, Ber. Verh. Sachs. Akad. Wiss. Leipzig Math-Phys Klass 74, 161 (1922).
24. S. Bhatnagar, P. Kapur, and M. Hashmi, J. Indian. Chem. Soc. 15, 513 (1938).
25. V. De Gaouk, R. J. W. Le Fevre, J. Chem. Soc. 1457(1939).
26. G. Lindemann, Zwiss. Photogr. 50, 347(1955).
27. C. Gheroghiu, Bull ecol polytech. Jassy, 2, 141 (1947).
28. N. Ebara, J. Chem. Soc. Jap. 82 (7), 941(1961).
29. a)N. Hendrics, O. Wulf, G. Hulbert, and U. Liddel, J. Am. Chem. Soc. 58, 1991(1936), b)C. Curran, and E. Chabit, J. Am. Chem. Soc. 69, 1134(1947), c)E. Bergman, E. Zimkin, and S. Pinchas, Rec. trav. Chim. 71, 168(1952).
30. 30. A. Weller, Naturwiss. 42, 175(1955), b)A. Weller, Z. Electrochem. 60, 1144(1956).
31. a)M. Cohen, Y. Hirshberg, and G. Schmidt, Bull. Res. Council Isr. 6a, 167(1957), b)M. Cohen, Y. Hirsberg, and G. Schmidt, Pure and Appl. Chem. 2, 83(1957).

32. 32. M. Cohen, Y. Hirshberg, and G. Schmidt, jn Hydrogen Bonding, D. Hadzi(Ed), London, N. Y., Paris Pergamon Press, 293(1959).
33. 33. M. Cohen, and G. Schmidt, in "Reactivity of Solids", Amsterdam, Elsevier, 560 (1961).
34. M. Cohen, and G. Schmidt, J. Phys. Chem. 66, 2442(1962).
35.)M. Cohen, G. Schmidt and S. Flavian, J. Chem. Soc. 2041(1964), b)M. Cohen, Y. Hirshberg, and G. Schmidt, J. Chem. Soc. 2051(1964), c)J. Bregman, L. Leizerovitz, and G. Schmidt, J. Cme. Soc. 2068 (1964), d)J. Bregman, L. Leizerovitz, and K. Osuki, J. Chem. Soc. 2086(1964), e) M. Cohen, Y. Hirshberg and G. Schmidt, J. Chem. Soc. 2060(1964).
36. a)M. Cohen, and S. Flavian, J. Chem. Soc. (B), 317(1967), b)M. Cohen, G. Schmidt, and L. Leizerovitz, J. Chem. Soc. (B)Phys. Org. (4), 329(1967) c)M. Cohen, S. Flavian, J. Chem. Soc. (B), Phys. Org. (4), 321 (1967), d)M. Cohen, and S. Flavian, J. Soc. Soc. (B), 334(1967), e)G. Schmidt, in "Reactivity of Photoexcited Org. Molecules" Interscience, London, 227 (1967).
37. M. Cohen, J. Chem. Soc. (4)373(1968).
38. B. Lang, M. Cohen, and F. Steckel, J. Appl. Phys. 36, 10, 3171(1965).
39. C. Hache, and H. Hancel, J. Phys. Chem. (DDR), 243, (1-2), 119(1971).
40. J. Delacre, and K. Nakatani in "Photochromism:Memories and Switches" Chem. Rev. 100(5) 1817 (2000).
41. a)R. Nurmukhametov, Yu. Kozlov, D. Shigorin, and V. Puchkov, Dokl. Akad. Nauk SSSR, 143, 5, 1145(1962), b)Yu. Kozlov, R. Nurmukhametov, D. Shigorin, and V. Puchkov, J. Fiz. Chim. 37, (11), 2432(1963), c)Yu. Kozlov, D. Shigorin, R. Nurmuzkhametov, and V. Puchkov in "Vodorodnaja Svjaz" ("Hydrogen Bond"), Moscow, Nauka, 223(1964)
42. a)O. Betin, R. Nurmukhametov, D. Shigorin, M. Loseva, and N. Chernova, Izv. AN SSSR, Fiz. 42(3), 533(1978). b)

R. Nurmukhametov, and V. Azarov, Optika i Spektr. 48(1), 30 (1980).

43. a)R. Nurmukhametov, O. Betin, and D. Shigorin, Dokl. Akad. Nauk SSSR 230(1)146(1976). b)A. Simonova, R. Nurmukhametov, and A. Prokhoda, Dokl. Akad. Nauk SSSR 230(4), 900(1976). c)O. Betin, R. Nurmukhametov, D. Shigorin, and N. Chernova, Dokl. Akad. Nauk SSSR, 227, 126, (1976).

44. a)B. Krasovitskii, V. Smeljakova, and R. Nurmukhametov, Optica i Spektr. 17, 558(1964), b)B. Krasovitskii, N. Mal'tseva, and R. Nurmukhametov, Ukr. Khim. Jurn. 31, 828(1965), c)N. Vasilenko, R. Nurmukhametov, and Ya. Pravednikov, Dokl. Akad. Nauk SSSR, 224, (6) 1334(1975).

45. a)O. Osipov, Yu. Zhdanov, M. Knyazhanskii, V. Minkin, A. Garnovskiii, and I. Sadekov, J. Fiz. Khim. 41(3), 641(1967). b)O. Osipov, Yu. Zhdanov, M. Knyazhanskii, V. Minkin and I. Sadekov in "Azometiny" , Publihed House of Rostov State Univer. 43 (1967).

46. M. Knyazhanskii, V. Minkin, and O. Osipov, J. Fiz. Khim. 41(3)649 (1967).

47. V. Minkin, M. Knyazhanskii, O. Osipov, Yu. Zhdanov, and V. Kurbatov, J. Fiz. Khim. 41(6)1383(1967).

48. Yu. Revinskii, M. Knyazhanskii, V. Minkin, and O. Osipov, J. Fiz. Khim. 48(4)933, (1974).

49. V. Minkin, B. Simkin, L. Olekhnovich, and M. Knyazhanskii, Teor. Eksper. Khim. 10(5), 668(1974)

50. a)V. Minkin, V. Bren', M. Knyazhanskii, andE. Malysheva, Reakt. sposob. Org. Soed. 5(4)978(1968). b)M. Knyazhanskii, M. Stjukov, and V. Minkin, Optika I Spektr. 35(5), 879(1972). c)M. Knyazhanskiy, M. Strjukov, V. Minkin, A. Ljubarskaja, and L. Olekhnovich, Izv. Akad. Nauk SSSR Fiz. 36(5), 1102

51. M. Strjukov, M. Knyazhanskii, V. Bren', V. Minkin, B. Simkin, and V. Usacheva, Teor. Eksp. Khim. 10(4), 520(1974).

52. M. Knyazhanskii, B. Simkin, B. Goljanskii, Teor. Eksp. Khim. 18(1), 108 (1982).

53. S. Aldoshin, M. Knyazhanskii, Ya. Tymjanskii, L. Atovmjan, and O. Djachenko, Khim. Fiz. 8, 1015(1982).
54. a)M. Knyazhanskii, O. Osipov, O. Asmaev, and Sheinker, J. Fiz. Khim. 42(2), 1017(1968). b)O. Asmaev, M. Knyazhanskii, V. Litvinov, O, Osipov, and Shejnker, J. Fiz. Khim. 46 (4), 902 (1972). c)V. Litvinov, MKnyazhanskii, O. Osipov, and V. Shejnker, J. Fiz. Khim. 47(6), 1366(1973).
55. V. Minkin, L. Olekhnovich, L. Nivorozhkin, Yu. Zhdanov, and M. Knyazhanskii, J. Org. Khim. 6, (2), 348 (1970)).
56. a)M. Strjukov, M. Knyazhanskii, O. Schipakina, T. Stul'neva, V. Bren', V. Minkin, and V. Orekhovskii, J. Fiz. Khim. 49, (11), 2924(1975). b)M. Strjukov, V. Orekhovskii, M. Knyazhanskii, B. Simkin, andV. Bren', J. Fiz. Khim. 52(4), 1075(1978).
57. a)M. Strjukov, M. Knyazhanskii, V. Bren', V. Minkin, and Zh. Bren', Opt. i Spektr. 35(6), 1051(1973). b)M. Strjukov, M. Knyazhanskii, V. Bren', V. Minkin, and B. Simkin, J. Fiz. Khim. 48(11), 2781(1974). c)M. Knyazhanskii, A. Ljubarskaja, and G. Paluj, Khim. Vysoc. Energ. 14(2), 130(1980).
58. V. Minkin, V. Shejnker, M. Knyazhanskii, and O. Osipov J. Fiz. Khim. 45(2), 221(1971).
59. a)B. Krasovitskii, O. Asmaev, M. Knyazhanskii, O. Osipov, N. Levchenko, V. Smeljakova, A. Nazarenko, N. Mal'tseva, and L. Afanasiadi, J. Fiz. Khim. 45(6), 1467(1971). b)M. Knyazhanskii, O. Asmaev, O. Osipov, and B. Krasovitskii, J. Fiz. Khim. 46(1), 178(1972). c)O. Asmaev, M. Knyazhanskii, B. Krasovitskii, and O. Osipov, J. Fiz. Khim. 46(3), 638(1972).
60. a)Ya. Tymjanskii, M. Knyazhanskii, Yu. Andrejchikov, G. Trukhan, and G. Dorofeenko, J. Prikl. Spektr. 25(2), 297(1976). b)M. Knyazhanskii, and Ya. Tymjanskii, Opt. Spektr. 43(2), 364(1977).
61. a)M. Knyazhanskii, P. Giljanovskii, V. Kogan, O. Osipov, O. Schipakina, and V. Litvinov, Opt. Spektr. 35, (6), 1083(1973).

b) M. Knyazhanskii, P. Giljanovskii, O. Schipakina, O. Osipov, and V. Kogan, J. Prikl. Spektr. 21(1), 183(1974). c)V. Minkin, L. Olekhnovich, M. Knyazhanskii, A. Ljubarskaja, J. Org. Khim. 10(4), 817(1974). d)P. Giljanovskii, M. Knyazhanskii, A. Burlov, V. Kogan, Yu. Revinskii, and V. Orekhovskii, Koord. Khim. 11(7), 889(1985).
62. W. Richey, and R. Becker, J. Chem. Phys. 49(5)2092(1968).
63. E. Hadjoudis, Mol. Cryst. Liq. Cryst. 13, 233(1971).
64. a)T. Kawato, H. Koyama, H. Kanatomi, and M. Issiki, J. Photochem. 28, 103(1985). b)T. Kawato, H. Ka-natomi, H. Koyama, and I. Igarashi, J. Photochem. 33, 199(1986).
65. E. Hadjoudis, I. Moustakly-Mavridis, and I. Xexakis, Isr. Journ. Chem. 18, 202(1979).
66. a)I. Moustakli-Mavridis, E. Hadjoudis and A. Mavridis, Acta Cryst. B34, 3709(1978). b)E. Hadjoudis, M. Vittorakis, and I. Moustakli-Mavridis, Mol. Cryst., Liq. Cryst. 137, 1(1986). c)E. Hadjoudis, J. Photoch 355(1981). d)I. Moustakali-Mavridis, E. Hadjoudis, and A. Mavridis, ActaCryst. B36, 1126
67. a)E. Hadjoudis, M. Vittorakis, and I. Moustakali-Mavridis, Chemotronics 1, 58 (1986). b)I. Moustakali-Mavridis, A. Terzis, and E. Hadjoudis, Acta Cryst. 43, 1389(1987). c) E. Hadjoudis, M. Vittorakis, and I. Moustakali-Mavridis, Tetrahedron, 43, 1345 (1987).
68. E. Hadjoudis, J. Photochem. 17, 335(1981).
69. a)V. Rybalkin, N. Makarova, V. Bren', A. Ljubarskaja, M. Knyazhanskii, and V. Minkin, J. Org. Khim. 22, 11, 2349(1986). (in Russian). b)M. Knyazhanskii, L. Sitkina, S. Aldoshin, A. Ljubarskaja, D. Dubonosov, O. Kozina, and L. Atovmjan, Khim. Fiz. 6(9), 1199(1987). (in Russian).
70. D. Higelin, and H. Sixl, Chem. Phys. 77, 391(1983).
71. S. Kawamura, T. Tsutsui, T. Saito, Y. Murao, and K. Khia, J, Am. Chem. Soc. 110, 509(1988).

72. E. Hadjoudis, P. Kondilis, I. Mavridis, and G. Tsoucaris, in "Advances in Inclusion Science", A. Hub-ler, J. Szejtli(Eds) Kluwer Academic Publisher, Dordrech, p. 119 (1988).
73. M. Cook, and A. Thomson, Chem. Ber. 20, 914 (1984).
74. a)G. Wettermark, and L. Dogliotti, J. Chem. Phys. 40, 1486(1964). b)A. Anderson, and G. Wettermark, J. Am. Chem. Soc. 87, 1433(1965).
75. a)M. Ottolenghi, and. D. McClure, J. Chem. Phys. 46, 4613(1967). b)T. Rosenfeld, M. Ottolenghi, and A. Meyer, Mol. Photochem. 5, 39(1973).
76. R. Nakagaki, T. Kabayashi, J. Nakamura, and S. Nagakura, Bull. Chem. Soc. Jpn. 50, 1909(1977).
77. R. Becker, C. Lenoble, and A. Zein, J. Phys. Chem. 91(13), 3509(1987).
78. P. Barbara, P. Rentzepis, and LBrus, J. Am. Chem. Soc. 102, 2786(1980).
79. J. Lewis, and C. Sandorfi, Can. J. Chem., 60, 1720(1982).
80. U. Grummiit, J. Prakt. Chemie, 327, 220(1985).
81. J. Ledbeter, J. Phys, Chem. 86, 2449(1982).
82. H. Lee, and T. Katagawa, Bull. Chem. Soc. Jpn. 59, 2897(1986).
83. J. Orbiz-Sanchez, R. Gelabert, M. Moreno, and J. Luch, J. Chem. Phys. 129, 214308(2008)
84. M. Zgierski, and A. Grabowska, J. Chem. Phys. 112, 14(2000).
85. M. Sliwa, N. Mouton, C. Ruckebuch, L. Poisson, A. Indrissi, S. Aloise, L. Porter, J. Dubous, O. Pozat, and G. Buntinx, Photochem. Photobiol. Sci. 9, 661 (2010). The articles /41-61/have been published in Russsian. There are translations on English.

Chapter 2

The Ground State Structural Transformations Of The Photochromic Anils... Solvato And Thermochromism.

Introduction

The OH and NH structures connected with the reversible Ground State Proton Transfer(GSIPT) reaction between the two reactional centers (O and N atoms) have the longwavelength absorption bands in the very different regions of the near UV-Vis spectra.

Therefore the OH⇔NH tautomeric equilibrium sensitivity to the external effects is the basis of the UV-Vis absorption spectra changes under influence of the media and the temperature (Solvato and Thermochromism).

The sensitivity of the equilibrium to the external factors depends of the energetic and the kinetic parameters of such an equilibrium and so it is a function of the molecular structure.

At the same time the position of the ground state OH⇔NH equilibrium in the molecular systems with the photochromic anil molecules determines their behavior in the excited state and therefore it is one of the principle factor that influences the system light's sensitivity to the change of the color (photochromism).

And so the study of the molecular structure of the OH and NH forms and its effect on the Ground state equilibrium under various conditions is the general problem connected with the anil's photochromism.

2. 1. The topography of the Ground state PES for Salicylidene molecule. The main structures, and transition states.

The topography of S_0PES for molecule SA has been studied in detail with help of the quantumchemical calculations (semiempirical PM3 method) /32/on the base of the gradient lines csonstruction.

The geometry and energy parameters of the thermodynamic stable structures and the transient states, the frequencies and the shapes of the active normal vibrations have been calculated (Sch. 2. 1 and 2. 2).

The OH(Enol-trans around C=N)structure (ESotr C=N) is the most stable one. The NH, Keto-cis (around C=C)structure (K SocisC=C) is less stable then ESotrC=N one (ΔH=8. 6kcal/mole). The reversible reaction of isomerization (GSIPT) ESotrC=N <=> KsocisC=C passes along the vector of the normal vibration $\sqrt{}$=3060 cm^{-1} via six center quasicyclic transient state T(EK) with the calculated value ΔH=27kcal/mole that is seems obviously overestimated and does not really exceed 10 kcal/mol. In any case it is evident the tautomeric equilibrium ESotr⇔KSocis under the common conditions is shifted completely towards E-side that consistents completely with experimental data.

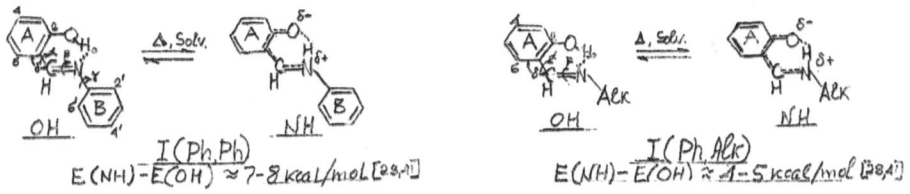

Scheme 2. 1
Tautomeric equilibrium as a result of the GSIPT
(GSIHT) for the molecules SA and SALK

Table 2.1
Selected Interatomic distances (Å) and selected angles(deg) in the molecules I(Ph,Ph) and I (Ph,Alk)
(see Sch.2.1)

Tautomer. Struct.	Distances and angles	SA (I Ph,Ph) Methods and References					Salk (IPh,Alk) Methods and References				
		Cal/8/	Cal/27/	Cal/39/	Cal/40/	Exp[1]/16/	Cal[2]/42/	Cal[3]/28/	Cal[1]/41/	Cal[3]/6/	Exp[4]/6/[A]
OH Struct.	C_1-C_2	1.403	1.41	----	1.403	1.384	----	1.396	1.395	1.408	1.377
	C_1-C_7	1.488	1.46	----	1465	1.453	----	1.464	----	1.467	1.448
	C_7-N	1.278	1.30	----	1.262	1.294	----	1.261	1.256	1.287	1.284
	C_2-O	1.375	1.35	----	1.331	1.347	----	1.352	1.332	1.367	1.369
	O-H	----	0.97	0.998	0.958	----	0.958	0.985	0.958	----	----
	N-H	----	1.83	1.721	1.896	----	1.688	1.789	1.891	----	----
	O-N	----	----	2.622	----	----	----	----	----	----	----
	α	----	0	----	0	0	----	0	0	0	0
	δ	----	119	----	----	139/69a/	----	120	----	----	121.6
	γ	36/69a/	----	33.1	44,09	49	----	----	----	----	----
NH Struct	C_1-C_2	----	1.47	----	1.463	----	----	1.448	1.437	1.462	----
	C_1-C_7	----	1.39	----	1.374	----	----	1.384	----	1.389	----
	C_7-N	----	1.37	----	1.37	----	----	1.328	1.314	1.347	----
	C_2-O	----	1.24	----	1.24	----	----	1.311	1.228	1.252	----
	O-H	----	1.83	1.639	1.83	----	1.866	1.752	1.890	----	----
	N-H	----	1.02	1.053	1.02	----	1.009	1.020	1.005	----	----
	O-N	----	----	2.558	----	----	----	----	----	----	----
	α	----	0	----	0	----	----	0	0	0	----
	δ	----	121	----	----	----	----	----	----	116.5	----
	γ	----	----	7.4	28.81	----	----	----	----	----	----

1/X-ray struct. Analysis Alk: 2/ ,CH_2Ph ,3/CH_3 ,4/C_9H_{11} .

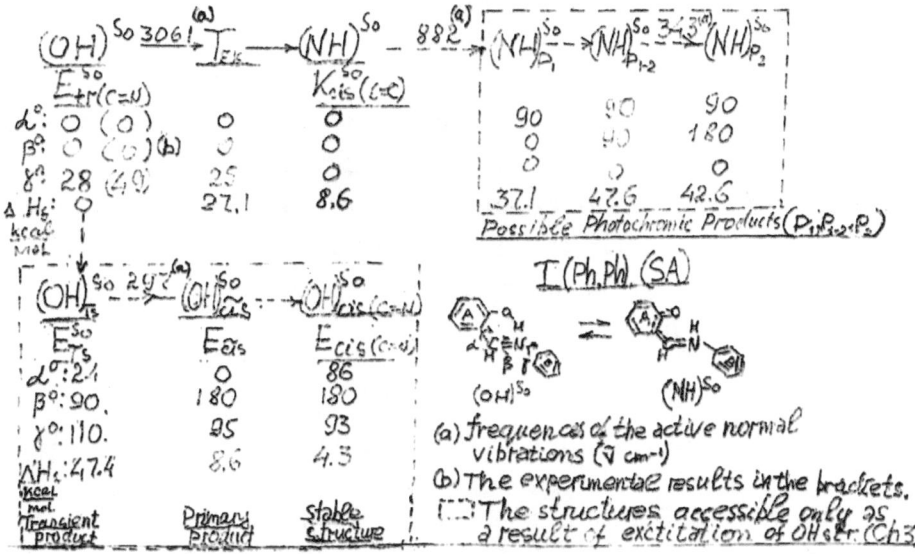

Scheme 2.2
The geometry and the energetic parameters of the stable and the transient structures in the ground state (calculated by PM3 method).

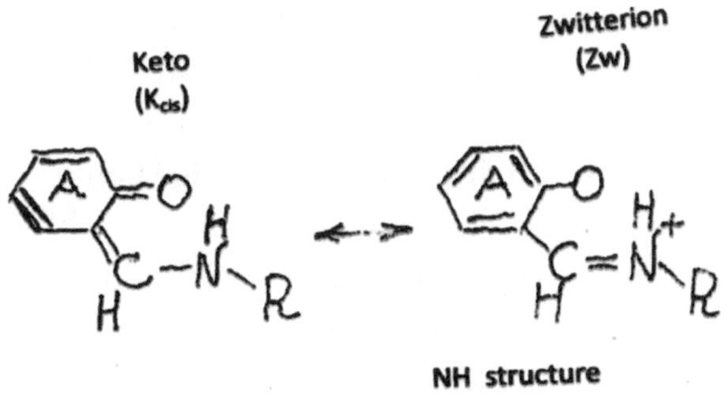

Keto (K_{cis}) ↔ **Zwitterion** (Zw)

NH structure

Scheme 2. 3
The general form of the NH structure with variation of the environment

It is obviously the metastable structures (NH)p1 and (NH)p2 with high energy are unattainable when moving along SoPES (see discussion about the photocolored structure, Ch3). On other hand the trans--cis isomerization about C=N bond (ESotrns<==>ESocis) can be realized as a result of the complicated moving (conversion + rotation) via the unstable nonplanar transition states (ESotr and ESocis) as for the related molecules (i. g. arylethylenes) (seeCh. 3).

Therefore the formation of ESocis structure when moving along SoPES is hardly possible in the condensed media and can be accessible only via excited state (see Ch. 3). Hence the SA molecule exists mainly in the OH form as ESotr structure in the gas phase, inactive media, and the crystal state. The change of the conditions and the molecular structure can lead to the equilibrium shift with the appreciable content of the NH structure.

2. 2. The investigation of the OH and NH structures.

The study of OH form for the simplest compounds I(Ph, Ph) and I(Ph, Alk) carried out mainly by the X-ray analysis, NMR, IR, and Raman spectroscopy and by the semiempiric and ab-initio quantum-chemical calculations (tables 2. 1 and 2. 2).

The data of Xray analysis and the quantum-chemical calculations of the structure parameters for the simplest structures without substituents have been obtained mainly in the works /6, 8, 16, 27, 32, 40, 69a/ for SA (IPh, Ph) and /6, 28, 41, 42/ for SALK I(Ph, Alk)*).

As it follows from the table 2. 1 the calculated bond lengths exceed a little the experimental findings with preservation of their correlation. The C=C bond in the ring A are markedly less in SALK than in SA. At the same time there is tendency towards lengthening of the N-H distance, CO and CN bonds in SALK as compared with SA. The aldehyde structural fragment (with ring A) is planar ($\alpha = 0$ deg) in both SA and SALK structures while N-Ph moiety in SA is twisted ($\gamma = 33\text{-}49$ deg).

The vibrational spectroscopy has been used in the works. The resonance Raman (RM) and IR spectra have been studied /1/ in the crystalline state and in the media with the various polarity. The bands' assignments have been carried out with help of 15N and deuterium

---

* The X-ray analysis is impossible for the simplest I(Ph, CH3) molecule which exists in the liquid phase at the room temperatures.

substitution and lead to the conclusion that the main bands correspond to the vibrational modes typical for the E-structure(table 2. 2.).

Authors /5/ came to the analogous conclusion when studying of the transient IR spectra in the solutions with use of the isotopic 15N and 18O substitutions. At the same time one of the typical vibrational modes($\sqrt{}$ =1577cm-1 Rm /1/, $\sqrt{}$= 1514cm-1 IR/5/) (table 2. 2) is interpreted differently. It is the most intensive in the Raman spectra under excitation in the longwavelength electronic absorption band(λ^{max} =340nm)and assigned in /1/to deformations of the structural group HOCCC=N that can facilitate considerably the ESIPT(seeCh. 3).

However in/5/ this band is assigned to the C=N stretch vibrations that is less likely. The same authors have produced the chemical shifts in NMR spectra for the sensitive H/1/ and C//19, 22/atoms of the phenyl ring and OH-proton/28/. The most typical data are shown in the table 2. 2 for OH form of SA and SALK structures. The little chemical shift to the low fields for OH proton in the SALK structure as compared with SA one points to the strengthening of OH...N bond owing to the increase of N atom basicity with localization of the n(N) electrons with remove of the N-Ph substituent.

Table 2.2
The main spectral data of the different authors for photo-thermochromic
SAs,I(Ph,Ph) and SALKs,I(Ph,Alk)

Tautom. form	Assign.	ν cm^{-1}	Method	Medium	Ref	Taut. form	Assig	λ^{max} nm	Medium	ε	Ref
O H	C$_7$=N Stretch	1621	Rm	Crystal Hexane	/1/	OH	E$_{\pi\pi}$*	332	Calculated	0	/8/
							E$_{\pi\pi}$*	345	Crystal,Hexan.	1.9	/1/
		1615	IR	CCL$_4$	/2/		E$_{\pi\pi}$*	340	Hexane	1.9	/39/
							E$_{\pi\pi}$*	336	Ethanol	24.3	/21/
		1620	IR	Acetone D$_3$	/5/		E$_{\pi\pi}$*	336	Methanol	32.6	/9/
							E$_{\pi\pi}$*	336	CH$_3$CN	37.5	/8, 43/
	HOCCCN Deform.	1577	Rm	Crystal Hexane	/1/		E$_{\pi\pi}$*CT	333	IIP	---	/16/
							Zw$_{CT}$	425	Trifluoroetha-nol(TFE)	16.7	/1/
		1574	IR	Acetone D$_3$	/5/	N H	K$_{\pi\pi}$*	430	Ethanol	24.3	/21/
							Zw$_{CT}$	400	HFIP	26.7	/1/
	C-O Stretch	1281	IR	Acetone D$_3$	/5/		K$_{\pi\pi}$*	432	Methanpl	32.6	/9/
							K$_{\pi\pi}$*	440	Methanol	32.6	/1/
	A< Deform.	1484	IR	Crystal Hexane	/1/		K$_{\pi\pi}$*	400	CH$_3$CN	37.5	/43/
							K$_{\pi\pi}$*	450	crystal	----	/22/
	C=N-C$_{Ph}$	852	Rm	Crystal Hexane	/1/		Zw$_{CT}$ *)	430	Zeolit(Zeol)	----	/1/
N H	H C=N$^+$-C$_{Ph}$ Stretch	1641	Rm	Hexa-fluoro-2-propa-nol (HFIP)	/1/	*)Diffuse reflectance spectra (DFS). SALK					
						Taut. form	R	λ^{Max} nm	Medium (solv.)	As-sig.	Ref.
		1635	IR		/5/	OH	CH$_3$	317	IIP	E$_{CT}$	/28/
	C$^+$-O$^-$ H	1543	IR		/5/		(CH$_2$)$_{17}$CH$_3$	320	IIP	E$_{CT}$	/28/
	C=N$^+$ Stretch	1547	IR		/1/	NH	CH$_3$	390	IIP	Zw	/28/
							(CH$_2$)$_{17}$CH$_3$	390	IIP	Zw$_{CT}$	/28/

(C) IIP-Isopropanol-isopentane 1:4 (77K)

NMR spectra

	SA ,OH form			SALK,OH form		
Assign	δ ppm	Medium	Ref.	H(O) R	δ ppm	Medium	Ref.
H$_1$	6.88d 6.99 d	TFE HFIP	/1/ /1/	CH$_3$	13.38	Toluene D$_8$	/28/
H$_5$	6.84 t 6.81 t	TFE HFIP	/1/ /1/				
H$_{7or9}$	7.18 d 7.33d	TFE HFIP	/1/ /1/	(CH$_2$)$_{17}$CH$_3$	13.61	Toluene D$_8$	/28/
C	160.9 162.3	CDCl$_3$ Cryst	/19/ /22/				
H$_{(O)}$	13.24	Toluene D$_R$	/28/	------	------	----	

The all obtained data of X-ray analysis, vibrational, NMR spectroscopy and quantum-chemical calculations lead to the unanimous conclusion that OH form of the molecules SA and SALK has the Enol(E) trans (about C=N) structure(Etrans) with planar aldehyde fragment including the strong IHB while the N-Ph moiety is twisted considerably in SA molecule.

At the same time the structural peculiarities of SALK molecule described above must be favorable to the formation of the K-structure and to the shift of the tautomeric equilibrium OH⇔NH towards K-form that is really observed experimentally/9, 28/. The typical longwavelength band in the UV-Vis absorption spectra of SA and SALK E-structure (table 2.2)is caused by the $S_o—S_1(\pi\pi^*)$ transition of the same nature including the ICT from OH-Ph group to C=N one with N-Ph moiety playing the role of the electron acceptor substituent.

Thus the observed "blue" shift of the transition from λ^{max}=325-345nm(SA) to λ^{max}=310 317nm(SALK)(table 2.2) is caused by an increase of the electron density on C=N group of SALK structure and of the observed shortwavelength band shift with the rise of the solvent polarity can be explained by the decrease of the dipole moment value($\mu(e) < \mu(g)$)in the excited S1*state of the ICT nature. *)

The information about NH form structure of the simplest molecules SA and SALK is less definite. Such a situation is the consequence of the several reasons:

1. The direct X-ray analysis of the crystal is hardly be used since the NH-form is not separable by crystallization under the proper conditions (see however Ch. 3).
2. The use of the NMR spectroscopy is ineffective due to a low concentration of NH-form in the crystals and the solutions under usual conditions while its rise with temperature is connected apparently with modification of NH-structure and mechanism of its generation (see "thermochromism").
3. The indirect methods with use of the variation of the medium and the molecular structure fail also because of their influence on the NH-form nature. In this connection the most efficient ways of the NH-structure study are the use of the sensitive spectroscopic methods and the quantum-chemical calculations.

---

* See e. g. N. Bakhshiev, M. Knyazhanskiy et al., Usp. Khim. 38(9)1644 (1969) (in Russian).

The comparative data for the calculated structural parameters allow to judge about their variations as a result of the OH-NH transformation in the simplest anils(tab2. 1)/6, 8, 16, 27, 28, 39-42/. The transformation leads apparently to the inversion of the interatomic distances O-H and N.. H, and to the forming of the N-H...O bond that is weaker than O-H...N one in the E-structure. The lengths of the C_2—O and C_1---C_7 bonds decrease considerably and those of the C_1—C_2 and C_7—N bonds increase owing to the corresponding changes of the bond orders.

The changes are more strong for the SALK molecule (I Ph Alk) than for SA one(I Ph, Ph)and connected with formation of the NH structure destabilized relative to the OH(ESotrans) structure($\Delta H \approx$ 4-8kcal/mole)/41, 42/ with a sufficiently uniform charge distribution(the dipole moment is small, $(\mu(g) \approx 3, 3D)$). This structure is responsible for the longwavelength absorption band in the UV-Vis spectral range with $\lambda^{max} \approx$ 390nm/28/and $\lambda^{max} \approx$ 400-420nm/1, 9, 21, 22, 43/ for molecule SALK and SA respectively (table 2. 2). Such a NH-form is very similar to the Keto–structure(Kcis)(Sch. 2. 3), that is supposed as the most probable for the participation in the Ground state OH$\Leftrightarrow$NH equilibrium in the nonpolar media and the crystal i. e. as the structure responsible for the color change in such an environment.

However with change of an environment and a temperature the equilibrium can be shifted completely to the NH-form that gains the Zw structure(Sch. 2. 3)with the keto-like absorption spectra (λ=390-400nm.

Such a situation may be realized when strong temperature decreasing as it supposed /50/ and in the polar media as it has been shown by the vibrational spectroscopy/1, 5/(table 2. 2).

These phenomena will be discussed below.

2. 3. The OH <=>NH equilibrium, and the mechanism of the Ground State Intramolecular Proton Transfer (GSIPT).

The kinetics of the tautomeric equilibrium has been studied theoretically /15/for the two model thermochromic(in the crystal)compounds –N-tetrachlorsalicylideneaniline and N-tetrachlorsalicylideneaniline-1pyrenilaniline. The calculations have been conducted by the semiempirical SCFMO method with the energy gradient technique. It has been shown that thermochromic reaction by the GSIPT OH→NH proceeds via the six –membered ring transition state with change of the π-electron configuration, over the potential barrier(about 20-21 kcal/mol)under effect of the intermolecular interactions. The results obtained are similar to those of /32/and also overestimate the barrier height by 6-10 kcal/mol. The very important conclusions about the kinetics and the mechanism of the GSIPT in the structures I(Ph, Ph)-(SA), and I(Ph, Alk)-SALK, have been made with help of the ab –initio quantum-chemical calculations by the Hartree-Fock method on the 6-31G*basis set/40-42/ and recently by the Density Functional Theory (DFT) method /42b/.

The data obtained do not differ in the value on the whole for the both structures. The cisNH-structure is destabilized by about 5-8 kcal/mol relatively OH(E) one, the reverse proton transfer may occur via calculated barrier of 0, 9-9kcal/mol and for direct GSIPT it is about 6-15kcal/mol that is very close to the data of /15, 32/. The calculated rate constant(T=293deg) for GSIPT, $k_{so}=(0.5-1.4)\ 10^{11}s^{-1}$ has been obtained with the effective frequencies of the vibration mode $\sqrt{(E)}=3129 cm^{-1}$ and $\sqrt{(K)}=2594 cm^{-1}$ for the Enol and Keto structures respectively which correspond to O-H and N-H bond stretch vibrations. Such a high rate of the GSIPT is caused by taking into account the coupling with low frequency so called "transverse "modes of the antisymmetric deformation vibrations C_2-O—H, C_1-C_7-N, C_2--C_1-C_7 with the frequencies of $\sqrt{}=490, 573, 652$ and $1055 cm^{-1}$(

the C-N stretching mode in the SALK structure) which decrease the O↔N distance closing the reaction sites, and promotes the tunneling mechanism of the GSIPT. Taking out the account of the transverse modes leads to the sharply fall of the proton transfer rate, almost by the order of magnitude($k_{so} = 3 \times 10^{10}$ s^{-1}) that can point to the essential role of the tunneling mechanism in the GSIPT.

The experimental investigation of the OH⇔NH equilibrium for the model molecules of the I(Ph, Ph) type structure has been conducted /2-4/ with use of the Nuclear Quadrupole Resonance (NQR) spectroscopy method.

The temperature dependence analysis for ^{14}N /3, 4/ and ^{17}O /3/ NQR frequencies with approximation of a very small probability (p) of finding of a molecule in the NH configuration allows to estimate the energy differences (ΔH) between the OH and NH forms of anils with various substituents in the A and B rings. Being based on the date obtained the authors supposed that OH⇔NH equilibrium is connected with a fast proton exchange between inequivalent states of the weakly asymmetric OH...N double minimum of H-bond PES.

The time averaged O-H distance decreases with fall of temperature. At 300 K the proton spends only from p=0.4 to 10% of its time in the O...H-N configuration in spite of its fast movement along the IHB and a very small energy difference between NH and OH structures ($\Delta E = 1.38$-2.39 kcal/mol (table 2.3) (see also sec. 6). The experimental energy differences are too small as compare with the calculated values /15, 32, 40/. On the other hand the calculated values of the potential barriers are very high to provide the fast H exchange observed in the experiment. At the same time the calculated rate (k_{so}) of the GSIPT with taking into account of the active and transverse vibrations promoting the tunneling /40/ conforms with the experimental data perfectly.

Thus the experimental findings can be explained good if to take into consideration the tunneling that goes along the active stretch O-H and N-H vibration with assistance of the transverse ones which provide in common the lowering and contraction of the barrier along the reaction coordinate of the GSIPT.

Being based on the comparative data presented above it can be believed that in spite of some kinetic differences the GSIPT mechanism is not depended essentially on the nature of the N-R (R:Ph or Alk) structural fragment in the simplest I(Ph, Alk) and I(Ph, ALK) compounds. At the same time one may be expected that a situation can be some altered with the strong structure changes.

2. 4. The structure effect on the OH⇔NH equilibrium.

With help of the ab-initio quantum-chemical calculations by the restricted HF method on the basis STO-3-27G and STO-6-31G(d, p) authors /30d/ have shown that the E<=>K equilibrium dependence on the π-system size or the substituents' nature in the aldehyde moiety can be predicted by the length (bond order) of the ring C_1-C_2 bond adjacent to the OH-group of the E-tautomer. It is analog of the so named "Erlenmeyer rule", and on the base of the obtained findings the orbital nature of this bond has been studied by the authors in the wide series of the anil molecules and their analogs.

The C_1-C_2 crystal bond length has been estimated with the value $L(C:::C)=1.397Å$. So that above and below this value the K or E structure is stabilized respectively. It allows to predict the equilibrium position with the change of the molecular structure and even to explain the equilibrium shift with the change of environment (see below).

2. 4. 1. Effect of the substitutions in the phenyl rings.

The findings of the substituents' effect on the equilibrium of the OH and NH structures have been obtained in the above mentioned work /3/ by the NQR method and produced in the table 2. 3. It is very interesting that the energy differences between the reaction sites (ΔE) are very small*) and for the planar thermochromic compounds (2-8)

---

* In the table 1 of /3/ the misprint is apparently committed in the designation of the energetic units: "Mev" must be replaced.

are very close(ΔE=1.29-2.40kcal/mol) to the difference between the effective frequencies of the O-H and N-H bond stretching vibrations($\sqrt{}$=1.53kcal/mol)/40/ responsible for the GSIPT together with the low frequency transverse modes (see above). At the same time this value for nonthermochromic(photochromic)structure (1) is much more (ΔE=3.23kcal/mol).

Table 2.3
Dependence of OH⇔NH equilibrium on substituent /3/ (crystal).

N_o	Substituent in rings		ΔE *)	P%	Type
	A	B	meV(kcal/mol)		
1	H	2'Cl	140(3.23)	0.4	Photochromic
2	5Cl	H	73(1.68)	5.6	Thermochromic
3	H	4'Cl	91(2.10)	2.9	Thermochromic
4	3-OCH$_3$	2 N-H**)	60(1.38)	8.8	Thermochromic
5	3-OCH$_3$	2N-6'CH$_2$	86(1.98)	3.3	Thermochromic
6	5-Br	2N-6'CH$_3$	104(2.40)	1.8	Themochromic
7	3,5-Cl	2N-6'CH$_3$	56(1.29)	10.3	Thermochromic
8	3,5-Cl	CH$_2$-Ph	63(1.45)	8.1	Thermochromic

*)Percent age error:6-10%.   **)2N-2-aminopiridine.

Table 2.4
Comparative experimental (exp) and calculated (calc) bond length(A) for the 3OH substituted and unsubstituted Alkilimines (see schemes 2.1 and 2.4).

Structure->	1(3OHPh, Alk)			1{Ph, Alk)		
Alkil-->	CH(CH3h	CH3	CH3	C9H11	CH3	CH3
Bond. J,	exp	E(calc)	K(calc)	exp	E(calc)	K(calc)
C1-C2	1.433	1.403	1.452	1.377	1.408	1.462
c,---C7	1.412	1.468	1.389	1,448	1.467	1.389
CrN	1.301	1.287	1.346	1.284	1287	1.347
C2-O2	·1.294	1.372	1.255	1.369	1367	1.252

The thermochromic compounds (2-8) (table 2.3) can be roughly grouped in the two types:

  i. With the high energy difference ΔE and the low NH structure population P% (molecules 3, 5, 6),

ii. With the low ΔE and the high P% (molecules 2, 4, 7, 8).

Authors /9/ suppose also that introduction of 5 Br and 5 OCH3 substituents (in the ring A) leads to the stabilization of the E structure that makes for the absorption band longwavelength shift. It is obvious that there is the complicated dependence of ΔE, and hence P%, values on the competition of the electronic and the steric factors, and the GSIPT can be regulated, as authors /3/note correctly, by the very delicate mechanisms which need the special careful studies.

The phenyl ring electrondonor-acceptor substituents' effects have been carried out in the series of the works. The relationship between IHB strength, H-transfer efficiency, the position of the OH⇔NH equilibrium on the one hand and the molecular structure, electron density distribution including OH group acidity and N atom basicity on other hand, has been studied for the series of anils/18/with help of multinuclear(H, ^{13}C, ^{15}N) NMR spectroscopy, measurement of the dipole moments with the quantum–chemical analysis (AM1, PM3, ab-initio)/58/and the studies of the thermodinamical parameters of the tautomeric equilibrium /63/.

The results obtained demonstrate the strong influence of the substituents on the NH structure relative stability (within the range ΔH=-1. 2 to-3. 4 kcal/mol)and the great contribution of the negative entropy factor (ΔS=-0. 0067 to-0. 011kcal/mol) into the GSIPT reaction.

The analogous studies have been conducted for four crystalline 3, 5 dichlor and 3 nitro salicylideneanilines /14/and 5-Nitro –N salicylidene ethylamine /57/ by X-ray diffraction and solid state ^{13}C NMR. For such compounds the significant contribution of the Zwitterionic NH structure in the crystal makes the bond lengths very close to the E structure ones. The unusual ionic intramolecular O-...H-N+ bond in $5NO_2$ substituted SA, I($5NO_2$, Ph, Ph), is realized due to a double through resonance effect and an increase of OH group acidity, and such a bond has been firstly observed by authors. In spite of the strong intramolecular O...H N bond this compound forms its lattice by several another different types of the weak intermolecular O-...H-N+, N-H..

O, and C-H...O hydrogen bonds. This lattice is characterized by a coplanar arrangement of the molecules with the aldehyde moieties of the adjacent molecules separated by a long distances ("head" to "tail" packing that is not favorable to the thermochromic properties)(see below).

At the same time it is shown /48/with help of the X-ray crystallography and the UV-Vis spectro-scopy that 3NO2 substituted molecules, I(3NO2, Ph), exists as mixtures of the OH and NH forms with the early equal populations in the crystal.

The effect of substituents on the GSIPT for anils in the solvents was studied by the UV-Vis absorption, emission, 13C NMR spectroscopy and the semiempirical quantumchemical(AM1) calculations /25, 26/. It has been shown that the E<=>K equilibrium is shifted towards the K-structure owing to the strengthening of IHB and the increase of the GSIPT efficiency under influence of substituents.

On the base of the structural and spectral data especially with use of the fluorescence excitation spectra authors /52, 53, 64/ supposed that the strong shortening of the IHB OH...N occurs due to a steric effect of the chlorine substituents in the molecules I(3-6ClPh, Ph) and I(3-6CLPh, Perylen)with the complete substitution in A-ring by Cl-atoms. Under the influence of the Cl substituents especially in I(3-6ClPh, Perylen)(sec. 2. 6, Sch. 2. 15), the long wavelength absorption band of the E-structure shifts strongly towards red side (λ^{max} =650nm) having CT-nature and overlaps with the NH-structure absorption band(λ^{max}= 625nm). In addition the equilibrium is shifted considerably towards the latter even at T=77K and thermochromism is very weak (see also below, sec. 2. 6a).

The systematic investigation of the OH⇔NH equilibrium in terms of of E⇔K one in N-salicylidene(4')anisidine, I (Ph, Anisidine), in methanol H-bonded complexes was carried out /46/ with help of the fluorescence excitation spectroscopy in the methanol solution at the temperature range 260-330K. The 4' –substituent effect (4'-Me, OMe, NMe2 as electron donors and COMe, CN, NO2 as electron acceptors)on the spectra, tautomeric equilibrium constant(Keq), and standard thermodinamic parameters $\Delta H=H(K)-H(E)$, $\Delta G=G(K)-G(E)$, $\Delta S=S(K)-S(E)$ has been studied by correlation analysis with the Hammet parameters (δ, p)and by quantum –chemical semiempirical (AM1)and ab-initio calculations.

The red shift(up to $\Delta\lambda=40$nm)of the longwavelength absorption(excitation)bands of the Eand K structures with the insertion of the electron donor NMe2 substituent is evidence of the large contribution of the CT from N-Aryl ring in the first electronic transition. At the same time the low sensitivity to the 4'-effect on the equilibrium parameters (small p-constant)can be explained, as it can believed, by acoplanarity of the aniline ring. However the rather distinct linear correlations with δ parameters connected with the electronic substituent effect show that ΔG(6. 78-9. 64 kJ/mol), ΔH(3. 80-7. 10kJ/mol) increase and Keq(6. 48-2. 19)falls when changing of the substituents from electron donor to electron acceptor. Such a behavior of the molecular systems can be explained naturally by a corresponding alteration of the electron density on the C=N group.

At the same time the E⇔K equilibrium has an important entropy contribution at293K(TΔS=1. 94-3. 08 kJ/mol, i. e. 20-40%ΔG). This contribution is higher for the molecules with the more polar substituents(4'NMe2, NO2) and correlates with the changes of the dihedral angle $\Delta\gamma = \gamma K-\gamma E$ and the dipole moment $\Delta\mu=|\mu K|-|\mu E|$ i. e. with variation of the molecular structure in the GSIPT reaction depending on the substituent nature. Authors suppose the Entropy change value is measure of the reorganization of methanol molecules around the solute as a result of changes in the solute molecular parameters when E=K reactions take place. It should be noted that authors' conclusion /46/ are true only with the full shift of the K*⇔E*equilibrium towards the K* structure in the excited state i. e. when the back ESIPT reaction(K*-E*) is absent. In spite of the importance of this question it has not been discussed in /46/ but it will be considered below (Ch. 3).

Besides that the mechanisms of the GSIPT and ESIPT reactions responsible for the generation of the fluorescent keto-srtructure (K)in the complexes I(4')-methanol with the strongly weakened or even broken IHB OH...N has not been also discussed by authors /46/. The strong effect of 3OH substituent on the OH⇔NH equilibrium in the crystals and the solvents has been studied. The structures I(3OHPh, Ph)and

I(3OHPh, Alk) with the electron donor and withdrawing substituents (R')(Sch. 2. 4a) have been studied by the X-ray analysis/6, 12/, the solid ^{13}NMR and IR spectroscopy/6, 55/ and also by the semiempirical calculations/55/.

In the crystalline state all compounds are associated in the H-bond dimmers I(AB)(Sch. 2. 4b) with 3OH NH tautomerism is realized by the fast proton exchange. OH group forming intermolecular ten-membered pseudocycle with preservation of IHB O-H.. N in which the OH⇔NH. There are three dimer types as (E-E), (E-K) and (K-K) structures. The compounds with aliphatic amines I(Ph, Alk) are characterized by dominant K-structure ((K-K) and (E-K) dimer structures.

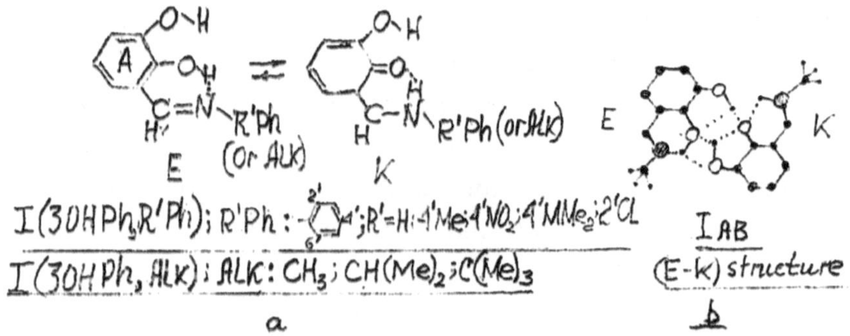

Scheme 2. 4
The effect of the 3OH substituent on E⇔K equilibrium.

a) The structure in the solvent (see text). b) H-dimmer structure in the crystals.

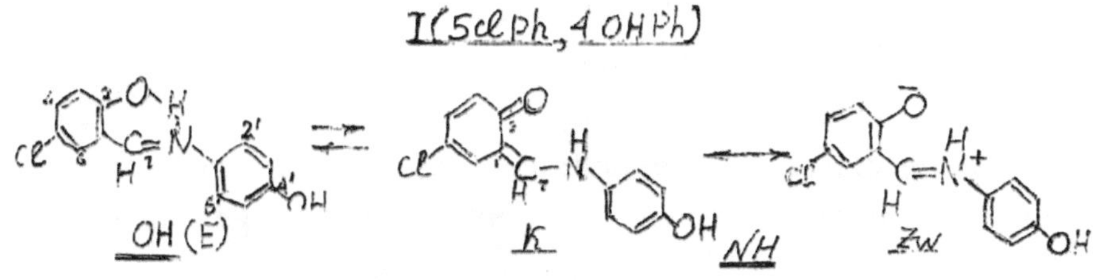

Scheme 2. 5
The effect of 4'OH substituent. The NH structure can
be more stable then OH one forming aggr.

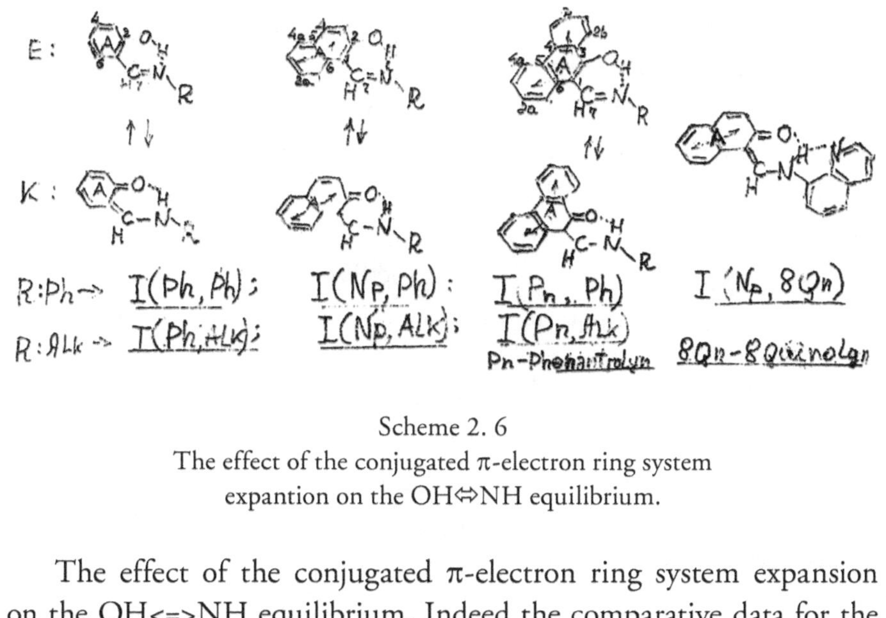

Scheme 2. 6
The effect of the conjugated π-electron ring system
expantion on the OH⇔NH equilibrium.

The effect of the conjugated π-electron ring system expansion on the OH<=>NH equilibrium. Indeed the comparative data for the typical bond lengths (Table2. 4) show that experimental values (X-ray analysis) are much close to the calculated ones of E and K structures for the compounds I(3OHPh, Ph) and I(3OHPh, Alk) respectively. At the same time the experimental C_1-C_2 bond lengths are more and less for I(3OHPh, Ph) and (Ph, Alk) correspondingly than the critical that (l=1. 39 Å, see above /30d/). It is indicative of the significant shift towards K-structure in the I(3OHPh, Alk) compounds. For the compounds I(3OHPh, Ph) the position and the nature of the substituent may favor

E or K form but always with a significant amount of the K species ((E-K) and if it is possible (K-K) dimers).

The electron acceptor and electron donor substituents favor stabilization of the K or E tautomers respectively. The tautomeric equilibrium mixture is greatly influenced by the aggregate state of the compounds and is different in the solid state and in the solvents and not always correlates each with other. The crystal structure of the compounds I(3OHPh, ALK)is favorable to thermochromism(see below).

A situation becomes complicated also by introduction of OH-group into the 4'position of the imine ring. So according to/47, 49, 51/the molecules I(5ClPh, 4'OHPh)(Sch. 2. 5)also form H-bond aggregates with participation of 4'-OH group. The NHZw structure is strongly stabilized in aggregates by dipole-dipole interactions and can become more stable than OH one(see also 2. 6). Thus the NH form is predominantly zwitterionic (Zw) in crystal and predominantly quinodial in gas phase and nonpolar media/49/.

The steric and electronic influence by inductive effect of the bulky (tert-butyl, alkyl)substituents in the rings A, B has been studied in /7, 37/. Insertion of such substituents leads to the strengthening of the IHB, and to the marked shift of the OH⇔NH equilibrium towards the NH structure. It is displayed by the shift of OH proton PMR signal to the more weak fields as compare with unsubstituted molecules (14. 3 ppm as against 12-13 ppm), some weakening of the C=N bond(shift of the stretch vibration frequency to 1614-1625cm^{-1}), and appreciable increase of the K-structure absorption band ($\lambda^{max}\approx$440-450nm) intensity.

The comparison of the reflectance spectra of the powders of the compounds without (a) and with (b) tert-butyl substituents in the aldehyde ring shows that these structures can be classified by the two types according to the change of the relative intensities of the K-structure band with the insertion of the tertbutyl substituents (a→ b). In the first type the intensities of the K-band almost the same for a and b structures meanwhile in the second one the optical density decreases in the structures b. The origin of this spectral change may by interpreted in terms of the change in the molecular shape /7/.

2. 4. 2. Effect of the ring π-electron system extension.

The influence of the conjugation on the GSIPT has been studied by the quantum-chemical methods (HF CIS)/62/. The results show that GSIPT barrier ΔE(GSIPT) decreases considerably: ΔE(GSIPT)=19. 3,

17. 3, and 9. 2 Kcal /mol for one two, and three conjugated cycles in the aldehyde ring i. e. (Ph, Ph), I(Np, Ph), and I(Pn, Ph) structures correspondingly(Sch. 2. 6.), and changes a little for the larger conjugated systems (from three to five phenyl cycles). Meanwhile the effect of the electron donor and acceptor substituents on the GSIPT barrier is very small.

At the same time the calculated and experimental values of the C1-C2 ring length in the structures I(Np, Ph), I(Pn, Ph) (see e. g. table 2. 5.) satisfy to the criterion $l(C_1-C_2)$ <1. 397 which favors formation of the K-structure /30/. The expansion of the A rings' π-system for example in the series I(Ph, Alk), I(Np, Alk), I(Pn, Alk) (Alk:n-propyl) results in the considerable strengthening of IHB OH...N(δH=14. 4ppm), shortening of C1-O bond (l_{co}=1. 254Å) as compared with the structure I(Ph, Alk)(table 2. 1), and the H...N distance (2. 403 Å)/13/. Such structural parameters of the IHB are favorable to the lowering of the GSIPT reaction activation energy and to the shift of the tautomeric equilibrium E⇔K towards K-structure.

The above mentioned data are agreed with the findings of the UV-Vis absorption spectra obtained earlier/29a/ and much later /30a, b, c/ which have shown that E⇔K equilibrium for I(Np, Ph) compounds and their structural analogs is strongly shifted towards the K-structure, very likely via dimerization. At the same time the detailed investigations with use of NMR (1H, 13C), absorption and fluorescent spectroscopy in the solvents and the crystal state and semiempiric quantum-chemical calculations (AM1) have been conducted recently /19-22/. The most sensitive to the E⇔K equilibrium parameters are: the NMR chemical shifts (δppm)for the 1H, $13C2_2$, 13C7 atoms directly included in the active structural fragment $HO-C_2-C_1-C_7-N$, the distances between these atoms (bond length (Å)), and the stretch band vibration frequencies in this fragment (e. g. C_2-O, C_7-N), and also the ratios of the electron absorption band intensities of K and E structures, Iabs. 440nm/Iabs. 340nm, and corresponding fluorescence ones, I flu. ex. 440nm/Iflu. ex. 340nm, (table 2. 5).

For the most of the compounds studied the behaviors in the crystal and the solvents are very similar. The distinct E⇔K equilibrium shift towards the K-structure takes place with the consecutive ring addition(annelation)in the "aldehyde" moiety in the series I(Ph), I(Np), I(Pn)(see Sch. 2. 6). Indeed, the calculated values of the free energy differences between K and E structu-res (Δ(K-E)) are agreed by the order of magnitude with the experimental data/2-4/showing the increase of the K-structure stability.

In this case the equilibrium is completely shifted towards E or K structures in I(Ph)and I(Pn) respectively, and the corresponding characteristics (e. g. relative intensities of the absorption bands or the 13C chemical shifts) take the extreme values typical for each of these structures (e. g. $\delta 13C_2$=160 and 180ppm for E and K structures respectively) which don't change with the temperature decrease practically (e. g. $\delta 13C_2$=180. 7 and 181. 4 ppm at T=315 and 265K respectively). On the other hand the compound I(Np, Ph) exhibits significant amounts of the both E and K structures in the equilibrium(Keq=1. 8 at the ambient temperature) which is caused by the fast proton exchange OH<==>NH, and strongly depend on temperature.

At the same time the study of the 13C, 14N residual coupling effects(see also /23/)allows to distinction to be made between dynamic proton transfer systems on the one hand, I(Np, Ph), and the structure which is significally shifted towards one of the tautomers on the other hand, I(Ph, Ph), I(Np, Ph).

The similar conclusions about the mechanism of the E⇔K tautomeric equilibrium have been made /24/on the basis of the study of the substituted compounds I(Ph, Ph), I(Np, Ph)and I(Pn, Ph) with help of the method based on the correlation between the Deuterium induced chemical shift (DIS) differences in the C13 and the chemical shift of the exchangeable proton (PS).

Unlike /19-22/it has been established in /24/ that the fast proton exchange is absent only in I(Ph, Ph) molecule in which the tautomeric equilibrium is strongly displaced in favor of the E-structure. This conclusion arises from the linear plot of log(DIS) vs (PS) for this molecule.

However such a correlation is not kept for I(Np,Ph) and I(Pn,Ph) molecules in which the fast proton exchange between O and N atoms is realized, and these molecules were found to be tautomeric mixtures dominated by the E-form.

Table 2. 5
Changes of the typical spectral and structural parameters of Anils with the ring addition in the aldehyde moiety /20-22/ (see scheme 2. 6)

Structure	UV-Vis spectrsc.			NMR ^{1}H, ^{13}C spectroscopy δ ppm						IR spect.cm^{-1} Stretch		X-ray analysis Bond length Å			5) Δ_{K-E} Kcal/mole
	Ethanol		Cryst	$^1H_{(O)}$	$^{13}C_7$	$^{13}C_2$	$^{13}C_7$	$^{13}C_2$ Extr 3)						4)	
	I_{NH}/I_{OH} absorp	I_{NH}/I_{OH} exc.flu 6)	I_{NH}/I_{OH} exc.flu	CDCl$_3$	Cryst	CDCl$_3$	crysta	E	K	C$_2$-O	C$_7$-O	C$_2$-O	C$_7$-N	C$_1$-C$_2$	
I(Ph,Ph)	0.03	0.36	0.06	13.38	164.2	160.9	162.3	160.0	----	----	1615	1.34	1.28	1.41	4.3
I(Np,Ph)	1.3	1.18	0.64	14.4 1)	144.8	171.2	180.0	155.0	----	1595	1622	1.30	1.29	1.39	-1.6
I(Pn,Ph)	3.8	1.99	2.54	14.8	146.8 2)	181.0	181.0	150.0	180.0	1622	----	----	---	1.38	-6.1

1) in crystal. 2) In CD Cl$_3$ and crystal. 3) The approximate values of the extreme $^{13}C_2$ shifts for tautomer E and K /20,22/. 4) From /27/. 5) Obtained from the table 2.3 /20/. 6) The method is correct only if in the S$_1$ state O⇔H equilibrium shifts towards NH structure completely i.e. the reverse ESIPT is absent.

The (DIS) also revealed that I(Pn, Ph) exists in the solution as a dimer (see also /30a-c/).

The proton transfer equilibrium has been studied also /68/ by 1H, 13C, 15N NMR spectra and DIS of 13C, 15N at variable temperatures (200-300 K) in different solvents with use of the 1J(15N, 1H) and 3J(15NH, 1H) coupling constants. It has been shown that all compounds of I(Np,) type with the different R substituents exist mainly as NH (Keto) tautomer, and the equilibrium in I(Np, 8Q) is shifted completely to the NH form stabilized to the IHB (Sch. 2. 6)/69/. The mechanism of the tautomerism has been studied also in /60/ with help of 17O and 13C NMR spectroscopy in the different solvents. According to the data obtained the compound I(Ph, Ph) exists mainly in the E form and I(Ph, Alk) one as the E⇔K equilibrium mixture.

Alkylimines – I(Ph, Alk), I(Np, Alk) and Arylimines-I(Ph, Ph), I(Np, Ph) give the shift towards (40-50%) and OH(70-80%) structures respectively, and the tautomeric equilibrium are shifted towards OH-form in the nonpolar solvents and with temperature increase. At the same time according to the author's viewpoint the values of the NH structure percentage in some Naththalen derivatives, I(Np, PhR), produced in /19/ are overestimated due to incorrect choice of the standard C_2 chemical shift values for E and K structures.

The two new sensitive methods have been suggested /59/ for the investigation of the OH⇔NH anil equilibrium in the ultrathin solid amorphous films prepared by vacuum evaporation which are based on the interpretation of the Near–Edge X-ray Absorption Fine Structure(NEXAFS) spectra at N and O K-edges and the X-ray Photoelectron Spectra (XPS) in the N and O 1s regions. Such methods can give a very reliable and detailed information about mechanism of equilibrium and according to the data obtained for the molecule I(Np, Py)(2-Hydroxy-Naphthylidene Pyrenylaniline) the E-structure predominates (75-80 %) in such conditions that conforms weto the data presented above/60/.

Thus although the tautomeric equilibrium by the fast proton exchange is established for anils in /2-4/, /19-22/, /24/, /59, 60/ the contradictions exist between the results obtained by the different methods and under different conditions. So such an equilibrium is observed for I(Ph, Ph) only by NMR-DIS method in the solvents /24/. In above structures the fast proton exchange equilibrium has not been observed by the NMR and the UV-Vis spectroscopy in the crystal and the solvents /19-22/. Such a situation needs the additional special investigations.

The influence of the ring π-system expansion in the aldehyde moiety of the anil molecules by the ring addition to the 3, 4 or 5 phenyl ring positions with the electron donor and electron acceptor substituents in the 4' position of the aniline ring-I(Np3, 4 ;Ph) and I(Np5, 6;Ph) (see Sch 2. 7)has been studied (/44, 45a/ and the referents there)by the UV-Vis spectroscopy in the solvents of the different nature and the quantum-chemical calculations.

There are the two main types of the spectral changes as compared with I(Ph, Ph), i. e. (SA), molecule irrespective of 4'substituent (see table 2. 6).

1. The longwavelength shift of the first electronic transition of the CT nature in the E and K structures in the nonpolar and polar solvents by $\Delta\lambda \approx 40$ and 20 nm for the E and K structures respectively due to the fall of the ionization potential of the elecdonor "aldehyde "moiety with the expansion of the π ring electron system.
2. The shift of the OH⇔NH equilibrium towards NH structure with the corresponding Increase of the NH absorption band intensity that conincides with the data presented above.

The valie of the "red" absorption band shift in the both forms increases irrespective of the substitutes nature but changed sharply with inclusion of 4' NMe_2 group. The small equilibrium shift in the nonpolar media is caused by the structural stabilization of the NH form.

At the same time the strong shift in the polar solvents at ambient temperature is exhibited for the both structures, and intermolecular factors are coupled with effect of the entropic term TS providing the better solvation of the NH form.

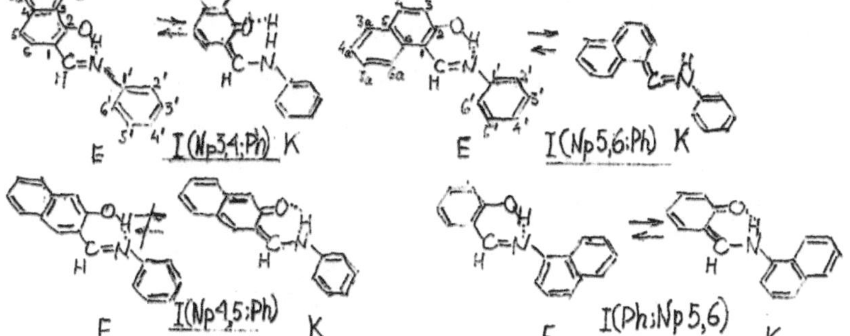

Scheme 2. 7
The anil structures with the π electron ring system expansion and the donor and acceptor substituents

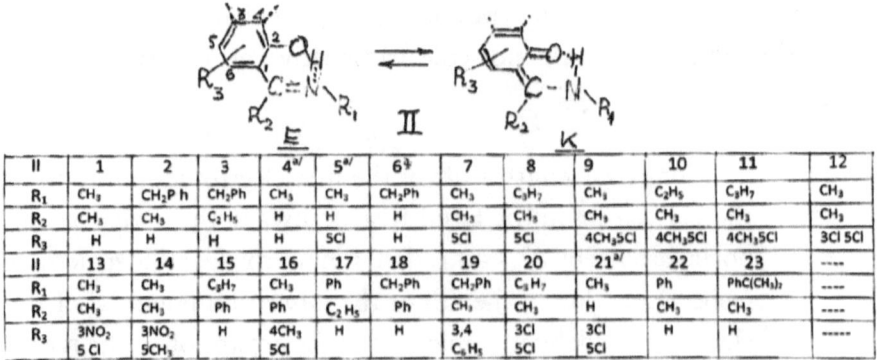

II	1	2	3	4[a]	5[a]	6[+]	7	8	9	10	11	12
R$_1$	CH$_3$	CH$_2$Ph	CH$_2$Ph	CH$_3$	CH$_3$	CH$_2$Ph	CH$_3$	C$_3$H$_7$	CH$_3$	C$_2$H$_5$	C$_3$H$_7$	CH$_3$
R$_2$	CH$_3$	CH$_3$	C$_2$H$_5$	H	H	H	CH$_3$	CH$_3$	CH$_3$	CH$_3$	CH$_3$	CH$_3$
R$_3$	H	H	H	H	5Cl	H	5Cl	5Cl	4CH$_3$5Cl	4CH$_3$5Cl	4CH$_3$5Cl	3Cl 5Cl
II	13	14	15	16	17	18	19	20	21[a]	22	23	
R$_1$	CH$_3$	CH$_3$	C$_3$H$_7$	CH$_3$	Ph	CH$_2$Ph	CH$_2$Ph	C$_3$H$_7$	CH$_3$	Ph	PhC(CH$_3$)$_2$	----
R$_2$	CH$_3$	CH$_3$	Ph	CH$_3$	C$_2$H$_5$	Ph	CH$_3$	CH$_3$	CH$_3$	H	CH$_3$	----
R$_3$	3NO$_2$ 5Cl	3NO$_2$ 5CH$_3$	H	4CH$_3$ 5Cl	H	H	3,4 C$_6$H$_5$	3Cl 5Cl	3Cl 5Cl	H	H	-----

a/ The model structures I have been designed in the structures II with the same substituents respectively

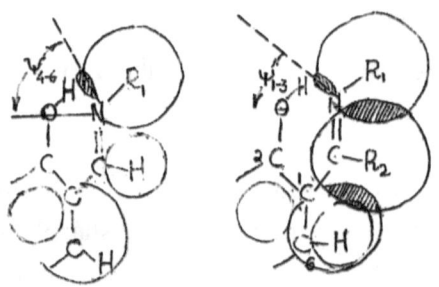

Scheme 2.10
The role of the polar resonance structures in the OH⇔HN equilibrium of the structure II.

(OH)⇔(NH) = (E<->ZwOH)⇔(K <->ZwNH)
E,K-Enol,Keto structures
ZwOH=(O$^+$-H...N$^-$), ZwNH=(O$^-$...HN$^+$).

Scheme 2.9
The influence of the imine bridge substituent R$_2$ on the Enol structure II and the NH one stabilization.

At the same time the electron donor and acceptor substituents preferentially stabilize the K and E tautomer respectively. The sharp equilibrium shift with cooling of the solutions (below 200K) independently of the nature of the solvent is explained /44/ by the change o the thermodynamic parameters at the indicated point caused by an unusual solvent-solute interactions (see below).

In accordance with the data obtained the authors /44/ believe tautomerization at ambient and low temperatures involves both electronic rearrangements and an only very short distance proton transfer process with the important role of the environment.

However too large electronic and structural transformations requiring a lot of energy is unfavorable to the K-structure formation. For this reason the addition of the ring to the 4, 5 positions in the structures I(Np4, 5;Ph) does not lead to the shift towards the NH tautomer unlike structures I(Np3, 4;Ph) and I(Np5, 6;Ph) since the desaromatization of the two rings in the K-structure is en-ergetically unfavorable (Sch. 2. 7). Recently the analogous results have been obtained /54 a, b/ for the compound I(Np5, 6;Ar) by the steady-state and transient absorption, fluorescence emission, and excitation spectroscopic methods.

It is shown however that the analogous π-system expansion in the imine moiety in compound I(Ph, Np5, 6)(Sch. 2. 7) does not result in any significant variations of E⇔K equilibrium shift under any conditions(unlike the SA molecule). Thus the essential role of the "aldehyde "structural fragment in the E⇔K equilibrium under various conditions is evident.

2. 4. 3. The role of the substituents in the imine bridge

The large series of azomethine of the type II structure with the bulky substituents R_2 joined to the C=N-group (Sch. 2. 8) has been investigated in the great number of works/33a-n/.

The various methods of the UV-Vis, IR absorption, NMR spectroscopy, X-ray structure analysis and the quantum-chemical calculations have been applied. It has been shown on the base of the comparative studies of the compounds II4-II6 and II1-II3/33c/that distortion of the structural fragment including IHB occurs.

Due to the steric C-C interactions(E str) the n(N) electrons' lone pair direction' (angle ψ) change, the considerable shortening of O....N distance and the relative stabilization of the NH-structure (ΔE)take place (Sch. 2. 9, tab. 2. 7). The most significant role is played by the interactions between the R_2 substituent and the C_6 ring atom (compare II1 with II4 and II3, II6 with II6)(Sch. 2. 8). Meanwhile the R_1-R_2 steric interactions are of less importance (compare the series of the structures I1, I2 and I3).

At the same time the steric repulsion as it has been shown by the calculations of the adiabatic potential for the proton motion /33c, g/not only decreases the energy for O-...HN+ state but makes the potential pit more shallow that results in the proton delocalization. The all above structural and energetic changes stipulate the strengthening of IHB, the rise of H-transfer probability and the shift of the equilibrium towards NH-structure(λ_{abs}=390-400nm)

Table 2. 6
Spectral (in ethanol) and thermodynamic parameters of E⇔K equilibrium of 4'substituted of 2-hydroxy nathtalidene aldehyde I(5, 6 Np) in ethanol and CH_3Cl solutions /44/.

4'-substient	Observed λ^{max} nm		Equilibrium constant k_t (200K)			ΔH_{K-E} **) Kcal/mol
	E	K	C_2H_5OH	$CHCl_3$	Calc.(Ab-initio)	
NMe_2	412	470	1.267	0.667	0.7	1.595
CH_3	378	440,461	1723	0.555	----	2.061
OCH_3	383	445,467	1.673	0.754	1.0	2.089
H	376	438,458	1.110	0.522	0.7	1.848
Cl	376	441,460	0.745	0.515	1.3	1.225
Br	378	442,461	0.805	0.351	----	1.298
I	381	444,464	0.814	0.492	----	1.586
CN	387	448,466	0532	0.416	1.3	---
NO_2	390	456,480	0.562	0.413	----	---
I(Ph,Ph),SA')	336	410	0.08	<10^3	----	4--8

*)See /2/ and sec 2.3.**)Percentage error 6-17 %

Table 2. 7
Changes of the selected parameters in the structures II as a result of the steric interactions(see scheme 2. 9) /33/

\Para- \ me- II \ ter	O-N Å		Ψ degree	Estr Kcal/mol	ΔE Kcal/mol
	X- Ray	Calculated			
1	2.459	2.484	24.8	0.97	2.77
2	2.497	2.490	25.9	1.01	2.76
3	2.494	2.492	24.9	1.18	2.29
4	------	2.617	32.3	-----	5.00
5	2.559	2.664	32.2	0.42	4.57
6	2.587	2.619	31.5	036	4.87

Table 2. 8

The effect of the sabstituents on keto-enol equilibrium of compound II

A	B
Effect of single and double substitution on the E K equilibrium and the **O-N** distance /33g/.	Effect of double substitution on the relative stability of the E and K forms, ⅡG, their equilibrium, and the strength of H-bond, 6(H), /33n/.

Str. II	O-N (A) X--ray	K_{eq}		Str II	ΔG (K-E) Kcal/mol	ΔH ppm	K_{eq}
5	2.559	0.28		1	1.16	16.46	0.14
7	2.490	0.54		12	-1.09	17.76	1.55
8	2.487	0.47		20	-0.97	17.60	1.18
9	2.487	0.52					
10	2.491	0.43					
11	2.494	0.33					
21	2.574	0.89					

400 nm) clearly observed by the experiment especially in the polar proton solvents (e. g. compounds II3, II17, II18)/33d, e, f, I, j/. However a note should be taken that the origin of the NHzw band (λ=400nm) in the alcohol solvents is explained by the Intermolecular HB complex with the solvent in spite of the significant strengthening of IHB in such structures(δH(OH)=15. 2-16ppm/33d, e, i/ and convincing arguments of the Intramolecular nature of this bond under such conditions /1/.

In the stericly deformated structures II the effect of the aryl ring substituents on the OH$\Leftrightarrow$NH equilibrium becomes more strong (tab. 2. 8A). The raising of the OH group activity when inserting the single withdrawing 5-CL substituent (structures II5-II10)/33g/ shifts the equilibrium towards

NH –structure even with the sufficiently long O.... N distance (structure II5) but in the strained structure II7 the distance is shortened and the equilibrium is shifted still more.

On other hand the competition of the substitution effects (R_3=4CH_3 and 5Cl)(II9) gives a more intermediate proton position, and the lengthening of the N-chain(from CH3 to C2H5) results in subsequent shift of the equilibrium towards the E-structure(II9-II10).

When the double substituted by 3, 5Cl atoms(II12, II20, II21)/33n, 33g/(tab. 2. 8A, B)the NH-structure stability is increased drastically as compared with the unsubstituted structure II1(ΔG(NH-OH)is negative !), the IHB is strengthened (δ(XH)ppm shifts to the low field)and equilibrium constant Keq=[NH]/[OH] increases by the order of magnitude (tab. 2. 8B). Such a situation arises in the structures II12, II20, II21 owing to the combined effect of the two factors (tabl. 2. 8A, B):

i. The considerable growth of the phenol moiety acidity (for series II1, II5, II21 Keq =0. 14, 0. 28, 0. 89 respectively)and
ii. The O...N distance shortening with inclusion of R_2=CH_3 substituent(Keq=0. 28, 0. 54 for the structures II5, II7 correspondingly and Keq =0. 89 and 1. 55(1. 18) for the structures II21and II12(20) respectively).

Thus both a structural deformation with inclusion of $R_2=CH_3$ group and the double substitution by two (3, 5 Cl) acceptor atoms play the equally important role in the equilibrium shift.

The structural deformations with inclusion of R_2=Alk substituent makes the molecule more sensitive to the equilibrium shift towards NH structure by the expansion of the aromatic π-system in the A-ring moiety(struct. II19)/33k, 33m/.

One can conclude from the comparative findings (tab. 2. 9) the expansions of the π-system (compare II2 and II19) leads only to the changes of the interatomic distances (O.... N, C-O and C-N)while with the introduction of the $R_2=CH_3$(compare I(Np, Alk)and II19) not only the interatomic distances change but the IHB is strengthened considerably (the strong O-H proton shift to the low field in the PMR spectra (tab. 2. 9).

As a result of the structural variations the significant shift towards NH structures is observed by the strong increase of the relative intensity of the NH and OH absorption band λ=400 and 325nm (INH/IOH). From the data of the UV-Vis, IR spectra dependence on the temperature, solvent polarity and the calculations of the force constants for the normal vibrations including C=N, C=O, OH, NH stretching modes can come to conclusion about the tautomeric form structures /33/. The situation can be described in general form from the viewpoint of the equilibrium of the resonance structures (Sch. 2. 10).

In some cases (struct.II19 in the alcoholic solvents) the OH structure can exist in form of E and Zw (O-...NH+) structures and the latter can be turned into anion when interacting with alcohol. It has been shown also that δ(C=N-H)vibration mode gives the important contribution in the no-rmal vibration connecting with the GSIPT reaction in this structures.

In the crystal state of the compound II the influence of the structure deformation is manifested with a lesser degree (e. g. for the compound II12 Keq=1. 55 and 0. 69 for the solvent and the crystal respectively)/33h/. Even the substitution of Cl atom for the more strong acceptor NO_2-group does not change significantly the contribution of the NH structure in the crystal state(Keq=0. 69, 0. 75, 0. 70 for the compounds II12, II13 and II14 respectively)/33h/. The inclusion of the bulky (R_2=Ph) substituent (compounds II15, II16) results in the decrease of the NH-structure amount (e. g. compare Keq=0. 32, 0. 36 for the compounds II15, II16 respectively and 0. 54, 0. 56 for II7, II9 ones respectively) apparently owing to the fall of the N-atom basicity /33h/in spite of the pos-sible steric effect.

From the data presented above as it was expected by authors/33/ the heightened ability for the thermochromism must be observed for the compounds II. However unfortunately any concrete information concerning thermochromism in the crystal has not been given.

Furthermore contrary to the expectations no photochromism in crystal has been detected for the structures II22, and even II23 (see /7/) and also for other derivatives of hydroxyacetophenon /7, 34, 35a, b/ that has been explained unlike /33/by unfavorable steric effect of C-Alk group(see also below).

2. 5. The solvent effects. Solvatochromism.

In the polar environment unlike the nonpolar solvents along with the absorption band about λ^{max}=333-340nm which belongs to the E-structure, the new band appears about λ^{max}=400nm.

Thus the changes of the spectral distribution in the visible range of the absorption spectra depend on the solvent nature and can be caused by the two factors:

i. The spectral shift of the above bands connected with the nonspecific and specific interactions with the solvent and
ii. The alteration of the relative intensities of the bands as a result of the OH⇔NH equilibrium change under influence both the nonspecific and specific interactions with the solvent.

The small solvent spectral shifts of both bands with the change of the solvent nature(tab. 2. 2c) don't play a significant role in the solvatochromic effect. However these shifts are important for the understanding of the nature of the electronic S_0—S_1^*transitions in each of the structures and will be discussed later.

The second factor has the decisive importance and is connected with understanding of the mechanism of generation and the structure of the form responsible for the corresponding absorption band.

The data obtained before the nineties did not allow to interpret definitely the band and especially to assign it to the low polar K-structure. The detailed study has been carried out in the liquid solvents and the solid (Dehydrated sodium–exchanged zeolite Y)polar media(tab. 2. 2)/1/.

The typical stretching vibration C=N frequency ($\sqrt{}$=1641cm-1)(tab. 2. 2), the short wavelength shift of the electronic absorption band with the increase of the solvent polarity (λ^{max}=425 and 400nm for TFE and HFIP correspondingly, tab. 2. 1c), interpretation and shifts of the PMR signals ((tab. 2. 2b) testify to the high –polar (Zw) NH-structure(Sch. 2. 12)which is responsible for the corresponding absorption band.

The correlation between the dipole moment values(μ=1. 65-2. 05 D)and the acidity(pK till 9. 3) of the solvents on one hand and the Zw-structure concentration on other one, are observed. This correlation points to the important role of the electron density distribution into the active structural fragment OH-C=C-C=N.

However according to /5/ such a correlation for IR spectra of I(Ph, Ph) molecule is although kept in HFIP (μ=2. 05D) but is broken in acetonitrile-d3((μ=3. 5D))in which the IR spectra correspond very well to the E-structure(like crystal or hexane) that is hard to explain. The Zw-structure, similar to that in the polar protic solvents, is also formed in the highly polar internal nonacid (Brönsted) cages of the dehydrated zeolite with Na+cations(NaY) and manifested by the intensive band (λ^{maxx}=430nm) in the Diffuse Reflectance spectra (tab. 2. 2)/1/.

The formation of the NH Zw structure under such conditions indicates that role of Brönsted acidity (i. e. Intermolecular H-bond and GSIPT)can be unimperative influencing the OH⇔NH equilibrium.

Authors believe polarity of the surrounding medium plays the major role in the solvent also in the promoting Zw formation by the GSIPT (Sch. 2. 11) /1/.

However one can believe from our point of view it is necessary to consider also an additional possibility of the NH Zw structure origin in Zeol as the D+A-complex (Na+-Anil-) formation. In the work /33l/ authors observed solvatochromism of molecule I(Ph, CH_3) in the solvent series CCl_4, CH_2Cl_2, CH_3OH with ε=2. 2, 8. 9, 32. 6 correspondingly. The concentration of the Zw↔K resonance form with the absorption in the range of λ=400nm, the dipole moment μ=5.2D and the Zw structure contribution about 40-50% grows with the increase of the solvent polarity.

In spite of the results obtained earlier /1, 5/the solvatochromism of the molecules I(Ph, Ph) and I(Ph, CH_3)in the alcohol solvents (i. e. the appearance of the absorption band at 400-430nm)is considered /9/from viewpoint of the equilibrium of the solely low polar Enol(E) and Keto(K) structures E<=>K without any discussion of the above mentioned findings.

At the same time the solvatochromism is manifested as a result of the OH⇔NH equilibrium shift not only with change of the solvent polarity but also the acidity (or basicity).

This phenomenon has been studied by authors /61/with use of the steady-state UV-spectroscopy in various solvents and their mixtures. The full equilibrium shift to the ZwNH structure occurs with addition of the salts (e. g. $CaCl_2$) dissolved in methanol. It gives an opportunity to determine the extinction of the absorption band, full spectra, concentration of the NH form, and the equilibrium constant in various solvents which increases with growth of the solvent acidity.

In the series of the works the effect of the molecular structure on solvatochromism has been studied. The dependence of the OH⇔NH equilibrium shift towards NHZw form on the electronic structure of various Schiff bases has been analyzed/56/by means of the deuterium isotope effect on the ^{15}NMR, UV-Vis spectroscopy in the different solvents and by the semiempyrical quantum-chemical (PM3) calculations.

According to /46/ the position of the So-S_1^* electronic transition in the E structure of the molecule I(Ph, 4'Ph) does not depend on the solvent nature. Meanwhile the longwavelength low intensive absorption band (λ^{max}=450nm) responsible for the solvatochromic effect in methanol has been assigned unlike /1,33k,44/ to the So-S_1^* transition in the nonpolar NHKeto(K) structure in the solute –solvent H-bonding complexes without any discussion of the above cited results.

According to /44/ (and references there) and to /54a, b/ the expansion of the ring π-system only in the aldehyde moiety, I(Np, Ph or Alk), increases sharply the sensitivity of the OH<=>NH equilibrium to the solvent polarity. The quantitative study /44/ has shown the equilibrium to be shifted especially strongly towards NH form in the protic polar solvents (K=[NH]/[OH]=1. 4 and 3. 3 for ethanol and H_2O correspondingly) owing to the rise of the relative stability of the polar NH –structure (for I (Np, Alk), OH and NH structures $\mu(g)$=2. 31 and 3. 15 D respectively).

There are the linear dependences: k(T) vs $\Delta\mu$ ($\Delta\mu=\mu(g)OH-\mu(g)NH$) for molecules with various substituents and k(T) vs ε for the solvents of the different polarity. For the protic solvents (H_2O, EtOH, MeOH) the deviation from the linear correlations is observed due to the specific interac-tions. However in such solvents the abnormal k(T) increase takes place as a result of the GSIPT with an assistance of the specific interactions without rupture of the the OH…N IHB (see the analogous interpretation in /1/ and the corresponding comments).

The strong E⇔K equilibrium shift at room temperature towards the K structure is observed in the polar aprotic (acetonitrile) and especially in the polar protic (ethanol) solvents in the series of the compounds I(Np, Ar or Alk)(1-5)(Sch. 2. 11, 2. 12)/54b/. At the same time the NH-band is shifted to the red in the series 1, 2, 3(λ^{max} =456, 474, 482nm correspondingly) indicating the influence of the aromaticity with increase of the ring number, but in 4, 5 the intensity of the NH band (λ=420nm) increases strongly without shift.

The increase of the ring number in the B moiety only I(Ph, Np) does not lead to the equilibrium shift and the compound 6 is not solvatochromic. Thus the solvatochromism of the o-hydroxynaphthaldehyde derivatives is stipulated mainly by the strong sensitivity of the OH⇔NH equilibrium (the GSIPT probability) to the solvent nature in these compounds. The sensibility is caused by the considerable increase of the electron density in the C=N group as a result of the influence of the A–ring conjugation and of the concerted action of the A and B ring π-electron systems.

Thus it should be supposed that the solvatochromism is caused mainly by the shift of the equilibrium OH⇔NH towards the polar NH structure in the polar and /or protic solvents due to the GSIPT. The polar NH structure has rather resonance form (K↔Zw) and the heightened stability under such conditions due to a charge separation. The contribution of the utmost Zw structure (the extreme charge separation) is varied within wide range (from several to 60%) in depen-dence on the molecular structure, external conditions and on nonspecific (Sch. 2. 11a) and speci-fic (Sch. 2. 11b) interactions with the solvent in particular (e. g. H-bond formation. Sch. 2. 11d). In any case the intrinsic GSIPT is the basic mechanism of the NH-structure generation even in strong protic solvents in which it is assisted by formation of the solute-solvent complexes (Sch. 2. 11d).

2. 6. The temperature effect on the OH⇔NH equilibrium. The problems of the thermochromism in the crystals and the solutions.

The thermochromism is the distinctive manifestation of the So state OH⇔NH equilibrium dependence on the temperature. It makes for the coloration connected with the formation and the change of the intensity of the absorption band in the visible spectral range of the NH structure absorption spectra when changing of the temperature.

2. 6. 1. The positive thermochromism (crystals).

2. 6. 1. 1. The crucial role of the aldehyde moiety and the classification of the crystal structures.

The positive thermochromism is displayed with the temperature increase in the ordered crystal structures only. Therefore it is obviously that the positive thermochromism is connected with the specific crystal lattice arrangement making for the required intermolecular interactions between the adjacent molecules.

The recent studies have shown that such interactions of the different structural fragments of the anil molecules in the crystal play the role of the different kinds. One can suggest with the high degree of probability that interaction of the just aldehyde (Ald) fragments (Ar-C=N) of the adjacent molecules plays a decisive role in thermochromism.

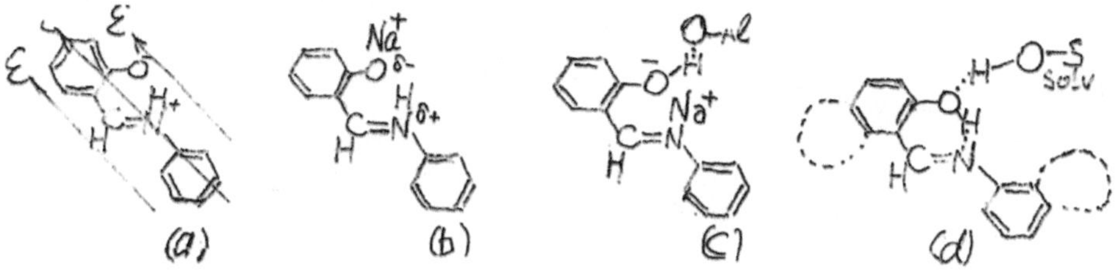

Scheme 2.11
The influence of the environment on the formation of the NH structure.
a) Polarity of the solvent, b) D*-A complex formation, c) Intermolecular H bond with the DA complex, d) Intermolecular H bond with the solvent.

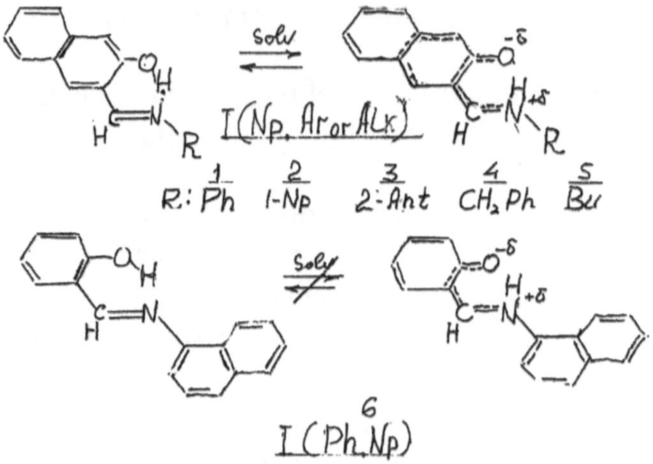

Scheme 2.12
Solvatochromic (INp,Ar or Alk imines), and nonsolvatochromic (IPh,Np)naphthalene derivatives of anils.

Table 2.9
The influence of the A-ring π-system expansion on the structural and spectral parameters and OH⇔NH equilibrium of the molecules II.

Structures	Atom distances (Å)			NMR δ(H)ppm CDCl$_3$	Absorp.spectra I_{NH}/I_{OH} Ethanol
	O--N	C--O	C--N		
II '19 [1)]	2.449	1.217	1.307	16.0	2.54
I(Np,Alk)	----	1.310 [2)]	1.300 [2)]	14.4 [3)]	1.3 [3)]
II2 [4)]	2.497	1.335	1.286	16.46 [5)]	1.04

1)All data for II19 have been taken from /33k,33m/. 2)From /14/.3) From /20,22/.4)All findings for II2 excluding δ(H) have been taken from /33c/.5) From/33n/for the related molecule II1

Table 2.10
Thermochromism and photochromism of anils I(Ph,Alk Ph)and I(Ph, Alk Pr) (Sch.2.17)/9/.

No	A bsorp.longlengthwave band max,nm			Rigid glass, 77K			Crystal
	MCH	MET		MCH	EPA	E/M	
1	345	338	(432)	Ph	P	P	P
2	322	316	(400)	T⁻ Ph	P	P	T⁺ P
3	318	313	(400)	T⁻	P	T⁻ P	T⁺ P
4	314	310	(380)	T⁻	T⁻	T⁻	T⁺ P
5	340	--------		P	P	P	P
6	334	328	(415)	P	P	P	P
7	340	334	(425)	T⁻	T⁻	---	T⁺
8	340	333	(423)	T⁻	T⁻	---	T⁺
9	343	338	(430)	T⁻ P	P	P	T⁺ P
10	326	322	(410)	T⁻	P	T⁻ P	T⁺ P
11	335	334	(425)	T⁻	T⁻	T⁻	---
12	335	333	(423)	T⁻	T⁻	T⁻	---
13	350	-----------		P	P	P	T⁺
14	320	-----------		P	P	P	P
15	311	307	(400)	T⁻ P	--	---	T⁺ P

Wavelength in the parentheses belongs to max. of the solvatochromic band.
P-- Photochromism.   MCH--methylcyclohexane
T⁺ ¨positive thermochromism.   E-Ethanol,M--Methanol.
T⁻ negative thermochromism.   EPA--Either: Isopentane:Ethanol 1:5:5

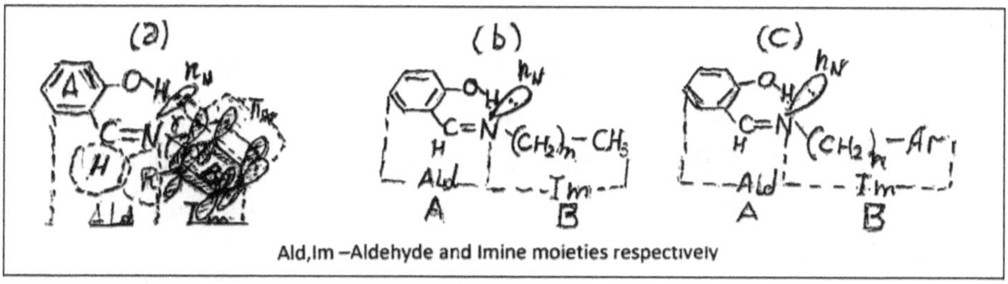

ESIPT Photochromism   59

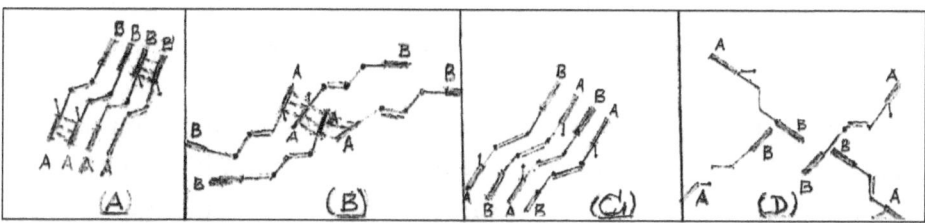

Scheme 2.13
Modification (a) and elimination (b) (c) of the n_N-π_{Ar} interaction in the anil structure.

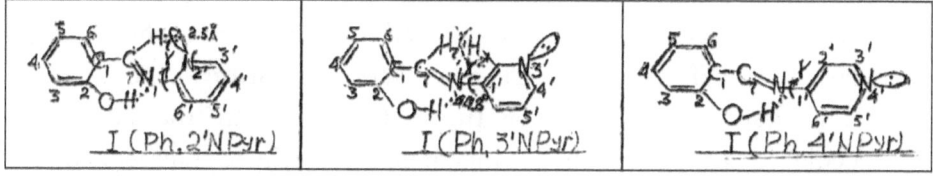

Scheme 2.14
The classification of the crystalline packing within the separate domains of the anil structure

| I (Ph, 2'NPyr) | I (Ph, 3'NPyr) | I (Ph, 4'NPyr) |

Scheme 2.15
The correlation between efficiency of thermochromism and the value of the dihedral angle γ in the crystalline pyridine derivatives.

This interaction is the necessary condition and the main driving force for the coordinated GSIPT reactions in all molecules included in the "crystalline quasiaggregate" (see below, sec. 2. 6. 2. 2.).

The imine moiety N-R plays the double role due to

(i). the electronic interactios between C=N group and N-Aril structural fragment (n-π interaction) and

(ii). the steric interactions between the imine moieties of the adjacent molecules in the crystalline packing.

The delocalization of n(N) electrons and the appropriate decrease of the N (C=N) atom basicity are provided by the n(N)-π(Ar) interaction. Thus the decrease of the latter leads to the stabilization of the NH-form (ΔE) and to the increase of the GSIPT reaction efficiency as compared with the nonthermochromic crystal structure (tab. 2. 3)(see /3/and /74/).

The n-π interactions can be modified by various ways (Sch. 2. 13): introduction of the bulky substituents to the 2' position of the phenyl ring (increase of the dihedral angle γ and n-π interaction), the flattening of the N-Ar moiety in the crystal packing (decrease of the γ angle and n-π interaction)(Sch. 2. 13a) or can be eliminated due to the substitution of the N-Aril by N-Alk radical(Sch. 2. 13b)or by the inclusion of the Alkyl spacer (Sch. 2. 13c). The role of the n(N)-π(Ar) interaction can be illustrated by comparison of the findings for the structures 7and 8 (tab. 2. 3) with the identical substituents in A ring which have almost equal ΔE values. At the same time the thermochromic compounds have been grouped in the two categories/3/ –with the high (3, 6) and the low (2, 4, 7, 8) ΔE values that may be assigned to the variations of both the planarity of the imine moiety and the substituent nature in the A-ring.

The steric interactions of Imine fragments in the adjacent molecules influence also the arrangement of Aldehyde fragments and therefore modify their interactions. The supposed mechanism of the thermochromism in the crystals based on the conception of the decisive role of the interaction of aldehyde structural fragments belonging to the adjacent anil molecules allows to propose the classification of the crystal structures according to the symbolic scheme(Sch. 2. 14).

Type A. The close stack-like packing of the flat molecules with "head –to-head" of the adjacent ones is realized with a high probability within each domain. Such a structure provides the maximum approaching of the Aldehyde fragments and the steric inhibition of the conformational transformations in the Aldehyde fragments. Such a crystal structure is typical for N-Aryl molecules (Sch. 2. 13a) and has to show the thermochromic properties only (Sch. 2. 14A).

Type B. The crystal structure provides the closing of the Aldehydes fragments, and their efficient inductive interactions. The strengthening of the IHB and the weak steric or specific inhibition of the conformational changes in the Aldehyde fragment are caused by the peculiarities of both molecular and crystal structures. Such a crystal packing is typical for the molecular structures (b)and (c) (Sch. 2. 13), and can be both thermo and photochromic. However they can be only thermonochromic if the steric or specific interactions in the crystal lattice prevent the changes of the conformations connected with the formation of the photocolored structure(see bellow).

Type C is characterized by the crystal lattice in which the Aldehyde fragments of the adjacent molecules cannot interact owing to the peculiarities of the molecular structure or(and) too long distance between them. At the same time the steric interactions or the specific molecular structure exclude the conformational changes in the Aldehyde fragments. The crystal lattices meeting such conditions can be formed by the all, (a), (b), and (c) molecular structures (Sch. 2. 13), and have the various structures (including the packing from Sch. 2. 14). The crystals of such structures do not manifest both thermo and photochromism (see below).

In the D type crystalline structure the Aldehyde fragments of the adjusting molecules are also located unfavorablfor the interactions but the conformational changes in this moiety cannot be prevented by any way. All molecular structures (a)-(c) (Sch. 2. 13) can form the crystals of this type and will be photochromic but not thermochromic under such conditions.

The A structure type has been studied in detail in the earliest studies(/74, 75/and the references there including the classical works) (see Ch. 1). According to the data obtained (see also /76/) a flattening of the acoplanar bare molecule (γ=30-60°, tab. 2. 1)to form a crystal packing of the type A requires not more than 2. 6 kcal/mol and the interaction energy of the adjacent molecu-les is about 10kcal/mol. As a result the flat molecules can form the face-to-face stack-like close crystal packing with the short intermolecular contacts between the aldehyde fragments (about 3. 4kcal/mol)of the adjacent molecules in the stack and the colored NH form is stabilized by the dipole-dipole interactions. These investigations were continued in 70-90th. The novel thermochromic structures of A-type have been synthesized and studied in detail /74, 77-79/. The principle idea consists of the creation of the flat stable molecules in the crystal by introduce of the N-heteroatom into the 2'aniline ring position for the exclusion of the CH(ring)--CH(C=N) steric interactions(Sch. 2. 15).

Such a structural modification not only excludes the steric interactions of H_7 and H_2' atoms but also stabilizes the flat structure owing to formation of the weak C_7-H_7...N_2' bond /77/. Such molecules form the stack-like close packing with interacting Aldehyde fragments with excluding of the conformational transformations typical for photochromic reactions. Thus such crystals show only thermochromism and belong to the A type.

The inclusion of the N-atom into meta (3') position, I(Ph, 3'NPyr), leads to a little acoplanarity of the pyridine ring ($\gamma=15°$) and deformation of the amine fragment even in the crystal /74, 79/. These crystalline compounds have only a weak thermochromism. From author's view it points to the monotonous correlation between efficiency of thermochromism and the value of the dihedral angle γ (Sch. 2. 15). Really, such a correlation can apparently reflect the dependence of the IHB strength on the n-π interaction (see Sch. 2. 13a). Thus these structures can be attributed to the A type also.

The effect of the n-electron lone pair of the pyridine ring does not count in the molecules of the para-substituted aminopyridines, I(Ph, 4'NPyr), and their behavior in the crystal is similar to that of I(Ph, Ph) molecules. Therefore the thermochromism and photochromism for such molecules have been found to be mutually exclusive properties and the thermochromic crystals appertain to the type A. The novel data on the different types of the crystalline structures have been obtained for the N-tetrachlorosalicylideneaniline derivatives /64, 65/(Sch. 2. 16).

The first crystalline compound, I(3-6ClPh, Ph), has the typical stack-like close packing with the short distance (about 3. 4Å) between the adjacent molecules. In spite of such a crystalline structure the usual thermochromism is manifested very weak although extremely short IHB has been found and photochromic properties are not observed.

In our view such a behavior is caused by a strong modification of the Aldehyde moiety interaction due to the full substitution of Cl atoms in the A-ring, and this structure can be designated as the A type with the thermo-chromism weakened by the peculiarities of the molecular structure. The second derivative of this aldehyde, I(3-6Ph, Prn), belongs to the crystal structure of the C-type (see below).

The B-type structures are constructed by the compounds that show both thermochromic and photochromic properties and have been revealed among the molecular structures of C type (see Sch. 2. 13 and 2. 17)/9/(Tab. 2. 10, structures 2, 3, 4, 9, 10, 15). However some of them display only thermochromic properties(Tab. 2. 10, structures 7, 8).

One can expect just B-type crystal packing must be realized for all above molecular structures (with exception of the structure 13). Indeed the crystalline packing of the typical molecule 10 (Tab. 2. 10) which crystal has the both properties, satisfies exactly to the signs of the B type.

This type of the crystalline packing is typical also for the compounds I(Ph4OMe, AlkTp)(struc. 16) with both properties in which the Ph or Pyr rings are replaced by the five membered triennial ring /80/. These findings testify also to the decisive role of the interactions of the Aldehyde fragments in the thermochromic mechanism. The convincing example in favor of such a notion about the Aldehyde fragment's role is the clear thermochromism of the crystals of the 3-OH substituted molecules, I(Ph3OH, CH_3)/6/(see also sec. 2. 4. 1). The favorable arrangement of the Adehyde fragment of the adjacent molecules in the crystals is ensured by the formation of the E-E (or E-K) H-bonded dimers with the participation of the 3-OH-groups (see Sec. 2. 4. 1 and Sch. 2. 4a, b). In such dimers the intermolecular distance O_2-O_3 is very short (2. 70 Å), and in the pseudocyclic chair conformation the distance between the two molecular planes(0. 87 Å)provides the strong Aldehyde fragment interaction. This interaction of the H-bonded Aldehyde fragments causes both the thermochromism and the lack of the photochromic transformations.

The type C packing has been found for the model molecular structure I(Ph], Ph) (Sch. 2. 18) with the excluded rotation (or twisting) of the A-ring /36/. The not very compact "head to tail" crystalling packing formed by the acoplanar molecules (γ=51°) with the distance between the main molecular planes about 3. 5-4. 0 Å provides the great space between the Aldehyde fragments of the adjacent molecules.

As one should be expected the thermochromism and photochromism are absent in spite of the strong IHB (O . N distance is 2. 64 Å). Another compound, N-tetrachlorosalicylidene-1-pyrenylaniline, I(3-6ClPh, Pn), /64/ (Sch. 2. 16) forms the crystalline packing of the planar molecules ("head-to-tail") with the short interplanar distances (3. 37-3. 39Å) but the large space between Aldehyde fragments of the adjacent molecules. Thermochromism is absent but the intermolecular charge transfer interaction takes place(see also 2. 4. 1). The structures unfavorable for the Aldehyde moieties interactions ("head –to-tail" packing) is realized also in the nonphotochromic crystals of the molecules of 5-nitro salicyliden-Nathilimine/57/as a result of the weak intermolecular H-bonds. (see above, sec. 2. 4. 1).

Thus all the crystal packing of the C-type can be considered as the model structures conforming the idea about of decisive role of the Aldehyde fragments' interaction in the mechanism of the thermochromism in crystals.

The D-type structures has been studied in detail like A-type ones in the earliest investigations.

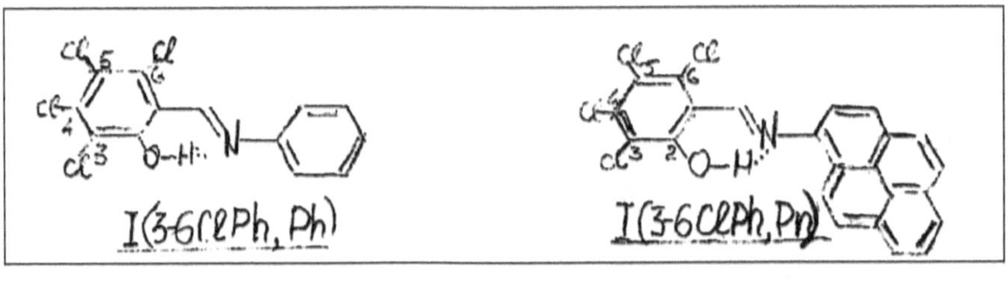

Scheme 2. 16
TetrachloroSA derivatives and photochromism.

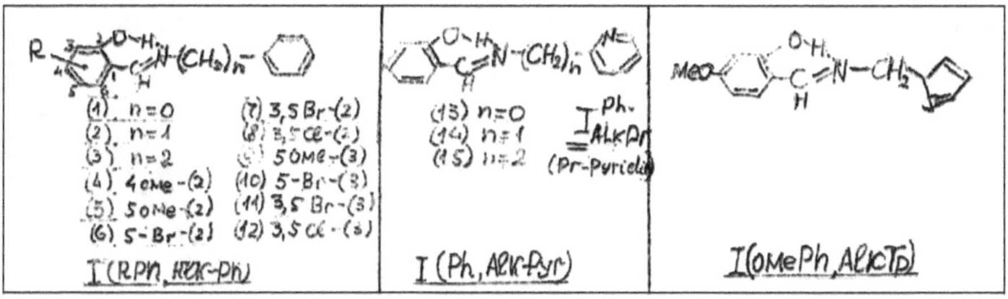

Scheme 2. 17
The molecules with (c) molecular structure (Sch. 2. 13) and with B type crystalline structure which can show both the photochromic and thermochromic properties /9/ (see table 2. 10)

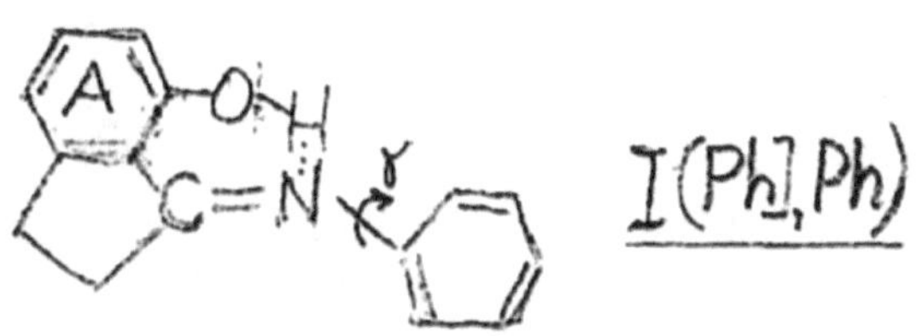

Scheme 2. 18.
The molecule with C-type crystalline packing that shows no thermochromic nor photochromic properties /75, 76/.

Recently side by side with the synthesis of the new photochromic crystalline anils the much attention has been given to the method of the change of tshe thermochromic crystal into the photochromic ones /81, 82/. These methods can be considered as a purposeful destruction of the crystal structure with intermolecular interactions that cause the thermochromism, and the creation of the crystal structures with a large free volumes giving a possibility for the conformational transformations. The results obtained by such methods also confirm the decisive role of the Aldehyde fragments' interactions in forming of the thermochromic properties in the crystals. The works connected with the investigations of Photochromism will be considered in detail in the following chapter. It follows from the findings presented above the classification of the packing arrangement by the types A-D is evidence for the decisive role of the nonspecific interactions between the Aldehyde fragments of the adjacent molecules in the anil crystals.

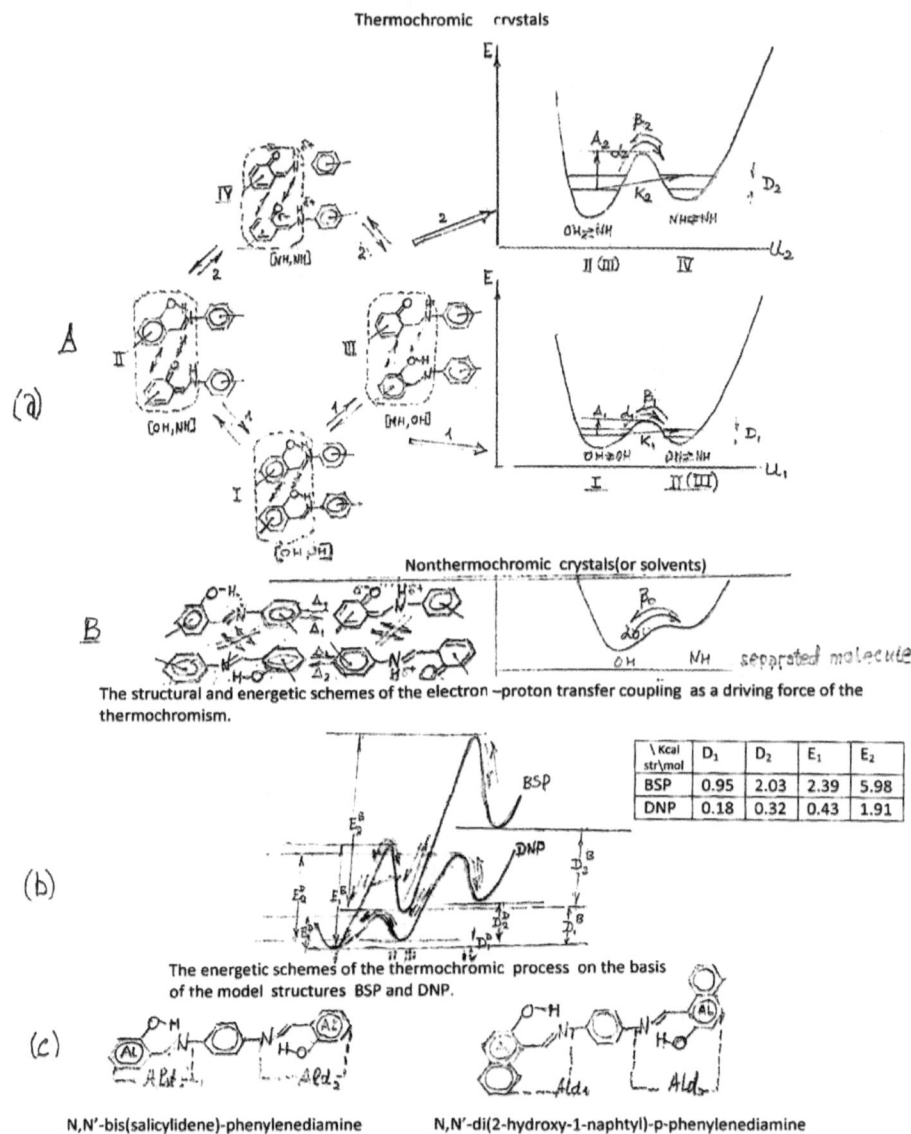

Scheme 2.19
Semiquantative view of the mechanism of the thermochromism in the crystal based on the study of the GSIPT of the model structures BSP and DNP (Taken from /70-75/, adopted and modified.(See text).

Thus the topochemical factor /80, 81/ is connected with realization of the favorable sitions of the aldehyde moieties of the adjacent molecules. With such a location their interaction becomes sufficiently appreciable for the manifestation of the cooperative effect of the OH<=>NH equilibrium shift with the temperature increase.

Thus there are two necessary conditions for the formation of the positive thermochromism:

i) The N(C=N) atom basicity increase due to diminution or exclusion of the n(N)-π(N-Ar) interaction as a result of the flattening in the crystalline packing or the modification, separation(or removal) of the N-Ar ring in the special molecular structure,
ii) the realization of the cooperative effect ensured by the specific interactions of the Aldehyde fragments of the adjacent molecules which are possible in the crystal packing only.

In this connection it should be note that the numerous attempts of the realization of the thermochromic crystalline anils on the basis of optimization of only intramolecular factors (strengthening of the H-bond, expansion of the π-ring system, flattening of the structure)/33, 44, 45/ without prediction of the possible crystalline structure seems to be not efficient.

2. 6. 1. 2. Semiquantitative conception of the thermochromic process development.

It results from the above that within the limits of each domain in the thermochromic crystal the arrangement of the Aldehyde moieties of the adjacent anil molecules in the crystalline packing is favorable to the interaction by their quasi-π-conjugated network involving the π-ring systems and Hydrogen bond (HB) cycles.

The BSP and DNP molecules (Sch. 2. 19 c) consisting from the two Salicylideneaniline, I(Ph, Ph), or Hydroxynaphthalideneaniline, I(Np,

Ph), moieties respectively and interacting via the π-aromatic ring system may be a proper models for the discussion of the semiquantitative conception of the anil thermochromism in the crystals.

The discussion conducted below is based on the assumption that interactions between Aldehyde moieties in the model BSP or DNP molecules are analogues to those of stacking of molecules in the crystal lattice.

The temperature dependence of the molecular and crystal structures and OH<=>NH tautomeric equilibrium under influence of the interaction between the Aldehyde moieties of BSP and DNP molecules (and of corresponding 1, 6 Pyrene derivative /71/) has been studied comprehensively by the spectral (IR, UV-Vis), the low temperature X-ray structural analysis, and the combi-nation of the ^{1}HNMR relaxation, and low temperature (~26K and 130-400K) ^{15}N-NMR chemical shift measurement /70-73/. The temperature dependence of the spin-lattice relaxation rate of the proton and ^{15}N chemical shift in the BSP and DNP thermochromic crystals are associated /72, 73/ with the proton transfer (GSIPT) in the two interacting Aldehyde moieties (Al_1 and Al_2) (Sch. 2. 19c).

Such comprehensive measurements have allowed to determine the relative energy differences (D_1, D_2), the activation energy (A_1, A_2), and the classical and tunneling rates (α_1, α_2 and β_1, β_2 respectively) for four sets of the interacting structures /72, 73/.

According to the above assumption of the model structures the discussion below can be related to the divided stacking interacting anil molecules in the domains of the thermochromic crystals.

Therefore in the Sch. 2. 19c the interacting halves of the structures BSP and DNP have been replaced by the sets of the corresponding anil structures interacting by their Aldehyde or Keto moieties in the quasiaggregates, and the following discussion applies equally to the latters.

The electron-proton coupling in the frames of the adjacent Aldehyde moieties is the charge-proton transfers' interaction that can be considered as a peculiar kind of the electron-"phonon" coupling in the crystal where the proton transfer along with the assisted low frequency vibration modes can operate as a kind of "phonon". Such a coupling means that GSIPT in one Hydrogen Bond induces the modification of the potential function for the GSIPT in other Hydrogen Bond and vise versa. The GSIPT OH⇔NH in the set of the quasiaggregates I ⇔(II⇔III)⇔IV proceeds by both the tunnel (k_1, k_2) and overbarrier(αi, βi) mechanisms (Sch. 2. 19a, A). The height and the shape (variation of the width) of the potential barrier for the GSIPT in the quasiaggregates depend strongly on the interaction between the Aldehyde moieties (OH-form) of the adjacent molecules; it is the necessary factor by which the crystalline thermochromism is caused.

The another very important condition is the marked stabilization of the colored NH structure as a result of the dipole-dipole interaction between the "keto⇔zwitterion" structural fragments of the adjacent NH structures in the quasiaggregates in the thermochromic crystals.

Such interactions can play important role in the crystals with the double (thermophotochromic) properties /9/ (B-type crystal structures, Sch. 2. 14).

The shift of OH⇔NH equilibrium towards NH structure (especially in IV) is dependent strongly on their stabilization in the quasiaggregates.

The Zw nature of the NH-form depends strongly on the molecular structure, and can be stabilized also in the polar solvents (solvatochromism, see above) and at very low temperature (see the negative thermochromism). In the liquid nonpolar solvents and nonthermochromic anil crystals the NH structure is usually unstable and decay fast to the OH form (Sch. 2. 19a, b).

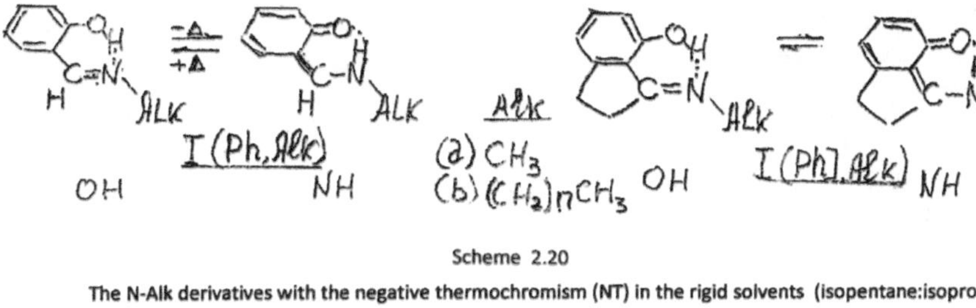

Scheme 2.20

The N-Alk derivatives with the negative thermochromism (NT) in the rigid solvents (isopentane:isopropanol 4:1 (IIP), 77K).

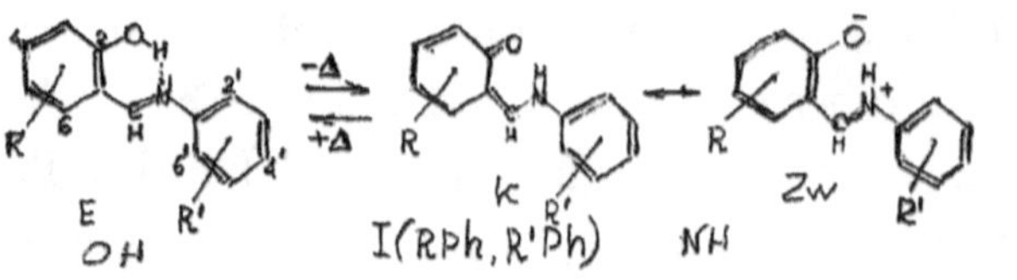

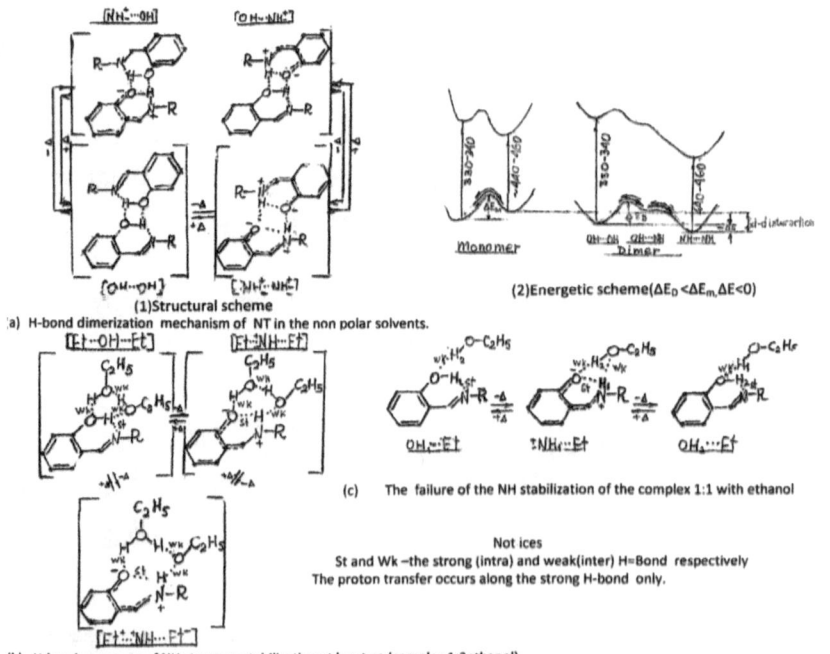

Scheme 2.21
The SA derivatives with the NT in the nonpolar solvents (75-300K)/50/.

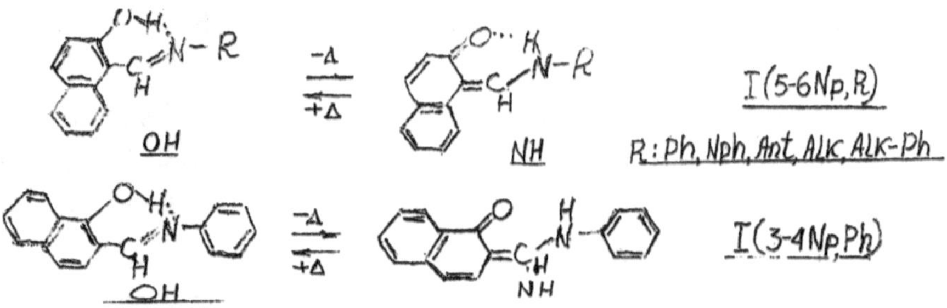

№	Position			
	3	6	2'	6'
1	H	H	H	H
2	H	Cl	H	H
3	H	OMe	H	H
4	H	NO_2	H	H
5	H	NO_2	CH_3	CH_3
6	Tert-butil		H	H

Scheme 2.22
The expanded π-system in anil structures with the NT in the nonpolar solvents.

(a) H-bond dimerization mechanism of NT in the non polar solvents.
(1) Structural scheme
(2) Energetic scheme ($\Delta E_D < \Delta E_m, \Delta E < 0$)

(b) H-bond aggregate of NH structure stabilization at low t-re (complex 1:2 ethanol)

(c) The failure of the NH stabilization of the complex 1:1 with ethanol

Not ices
St and Wk – the strong (intra) and weak(inter) H=Bond respectively
The proton transfer occurs along the strong H-bond only.

Scheme 2.23
Intermolecular mechanisms of the Negative Thermochromism

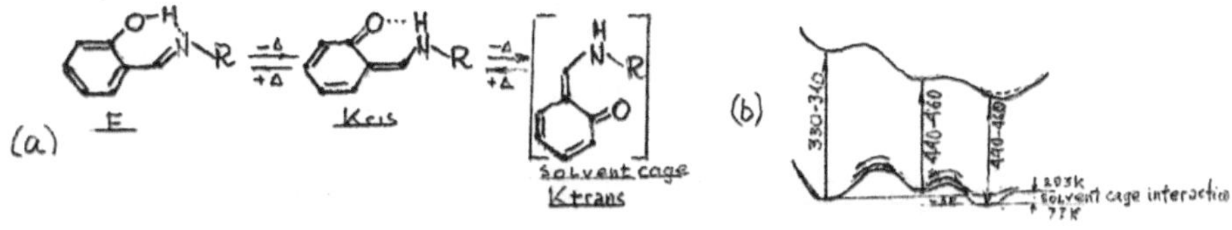

Scheme 2.24
Quasimonomolecular mechanism of the Negative thermochromism in the nonpolar solvents

Thus the thermochromic process in the crystal proceeds in the direction of the shift of the tautomeric equilibrium OH<=>NH towards NH-structure, i. e. the set of the quasiaggregates I, II-III, IV towards II(III) and especially IV with the increase of the temperature(Sch. 2. 19b).

The activation energy of the formation of the quasiaggregates with colored NH structures falls sharply and their stability increases with the expansion of the π-system and hence with intensification of their interaction (Sch. 2. 19b, compare BSP and DNP).

So that the thermochromism of the anil crystals(within the temperature range 70-300K) is simulated perfectly by the DNP structure (with the expanded π-system).

The intensification of the π-π and dipole-dipole (in the NH-structures)interactions can be promoted by the extension of the aggregates with the rise of the amount of the coupling molecules within the crystalline domains.

Thus the thermochromism (unlike photochromism)can be considered as a bulk crystalline, collective-like phenomenon which depends however on the shape and the size of the domains especially near the crystalline surface. In such a situation the photochromism is prevented by thermochromism due the strong electron-proton coupling and the steric interactions of adjacent molecule near crystalline surface.

Therefore the thermo-photochromism can also be considered as a nanoscale size dependent penomena in the crystal (see also below).

2. 6. 2. The Negative thermocromism –"cryochromism"*) (solution).

2. 6. 2. 1. Review and discussion of the experimental findings.

The Negative Thermochromism (NT) is manifested as an appearance and increase of the long-wavelength absorption band (λ^{max}=400-440nm)intensity with decease of the solution temperature below 150K and close to the rigid-glass formation (usually 77K).

The studies of the Cryochromism have been carried out in the series of the earlier works/11, 26b, 74, 76, 82/and especially recently/28, 33k, 33n, 45, 54/.

In spite of the likeness of this absorption band to the thermochromic and also solvatochromic one in the crystal and the polar solvent respectively it is impossible obviously to explain its appearance and growth with the drop of the solution temperature by a trivial shift of the OH⇔NH equilibrium towards the less stable NH-structure which may be either little polar or zwitterionic one depending on environment(see sec. 2. 5). Although the belonging of this absorption band to the NH-structure is not a subject to doubt, the mechanism of the equilibrium shift requires the complementary studies.

For the simplest structures (the (b) type(Sch. 2. 13), I(Ph, Alk), and their "fastened" structural analogs, I(IPh], Alk)(Sch. 2. 20))the equilibrium dependence on temperature in isopentane-isopropanole mixtuture (4:1)(IIP)has been studied /28/by UV-Vis spectroscopy (at 77 and 290 K) and quantum-chemical (semiempirical and ab-initio) calculations. The equilibrium shift towards NH(K)structure with the decrease of the temperature is observed clear for all I(Ph, Alk) and I(Ph], Alk)compounds unlike corresponding Aryl derivatives, SA, I(Ph, Ph)

*)This term has been proposed in /9/and more exactly corresponds to experimental findings. (Tab. 2. 11). It depends strongly on the molecular structure: the sharp shift is observed for the molecule with the short Alkyl radical R=CH_3 or for the rigid model structure I(Ph], Alk). The combination of the both structural factors, I(Ph], CH_3), leads to the very strong shift towards the NH-structure. The concentration

effect (10^{-5}-10^{-2}M) on the equilibrium and its temperature dependence are not observed, and quantum–chemical data do not indicate existence of any stable aggregate (or dimer) structure. Therefore authors /28/ have been forced to assume the "quasi" monomolecylar explanation for this unusual phenomenon.

The quantum–chemical study has not allowed to find out any transient conformation that can be responsible for the inversion of the E(So) and K(So) energy position.

Meanwhile the assumption about molecular structure distortion in the solvent cell at the low temperature allows to point to the simple criteria (see Sec. 2. 4) for the bondlength C_1C_2 (l=1. 397Å) above which the NH structure becomes more preferable. However there is not evidence for such a low temperature distortion.

At the same time the similar interpretation based on the thermodynamic considerations is discussed also by authors /44, 45/ for the molecules with the expanded A-ring π-system (see below). The Negative thermochromism for solutions of the more complex, C-type, structure. I(Ph, Alk-Ph), (Sch. 2. 17) in nonpolar (MCH, IP) and polar, protic (EPA, E/M) solvents have been studied /9/ (Tab. 2. 10, Sch. 2. 17 and the structural numeration there).

The cryochromism is manifested by the majority of them in the nonpolar solvents (2-4, 7-12,15) and by only part of them (3, 4, 10-12) in the solvents of the both types. Some (5, 6, 13, 14) do not show cryochromism at all.

The typical features of the compounds I(Ph, Alk-Ph) is following:

1) All Negative thermochromic compounds are also Positive thermochromic in the crystals.
2) The n(N)-π(Ar) interaction, i. e. the delocalization of the n(N) electrons along the π-Arylimine system prevents the Negative thermochromism (e. g. compounds 1, 13).
3) The relative intensity of the thermochromic absorption band, J(Th, rel) grows clear with the decrease (but not increase!) of the concentration in the nonpolar solvents (e. g. MCH).

4) In the nonpolar solvents the I(Th, rel) increases sharply with the rapid cooling as compared with the gradual cooling.
5) The features 3) and 4) are absent in EPA and E/M.

The above findings are the evidence of the decisive role of the intermolecular interactions in the mechanism of the equilibrium shift towards NH-structure. This mechanism involves the formation of H-bonded dimers in which the molecules are packed by head-to-head in the nonpolar solvents. The H-bonded complexes with the molecules of the protic solvent prevents the formation of dimers and hence the NH-structure generation (item 5).

However the difficulties arise with the explanation of the reverse concentration dependence in the nonpolar solvents (item 3). The impression arises that formation of the dimers is not fovarable to the effective production of NH-structure at low temperature (this is natural if the dimers are the H-bonding complexes). However such a situation is hard to explain the positive (but not negative) effect of the cooling speed (item 4). Besides it a dimers' decay cannot explain why the equilibrium is been shifted towards the less stable monomeric NH-structure with the temperature decrease.

At the same time the concentration effect has not been found for the structures 7, 8 in both MCH and EPA in the earlier work /74/. And furthermore in the mixture IP:MCH (3:1) the Negative thermochromism has been also observed in the liquid solvent (-80 and -120 °C) when the thermochromic form was retained for a long time indicating to its high stability. At the same time it has been found out /50/ that all structures of the type I(PhR, PhR')(Sch. 2. 21) besides the compound 6 in the nonpolar solvents (isopentene and isopentene methylcyclohexane mixture) display a strongly pronounced Negative thermochromism at the concentrations within $2.2-4.6 \times 10^{-5}$ M with the almost full shift of the OH$\Leftrightarrow$NH equilibrium towards NH-structure when decreasing of the temperature from 300 to 77 K.

Such results in the nonpolar solvents at low temperatures suppose the aggregates are formed by the salicylidene molecules in which the

NH structure is strongly stabilized. The evidences of the aggregation as follows:

1. The NH form appears at higher temperatures in the solutions of the higher concentrations.
2. The Negative thermochromism is lack for the compounds 1-5 in EPA where aggregation is in-hibited by the intermolecular hydrogen bonding with ethanol.
3. The aggregation of the molecules 6 is prohibited by the bulky tert-butyl groups in the A-ring.

According to /50/ the Zw structure makes the considerable contribution in the NH-form being strongly stabilized in the aggregates by the electrostatic intermolecular (dipole-dipole(d-d))interactions.

At the same time it is necessary however to note that:

i) Such huge variations of the equilibrium in salicylidenephenylimines under the same conditions have not been observed earlier or reproduced later by another researchers and
ii) With the full shift of the equilibrium towards NH form the photochromism cannot be manifested in the glassy solvents at 77K due to the photostability of the NH structure(see below), and this is contradiction to the numerous experimental findings.

Meanwhile the findings /50/testify to the realization of the aggregate mechanism.

The strongly pronounced Cryochromism has been found in the solutions of the compounds with the expanded ring π-system I(5-6 Np, R) and I(3-4Np, Ph) (Sch. 2. 22) in the polar and nonpolar solvents (/44, 45/ and /54a, b/)by the fluorescence excitation spectra(see also sec. 2. 4. 2.)

It results from the table 2. 12 the percentage (X%) of the NH-structure increases considerably in both nonpolar (cycohexane, CH, metylcyclohexane, MCH, toluene, Tn)and polar(ethanol, Eth) solvents as compared with the gas phase at room temperature and reaches 100% at low temperatures (T≤100K). Any spectral evidence for dimerization has not been found by authors at the concentrations up to 10^{-4}M. Furthermore the spectral variations in the MCH/Tl mixture and Eth are identical despite the aggregation(dimerization) is unlikely in the latter owing to the specific solvation. According to /44, 45/ the both observation are the strong support for the quasimono-molecular nature of the Negative thermochromism of these compounds.

Although the authors do not produce any evidences they are inclined to the opinion that "there is indeed a shift in the thermodynamic parameters around 200K in MCH-Tn mixture which in turn could imply that cooling and compression of the solutes are forced in different configuration".

It is interesting that in /28/authors also come to such a conclusion for the compounds I(Ph, Alk)(see above). However in the work/54a/ the Negative thermochromism of the compound I(5-6Np, Ar) in the nonpolar solvent caused by the analogous equilibrium shift towards NH-structure with the lowering of the temperature is explained unlike /44, 45/ by formation of the dimers.

In the compounds with the Alkyl substituents in the imine bridge-CR_2=N-(R_2=Alk)/33/ (see above, sec. 2. 4. 3., Sch. 2. 8)the Negative thermochromism is observed /33n/ within the temperature range 298-110K even in the liquid CH_2Cl_2. Under such conditions the equilibrium constant k=[K]/[E] and K-structure stability(ΔG) increase considerably, and the IHB(O-H ...N) (chemical shift δ)is strengthened markedly(Tab. 2. 13). The same NH-structure band appears in the fluorescence (λ^{max}=500nm)excitation spectra (λmax=400nm) in the glassy polar solvents (77K) for the related structure II_3(R_1=CH_2Ph, R_2=C_2H_5)/33d, f/.

The data produced above point to the strong equilibrium shift towards NH –form caused by the stabilization of this structure with H-bond strengthening owing to the proton displacement to the N-atom under temperature decrease. However the role of any structural factors or intermolecular interactions including a possible dimer (aggregate) formation in the equilibrium shift has not been studied in /33n, d, f/, and the-refore the mechanism of the Negative thermochromism for these compounds remains unclear.

As it was discussed above (Sec 2. 4) the influence of the substitution can change considerably a spectral behavior of molecule in the crystals with variation of the temperature. However the first observation of the crystal structural change for the thermochromism of the substituted N-salicylideneanilines with use of variable temperature X-ray analysis and UV spectroscopy of the crystalline I(5ClPh, 4'OHPh) has been performed in /47, 49/(see also above, Sec. 2. 4. 1. Sch. 2. 5). According to the data produced such crystals show an unique behavior: UV spectra show that crystals are negative(!) thermochromic.

Table 2.11
The dependence of the equilibrium on the temperature for Alkylimines, I (Ph,Alk) (see Sch.2.20)

Compounds	UV spectra (in IIP)			K=[NH]/[OH]		PMR
	λ^{max}_{OH} nm		λ^{max}_{NH} nm			δ (H$_{(O)}$)
	290K	77K	77K	290K	77K	ppm
I(Ph,Alk) (a)	317	320	390	<0.01	7.0	13.38
I(Ph,Alk) (b)	317	320	390	<0.01	0.32	13.61
I(Ph],Alk)(a)	310	315	390	<0.01	20.0	---
I(Ph],Alk)(b)	311	313	400	<0.01	2.8	11.55
I(Ph,Ph),SA	340	342	430	<0.01	<0.04	12.24

Table 2.12
The negative thermochromism of the structures with the expanded π-ring system /45,47/.

compounds	Solvent	Temperat. Tk	X_{NH} %	UV absorption, nm		X% (Gas) [2]	$E_{NH} - E_{OH}$ Kcal/mol
				OH-form	NH-form[1]		
I(5-6 Np,Ph)	CH	300	60	424	492	8.2	1.848
	MCH/Tl	100	100	424	542,500		
		300	48,5	376	465		
	Eth	100	100	376	440		
I(3-4Np,Ph)	CH	300	8.3	376	460	16.7	0.89
	MCH/Tl	100	100	376	460,440		
		300	59.8	383	480,453		
	Eth	100	100	383	482,455		

1)The bands appearing as shoulders are not shown. 2)X%=k$_T$/1+k$_T$ x100 k$_T$ is taken from table 2.2 /45/.

The absorption band at λ^{max}=485nm which assigned to the NH form, appears at the room temperature and increases in intensity with the lowering of temperature.

The similar results have been obtained by X-ray analysis (Tab. 2. 14). Realy the C_2--O and C_1--C_7 bonds are shortening and C_1--C_2, C_7--N ones become longer with the temperature fall from 375 to 90K that corresponds to the increase of the NH structure contribution in the OH⇔NH equilibrium under such conditions. At the same time even at 90K the above bond lengths differ strongly from those expected for a typical "Keto" (K)-structure(Sch. 2. 5 and Tab. 2. 14). It means that the NH-form in the crystalline I(5ClPh, 4'OHPh) has the character of the Zw mainly. Thus the UV-spectra and X-ray analysis findings obtained /47, 49/ lead to the conclusion that there is an equilibrium OH⇔NH in the crystal, and the percentage of the NHZw form increases with lowering of the temperature.

Contrary, to the crystalline state the compound I(5ClPh, 4'OHPh)exists exclusively as the OH structure in the solution, and the stabilization of the NH-form in crystal is therefore ascribed to intermolecular interactions. Really the molecular packing of the crystal I(5ClPh, 4'OHPh)reveals the occurrence of the intermolecular Hydrogen bonding $O_2 \ldots H-O_4'$, and stabilization of the NH-form with the lowering of the temperature (Negative thermochromism) results primary from the intermolecular bonding in the crystals. Sss

These results have the general meaning and from our viewpoint can throw light on the Negative thermochromism in the solutions also. At the same time it should be noted that the phenomenon described above is the sole case of the Negative thermochromism for the crystalline anils that unfortunately could not be reproduced by another studies /67/.

Unfortunately the results discussed above illustrate the absence of the coordinated experimental findings for the related compounds. And so, the notions of the different authors about the nature of the Negative thermochromism differ very much (see above and Tab. 2. 15). Realy the observed concentration dependence of the equilibrium shift and consequently the role of aggregation in these systems are discussed for the solutions /9, 50, 54a/ and for the crystal of the specific structure /47/.

However in /50/ the Negative thermochromism and concentration dependence are observed only in the nonpolar solvents unlike /9/where the Negative thermochromism is observed in both solvent types but a concentration dependence has been found only in the nonpolar solvents and it has the reverse direction. At the same time in /28, 44, 45/the concentration' dependence has not been found even in the nonpolar solvents, and another mechanism is discussed which is connected with the deformation of the molecular structure in the cell of the cooling solvent. Although the concentration dependence has not been studied in /33/but the manifestation of the Cryochromism in the polar solvents also testifies against the mechanism caused by the dimer formation. Such a situation can testify to the differing mechanisms of the phenomenon that is sensitive to both molecular structure and environment nature. Thus the mechanism of the Negative thermochromism (Cryochromism) requires subsequent more detailed investigations.

Table 2.13
Thermochromic behavior of the compound II$_1$ (see scheme 2.8).

T$_K$	k	ΔG kJ/mol	T$_K$	δ (H) ppm
298	0.14	4.87	293	16.46
274	0.20	3.67	273	16.68
256	0.25	2.95	253	16.90
237	0.29	2.44	233	17.13
217	0.37	1.80	213	17.34
185	0.50	1.07	203	17.44
179	0.63	0.69	193	17.55

Table 2.14
The length (Å) of the selected bonds of the NH form in the crystalline compound I(5ClPh,4'OHPh) compared with standard lengths for the keto –structure (Schem.2.5)

T$_k$	C$_2$:::O	(C$_2$=O)$_{keto}$	C$_1$:::C$_2$	(C$_1$=C$_2$)$_{keto}$	C$_1$:::C$_7$	(C$_1$–C$_7$)$_{keto}$	C$_7$:::N	(C$_7$–N)$_{keto}$
375	1.320	1.222	1.414	1.464	1.434	1.340	1.288	1.355
90	1.310		1.433		1.425		1.308	

It is seen good from the table that the NH structure becomes closer to keto one when the temperature falls down.

Table 2.15
The main types of the compounds with the Negative thermochromism (NT) of different nature including the role of molecular aggregates

Compounds	NT in solvents 1/ Nature of the solvents		Aggregate formation 2/ Nature of solvent		References
	Polar	Nonpolar	Polar	Nonpolar	
I(Ph,Alk)	?	+	--	--	/28/
I(Ph,Alk-Ph)	+	+	--	+	/9/
I(Ph,Ph)	--	+	--	+	/50/
I(5,6Np,Ph)	+	+	--	--	/44/
I(3,4 Np,Ph)	+	+	--	--	/45/
I(5,6 Np,Alk)	+	+	+	+	/54/
II(Ph,Alk)	+	+	--	--	/53/
I(5Cl Ph,4'OHPh)	Negative thermochromism in the crystal !				/47,49/

1/+ NT is clearly observed ,-- NT is not observed.2/+Formation of aggregates(concentration dependence is clear),--Aggregate formation is doubtful.

2. 6. 2. 2. The comparison of the Negative and Positive thermochromism natures, and the possible mechanisms of the Negative thermochromism.

In spite of the considerable vagueness in the experimental data and in the understanding of the nature of both Positive (PT) and especially Negative (NT) thermochromism there are the sufficiently clear similarities and differences between these phenomena.

The similar features are following:

i. The both PT and NT are caused by the shift of the tautomeric equilibrium OH⇔NH towards NH-form as a result of the fast reversible GSIPT along the very strong intramolecular H-bond OH…N and therefore they lead to the almost identical spectral variations with change of temperature.
ii. In both phenomena the equilibrium shift is regulated by the similar intramolecular factors which influence the H bond strength and the GSIPT efficiency. Therefore the PT and NP are maninfested mainly by the same molecular structures (see e. g. Tab. 2. 10).
iii. In spite of their intramolecular nature the PT and NP can be realized also under the influence of the intermolecular interactions. The both "quasimonomolecular" processes can be stimulited by the intermolecular interactions and depend on the specific environment favorable such interactions.

The main differences are following:

i. The relative NH structure stabilization energy $G(NH)-G(OH) = \Delta G(NHrel) > 0$ (like a bare molecule) and $\Delta G (NHrel) < 0$ for the PT and NT molecular systems correspondingly (Sch. 2. 23a, 2. 24b) leads to the change of the thermodynamically caused NH-structure population i. e. its increase or fall with the temperature growth for PT and NT respectively.
ii. Such a radical difference between the NH forms' relative stability is caused by the principal distinction between the intermolecular interactions in the crystals and the solutions for the PT and NT systems correspondingly. In the PT the OH⇔NH equilibrium shift towards NH structure with the temperature increase are provided by the nonspecific

electrostatic interactions between the aldehyde moieties of the adjacent molecules in the crystalline packing (Sch. 2. 14A, B) (see above). Such interactions promote the higher efficiency of the NH structure formation when temperature increasing and necessary condition for the PT. Unlike the PT the necessary condition for the NP is the strong stabilization of the NH structure relatively OH one that provides its accumulation when lowering temperature.

The most admissible mechanism of the NH structure stabilization can be connected with dimerization of the anil molecules in the nonpolar solvents. However the contradictory of experimental findings concerning the intermolecular interactions(including dimerization) gives possibility to suggest another mechanism. Competition between those mechanisms is determinated by the molecular structure and the nature of the medium.

The possible mechanism in the nonpolar solvents can be connected with formation of the dimers of the H-bonded Enol(E))molecules in the solvent cage at low temperatures(Sch. 2. 23a_1,) The weak intermolecular H-bonds ($N_1H_1...O_2$, $N_2H_2...O_1$) promotes the GSIPT ($O_1H_1 \rightarrow N_1H_1$ and $O_2H_2 \rightarrow N_2H_2$)that can proceed by step-by-step or by the connected double proton transfer mechanisms with formation of the two NHZw structures stabilized by the strong dipole-dipole(d-d) interaction. The concentration and the cooling rate dependences /50/is a manifestation of such a mechanism.

The dimer formation in the polar solvents is inhibited as a result of the insulation of the molecules by the salvation shells, and the NT caused by the dimer mechanism is not realized.

At the same time in the protic solvents the NHZw structure can be generated in the aggregate of the three molecules (two solvent and one solute)as a result of the GSIPT promoted by the weak Intermolecular O-H...O bonds with the solvent molecules (Sch. 2. 23b)and stabilized by the Coulomb interactions in that structure in the solvent cage. The generation of the most stable trimmer structure is promoted by the strong Coulomb interactions of the solute MHZw± with the two opposite solvent ions (Et⁻⁺NH⁻⁺Et) which are generated as a result of the H-transfer-O-H-->O between the solvent molecules (see Sch. 2. 24b). With such a mechanism the stable NHZw form can be generated in the protic solvent without a concentration dependence and with a small efficiency /9/.

Unfortunately the simple mechanism with the H-bond solute-solvent 1:1 complex (Sch. 2. 23b) cannot produce the stable NH structure due to the intermolecular double proton transfer (±NH Et→OH...Et) with the production of the initial molecular structure. Such a situation can be realized for the anil molecules in ethanol where the NT is not observed /50/. The quasimonomolecular mechanism of the NT can be connected with the generation of the NH structure which stability can exceed that of the OH form (E(NH)-E(OH)<0) when lowering temperature due to the interaction with the solvent cage.

Such a NH structure can be formed as a result of the additional structural transformation of the initial OH form with the sufficiently high relative stability not exceeding however the OH that (E(NH)-E(OH)>0) (Sch. 2. 24a). In these molecules the tautomeric equilibrium OH⇔NH is shifted strongly towards NH form. It especially concerns anil structures with the heightened ability for the GSIPT OH->NH(e. g. N-Alkylimines /9, 28/, Naphthylimines/44, 45, 54a, b/or with the bulky substituents in the imine bridge /33/). The cis –trans isomerization about one of the weakened (C=C or C=N) double bonds is the most probable additional structural transformation, and since the C=N isomerization is only possible way for the model structure I(Ph], CH_3) with the clear expressed NT properties /28/ the formation of the trans-isomer around C--N bond must be the NH structure responsible for the NH. As such an isomerization may proceed with the very high efficiency around the ordinary C-N bond, the low polar Keto-trans (K_{trans}) is the most probable low temperature colored form especially in the nonpolar solvents unlike the high polar ZwNH structure with the high C=N double bond order. The supposed K_{trans} structure is obviously extremely kinetically unstable at ambient temperatures but can be sharply stabilized in the solvent cage when temperature lowering.

With the cis-trans isomerization about C-N bond the chromophore with the Keto ringresponsible for the lowest electronic transition does not undergo to any marked variations and no spectral shift of the thermochromic band must be observed when lowering temperature in accordance with the experimental findings. Thus the quasimonomolecular mechanism of the NT involves the shift of the equilibrium Enol⇔Keto cis ⇔Keto trans towards the latter with the lowering of the temperature (Sch. 2. 24 a, b).

No concentration dependence must be observed with such a mechanism(see /28, 44, 45, 33, 54/).

Furthermore the monomolecular reaction of the cis-trans isomerization is inhibited in the dimer structures and its efficiency has to become higher with the concentration fall according to the experimental findings /9/.

Thus the experimental data on the NT can be explained qualitatively sufficiently good by the two competing mechanisms discussed above. However for the clear understanding of the many important details the investigations are required. In particular the NT efficiency dependence on the molecular structure and the nature of the environment must be studied, especially in different polymer matrixes and amorphic materials which NT properties did not unfortunately studied earlier although they can be very prospective for the low-temperature sensors and another technical applications.

Conclusions

The principle results of molecular and the crystalline structures of anils in the ground state have been obtained within last two decades by the modern spectral, structural and quantum-chemical methods.

According to the notions based on the earlier, later, and recent findings the main trait of anil molecules is the strong intramolecular Hydrogen bond O-H...N in the Enol(OH) structure resulted in the tautomeric equilibrium (teq) OH⇔NH (Enol(E)⇔Keto(K)) that is highly sensitive to the molecular structure, media nature, and temperature.

The variation of the teq in dependence on the solvent nature or temperature results in the teq shift towards NH(K) "colored" structure (absorption band λ^{max}=420-440nm) which causes the solvato or thermo chromism correspondingly. The intermolecular interactions play a clue role in both phenomenon. The solvatochromism is caused by a teq shift towards NH structure under influence of the dipole-dipole or (and) H-bond interactions with the solvent molecules.

According to the recent notions the "Positive" thermochromism in the crystals is caused by the collective (or concerted) process of the teq shift towards NH (Keto) structure with the increase of the temperature stimulated by the electron-proton coupling in the C-hydroxyphenyl moieties of the adjacent molecules in the quasiaggregates within the crystalline packing. In some cases such a thermochromic crystalline structure does not tight packing with the parallel adjacent molecules usual for the typical (classical) thermochromic crystal and has a sufficient room for the structural transformations responsible for photochromism.

Thus the combined thermophotochromic properties can be provided by such crystal structures which really have been obtained recently. This is the evidence of the new, more accurate notion of the thermochromic mechanism in the crystals. The Negative thermochromism (or cryochromism), the new phenomenon, discovered recently, is caused by the teq shift towards the NH structure with the temperature drop up to 77K. It is apparently, the teq OH<==>NH(E<==>K) shift towards NH structure is caused by its stability rise as a result of the solute-solvent and (or) solute –solute interactions which are modified with change of the properties (e. g. microviscosity and microrigidity) with the temperature decrease.

Unfortunately the problem of the nature of the Negative thermochromism in the solvent of the different natures have not still be solved finally up to now and needs subsequent discussion. The solvate and thermochromic NH structures are fluorescent, photostable and cannot be transformed by the excitation into the photocolored or the initial E structures with the longer or shorter longwavelength respectively.

In addition the strong intermolecular interactions including the sterical those in the thermochromic crystals in the ground state don't favorable the structural transformations responsible for the photochromism.

Thus both in the crystals and in the solvents the thermochromic structure is not favorable for the photochromic activity, and photocromism is depressed by thermochromism.

References

1. W. Turbeville, and P. Dutta, J. Phys. Chem. 94, 4060(1990).
2. F. Milia, E. Hadjoudis, and J. Seliger, J. Mol. Struct. 177, 191(1988).
3. J. Seliger, V. Zagar, R. Blinc, E. Hadjoudis, and F. Milia, Chem. Physics, 142, 237 (1990).
4. E. Hadjoudis, F. Milia, J. Seliger, V. Zagar, and R. Blinc, Chem. Physics, 156, 149 (1991).
5. T. Yuzawa, H. Takahashi, and H-o Hamaguchi, Chem. Phys. Lett., 202(3, 4), 221, (1993).
6. M. Carles, F. Mansila-Koblavi, J. Tenon, Th. N'Guessan, and H. Bodot, J. Phys. Chem. 97, 3716(1993).
7. a) T. Kawato, H. Koyama, H. Kanatomi, H. Tagawa, and K. Iga, J. Photochem. Photobiol. A:Chem. 78, 71(1994). b) K. Animoto, H. Kanatomi, A. Nagakari, H. Fukuda, H. Koyama and T. Kawato, Chem. Comm. 870(2003).
8. K. Kownaćki, A. Mordzinski, R. Wilbrandt, and A. Grabowska, Chem. Phys. Lett. 227, 270 (1994).
9. E. Lambi, D. Gegiou, and E. Hadjoudis, J. Photochem. Photobiol. A:Chem. 86, 241(1995).
10. S. Aldoshin, and I. Chue, Correlation, Transformations and Interactions in Organic Crystal Chemistry, Ed. by D. Jones, and A. Katnesiak, pp. 79-92, Oxford Univ. Press (1994).
11. L. Olechnovich, A. Lubarskaja, M. Knyazhansky, and V. Minkin, Zh. Org. Chim. 9, 1724(1973)(in Rus.)

12. F. Mansilla-Koblavi, J. Tenon, S. Toure, and N'Dede Ebby, Acta Cryst. C51, 1595(1995).
13. T. Hókelek, N. Gűnduz, Z. Hayvali, and Z. Kilic, Acta Cryst. C51, 880(1995).
14. K. Wozniak, H. He, J. Klinovski, W. Jones, T. Dziembowska, and E. Grech, J. Chem. Soc. Far. Trans. 91(1) 77(1995).
15. W.-H. Fang, Y. Zhang, and X-Z You, J. Mol. Struct. THEOCHEM, 334(1), 81, (1995).
16. M. Knyazhansky, A. Metelitsa, A. Bushkov, and S. Aldoshin, J. Photochem. Photobiol. A:Chem. 97, 121(1996).
17. G. Pistolis, D. Gegiou, and E. Hadjoudis, J. Photochem. Photobiol. A:Chem. 93, 179(1996).
18. J. Sitkowski, L. Stefaniak, T. T. Dziembowska, E. Grech, E. Jagodinska, and G. Webb, J. Molec. Struct. 381(1-3), 177(1996).
19. S. Alarćon, A. Olivieri, and M. Gonzalez-Sierra, J. Chem. Soc. Perkin Trans. 2. 1067 (1994).
20. S. Alarćon, A. Olivieri, G. Labadie, R. Gravero, and M. Gonzalez-Siera, J. Phys. Org. Chem. 8, 713(1995)
21. S. Alarćon, A. Oliviery, G. Labadie, R. Gravero, and M. Gonzalez–Siera, Tetrahedron, 51, 4619(1995).
22. S. Alarćon, A. Oliviery, A. Nordon, and R. Harris, J. Chem. Soc. Perkin Trans. 2, 2293 (1996).
23. A. Oliviery, J. Chem. Soc. Perkin Trans. 2, 85(1990).
24. A. Katritzky, I. Chivinga, P. Leeming, and F. Soti, Magnetic Res. in Chem. 34, (7), 578(1996).
25. G. Ledesma, G. Ibañez, G. Escander, and A. Oliviery, J. Molec. Struct. 415(1-2), 115 (1997).
26. S. Alarćon, D. Pagani, J. Basigalupo, and A. Oliviery, J. Molec. Struct. 475(2-3), 233(1999).
27. M. Kletskii, A. Millov, A. Metelitsa, and M. Knyazhansky, J. Photochem. Photobiol. A:Chem. 110, 267(1997).
28. M. Knyazhansky, A. Metelitsa, M. Kletskii, M. Millov, and S. Bezugliy, J. Molec. Struct. 526, 65 (2000).

29. a)M. Cohen, Y. Hirshberg, and M. Schimdt, J. Chem. Soc., 2060(1964). b)M. Cohen, Y. Hirshberg, and M. Schmidt, J. Chem. Soc., 2051(1964).
30. a)M. Knyazhansky, O. Osipov, O. Asmaev, and V. Sheinker, Zh. Fiz. Khim. 42(2), 1017(1968); b)O. Asmaev, M. Knyazhansky, V. Litvinov, O. Osipov and V. Sheinker, Zh. Fiz. Khim. 46, (4), 902 (1972);c)V. Litvinov, M. Knyazhansky, O. Osipov, and V. Sheinker, Zh. Fiz. Khim. 47(6), 1366(1973). d)M. Kletskii, A. Milov, M. Knyazhansky, and A. Metelitsa, Zh. Obsch. Khim. 69 (8), 1335(1999). (All the articles 30 a-d in Russian).
31. F. Hückel, Grundzüge der Theorie ungesättiger und aromatisher Verbindunger, Verlag, Chemie, Berlin (1938).
32. M. Kletskii, A. Millov, and M. Knyazhansky, Zh. Obsch. Khim. 68(10), 1700(1998)(in Russian).
33. a)A. Filarowski, A. Szemik-Hojniak, T. Glowiak, and A. Koll, J. Mol. Struct. 404(1-2), 67(1997); b)A. Filarowski, and A. Koll, Vibr. Spectr. 17(2), 123(1998);c)A. Filarowski, T. Glowiak, and A. Koll, J. Mol. Struct. 484, 75(1999);d) A. Mandal, A. Koll, A. Filarowski, D. Majumder, and S. Mukherjee, Specrochim. Acta P. A. 55, 2861(1999);e) A. Mandal, A. Filarowski, T. Glowiak, A. Koll, and S. Mukherjee, J. Mol. Struct. THEOCHEM, 577, 153(2002);f) D. Guha, A. Mandal, A. Koll, A. Filarowski, S. Mukherjee, Spectrochim. Acta P. A. 56, (14), 2669(2000);g)A. Filarowski, A. Koll, and T. Glowiak, J. Chem. Soc., Perkin Trans. (2), 835(2002);h)A. Filarowski, A. Koll, and T. Glowiak, J. Mol. Struct. 615, 97(2002);i)A. Koll, A. Filarowski, D. Fitzmaurice, E. Waghorne, A. Mandal, and S. Mukherjee, Spectrochim. Acta, P. A. 58, 197(2002);j)A. Mandal, D. Fitzmaurice, E. Waghorne, A. Koll, A. Filarowski, D. Guha, and S. Mukherjee, J. Photochem. Photobiol. A:Chem. 153, 67(2002)k)A. Mandal, A. Koll A. Filarowski, and S. Mukherjee, Ind. Journ. Chem. 41A, 1107, (2002);l)A. Koll, Int. Jour. Mol. Sci. (4), 434(2003);m)A. Mandal, D.

Fitzmaurice, E. Waghorne, A. Koll, A. Filarowski, S. Quinn, and S. Mukherjee, Spectrochim. Acta, P. A. 60, 805(2004);n) I. Król-Starzmoska, A. Filarowski, M. Rospenk, A. Koll, and S. Melikova, J. Phys. Chem. A. 108, 2131(2004).

34. a)T. Kawato, H. Kanatomi, H. Koyama, and T. Igarashi, J. Photochem. 29, 199(1986). b)T. Kawato, H. Koyama, H. Kanatomi, and M. Isshuki, J. Photohem. 28, 103(1985).

35. a)G. Dudeck and E. Dudeck, J. Am. Chem. Soc. 88, 2407(1966);b)Yu. Kozlov, R. Nurmukhametov, D. Shigorin, and V. Puchkov, Zh. Fis. Khim. 37(11)2432(1963)(in Russian).

36. M. Knyazhansky, S. Aldoshin, A. Metelitsa, A. Bushkov, and O. Filipenko, Khim. Fiz. 10(7), 964(1997) (in Russian).

37. H. Fukuda, K. Animoto, H. Koyama, and T. Kawato, Org. Biol. Chem. 1(9), 1578 (2003).

38. H. Koyama, T. Kawato, H. Kanatomi, H. Matsushita, and K. Yonetani, J. Chem. Soc. Chem. Comm. (5), 579(1994).

39. N. Otsubo, C. Okabo, H. Mori, K. Sakota, K. Animoto, T. Kawato, andH. Sekiya, J. Photchem. Photobiol. A:Chem. 154, 33(2002).

40. M. Zgierski, and A. Grabowska, J. Chem. Phys. 112, (14), 6329(2000).

41. M. Zgierski, and A. Grabowska, J. Chem. Phys. 113, (18), 7845(2000).

42. a)M. Zgierski, J. Chem. Phys. 115, (18)8351(2001);b)J. Orbit-Sanchez, R. Gelabert, and J. Lluch, J. Phys. Chem. A. 110, 4649(2006).

43. M. Zloteck, J. Kubicki, A. Maciejewski, R. Naskrecki, and A. Grabowska, Phys. Chem. Chem. Phys., 6, (19), 4682(2004).

44. L. Antonov, W. M. F. Fabian, D. Nedelcheva, and F. Kamounah, J. Chem. Soc. Perkin Trans. 2, 1173 (2000).

45. a)H. Joshi, F. Kamounah, G. Van der Zwan, C. Gooijer, and L. Antonov, J. Chem. Soc. Perkin Trans. 2, 2303(2001);b) H. Johi, F. Komunah, C. Gooijer, Van der Zwan, and L. Antonov, J. Photochem. Photobiol. A:Chem. 152, 183 (2002).

46. V. Vargas, and L. Amigo, J. Chem. Soc. Perkin Trans. 2, 1124 (2001).
47. K. Ogawa, Y. Kasahara, Y. Ohtani, and J. Harada, J. Am. Chem. Soc. 120, 7107 (1998).
48. 48. K. Ogawa, and T. Fujiwara, Chem. Lett. (7), 657 (2000).
49. K. Ogawa, J. Harada, I. Tamura, and Y. Noda, Chem. Lett. (5), 528 (2000).
50. K. Ogawa, J. Harada, T. Fujiwara, and S. Yoshida, J. Phys. Chem. A. 105, 3425 (2001).
51. K. Ogawa, and J. Harada, J. Molec. Struct. 647(1-3), 211 (2003).
52. T. Sekikawa, T. Kobayashi, and T. Inabe, J. Phys. Chem. A. 101, 644(1997).
53. T. Sekikawa, T. Kobayashi, and T. Inabe, J. Phys. Chem. A. 101, 10645(1997).
54. a)A. Oshima, A. Momotake, and T. Arai, J. Photochem. Photobiol. A:Chem. 762, 473(2004); b)A. Oshima, A. Momotake, and T. Arai, Bull. Chem. Soc. Jp. 79 (2), 305 (2006).
55. H. Pizzala, M. Carles, W. E. E. Stone, and A. Thevand, J. Chem. Soc. Perkin Trans. 2(5), 935(2000).
56. T. Dziembowska, E. Jagodzinska, Z. Rozwadowski, and M. Kotfica, J. Mol. Struct. 528, (2-3), 229 (2001).
57. T. Krygowski, K. Wozniak, R. Anulewicz, D. Pawlak, W. Kolodziejski, E. Grech, and A. Szady, J. Phys. Chem. A. 101, 9399(1997).
58. A. Koll, M. Rospenk, E. Jagodzinska, and T. Dziembowska, J. Mol. Struct. 552, 193(2000).
59. E. Ito, H. Oji, T, Araki, K. Oichi, H. Ishii, Y. Ouchi, T. Ohta, N. Kosugi, Y. Maruyama, T. Naito, T. Inabe, and K. Seki, J. Am. Chem. Soc. 119, 6330(1997).
60. J.-C. Zhuo, Magn. Res. Chem. 37, 259(1999).
61. R. Herzfeld, and P. Nagy, Curr. Org. Chem. 5, 373(2001).
62. D.-P. Wang, Sh.-G. Chen, and De-Zh. Chen, J. Photochem. Photobiol. A:Chem. 162, 407(2004).

63. I. Król-Starzomska, M. Rospenk, Z. Rozwadowski, andT. Dziembowska, Polish. J. Chem. 74, 1441 (2000).
64. T. Inabe, I. Gautier-Luneau, N. Hoshino, K. Okaniwa, H. Okamoto, T. Mitani, U. Nagashima, and Y. Maruyama, Bull. Chem. Soc. Jp. 64, 801(1991).
65. T. Inabe, New J. Chem. 15, 129(1991).
66. E. Hadjoudis et al., Mol. Cryst. Liq. Cryst. 221, 242(1994).
67. E. Hadjoudis, Private communication (2008).
68. T. Dziembowska, Z. Rozwadowski, A. Filarowski, and P. Hausen, Magn. Res. Chem. 39, 567, (2001).
69. K. Abbas, S. Salman, S. Kana'an, and Z. Fataftah, Can. J. Appl. Spectr. 41, 119(1996).
70. N. Hoshino, T. Inabe, T. Mitani, and Y. Maruyama, Bull. Chem. Soc. Jpn. 61, 4207(1988).
71. T. Inabe, N. Hoshino, T. Mitani, and Y. Maruyama, Bull. Chem. Soc. Jpn. 62, 2245(1989).
72. J. Takeda, H. Chihara, T. Inabe, T. Mitani, and Y. Maruyama, Chem. Phys. Lett., 189, (1), 13 (1992).
73. J. Takeda, T. Inabe, C. Benedict, U. Langer, and H.-H. Limbach, Ber. Bunsenges. Phys. Chem. 102(10), 1358 (1998).
74. E. Hadjoudis, M. Vittorakis, and I. Moustakali-Mavridis, Tetrahedron, 43, 1345(1987).
75. E. Hadjoudis, in Photochromism, Molecules and Systems, H. Dürr and H. Bounas-Laurent, Eds. Elsevier, Amsterdam, p. 685(1990).
76. M. Knyazhansky, and A. Metelitsa, Photoinduced processes in molecules of Azometines and their structural analogs. Publish house of Rostov State University, 207pp. (1992). (in Russian).
77. E. Hadjoudis, I. Moustakali-Mavridis, and I. Xexakis, Isr. Journ. Chem. 18, 202(1979).
78. I. Moustakali-Mavridis, E. Hadjoudis, and A. Mavridis, Acta Cryst. B34, 3709

79. E. Hadjoudis, M. Vittorakis, and I. Moustakali-Mavridis, Mol. Cryst. Liq. Cryst. 137, 1 (1986)
80. E. Hadjoudis, and I. Mavridis, Chem. Soc. Rev. 33, (9)579 (2004).
81. E. Hadjoudis, Mol. Eng. 5, 301 (1995).
82. E. Hadjoudis, J. Photochem. 17, 355 (1981).

Chapter 3

The Modern Ideas Of Photochromism And Accompanying Photoinduced Processes In Anil Molecules.

Introduction

The Chapter is devoted to the discussion of photochromism of the simplest imines of orthohydroxyaldehydes (anils) and it is the main part of this book.

The object consists in the detail consideration and discussion of the experimental and theoretical data obtained by the various research groups mainly for the recent two decades with the goal to get the grounded conclusions about the mechanisms of each stage of the photochromic reaction and of photochromism as a whole including the competing processes.

For that and for the discussion of the well known conventional mechanisms, and also of the novel nontraditional mechanisms and structures connected with the various consecutive stages of the photochromic process are considered. In particular the competition between the ESIPT and the trans –cis photoisomerization in the Enol structure, the structure of the state responsible for the fluorescence with the Anomalous Stokes Shift (ASS flu).

For the first time the special attention is devoted to the basing of the twisted structure role in the mechanism of the ASS flu quenching and the Photocolored (PC) form generation. The particular attention is given to the discussion of the mechanisms of the twisted ("post-TICT" or the post Conical Intersection, "post CI"), and also the planar PC structures' formation via the twisted structure.

In conclusion the general scheme of the photochromic process in the solvent is given.

At last the distinctions between the nature of photochromism and thermochromism in the crystals, the methods of the preparation of the photochromic crystals, and the peculiarities of the photochromism in the crystals as compared with that of the solutions are discussed briefly.

3. 1. The characteristics and deactivation of the electronic excited states of imines of o-hydroxyaldehyde Enol structure.

3. 1. 1. The excited states' deactivation and the structural transformations in the Enol form of Salicylideneanilines.

The Enol(E or OH) structure is only one from the several So state structures (Ch. 2)which is the sssssssssssssssinitial form of the following photoindused structural transformations connected with the Excited state Intramolecular Proton Transfer (ESIPT). Therefore the excited states' deactivation and the possible photo-transformations of the OH structure on the one hand and ESIPT followed by the subsequent structural transformations on the other hand are the competitive processes. For the flexible molecular structures (e. g. SA) the path-ways of the electronic deactivation depend strongly on the relative positions of the low electronic states which can be changed and even inverted with the photoinduced structural transformations (see /1/, Sch. 3. 1b).

The relative positions of the first $S_n^*(n=1-4)$ electronic states of the initial planar(Cs)Enol(E) structure(Y=0) calculated by the semiempyrical (INDO/S)/2/ and ab-initio /1/methods in comparision with the experimental findings /2/ (Sch. 3. 1)show that the experimental and calculated data are conformed satisfactory with data of INDO/S method for $S_n(\pi\pi^*)$ states. The $T_1(\pi\pi^*)$ energy values obtaned by various both experimental and calculated methods (Tab. 3. 2) are considerably scattered.

At the same time a remark should be taken also that overestimated values/3/ have been obtained for the structure I(Ph, CH_3)(see Tab. 3. 2), which differs from I(Ph, Ph)one, and the same conditions (excitation by the light with λ=320nm at 77K) favor a strong prevalence of the Keto(K) structure in the E⇔K equilibrium (see Ch. 2). It is obviously the key role in the photoinitiated processes are played by the lowest excited $S(π, π^*)$ and $S(n, π^*)$ states.

The energy of the $So-S_1^*$ transition of SA structure has been obtained by UV-Vis absorption spectra and the quantum-chemical calculations of the different levels (semiempirical and ab-initio)(Tab. 3. 1a). The experimental data are overestimated by the calculated those by $Δv=1500 cm^{-1}$.

It has been shown by calculations/12/(Tab 3. 4 with Scheme) that nature of the $So-S_1^*(π, π^*)$ transition is connected with the two Intramolecular Charge Transfers(ICT) of the opposite directions from the both aromatic rings into C=N group. Such a charge redistribution with the electron excitation results in the decrease of the charge separation and of the electric dipole moment value in the S_1^* state(μe) as compared with that in So state(μg), (Δμeg<0)/13/.

In such a case the change of the nonspecific interactions between the solute and solvent molecules leads to the "blue" shift of the absorption band with the solvent polarity increase /19/ from cyclohexne to acetonitrile (Δv=+700nm, see tab. 3. 1a).

However the "blue" shift of the absorption band caused by the decrease of the μ value in the S_1^* state is nontypical for the π—π* transition connected with the peculiarities of ICT in the SA structure. Unlike $S_1^*(π, π^*)$ state the energy position of the lowest nπ* state cannot be determined in confidence with absorption spectra/8/ due to the complete overlap of the very weak n—π* and strong π—π* transitions even in the very strong polar solvents owing to the unusual "blue" shift of the $So-->S_1^*(ICT, π-π^*)$ band(see above). But at the same time the qualitative comparative consideration of SA and the model structures (BA and MBA, Scheme with the table 3. 5)/20/ points at the higher

$S_1(n\pi^*)$ state energy than $S_1^*(\pi, \pi^*)$ one. Really even BA structure has no $S(n, \pi^*)$ state below $S_1^*(\pi, \pi^*)$ one.

The ICT $S_0 \to S_1^*$ transition of MBA (the structural analogue of SA without H-bond) is shifted towards long wavelengths by about 1200cm^{-1} without market shift of the $n\pi^*$ state. At last the H-bond formation l is stabilized S_0 state (i. e. increase of the $n\pi^*$ state). Thus in SA the energy gap between $S(n, \pi^*)$ and $S(\pi, \pi^*)$ states has to increase strongly in comparison with that of BA and MBA, and it is impossible to expect the lowest $S(n, \pi^*)$ state in such a case. The conclusion is corroborated by the most results of the quantum-chemical calculations of various levels for almost planar Enol (E) structure (~Cs symmetry) (Tab. 3. 5)/7, 9, 21/, and according to the recent data/1/(DFT/TDDFT method)the $S(n, \pi^*)$ state is the S_3^* one.

Therefore the weak longwavelength tail of the absorption band ($\lambda^{max} \approx 390$nm) in the highly polar ACN can hardly be assigned to $S_0 \to S_1^*(n, \pi^*)$ absorption band(according to /8/) but could probably originate from the Ground state Zwitterion NH tautatomeric structure(see Ch. 2).

However there are contradictory findings of the ab-initio calculations about the strong lowering /7/ or on the contrary strong heightening/1/ of the $S_1(n, \pi^*)$ state when increasing of the structure nonplanarity.

Such a situation is connected with the interaction between the lowest $\pi\pi^*$ and $n\pi^*$ states along the structural transformations and can substantially influence the notice about the excited state dynamics of the SA structure (see below). At the same time according to the calculated data of the different levels /27, 12, 13/ the excitation of the planar $S_1^*(\pi, \pi^*)$ state (Y=0, Cs symmetry) leads to the two very important changes in the electronic and geometrical structures of SA molecules.

i. The electron excitation from the HOMO(π) to the LUMO(π^*) (95%) pushes the electronic density from the phenol(80%) and aniline parts of the molecule onto C=N group (Tab. 3. 4). The changes modify the charge distribution on the H-chelate ring and thus increase acidic and basic character of the OH group and N-atom respectively.

ii. The excitation of the $S_1(\pi, \pi^*)$ state with almost planar (~Cs symmetry) structure leads to lengthening of the O-H, C_1-C_2 and C_7-N bonds and shortening of the C_2--O, C_1--C_7 and N-H ones introducing a considerable characteristics of the keto(K) structure (Tab. 3. 3).

Thus the both modifications favor strongly the ESIPT reaction by the lowering of the activation energy and (or) rise of the tunneling rate in the S_1^* excited state. On the other hand in the nonplanar S_1^* state structure ($\gamma \neq 0$, C_1 symmetry) the fall of the ICT efficiency from imine ring, decrease of the O-H length and lengthening of the C_2-O, C_1-C_7 and N-H bonds (Tab. 3. 3) does not the ESIPT but weaken the IHB O-H...N and along with the lengthening of the C_7=N double bond promote the trans –cis isomerization around the latter with break of the IHB. Thus each of the proximate S_1, $_2^*(\pi, \pi^*)$ states differing in symmetry (C_1 or Cs) can provide an advantage for the different photoinduced structural transformations (see below).

One of the most typical feature of the E-structure of the SA molecule is the extremely ineffective emission or the complete absence of it under any conditions. Really under sready-state excitation So->$S_1^*(\pi, \pi^*)$ ($\lambda^{max} \approx 340$-350nm, Tab. 3. 1a) in the solvent of the various polarity at different temperatures the resonance fluorescence of both SA and MBA molecules has not been observed in most investigations (see reviews /20, 21/). However the weak fluorescence attributed to the emission of the E-structure has been detected in several works /4, 6, 8, 10, 11/ (Tab. 3. 16). The weak steady-state and the time resolved emission ($\lambda \approx 420$-500nm) with very short lifetime (τ<1psec /43/ and 50fsec /116/) has been detected under excitation at 260-390nm in ACN /8, 10, 11/. It could be considered as an evidence of the ultrafast deactivation or the Enol S_1^* state including the ESIPT /10/.

Table 3.1a

The experimental and calculated values of the energy (E) and the nature of the $S_0-S_1^*$ transitions for the Enol structure of SA obtained by the various methods and authors.

E (cm^{-1})	Nature	Medium	Method	Referents	Notes
28986	$\pi\pi^*$	CH	UV-Vis	/4/ /5/	
32850	$\pi\pi^*$	-----	Semiemp.	/6/	
24400	$n\pi^*$	-----	Semiemp	/2/	
30100	$\pi\pi^*$	-----	Semiemp.	/2/	
29800	$\pi\pi^*$	ACN	UV-Vis	/2/	
30248	$\pi\pi^*$	----	Ab init(TDDFT)	/7/	C_S symm.
29240	$\pi\pi^*$	-----	Ab init(TDDFT)	/1/	C_1 symm.$\gamma=36°$
35571	$n\pi^*$	-----	Ab init(TDDFT)	/7/	C_S symm.
29700	$\pi\pi^*$	ACN	UV-Vis	/8/	"blue"shift
29673	$\pi\pi^*$	ACN	UV-Vis	/10/	
≈ 29700	$n\pi^*$	ACN	UV-Vis	/8/	
29412	$\pi\pi^*$	ACN	UV-Vis	/9/	
29762	$\pi\pi^*$	ACN	UV-Vis	/11/	
29764	$\pi\pi^*$(ICT)	------	Semiemp.	/12/	
30005	$\pi\pi^*$(ICT)	IIP(77K)	UV-Vis	/12/	
30961	$\pi\pi^*$(ICT)	------	Semiemp.	/13/	
30963	$\pi\pi^*$(ICT)	-----	Semiemp.	/14/	
22220	S_1--$S_n(\pi\pi^*)$	CH	Laser.Transient Absorp.Spectr.	/4/ /5/	$\tau_{rise} \approx \tau_{dfcey} < 200$ fs

Table 3.1b

The fluorescence and the excitation fluorescence spectral bands attributed to the Enol structure (for the model compound-2-METOXYbenzilideneaniline).

Fluorescence band		Absorption (excitation) band		Stokes Shift	Solvent	Refer.
λ^{max}_{flu} nm	ν^{max}_{flu} cm^{-1}	λ^{max}_{ex} nm	ν^{max}_{ex} cm^{-1}	cm^{-1}		
480	20833	(390)	(25641)	4808	ACN	/8/
325	30769	(270)	(37037)	6268	CH	/4/,/6/
~480	20833	337	29673	8840	ACN	/10/
420sh	23809	355	28169	4360		
480sh	20833	266, 355	37593,28169	7366	ACN	/11/
330	30303	266	37593	7290		

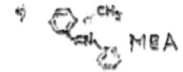

However the similar short-lived fluorescence with the marked intensity has been observed also in this region for the model molecule MBA (see Sch. and the Tab. 3. 16) with 2-OCH$_3$ substituent instead OH one /11/i. e. without OH.. N bond. Therefore if the quantum yields and lifetimes of the MBA and SA emission are close each other, the latter cannot be attributed to the "quasienol" fluorescent precursor of the excited cis-keto species /10/(see below and sec. 3. 2) but must be assigned to the emission from a very low stable excited (Enol)* state competing with an ultssrafast radiationless decay.

On other hand such as emission of SA can be overlap also with the fluorescence of the Zwitterion (Zw(NH))form traces in polar acetonitril (see Ch. 2).

Thus the additional information is needed for elucidation of the nature of the fast fluorescence at $\lambda \approx 420\text{-}500$nm. The fluorescence with $\lambda max \approx 325\text{-}330$nm of a marked intensity at the room temperature of SA and MBA in cyclohexane and acetonitril under steady-state excitation($\lambda=270$nm)/4, 6, 11/has been also attributed to the emission from the low Enol S_1^*state which geometry is different from that of the Frank-Condon S_1^*state and leads to the large Stokes Shift of fluorescence (Tab. 3. 16). Unlike SA, the absence of the competing ESIPT of MBA structure has to provide the increase of fluorescence relative efficiency for the latter as compared with the former.

But it is not corroborated by the experimental findings/11/. On other hand the existence of such a fluorescence has not been conformed later /8, 9, 10, 12/including the data of the author/4, 6/ themselves/5/. This fluorescence is also not manifested in any form in the transient spectra /11/ and is not included in the general scheme of the photochromic process /1, 11/.

Thus the subsequent investigations are needed for understanding of the nature of the fluores-cence attributed to the Enol structure of SA.

A very low fluorescence efficiency of both SA and MBA indicates to existence of the extremely fast radiationless process within the E structure competing with the emission and ESIPT in the S_1^* state.

The transient absorption spectroscopy with the ultrahigh time resolution (in pico-fem-tos time scale)is one of the most effective methods of investigation of such processes /4, 5, 6, 11/.

According to /4-6/ the broad band in the regions 400-500nm($\lambda max \approx 440$nm)that appears instantly during excitation ($\Delta \tau <100$fs, $\lambda \approx 360$nm, ACN)is assigned to the $S_1^*(\pi, \pi^*) \rightarrow S_n^*(\pi, \pi^*)$ transitions of the E tautomer. The ultrashort ($\tau<100$ps) change of the shape and intensity of the absorption band are connected with the efficient deactivation of the $S_1^*(\pi, \pi^*)$ state including both the radiationless process and ESIPT(τ(ESIPT)$\approx$210fs according to authors). The detail comparative studies of SA and MBA transient spectra after excitation pulse($\Delta\tau$=100fs) at different wavelengths (λ=266 and 390nm) in ACN have been conducted recently/11/. The broad negative (λ^{max}=330nm) and the positive (in the domain λ=350-600nm with $\lambda^{max}\approx$400nm) absorption bands appearing during excitation pulse (λ=266nm, $\Delta\tau$=100fs) are assigned to depopulation of the So and arising of the S_1^*Enol states correspondingly. The latter decay has the monoexponential kinetics with τ=550fs and reflects kinetics of the S_1^*Enol state deactivation competing with the ESIPT. With excitation at $\lambda\approx$355nm the relative efficiency of such an intraenol deactivation of SA in comparison with the ESIPT falls markedly (by$\approx$ a twice)(see also below).

Table 3. 2

The experimental and calculated values of $T_{\pi\pi^*}$ energy (cm^{-1})

Experiment	Method	Ref er.	Calculation	Method	Refer.
17422	T-Ttransfer	/15/	18148	SSP MOPPP	/16/
18390	So-T1abs. spect r.	/15/	18632	LCAO-MOSCF	/17/
19579 *)	Phosph. spectr.	/3/	21939	PPP	/18/
20407*)	T-Ttransfer	/3/	16530	PM3	/14/

*)For the structure J(Ph, CH$_2$Ph), (see below).

Table 3.3

The variation of the bond lengths $L_{S1}-L_{S0}= \Delta L$ (Å) with $S_0 \to S_1^*$ excitation of the different symmetry (C_S and C_I) of the Enol form structures of the anil molecules.*)
(molecule salicilidenaniline SA)

Bond	C_s-symm. /7/ ΔL	C_1-symm. /7/ ΔL	C_s-symm. /13/ ΔL	~C_s-symm. /6/ ΔL
O-H	+0.0091	-0.0097	+0.01	+0.01
C_2-O	-0.0132	+0.0188	-0.02	---
C_2-C_1	+0.0464	+0.0019	+0.05	---
C_1-C_7	-0.0605	-0.0128	-0.04	---
C_7-N	+0.0790	+0.0739	+0.05	---
N-H	-0.0078	+0.1821	-0.10	-0.11
Method of calcul.	Ab-initio		Semi-empiric.	

*) 1. Numering of C-atoms in the aldehyde ring is conducted counter-clockwise beginning from the atom C (C_1) bonding with the atom C_7 of the C=N group. 2/ C_S and C_I differ in the turning of the imine ring around C-N bond by angle y≈30° in the latter.

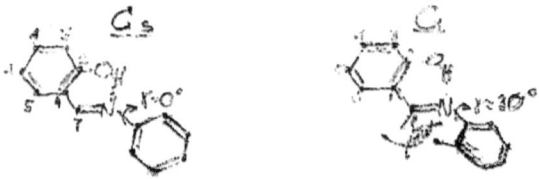

Table 3.4

Redistribution of electron charge under excitation $S_0 \to S_1^*$ excitation /12/
(molecule SA)

Fragments of the molecular structure, (Δq) ($S_1^*-S_0$)			
Aldehyde ring A	OH-group	C=N	Imine ring B
+0.075	+0.031	-0.140	+0.034

SA

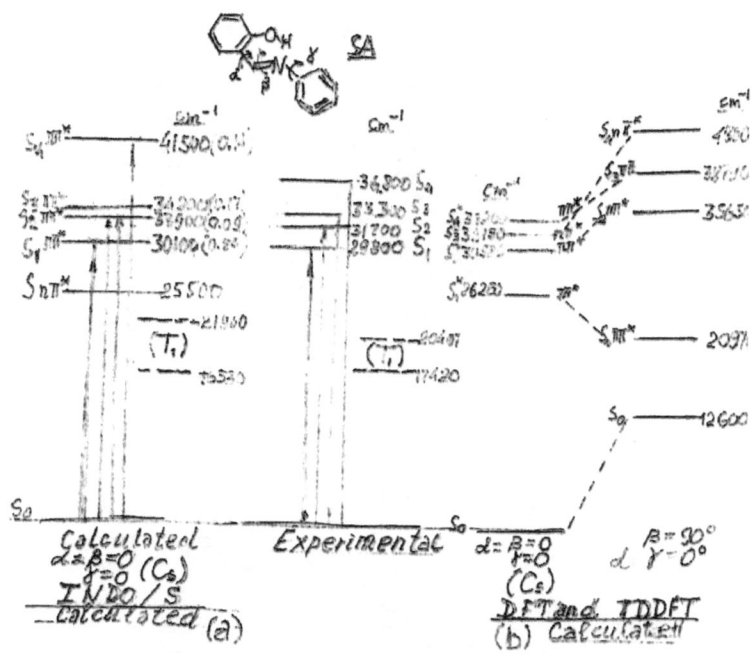

Scheme 3.1
The lowest electronic states of SA {l(Ph,Ph)}

a) The calculated (INDO/S) and the experimental (UV-Vis absorption spectroscopy) data of the relative positions of the S_n^* (n=1-4) excited states /2/ (calculated oscillator strength in the brackets) and the lowest T_1 state. For the T_1 state the limits of the obtained (by various authors) values are shown (see Tab.3.2).
b) The calculated (DFT/(TDDFT) relative positions of the electronic states (S_0 and S_n^* (n=1-4) and their changes with the variation of the molecular geometry (dihedral angle β)/1/.

Table 3.5
The relative positions of the singlet $n\pi^*$ and $\pi\pi^*$ states ($\Delta E_{n\pi-\pi\pi^*}$ 59) in the Enol structure of SA molecule.*)

$\Delta E(n\pi^*-\pi\pi^*)$ cm^{-1}	Method of calculation	Ref.
+2259	CNDO/S+CIS/6-31G*	/7/
+5323	TD/BSLYP/6-31G*	/7/
+2420	CNDO/S + CIS	/7/
+242	TD/B3LYP/6-31G*	/9/
-4598	INDO/S	/2/
+5890	DFT/TDDFT	/1/

*)Model structures- Benzalideneaniline and o-Metoxybenzalideneaniline

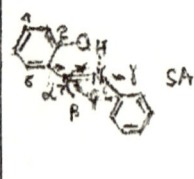

Table 3.6
The bond length (Å) and angles (°) of the transient (E_{ts}^{S1*}), (E_{ts}^{T1}) and final (E_{cis}^{S0}) structures for the reaction of the trans –cis photoisomeriztion with the respect to C=N bond in SA *) /14/
(see also Sch.3.2).

Bonds, Angles \ Structures	E_{ts}^{S1*}	E_{ts}^{T1}	E_{cis}^{S0}
O-H	0.955	0.957	0.950
C_2-O	1.359	1.364	1.367
C_1-C_2	1.411	1.415	-----
C_1-C_7	1.447	1.424	1.473
C_7-N	1.323	1.387	1.285
α	0	0	86
β	90	90	180
γ	23	180	93
φ	131	125	230

*) Numbering of C atoms see table 3.3.
Angels : α –turn of Ph ring around C_1-C_7
β—turn around C_7=N (trans-cis isomer.)
γ– turn of imine Ph-ring around bond N-C(imine ring)
φ- angle C_7–N—C(imine ring)

The additional transformations of the transient bands of SA and MBA are connected with the shift of the maximum towards λ^{max}=390nm followed by the increase of the intensity in the same spectral region ($\tau\approx 1.5$ps) and recovering of the initial spectra ($\tau\approx 16$ps) together with the very slow disappearance of the bleaching negative band ($\lambda max\approx 350$nm, $\tau\approx$ms). The analysis of these spectral transformations based on the data of the quantum-chemical calculations leads to the notions about mechanism of the intraEnol processes competing with the ESIPT and photocoloration.

The series of other methods suitable for the estimation of the rate of the ultrafast processes has been also used for the study of the S_1^*Enol state deactivation of SA including the ESIPT.

The known correlation $\Delta E \times \Delta \tau \leq h/2$ can be useful for the rate estimation of the above processes by Laser Induced Fluorescence(LIF) excitation spectra of the jet –cooled SA /9/. The broad vibrational band in the region of the 28000cm^{-1}of LIF excitation spectra ascribed to the $S^*n > S_1^*$transitions can be caused by the OH stretching mode with the homogenous width about $\Delta\nu \approx 100$cm^{-1}. However according to the findings obtained recently /4, 5, 10, 11/the ESIPT rate $\tau \approx 50\text{-}200$fs(see below) corresponds to the homogenous width $\Delta\nu \approx 30\text{-}50$cm^{-1} which is considerably less than experimental that(≈ 100cm^{-1}). Hence a fast radiationless process of S_1^*state with rate not less than of ESIPT rate can be realized. The femtosecond time resolved Resonance –Enhanced.

MultiphotonIonization(REMPI)spectroscopy/23/ has been also utilized for the study of the ultrafast deactivation process from S_1^*Enol state. The ultrafast decay component($\tau < 300$fs)is attributed to the decay of the $S_1^*(\pi\pi^*)$state of the Enol structure (λprobe=790nm)as a result of the ESIPT and Internal Conversion(IC) from the Enol S_1^*state. According to /23/ the IC can be caused by the $S_1^*(\pi\pi^*)\text{--}>S_1^*(n\pi^*)$ ($\tau \approx 500$fs) followed by the decay of the $S_1^*(n\pi^*)(S_1(n\pi^*\text{--}>So)(\tau=33$ps). The finding are agreed well with those of the transient spectroscopy.

Thus the discussed above experimental findings are the evidence of the ultrafast deactivation process that competes with the ESIPT and can be caused by the structural transformations within the E structure.

There are the two principle conceptions based on some differences of the experimental data and their interpretation on the base of the quantum-chemical calculations.

The first semiquantitative scheme of the structural transformations (Sch. 3. 2) is constructed on the base of the data of the spectral and kinetic investigations along with their interpretation based on the force-field ab-initio calculations /7/ with utilization of some results of the PM3 method with use of the gradient vectors in the S_1^*PES /14/. The planar ($\alpha=\beta=\gamma=\varphi=0°$) Enol $S_1^*(\pi\pi^*)$structure (Cs symmetry)is the only lowest state favorable to ESIPT(Tab. 3. 3, 3. 5)/7, 13, 14/. The excited state involving n orbital, so called $S(\pi, n^*)$ state /7/, lies more higher and does not take part in the ESIPT. The planar $S_1^*(\pi\pi^*)$ state is instable with respect to torsion vibration and of the Aldehyde and Imine rings with the very low frequency(f=22cm^{-1}/32/, 58cm^{-1}/7/) involving in the concert changes of the angles α, β, and γ. The conical intersection between the states of the $n\pi^*$ and $\pi\pi^*$ natures caused by such deformations is accompanied by the changes of orbital nature of the lowest state from $S_1^*(\pi\pi^*)$ for $S_1^*(n\pi^*)$ one in the strongly twisted conformation $E(S_1^*, tw)$(Tab. 3. 6) in which the ESIPT is impossible /7/.

On the S_1^* PES minimum with $E(S_1^*, tw)$ structure of the C_1 symmetry the energy gap between S_1^* and S_o states ($\Delta E \approx 15000$-$16000 cm^{-1}$) corresponds to the maximum of the high potential barrier on the SoPES in the region between the Etrans and Ecis structures(Tab. 3. 6).

On other hand in this area with very small S_1^*-T_1 splitting ($\approx 2000 cm^{-1}$, Tab. 3. 2) the ISC S_1^*-->T_1 is highly effective /14/. Unlike So and S_1^* states the T_1 PES is smoothed in this region and the isomerization via a transient structure $E(T_1, ts)$(Tab3. 6) can proceed almost without a potential barrier/14/. The both $E(S_1^*, tw)$ and $E(T_1, tw)$ structures are "ready" for the next step of the isomerization around C=N bond (see Tab. 3. 6). The final step of the isomerization with the additional twist of both A and B rings with generation of utmost twisted structure $E(So, cis)$(Tab. 3. 6)(see also /20/, /24/) with the low intensive "blue" shifted absorption band. Thus there are the competing path-ways both direct ($E(S_1^*, tw) \rightarrow E(So, cis)$ competes with $E(S_1^*tw) \rightarrow E(So, trans)$) and via triplet T_1 state ($E(S_1^*, tw) \rightarrow E(T_1, tw) \rightarrow E(So, cis)$ competes with $E(S_1^*, tw) \rightarrow E(T_1, tw) \rightarrow -E(So, trans)$). The path-ways to the initial $E(So, trans)$ structure are equivalent to the radiationless deactivation $E(S_1^*, trans)$-->$E(So, trans)$.

The existence of the path-way via T_1 state competing with the ESIPT(photocoloration) is supported qualitatively by the fall the relative efficiency in the series SA and its 5CL, 5Br, and 5I substituted (1. 00, 0. 15, 0. 14, and 0. 08 respectively) as a result of the increasing of ISC efficiency due to the heavy atom effect in this series /20, 25, 26/. In addition such an isomerization can also been sensitized by the triplet energy donor with the triplet energy E(T)>46kcal/mole. Diacetyle (E(T)≈56kcal/mol)/20/.

The strained nonplanar structure E(So, cis) is stabilized by the rigid matrix at 77K and can be detected by the typical change in the absorption spectra without photocoloration(see above). The reverse photoreaction E(So, cis)-hv->E(So, trans) can also be observed under these conditions with radiation at ≈313nm.

Thus the trans –cis isomerization about C=N bond is one of the final step of the ultrafast structural transformation competing with the ESIPT.

The second conception(Sch. 3. 2b) of radiationless deactivation as a result of the structural transformations in the E-form can be based on the experimental investigation of the ultrafast photoinduced process by the steady-state and transient (nano and femtosecond time scales) absorption and fluorescent spectroscopy with the Multivariate curve resolution (MCR) analysis /11/ together with the quantum-chemical analysis on the base of the ab-initial calculations with use of the Density Functional Theory (DFT) and time dependent DFT(TDFT)/1/.

In the present discussion the attempt has been undertook to coordinate the data of the experiment and the calculation in the common scheme.

According to the experimental /11/ and calculated /14/ data there are two excited S_1^* states of the $\pi\pi^*$ nature differing by symmetry and energy. These $S^*(\pi\pi^*)$ electronic states with $Cs(\gamma=0)$ (E^*Cs) and $C_1(\gamma\neq0)$(E^*C_1) symmetries are excited by $\lambda=335$nm and $\lambda=266$nm correspondingly/11/ having the structures which is favorable (Cs) and unfavorable (C_1) for the ESIPT. Really the efficiency of the ESIPT(photocoloration) $\Phi col=0.12$ and 0.14 and radiationless deactivation $\Phi d=0.23$ and 0.11 for excitation with $\lambda=260$nm and $\lambda=355$nm respectively.

Meanwhile according to/1/unlike /7/(Sch. 3. 2) the lowest $S(n\pi^*)$ state in the initial planar structure corresponds to the third excited singlet state (S_3^*) which lies ~150kcal/mol above the Frank-Condon point on the S_1^* state and rises up to ~60kcal/mol above $S_1^*(\pi\pi^*)$ with twist around C=N bond and therefore cannot take part in the deactivation process.

On the other hand the excitation at λ=266nm the PESs of the $S_1^*(\pi\pi^*)(C_1)$ and So states approach sharply lowering and rising correspondingly when twisting around C=N bond. Such a structural transformation towards the twisted Enol structure (Etw_1) with the broken O-H...N bond proceeds with τ<100fs (Φ=0. 23)/11/ or τ=37. 7fs/1/according to the experimental and calculated findings respectively.

The calculated energy difference between both states at the torsional angle around C=N bond β=90°less than 0. 2kcal/M and corresponds to the transient structure on SoPES with the calculated energy$\approx$50kcal/mol/1/.

It can be supposed that analogous twisted structure (Etw_2) can be formed with excitation by λ=355nm from the planar $S_1^*(\pi\pi^*)Cs$ state but considerably slower (τ=1. 5ps)along a less slopping PES that provides a successful competition of the ESIPT and even of emission with such a twisting transformation in the $S_1^*(\pi\pi^*)(Cs)$state. The Etw_2 structure can be generated also by excitation of the $S_1^*(\pi\pi^*)$ C_1 state (λex=266nm)/11/via path-way E^*C_1-->E^*Cs-->Etw_2 or by the transformation Etw_1-->Etw_2.

In spite of some doubts about attribution of the observed fluorescence to the Enol structure (see above)the emission bands with $\lambda max\approx$420-480nm/10, 11/and λ=330nm/11/are inscribed in the scheme and attributed to the $S_1^*E^*Cs$ and $S_1^*E^*C_1$ states correspondingly.

In the both cases the observed considerable Stokes Shifts of the fluorescence bands ($\Delta v\approx$7000cm^{-1}) can be explained by the marked changes of the structure in the $S_1^*E^*Cs$ and $S_1^*E^*C_1$ states towards geometries favorable for the ESIPT and the twist respectively/14/(see above). In the former case the fluorescence band decay after impulse excitation could be used for the estimation of the ESIPT rate /10/(see, however, above and sec. 3. 2).

The transient structures $Etw_{1,2}$ in So state are converted by the two path-ways into both the initial E trans structure (τ=16ps) with the absorption band at $\lambda^{max}\approx$336nm and the metastable utmost twisted Ecis like structure with a blue shifted absorption band. The latter turns slowly ($\tau\leq$1s)into the initial $Etr((C_1)==(Cs))$structure in the liquid solvent but can be stabilized strongly by a rigid media, and under appropriate conditions the photochromism can be accompanied by the marked "blue" shift of the initial absorption band ($\lambda^{max} \approx 336$nm).

Thus the competition between the ESIPT and the radiationless deactivation within the Enol structure can be described by the two different scheme including the same structural transformations but with absolutely different energetic of those. The conical intersections of the lowest $S_1^*(n\pi^*)$ states on one hand and that of the lowest $S_1^*(\pi\pi^*)$ and So states on another hand play the principal role in the first and the second schemes correspondingly.

The kinetic peculiarities of the processes in the solvents are described very well in the second scheme with taking into account of the two excited S_1^* states with different symmetry.

However the overestimated calculated instability of these states does not allow to explain the observed fluorescence with the marked efficiency. Within the limits of the second scheme unlike the first one it is impossible also to explain the experimental data about the participation of the Triplet state in the structural transformations.

The both schemes present the conformed combinations of the experimental and calculated data on the base of the conceptions stated in/7, 27/and/1, 11/. The produced schemes supplement each other, and reflect the present situation for the discussed problem.

3. 1. 2. Effect of the molecular structure and the medium on characteristics and deactivation of the excited state Enol form.

3. 1. 2. 1. Influence of the substituents in the phenyl ring

The experimental findings show that the substituents in the "Aldehyde" ring (A) influence stronger considerably the Enol structure excited states and their deactivation than corresponding ones in the "Imine "ring (B).

It is connected with the flattened structure of the "aldehyde' fragment with the IHB and the primary localization of the first electron transition on that fragment on the one hand and the twist of the imine ring with the n(N)->π(B ring) electron delocalization (see Ch. 2) on the another hand.

The studies of the effect of the strong electron donor 4-NMe_2 and acceptor 4-NO_2 substituents have been carry out by the methods of the steady-state absorption and fluorescent spectroscopy in the glassy solvent (IIP) at 77K with use of the 4-$N(Me)_2$ Benzylideneaniline (BA) as a model structure /20/ (see BA structure in the Tab. 3. 7). When inserting the NMe_2 group into the A ring the sharp energy decrease of the S_o->S_1^* transition (increase of λ^{max}_{abs}), the appearance of the fluorescence band (λ^{max}_{flu}) with the marked quantum yield (Φ_{flu}) and a not big Stokes shift ($\Delta\nu_{a-f}$) are observed.

As it is obvious from the findings produced the experimental charactetristics and the parameters calculated from the spectral data for the first electronic transition-the rates of the emission (k_{flu}) and radiationless (k_d) deactivation, transition dipole moment (D) and the electric dipole moment difference between S_1^* and S_o states /$\Delta\mu_{eg}$/ (obtained by the method of spectral shifts) are very close for the both NMe_2 derivatives of SA and BA (Tab. 3. 7).

The structural shapes of the absorption and fluorescence bands are also very similar. The large values and the proximity by the order of magnitude of the values /$\Delta\mu_{eg}$/ and D testify to the same ICT nature of the transitions in both SA and BA structures /31/.

Thus the all above data point out the same localization and the same nature of the longwavelength transitions in both molecules connected with the ICT from the A ring with NMe_2 group towards C=N one. This is at the bottom of the low probability of the ESIPT and absence of the photochromic reaction (see below).

At the same time the efficiency of the phototrans-cis isomerization about C=N-bond for SA is two and half times as low corresponding BA due to the stabilization of the planar structure by IHB, but is comparable with that of other derivatives of the SAs.

The strong withdrawing 3, 4 or 5NO_2 groups shift the E⇔K equilibrium towards the inactive Keto(K) structure with the marked emission (see also Ch. 2) that impedes considerably the study of the deactivation processes in the E-form /20/. However the excitation in the E structure absorption band allows to observe the reaction of the phototrans-cis isomerization around C=N bond under the steady-state (77K) and impulse (liquid solvent, 293K) excitation by the typical spectral changes (see above) /11, 20/.

The study of the thermochromic 5CL and 4'Me derivatives of SA in crystals has been carry out /29/ by the method of the steady-state absorption and fluorescent spectroscopy and femtosecond time resolved fluorescence spectroscopy at 293K and 77K (Tab. 3. 8). According to the data produced the energy of the S_1^* states is shifted to the low values for the all molecules (including SA) comparatively to those in the solvents and shifted in the same direction for the all molecules comparatively to SA. Such a situation can be explained by both the flattening of the molecular structure in the thermochromic crystals and the additional influence of the substituents. The fast component of the ASS fluorescence ($\tau \approx$2-3ps) has been assigned /29/ to the deactivation of S_2^* state or S_1^* state's high vibronic levels of the E structure.

Table 3.7
Effect of 4-NMe$_2$ group on the nature of S_0-S_1^* transition (see text)

Molecule	λ_{abs}^{max} nm	λ_{flu}^{max} nm	$\Delta\nu_{a\text{-}f}$ cm^{-1}	ϕ_f 3)	k_f 4) s^{-1}	k_{nr} 4) s^{-1}	---> 5) |D| D	-----> 6) |$\Delta\mu_{e\text{-}g}$|D
4-NMe$_2$SA	369	454 2) 470	5820	0.17±0.03	6.9x10^8	3.4x10^9	14.0±3.0	8.0±1.0
4-NMe$_2$BA 1)	376	450 2) 480	5800	0.09±0.02	3.0x10^8	3.0x10^9	13.0±3.0	7.5±1.0

1) BA-benzalydeneaniline. For numbering of C-atom in the benzaldehyde ring see tab.3.3. 2) The main max. is underlined.
3) Fluorescence quantum yield. 4) Rate constants of emission (k_f) and radiationless deactivation (k_{nr}). 5) Transition dipole moment (Debys).
6) Difference of the dipole moments of the excited (S_1') and ground (S_0) states (Debys).

Table 3.8

The comparative data for the Electronic states for SA and its derivatives.

No Comp.	Substituents	Crystal, E (S_1^* cm^{-1})		Solvent (Cyclohexane).Abs. band ν^{max}cm^{-1} /6/		
				$\pi \to \pi^*$ transitions		
		1> Flu. exc.	2> Abs. max.	$S_0 \to S_1^*$	$S_0 \to S_2^*$	$S_1^* \to S_n^*$ ($\tau_{\mu s}$)3>
1	5Cl	20133	23703	------	------	----
2	5Cl4'CH$_3$	21270	24840	------	------	----
3	4'CH$_3$	20730	24300	------	------	----
4	4C$_{12}$H$_{25}$O	----	--------	28571	37037	21276 (0.2-0.3)
5	H (SA)	----	28571 4>	29412 5>	37037	22727(0.2-0.4)

1>Estimated from the longwavelength edge of the excitation band at the intensity equal 1/20 th of the peak one/29/.
2>Estimated from the maximum of the longwavelength absorption band of the absorption spectra presented in /29/.
3> The transient absorption spectra with the decay time presented in the brackets .
4>The findings from /30/.
5>See also table 3.1.
6>For numbering of atoms C in the aldehyde ring see table 3.3 .The numbering of atoms C in the imine ring (n') is conducted clockwise from the atom C_1 bonding with atom N of the C=N group.

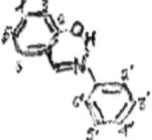

However such an interpretation of the fast fluorescence is extremely doubtful by the following reasons:

i. The excitation under the experimental conditions (T >293K, thermochromic crystal)at ~3. 2 eV can be assigned to the $S_0 \to S_2^*$ transition of the K structure.

ii. The considerable energy difference between the excitation (3. 22eV) and emission which is located below the longlength edge of the $S_0 \to S_1^*$ absorption band (<2. 4eV, $\Delta\nu$>6600cm^{-1}) testifies to a difference nature of the excited states responsible for the excitation and emission.

iii. The rate of the radiationless deactivation of the E –structure excited state higher at least by the order of magnitude the rate of the fast ASS fluorescence. The most probable such an emission and the variation of its kinetics depending on the substituents' nature belong to the processes occurring after ESIPT but before the appearance of the ASS fluorescence from the thermolized S_1^*K structure.

In the amphiphilic SA derivatives with the electron donor substituent having a very long alkyl chain ($4C_{12}H_2O$)in the A ring and the acceptor group (COOH) in the B ring /6/(Tab. 3. 8, Comp. 4), the longwavelength shift of the So->S_1^*ICT transition as compared with SA is observed(Tab. 3. 8).

The broad band in the transient absorption spectra with the decay time τ= 0. 2-0. 5ps is assigned to the S_1^*-> Sn*absorption like SA(see above /5, 7/). The longwavelength shift, the broadening and the luck of a vibrational structure of that band as compared with the SA one can be caused by the existence of the several conformers having a different orientation of the alkyl chains.

The interpretation of the fluorescence ($\lambda \approx 315$nm)as the S_2^*->So emission is doubtful like for SA(Tab. 3. 8). Unlike SA the amphiphilic compound 4 (tab. 3. 8)can form the Langmuir-Blodjet films and aggregates in cyclohexane. These aggregates with the absorption band about $\lambda \approx 285$nm give fluorescence ($\lambda^{max} \approx 400$nm, $\tau \approx 5$psec, Stokes shift $\Delta v \approx 10000 cm^{-1}$) which can belong to the monomer structure in the S_1^* excited state.

Such aggregates can't be formed in ethanol due to the specific interactions with the solvent. The steady-state absorption and fluorescence spectral investigations of 4'-OH substituted SA in the glass-forming solvent(Isopentane-Isopropanole-4:1, IIP) have been conducted/32/(Tab. 3. 9).

When including the 4'-OH group (B –ring)the absorption band of the So->S_1^*ICT transition is shifted towards lower energies ($\Delta v \approx 1600 cm^{-1}$)as compared with the corresponding transition of SA, and has unlike the latter the clear vibrational structure ($v=1300 cm^{-1}$) (Tab. 3. 9, str. i)which is smoothed out strongly in the band of the compounds iii, iv containing the tert-butyl group substituents in the B ring which inhibited torsion deformation and rotation vibrations of the OH group. Thus the electron donor nature and vibrations of the OH group determine the band position and the shape correspondingly.

At the same time the cis-isomer with respect to the C=N bond can be stabilized under influence of the 4'-OH substituent, and therefore the ge-neration of the cis-isomer can be observed easier especially in the glassy solvents at 77K as co-mpared with SA. Really the effective formation of the cis-isomer is observed by the steady-state irradiation of the compound i (Tab. 3. 9) under such conditions that is manifested in the strong drop of the intensity and loss of the vibration structure of the longwavelength absorption and fluorescence bands during the initial step of irradiation(τ<2min) without generation of the col-ored form (see above) (Tab. 3. 9).

The reaction is observed with almost the same efficiency in the structure iii with the tert-butyl substituents in the B ring only but it is absent in the structures ii, iv with the bulky substituents in the A ring.

It is obvious the structural transformations typical for the first step of trans-cis photoisomerization(see above) are inhibited by the bulky tert-butyl substituents in the ring A. Hence the above findings corroborate the idea about the trans-cis isomerization mechanism in the E structure competing with the photochromic transformations (see above).

The data about effect of the 4' donor and acceptor substituents (in B ring) have been obtained by the steady-state absorption spectra in the solvents of various polarity /33, 35/(Tab. 3. 10) and fluorescence excitation spectra in the crystal thin films /30/ which have been compared with the data of the semiempiric quantum-chemicalcalculat ions(INDO/S)/33, 34/(Tab. 3. 10).

Calculated findings for the absorption bands are shifted systematically towards the higher energy as compared with the experimental data but qualitatively correctly reflect the relative band positions. The longwavelength band's shift (λ^{max}_{abs}) accompanied by the increase of the oscillator strength (f_{osc}=33, 34/) with the drop of the ionization potential(the donor substituents) or the rise of the electron affinity(the acceptor substituents) testifies to the ICT transition nature. Such an interpretation is conformed with the "red" shift of the absorption bands in the polar solvent (ethanol)/33/ due to the nonspecific interactions with the solvent and the considerable increase of the dipole moment in the S_1^* excited state ($\Delta\mu_{eg}$)/19, 34) and also in the crystalline films /30/ owing to the flattened molecular structure in the thermochromic crystal (see above, Ch. 2).

The absent of the clear correlation between the changes of the dipole moments with the excitation ($\Delta\mu_{eg}$) and strengths of the donor and acceptors substituents is caused apparently by the complex redistribution of the electron density as a result of the competing of the oppositely directed ICT in the $So \to S_1^*(\pi\pi^*)$ transfer (see above) when twisting of the B ring under influence of the substituents. The anomalous behavior of the $So \to S_1^*$ transition connected with the ICT from the 4'NMe$_2$ group is expressed much less clear than in the molecule with 4NMe$_2$ one.

Thus the problem of the participation of the B ring in the ICT connected with the $So \to S_1^*$ in the E structure needs the detailed study by the experimental and quantum–chemical methods.

The investigation of the influence of the ring substituents allows not only to understand in detail the mechanism of the E structure deactivation that competes with the ESIPT but also to modify it advantageously to obtain the more effective photochromic anil structures in the crystals and the solvents.

3. 1. 2. 2. The effect of the imine moiety structure.

A. Salicylideneanilineiminopyridines.

The study of the combined influence of the bulky substituents in the aldehyde ring and the Nitrogen atom in the imine ring (Sch. 3. 3 c) is carried out /36/by both the steady-state and the transient absorption spectroscopy with use of the quantum-chemical calculations by the ab-initial, DFT, method. The findings about the influence of the bulky tert-butyl substituents in the aldehyde ring have been discussed above.

The replacement of the phenyl imine ring by the pyridine one brings the typical variations in the electronic and steric interactions of the aldehyde and imine moieties (Tab. 3. 11. e). The effect of the difference in positions of the

Nitrogen atom (2'or 4', Sch. 3. 3e)is insignificant, and the lowest transition So-->S_1^* is shifted of the both molecules, 2p and 4p by $\Delta E \approx 3000 cm^{-1}$ towards the low energy as compared with SA both in solvents and the crystals, and these data are reproduced by calculations very good.

However the significant flattening of the structure of I(Ph, 2'Pyr) (Struct. 2p)(Y=7°)as compared with I(Ph, 4'Pyr)(Struct. 4p)(Y=42°) (Tab. 3. 11e, Sch. 3. 3c)due to the exclusion of the steric interactions between C-H groups (Sch. 3. 3c)shifts the lowest electronic transition towards low energy $\Delta E \approx 670 cm^{-1}$ in the solutions. However no shift of that band is observed in the crystal with the change of the Nitrogen atom position from 4' to 2' with the exclusion of the weak steric interactions typical for both 4p and SA structures unlike 2p one.

The existence of the two lowest $S_1^*(\pi\pi^*)$ states of the different symmetries, $C_s(y=0°)$ and $C_1(y \neq 0°)$ caused by the weak steric interactions can be realizable only for 4p and SA structures unlike 2p one. Therefore the strong competition between the ESIPT and the twisting structural transformations within Enol form must be expressed more clear for 4p (like SA) in comparison with 2p. It may be suggested in this connection that ESIPT reaction with the corresponding ASS fluorescence in the thermochromic crystals (including 2p)are more efficient than in photochromic ones (e. g. 4p) (see also below).

Thus the molecules I(Ph, Pyr) are the perfect structural models for understanding of the radiationless deactivation nature in the Anil Enol structures.

B. Alkilimines and the related structures.

The two typical structures in which the N-aromatic ring is substituted (Sch. 3. 3a)or separated from the C=N group (Sch. 3. 3b) by the N-alkyl radicals with the various chain length have been studied by the experimental (steady-state absorption and fluorescent spectroscopy) by the quantum-chemical (semiempirical and ab-initial) methods /27, 28, 37-41/ (Tab. 3. 11 a-d).

Unlike SA the So->S_1^* transition has only a little contribution of ICT that can have even reverse direction (from the C=N group to the

Salicylidene ring /37/) due to the high electronic density on the N-atom already in the Ground state (Ch. 2)(see Tab. 3. 11c for SALKo). The calculated data of the transition energies are overestimated by ≈1000cm⁻¹ but reproduce the relative values correctly (see Tab. 3. 11 a, b /27, 28, 37, 40/).

The findings produced in the table 3. 11a show that the $S_o \rightarrow S_1^*$ transitions are shifted towards the high energies in both types of the molecules as compared with SA obviously due to exclusion (in SALKn i. e. I (Ph, Alkn))or a sharp decrease-(in SALKnPh i. e. I(Ph, Alkn Ph)) of the N-Phenyl contribution to the ICT transition.

In SALKn the shortwavelength shift does not depend on the length of the Alkyl chain /37/. Unlike SALKn in the molecules SALKnPh or SALKnPyr the structure of the N-Aromatic ring and its "distance" from the salicylidene moiety influence markedly the $S_1^*(\pi\pi^*)$ position lowering its energy when substituting Ph ring for Pyridine one and shortening of the Alkyl chain (Tab. 3. 11a) /39/. Thus in spite of the separation, the π-system sof the imine moiety makes an appreciable contribution into formation of the ICT nature of the $S_o \rightarrow S_1^*$ transition.

Due to the ICT nature, the position of the latter is sensible to even very little changes in the conformations of the both types of the molecular structures —the increase of an acoplanarity of salycilidene ring and C=N group with the insertion of the rigid Alkyl bridge (in SALKnAlk and SALKnAlkPh)leads to the S_1^* energy rise (600-1200cm⁻¹)/37/and, on the contrary, the increase of their coplanarity in the rigid matrix at 77K stipulates the little lowering (200-250cm⁻¹)of the $S_o \rightarrow S_1^*$ transition energy of the SALK molecules with the temperature drop (Tab. 3. 11a)/37/.

Thus even the little change of the conformation or variation of the environment can stipulate an alteration of the relative positions of the closely disposed excited states in the E structure.

According to the data of the quantum-chemical calculations of the various levels/37. 28, 37/ (Tab. 3. 11d)the structure of the S_1^* flattened state (~Cs symmetry)of the Enol structure for the both types of molecules gains partially the nature of the K form approaching to the structure of the transition state of the OH->NH reaction. Indeed the O-H, C_1--C_2, C –N bonds becomes longer and N-H, C-O, C_2-C ones are shortened. Such structural changes has to make for lowering of the ESIPT reaction barrier in the S_1^*state. The effect is expressed some more in SALKnPh than in SALKn and much less in SA than in two latters (compare the data in the Tab. 3. 11d and 3. 3).

Thus the rates of the Enol $S_1^*(\pi\pi^*)$ state decay due to the ESIPT reaction have to decrease in the order (SA)<<(SALKn)<(SALKnPh) (see also below in this Chapter). The lowest excited $S_1^*(n\pi^*)$ state of the compounds SALKnCH3Ph, that lies by $\approx$5000cm^{-1} lower $S_1^*(\pi\pi^*)$, has like SA the twisted conformation, and the changes of the bonds unfavorable for the ESIPT(Tab. 3. 11d)/28/. In the planar structure of SALKn the $n\pi^*$ state by $\approx$5000cm^{-1} higher than $S_1^*(\pi\pi^*)$ one and ctake part effectively in the deactivation of the latter.

Table 3.9
The spectral and the photochemical characteristics of 4'-OH and tert-butil derivatives of SA (iospentan,77K).

Comp. No	Substituent (see tab.3.3 and 3.8)					Long-wavelength absorption band's maxima, λ^{max} nm				Tr.-cis isomeris. around C=N bond
	3	5	3'	5'	4'					
i.	H	H	H	H	OH	330	345	<u>360</u>	380	+
ii	Tr.byt.	Tr.but.	H	H	OH	335	350	<u>370</u>	385	-
iii	H	H	Tr.but.	Tr.but.	OH	335sh	350	<u>360</u>	380sh	+
iv	Tr.but.	Tr.but.	Tr.but,	Tr.but.	OH	335sh	350	<u>370</u>	380sh	-
SA	H	H	H	H	H	342	----	----	----	----

The main maximum is underlined. + and – are the existence or the absent of the visible "sh" is band displayed like a shoulder. spectral display of trans-cis isomerization around C=N bond.

Table 3.10

Effect of the 4'-substituents in the imine ring (B) on the characteristics of the $S_0 \rightarrow S_1^*$ transition in the Enol (E) structure in various media.

Media / Spectr Refer / 4'-Substituent	λmax (E-structure) nm					Oscillator strength		$\Delta\mu_{e-g}(D)$
				Calcul.INDO/S		Calculation INDO/S		
	Cyclohex /35/	Ethanol /33/	Cryst. film /30/	/34/	/33/	/34/	/33/	/34/
H (SA)	342	336	347	334	306	0.193	0.32	1.8
NO$_2$	357	364	374	336	314	0.737	1.00	1.57
CN	352	---	---	332	---	0.326	---	1.99
CH$_3$	346	---	---	338	---	0.258	---	3.87
OCH$_3$	353	349	325	339	304	0.303	1.00	2.88
N(CH$_3$)$_2$	384	380	331	351	316	0.513	1.00	6.55
COCH$_3$	332	343	322	332	304	---	0.61	---

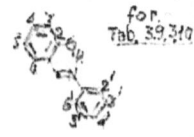
for Tab. 3.9, 3.10

Table 3.11
The energetic and spectral characteristics of the first electronic states of SA Enol structure with different imine moieties (See Scheme 3.3, Structures A,B,C)

a) S_1^* state energy (experiment)

Structure(A,B) Compound I(Ph,R)	Absorption spectra		Ref.
	Temp. K	Energy (cm^{-1})	
SALK$_0$	293	31941	/37/
	77	31250	/38/
SALK$_{17}$	293	31546	/37/
	77	31250	/37/
SALK$^{(Ph)}_{17}$	293	32154	333
	77	31949	
SA	293	29412	
	77	29240	//// /37/
SALK$_1$Ph		31056	/39/
SALK$_2$Ph		31446	
SALK$_1^{LPh}$Ph	293	32258	/37/
SALK$_0$ Pr		28571	
SALK$_1$Pr		31250	/39/
SALK$_2$Pr		32154	/39/
SALK$_1$CH$_3$Ph		31546	/40/

b) S_1^* state energy (calculation, DFT)

Structure (A) Compound I(Ph,Alk)	Energy (cm^{-1})	State Nature	Ref.
SALK$_0$	32950	$\pi\pi^*$ICT	/27/
	38773	$n\pi^*$	
SALK$_1$CH$_3$Ph	32236	$\pi\pi^*$ICT	/28/
	27165	$n\pi^*$	

c) Charge change(S_1^*—S_0)

Structure fragment	Charge change
OH	-0.09
Ph ring	-0.01
C$_7$	+0.08
N	+0.04
CH$_3$	-0.02

d) Length bonds and angles change(S_1^*-S_0)(A,B)

Bond	SALK$_0$		SALK$_0$CH$_3$Ph Ref /28/	
	Ref/37/	Ref/27/	S$\pi\pi^*$	S$n\pi^*$
O-H	+0.008	+0.031	+0.034	-0.01
N-H	-0.086	-0.182	-0.199	+0.23
C$_2$-O	-0.037	-0.035	---	---
C$_1$-C$_2$	+0.075	+0.093	---	---
C$_2$-C$_7$	-0.028	-0.053	---	---
C$_7$-N	+0.067	+0.041	---	---
α	-10	0	+0.1	+11.3
β	+171	+180	-178.6	+93.0

e) Spectra and structure of compound C /36/

Molecule No	Absorp.band, λ^{max} nm			Angle γ°	
	Experiment		Calc. DFT	X-ray	Calc DFT
	Cychex.	Cryst.			
2p	370	366	380	6.9	0
	313	309	313		
4p	361	368	365	41.8	37.8
	284	293	305		

Notes to the Tables 3.11a-e

(The conditions of experiments and the methods of calculations)

Tab (a) Experiment Tab (b) Calculations Tab (c) Calculations

/37/ Isopentane   /41/ TD/B3LYP/6-31 G* /37/PM3
/38/,39/Cyclohexane /42/B3LYP/cc-PVDZ S_0 state
/40/ Crystalline film /43/TDDTF CIS S_1^* state

Tab (d) Calculations   Tab(e) Experiment and Calculations

/37/STO—3,2 G-S_0state, PM3-S_1^*state /36/Cyclohexane, Crystal, Abs. X-ray
/27/ HF/6-31G*-S_0 state, CIS /6-31G*-S_1^* state /36/ DFT
/28/B3LYP/CC-PVDZ-S_0 state, TDDFT CIS-S_1^* state

I(Ph,Alk$_n$):SALK$_n$:   I(Ph,Alk$_n$Ar):   I(tbutPh,2'Pyr):
          X=CH:(SALK$_n$Ph):   (2p):
          X=N: (SALK$_n$ Pyr):

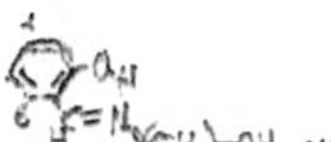

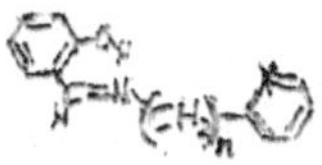

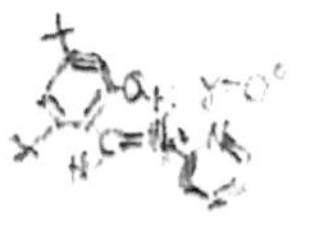

I(|Ph,Alk$_n$):SALK$_n^{LPh}$:   I(Ph,Alk CH$_3$ Ph) :
          SALK$_0$CH$_a$Ph :   I(tbutPh,4'Pyr):
                   (4p):

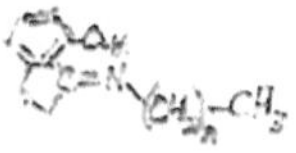

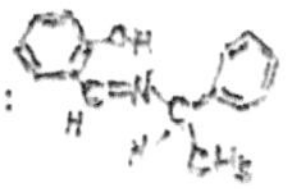

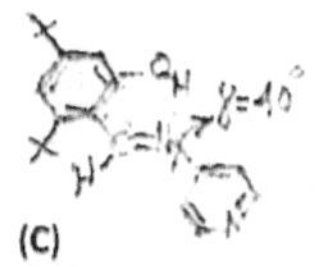

     (A)   (B)   (C)

Scheme 3.3
The studied SA structures with the various aril and alkil moieties (see tables 3.11 a-e).

Table 3.12

The influence of the ring conjugation the energy of the Enol structure's $S_1^*(\pi\pi^*)$ state obtained by the absorption (and fluorescent) spectra.

Compound 1)	Absorption band v^{max} cm^{-1}	Fluorescence band v^{max} cm^{-1}	Solvent 2)	Reference
SN	28511	---	3MP	/45/
	~28150	---	ACN	/2/
	28329	---	MCH	/47/
	28653	---	ACN	/47/
	28571	20000 3) 25000	ACN	/10/
5-6 NA (R=H)	27778	---	Eth	/30//42/
	26667	---	Cryst	/30/
	26667	---	Hex	/43/
	26596	---	MCH	/44/
	26596	---	MCH	/47/
	26596	---	ACN	/47/
3-4 NA(R=H)	26110	---	MCH	/44/
5-6NN	26042	---	MCH	/47/
	26178	---	ACN	/47/
	26667	---	MCH	/46/
PA	29070	---	Eth	/33/ /42/
SA	29070	---	MCH	/4/ /5/
	29762	---	ACN	/4/ /5/

1)See Scheme 3.4. 2)Solvents: 3MP-3Methylpentane, ACN-Acetonitrile, MCH-methylcyclohexane, Eth-Ethenol, Cryst-Crystal film. 3)Excitation-λ=385,415nm τ=40fs see text.

Scheme 3.4

The studied molecular structures with the polyphenil rings

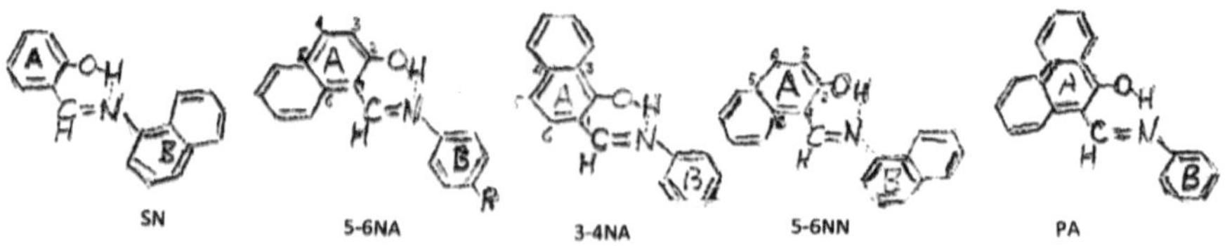

SN   5-6NA   3-4NA   5-6NN   PA

Table 3.13

The substitution effect on the $S_0 \to S_1^*$ ($\pi\pi^*$) transition energy (cm^{-1}) and intensity (oscillator strength)[4)]
in the structures with the different conjugation ring systems [1)] in various media.[2),3)]

R:	H		CH$_3$		N(Me)$_2$		OMe		COMe		Br		Cl		I		NO$_2$		CN		Refe-
Media \ Struc	Sol	Cr	Sol	Cr	Sol	Cr	Sol	Cr	Sol	Cr	Sol	Cr	Sol	Cr	Sol	Cr	Sol	Cr	Sol	Cr	rence
5-6NA	27778 (0.52)	26667	---	--	26596 (0.82)	26738	25773 (0.65)	26316	26525 (0.84)	31055	---	---	---	---	---	---	27027 (1.00)	26738	---	---	/30/ /33/ /42/
5-6NA	26596 (0.38)	---	26455 (0.43)	--	24272 (0.58)	---	26110 (0.56)	---	---	---	26455 (0.49)	26596 (0.51)	26247 (0.47)	---	---	---	25641 (0.64)	---	25840 (061)	---	/43/
5-6NA	26596	---	---	--	25641	---	---	---	---	---	---	---	---	---	---	---	25189	---	---	---	/44/
3-4NA	26110	---	---	--	26042	---	---	---	---	---	---	---	---	---	---	---	24814	---	---	---	/44/
PA	29070 (0.01)	29412	---	--	28511 (0.83)	28511	29240 (0.60)	---	27027 (1.00)	---	---	---	---	---	---	---	27700 (0.93)	---	---	---	/30/ /33/ /42/
SA	29762 (0.32)	28818	---	--	26316 (1.00)	30211	28653 (1.00)	30769	29154 (0.64)	31056	---	---	---	---	---	---	28248 (1.00)	26733	---	---	/30/ /33/ /42/

1)See scheme 3.4. 2) Sol.-Solvents: Ethanol/30/,/33/,/42/,/43/, Cyclohexane/44/. 3) Cr-Crystalline film. 4)Oscillator strength is calculated by the semi-empirical quantum-chemical method INDO/S (in the brackets)

Table 3.14

The dependence of the $S_0 \to S_1^*(\pi\pi^*)$ transition energy on solvent polarity for compound 5-6 NA (R=H)
(Sch.3.4)/43/

Solvent	Rel. permeability.	ν^{max}_{abs}
Hexane	1.88	26667
CCL$_{4r}$	2.24	26525
Acetone	21.4	26596
Acetonitrile	36.7	26667
DMS	46.7	26316

Table 3.15
The electron absorption and fluorescent spectral characteristics of ESA

Medium [4]	Temper.	Absorption		Fluorescence				References
		λ^{max}_{abs} nm	Attribut.[1] of bands	λ^{max}_{flu} nm [5]	λ^{max}_{exit} nm	Stokes shift (cm^{-1}) [5]	Attribution of fluor. bands	
n-Hex	room	325	E-cis and E-trans	365 530 ~420 [3]	~325 325	3143 11901	E=trans K-cis(ESIPT)	/49/
n-Hept	77K	325 ~440	E-cis K	530	325	11901	K-cis(ESIPT)	/50/
Eth	room	322 ~440	E-cis K	420-460 [2] 530	~360 322	~5000 11901	Anion,Zw.ion K-cis(ESIPT)	/50/
Mth	77K	—	—	530	322	11901	K-cis((ESIPT)	/50/
Wt EG	room	322 [2] 360	E-cis Anion	420-470 [2] 530	360 322	~5000 11901	Anion,Zw.ion K-cis (ESIPT)	/50/
Cryst.	room	—	—	530	322	11901	K=cis (ESIPT)	/49,50/

1)Trans and Cis Enol –see scheme 3.5 . 2)With edition of NaOH .3)With edition of Three Ethyl Amine(TEA)(see sch,3.5) .4) n-Hex-n-Hexane,n-Hept-n-Heptane,Eth-Ethanol,Mth-Methanol,W-Water EG-EthilenGlicole. 5)The ESIPT fluorescence is underlined.

Scheme 3.5
The various isomeric and interacting with the solvent the Enol and the Keto structures of the molecule 7Etyl SA (ESA)

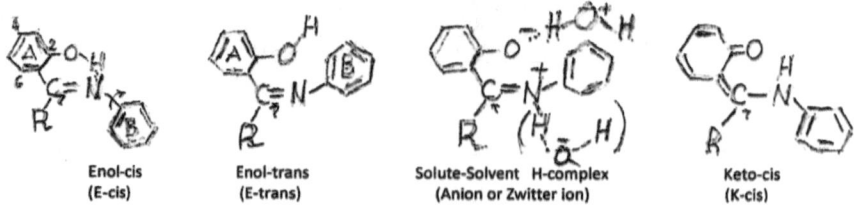

Enol-cis (E-cis)   Enol-trans (E-trans)   Solute-Solvent H-complex (Anion or Zwitter ion)   Keto-cis (K-cis)

Table 3.16
The electron absorption and fluorescent characteristics of ESALK
(see also sect.3.3.2.4 ,sch. 3.6)

Environment		Absorption		Fluorescence				Ref.
Medium 2)	Temp. K	λ^{max}_{abs} nm	Attr- 1) bution'	λ^{max}_{flu} nm	λ^{max}_{exc} nm 4)	Stokes Shift (cm^{-1})	Attrib.of fluores.	
MCH	room	322	E-trans	370	325	3742	E-trans	/54/
				400-420	350	~4120	NH-Zw	
				500	322	11056	NH-ESIPT	
Eth	room	322	E-cis	470	~330	9030	Zw-Solv	/52/ See 3.3.2
		395	Solute-Solvent H-complex	**500** Sh	330	10303	NH-ESIPT	
				500	395	~5300	K-cis	
Cryst.	room	---	---	**500**	325	10769	NH ESIPT	/51/
MCH	77	----	---	**490**	325	10360	NH-ESIPT	/51/
				500	400	5000	K-cis	
GLY	77	----	----	~420 3)	360	----	Zw T$_1$-S$_0$	/52/
				500	325	10769	NH-ESIPT	

1)See Sh 3.6 2)MCH-MethylCycloHexane,Eth-Ethanol,Gly-Glycerol..3)Phosphorescence,τ=0.7s 4)ESIPT – fluorescence is underlined.

Enol cis (E cis) — Enol trans (E trans) — Anion (A) — Zwitterion (Zw) — Keto cis (K cis) — Zwitterion with solv. (Zw –S)

Designations of the structures

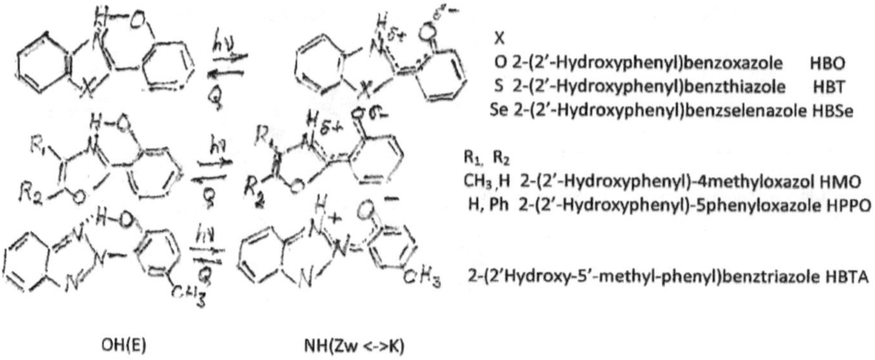

Scheme 3.7
The ESIPT and tautomeric structures of the SA related model molecules.

X
O 2-(2'-Hydroxyphenyl)benzoxazole   HBO
S 2-(2'-Hydroxyphenyl)benzthiazole   HBT
Se 2-(2'-Hydroxyphenyl)benzselenazole HBSe

R_1, R_2
CH_3, H 2-(2'-Hydroxyphenyl)-4methyloxazol HMO
H, Ph 2-(2'-Hydroxyphenyl)-5phenyloxazole HPPO

2-(2'Hydroxy-5'-methyl-phenyl)benztriazole HBTA

OH(E)   NH(Zw <->K)

At the same time the attempts were made to discover a fluorescence of the E structure for the both type of molecules in the different solvents at the room and low (77K) temperatures resulted in the conclusion of mainly radiationless deactivation of the $S_1^*(\pi\pi^*)$ state (like SA). At the same time the phosphorescence ($\lambda^{max} \approx 510$nm) has been found out in the Cyclohexane solutions of the SALKnPh at 77K that indicates the participation of the T_1 state in the deactivation of the Enol structure excited states.

The deactivation of the $S_1^*(\pi\pi^*)$ state is realized according to the Scheme 3. 2 c, d.

Although the high $n\pi^*$ states in the planar E form (Sch. 3. 2c) does not take part in the energy deactivation, the situation changes (unlike SA) with a distortion of the conformation only after very fast ESIPT(see below), and competition between the latter and the deactivation of the excited states in the Enol form is almost absent (Sch. 3. 2b). According to the Sch. 3. 2d /28/the lowest $n\pi^*$ state in the planar structure lies unlike SA lower the $S_1^*(\pi\pi^*)$ one but it does not havea stable minimum. Therefore after Internal conversion (IC) $\pi\pi^* \to n\pi^*$ in the planar geometry the $n\pi^*$ PES lowers sharply to the global minimum. In this region the PESs of S_1^* and So states with $T(\pi\pi^*)$ and $T(n\pi^*)$ ones between them(Tab. 3. 2)become very close and mixed. Therefore the fast deactivation ($\tau \approx 250$fsec/28/) can be realized in such structures with population of the T_1 state.

Table 3.17
The ESIPT in the related structures.
Parameters and conditions of the reaction (see scheme 3.7)

Structure	τ (fs) (k×10^{12})s^{-1}	Transfer promoting modes (TPM) cm^{-1}	Method Medium(solvent) Excitation, λ nm	Mechanism	Reference	Notes
HBTA	60-80 (12.5-17.0)	470 Inplane unharm.bend.	Pump-probe (trans.absorp.) CH 367	Barrier less +TPM	/55/	TPM –transfer promot. modes coupling
	500 (2.0)	Ring deformation	Pumt-probe(trans.absop.) CH 310	Role of TPM hasn't been analyzed	/56/	After ESIPT solvent assited Intramolecular vibration energy redistribution is possible
	~100 (~10.0)	1400 Elongation of ring	Pump=probe(trans.absorp.) CH 370	Barrierless +TPM	/57/	
HBO	60 ±30 (11-33)	147 Inplane ring bending 2650 OH-stretching 123,308 –inplan.ring 3270 OH-strething	Pump-probe (trans absorp) CH 310 Calculation ab-initio CIS /3-21G*	Small barrier Tunnel +TPM Almost barrierless+TPM	/58/ /59/	Analysis of Deuterium effect on vibration spetra.
HMO	<100 (>10)	----	Pump-probe (ASS-flu) Protein 325	Barrierless	/60/	
	132 (76)	~120 Ring-ring bending on plane	Calculation by two-state Parameters +TDDFT method of calculation	Barrierless, Assited byTPM	/61/	
HBT	160±20 (6.2±0.5)	50-250 In plane bend.with large amplitude of azole and phenyl	Pump-probe(trans absorp) TCF 310	Barrierless + TPM	/62/	ESIPT is not effected By Deuteration
	33 (3.0)	113 Bending	Pump-probe (trans absorp) CH 347	Barrierless +TPM	/63/	
	---	114 In plane ring bending	Calculation ab-initio CIS 6-31G * + TDDFT	Almost barrierless +TPM	/64/	
HPPO	~220 (4.5)	33 Out of plane tors 164 Out of plane deform.	Jet-cooled Gas 334	Barrierless Modulated by TPM and tunnelling	/55/ /66/	K^{ESIPT} decreases by factor 4-6 with change H by D
HBSe	40 (250)	380,710 Bending in plane and out of plane	Pump -probe (flu) THF 371	Barrierless Assisted by TPM	/67/	Probe –Intensity of E-structure fluorescence (λ=460nm) falls

Solvents:CH-cyclohexane,TCF-TetraChlorEthilene.THF-TetraHydroFurane

Table 3.18

The ESIPT rate values for molecule of SA determined by various experimental and calculated methods.

τ fs ($k \times 10^{12} s^{-1}$)	Method, [1] Probe type Registration, λnm	Excitation λnm	Medium [2] T_k	Remarks	Reference
50 (20)	SWFF 480 decay	385	ACN 293	No Deuterium effect	/40/
40 (25)	Time resolved mass spectroscopy	266	Gas phase	Deuterium effect has not been studied	/11/
45±5 (~22)	TABS positive decay 410	266	ACN 293	The kinetics of the positive and negative bands' decay are are different. Deuterium effect was not studied.	/11/
	TABS negative 340 and 620 (ASS flu) decay.	390			
49.6 (20.16)	Calculated by ab-initio TDDFT method	—	—	The barrierless ESIPT	/1/
210 (4.76)	TABS positive, 420	360	CH 293	The both probe bands (420,485) are attributed to the S_1^* NH state	/4/
380 (2.63)	TABS positive, 420	360	Eth, 293		/6/
180 (5.56)	TABS positive, 485 Decay and rise	360	CH, 293		/5/ /6/
<50 (>20)	TABS positive, 420 TABS negative, 620 SWFF decay, 480	390	ACN 293	Deuterium effect has not been studed	/8/
220 (4.54)	Calculated by ab-initio CIS/6 -31 G* method with instant approach.	—	-----	Barrier height is 5.7 kcal/mole Deuterium effect is $τ_D/τ_H$=12.3	/7/
1200 (0.83)				Barrier height is 7.2kcal/mole Deuterium effect is $τ_D/τ_H$=17.5	
<50 (>20)	LIF(see text) excitation spectra	—	Gas phase mixture: SA+Helium in the Jet cooled spectroscopy	Estimated by homogenous broadening of vibration band. (see text)	/9/
<750 (>1.33/	Femtosecond REMPI spectroscopy (see text) 395 decay	320 or 365	Gas phase mixture SA+Helium Vapor p=1atm 350	The absorption bands of $λ^{max}$= 395 (420) is attributed to S_1^*state of NH structure. Deuterium effect has not been detected. Potential barrier is small or absent.	/23/

<1> TABS –Transient Absorption Spectra. Short Wave-length.East Eluorescence.<2> Solvents:ACN-Acetonitril, CH-CycloHexane ,Eth- Ethamol.

Table 3.19

The change of the rise rate of the transient absorption bands attributed to the $S_1^* \rightarrow S_n^*$ transitions in the K-structure in the dependence on the solvent polarity (ε) and viscosity (η) /5/.

Solvent	τ(fs) (k x 10^{12} s^{-1})		ε (Polarity index)	η 25°C
	Absorption bands λ^{max}nm			
	420	485		
Cyclohexane	210	180	2.00	
	(4.76)	(5.56)	(0.00)	0.90
Acetonitril	240	200	37.5	
	(4.16)	(5.0)	(6.20)	0.34
Butanol	245	180	17.1	
	(4.08)	(5.56)	(3.90)	2.98
Cyclohexanol	280	360	18.3	
	(3.75)	(2.78)	(2.45)	2.45
Ethanol	380	500	24.3	
	(2.63)	(2.0)	(0.88)	1.20

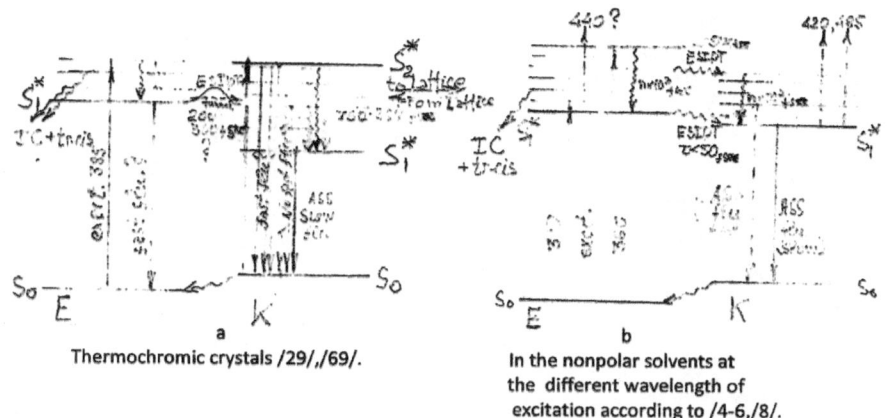

a
Thermochromic crystals /29/,/69/.

b
In the nonpolar solvents at the different wavelength of excitation according to /4-6,/8/.

Scheme 3.8

The ESIPT and deactivation of the electronic excitation in the thermochromic crystal and nonpolar solvents of the anile molecules.

Table 3.20
The participation of the vibration modes in the supposed ESIPT mechanism for SA molecule.

Vibration modes ν (cm^{-1})	Attribution of vibration modes	Method of investigation	Medium	Supposed mechanism	Refen. ces
1577	Skeletal vibrations in in the H-O-C-C-N structure fragment	Resonance Raman spectroscopy) λ_{ex}=350,7 nm	Crystal	?	/71/
?	Supposition about O-H stretching mode	Pump-probe(flu)	Crystal	Low-barrier Tunneling	/29/ /69/
?	Supposition about promotion by the low Frequency mode similar to HBT .	Pump-probe (TABS)	ACN	Barrierless	/8/
2884	Reaction mode				
1634i	Imaginary (i) mode				
*) 312 δ_s	C $_7$.C$_1$ -C$_6$: O-C$_2$ C$_3$.def.mode Phenyl ring wagging	Calculations		Tunneling	
558δ_s	C$_2$.O-H;C$_1$-C$_7$- N;C$_7$- NH; Ketyl C$_7$-C$_1$-C$_2$ Deform. modes.			98%	
652δ_s	CCC-ring ; C$_2$-OH; C$_1$-C$_7$-N; C$_7$-NH Deform. Modes.	Ab-initio		Barriers:	/7/
730δ_s	Ring CCC stretch (out of phase breath)	CIS /6 -31 G*		$\Delta E \approx 5.7$ kcal/mole or 7.2 kcal/mole	
853δ_s	Ring CC stretch. (in phase breath)				
898δ_s	Ring CC stretch.(in phase breath); N-C$_7$;C$_1$-C$_7$ stretch. modes				
94δ_a 244δ_a 537δ_a	Wagging and deformation modes				
?	Vibration broad band at electron transition band about 28200 cm^{-1} (see text)	Ultrasound Jet cooled LIF excitation	Gas mixture SA+He	Barrierless	/9/

*)δ_s and δ_a—Symmetric and antisymmetric transfer promotion modes respectively.

Nevertheless the $S_1^*(n\pi^*)$ state is deactivated almost completely by the barrierless ESIPT reac-tion with the rate by order of magnitude higher the radiationless deactivation/28/(see also 3. 2).

Thus the principle difference in the deactivation mechanism between N-Aryl and N-alkyl structures consists in the different competitive ability of IC as compared with the ESIPT which is clearly manifested in the N-Ph and practically absent in the N-Alk and N-Alk-Ph structures.

According to the conception (Sch. 3. 2b) discussed above such an effective competition can be promoted only in the molecular structure with N-Ph moiety by excitation of the two $S_1(\pi\pi^*)$ excited states with the different symmetry, one of which provides the ultrafast radiationless electronic deactivation.

In addition with the strengthening of O-H...N bond in the NAlkyl and N-Alkyl-Ph structures without n(N) lone pair delocalization along the imine ring (Ch. 2) the comparative ability of the ESIPT increases strongly as compared with the H-bond break, $\tau(ESIPT) \approx 30fs/27/$ or $15fs/28/$.

The variations of the relative rates of the sufficiently weak radiationless deactivation, the ESIPT reaction and the emission can promote the emission that competes with the ESIPT in such systems (see also Ch. 4).

3. 1. 2. 3. Effect of the rings' conjugation.

The distinctive feature of the compounds with the expanded π-system of the A or(and)B rings (see Sch. 3. 4) is the strong shift of the E⇔K equilibrium towards K structure especially at law temperature and in the polar solvent(see Ch. 2).

It result in the distortion of the longwavelength region of the Enol structure absorption spectra.

Therefore the difficulties arise in the determination of the band position for the So-->S_1^* transition, especially when elucidating of the substituents' influence. Under examination of the findings (Tab. 3. 12, 3. 13)/10, 30, 33, 42-46/the estimation of the So-->S_1^* transition energy of the E structure has been made by the position of the most shortwavelength band in the most shortwavelength region of the absorption spectra.

According to the data produced (Tab. 3. 12) the increase of the conjugated rings B in the acoplanar imine moiety (compound SN) results in the stabilization of the $S_1^*(\pi\pi^*)$ state in compari-son with SA by a little more than $\approx 1000 cm^{-1}$ whereas the expansion of the "Aldehyde "ring π-system (A)(compounds NA)leads to the lowering of the $S_1^*(\pi\pi^*)$ state by more $\approx 2000 cm^{-1}$, and addition of the second ring in the 3-4 position(3-4 NA) is more effective than that in the 5-6 position (5-6 NA).

The ring addition in the both moieties (5-6 NN) is a little more effective. At the same time the further expansion of the π-system with the addition of the third ring (PA) doesn't change the position of the $S_1^*(\pi\pi^*)$ state in comparison with SA on.

Thus the So->$S_1^*(\pi\pi^*)$ transition in the nonsubstituted compounds with the expanded ring π-system is localized mainly on the Aldehyde moiety with some contribution of the ICT from the ring A with OH group to the Imine moiety.

The π-systems of the rings modify only the electron donor and acceptor properties of A and B rings changing their ionization potential and electron ability correspondingly. The ICT character of the transition leads to the little increase of the excited state $S_1^*(\pi\pi^*)$ dipole moment that results in the trend to the longwavelength shift of the absorption band with the inc-rease of the solvent polarity(Tab. 3. 14)/43/.

The insertion of the substituents R modifies the ICT nature markedly: in the case of the strong electron donors the ICT gains the considerable reverse component (see above)and in the case of the strong electron acceptors the electron withdrawing nature of the imine moiety grows that results in both cases in typical increase of the oscillator strength of the electron transition and in some stabilization of the S_1^* state (Tab. 3. 13).

The differences in the variations of the above parameters with the insertion of the substituents in the structures NA, PA, and SA depends not only on the π-system size but also on the substituent effect on the coplanarity of the Imine and Aldehyde parts of the molecular structure studied.

The direction of the longwavelength absorption band's shift in the crystals in comparison with the nonpolar solvent(Tab. 3. 13)may possible serve as a criterion for the alteration of the structural fragments' acoplanarity with the formation of a tight crystalline packing. So the unsubstituted structures become more flattened in the crystal state (longwavelength shift of the absorption band), however with insertion of the substituents (with exception of NO_2) the structures can be less flattened in the crystal.

Such conformational changes are typical for both SA and the molecules with the two conjugated rings (NA) but with the more expansion of the π-system (PA) the conformation doesn't change practically with crystallization.

Obviously the change of the $S_1^*(\pi\pi^*)$ energy under influence of the ring conjugation, the substituents and the media variation for such molecular systems can play the essential role in the deactivation of the $S_1^*(\pi\pi^*)$ states in the Enol structure by the IC, ISC and ESIPT.

To the best of our knowledge the fluorescence of the Enol form has been found of the almost all compounds with the expanded π-ring systems discussed above especially because the study of such a very weak emission is hampered by its overlapping with the very intensive absorption band of the Keto-structure.

However the very weak ultrashort fluorescence in the short wavelength region (λ=400-500nm) of the ASSflu band has been found out for SN structure like SA one with excitation at $\lambda \approx 385$ and 415nm with decay time that is shorter than for SA (τ<40fsec).

This fluorescence is attributed (as for SA) to the "quazienol".

In any case the absence of the usual emission of the Enol form point to the mechanism of radiationless deactivation within the Enol structure which is similar to that of SA.

As long as in the structures with the expanded π-electron systems the $S(n\pi^*)$ states lies much more above the $S_1^*(\pi\pi^*)$ state than of SA ones, the deactivation can be connected like SA with structural transformations competing with the ESIPT and can be result finally in the reaction of the trans–cis isomerization with respect to C=N bond.

Really the typical for such a reaction spectral modifications have been observed in the solution at the room temperature by the signal of depletion at 300-400nm (disappearance of the E-trans structure in the So state) in the time ranges of $\tau \approx$ 1msec, 4msec, 0. 4msec in benzene, acetonitril, and ethanol correspondingly for 5-6 NA and 380nm for SN /47, 48/.

For the latter the typical spectral changes have been found also under the steady-state irradiation of the low temperature 3MP solution in the absorption band of the E trans-structure /48/. It results from the above findings that the cis-isomer is stable at 100 K and reverts efficiently to the trans-isomer only by heating to T>150K with activation energy about 60kJ/mol.

According to the supposition of authors /45/ the formation of the Enol cis-isomer, unlike SA, does not occur within E-structure but includes the ESIPT followed by the enolization of the twisted Kcis (Ktw) structure by means of the intermolecular mechanism. This supposition is however unlikely, and has not been confirmed by the following investigations (see above and /47, 48/).

The study of the radiationless deactivation of the excited 5-6NN with Naphthalene rings in the both moieties has been conducted by registration of the transient absorption spectra in the region 380-400nm in MCHX at room temperature /46/. It has been supposed that $S_1^*(\pi\pi^*)$ state decay time (τ~100psec) can be caused by both the ESIPT and IC.

However the study of their competition has not been conducted, and information about the trans-cis photoisomerization about C=N bond is absent. Thus the above discussed findings show that the $S_1^*(\pi\pi^*)$ state of the E structure in the compounds with the expanded ring π-system can be deactivated like SA by both the IC+ISC and the ESIPT.

However the rates of those are considerably lower than for SA that could be explained by the relative stabilization of the $S_1^*(\pi\pi^*)$ state, and especially by increase of the steric in-teractions, and modification of the vibrational modes which are involved in the ESIPT reaction (see below).

3. 1. 2. 4. The influence of the substituents in the azomethine bridge.

The insertion of the C substituents R(Sch. 3. 5) can cause the two competing effects: the considerable shortening (strengthening) of the H-bond (see Ch. 2) and the A-ring twist with break of the H-bond due to the steric interactions of the imine moiety (including B-ring) and the A-ring correspondingly (Sec. 2. 4. 3).

It is therefore expediently, from our point of view, to consider the two principal structure types differing in the steric interactions: the structural analogs of the SA molecules (Sch. 3. 5) and the SALK ones(Sch. 3. 6). The typical absorption and fluorescence data of ESA, the structural analog of the SA molecule are produced in the Tab. 3. 15.

Just as it has been expected, ESA unlike SA can exist in the liquid nonpolar solvents at room temperature in form of the two Enols, cis (Ecis) and trans (Etrans), structures, with the strong and broken IHB respectively(Sch. 3. 5).

The $S_1^*(\pi\pi^*)$ state of the Ecis structure corresponding to the Etrans one of SA lies higher the latter by $\approx 1400 cm^{-1}$ and like SA is responsible for the ESIPT with the ASS fluorescence (Stokes shift $\Delta\nu \approx 11000 cm^{-1}$) (see 3. 3).

The Etrans structure is less stable in the So state but more stable in the $S_1^*(\pi\pi^*)$ one which is responsible unlike SA for the marked Enol fluorescence ($\lambda \approx 370 nm$ with the normal Stokes shift $\Delta\nu \approx 3700 cm^{-1}$) in the nonpolar solvents at room temperature with very low quant. yield ($\sim 10^{-2}$) and very short lifetime ($\tau < 0. 5 nsec$).

Therefore the radiationless deactivation (IC+ISC) in the E*trans structure occurs with high rate and competes obviously with the very fast ESIPT in the E*cis structure also.

The sufficiently stable fluorescent anion or zwitterion(Sch. 3. 5) can be generated by the Etrans structure mainly from the H-bonded complex with the solvent or as a result of the H-atom abstraction by a base at room temperature in the protic solvents.

At 77K in the rigid glassy solvents and especially in the crystals(at room temperature and 77K) the equilibrium Etrans ⇔ Ecis in both ground and excited states is shifted completely towards the latter with the ESIPT fluorescence and without of the Enol fluorescence. Obviously the anion and Zwitterion structures is inhibited in such conditions also.

For analogs of the SALK molecules(ESALK, PSALK, MNALK see Sch. 3. 6)/51, 52/the some more shift of the equilibrium Ecis ⇔ Etrans towards Etrans especially in the $S_1^*(\pi\pi^*)$ state is observed due obviously to increase of the role of the steric interactions between the R substituent and the A-ring in comparison with interactions with the imine moiety.

For this reason the ASS fluorescence has considerably less intensity in the liquid solvents of ESALK than of ESA (see e. g. Tab. 3. 16).

At the same time the fluorescence of the E-trans structure (Stokes shift $\Delta\nu \approx 3700 cm^{-1}$) is observed, and absorption and fluorescence of the various polar structures which are formed in the Ground and Excited states of the solute (Etrans)-solvent(alcohol) complexes are observed also (Tab. 3. 16).

At the same time the practically complete shift of the equilibrium towards Ecis structure occurs in the glassy solvents at 77K and in the crystals (T=295, 77K).

However the phosphorescence of the polar Enol forms($\tau \approx 0. 7s$) arising in the H-complexes with the protic solvents is nevertheless observed in glycerin at 77K(Tab. 3. 16)/52/ showing participation of ISC in the radiationless deactivation.

The analogous phenomena are typical also for another C7 substituted alkilimines. For instance the appearance of the double –component low efficient fluorescence in the region of 300nm for PSALK ($\tau \approx 0. 8ns$, $\varphi \approx 2. 3 \times 10^{-3}$ and $\tau \approx 1. 5ns$, $\varphi \approx 2. 9 \times 10^{-3}$ in hexane)testifies to the existence of the two excited Enol structures deactivated mainly by the radiationless processes with high rates ($K_{nr} = 0. 22 \times 10 s^{-1}$ and $0. 82 \times 10 s^{-1}$ in glycerin)by way of torsion or bending motion which is hindered by the viscous drug /49/.

On excitation of the Ecis form of MNALK /53/ it also converted to the Etrans form in both the liquid solution and the low temperature solid matrix at 77K. Under such conditions the Zwitterionic form has been detected both in the Ground and Excited states. The shift towards Ecis structure occurs only in the crystalline matrix (T=295K).

Thus the most important distinction of the 7-subtituted azomethine analogs is the high probability of existence of the two, Ecis and Etrans, forms with and without H-bond correspondingly in the Ground and Excited states.

Unlike SA and SALK, the excited Enol structures of such compounds can be deactivated by the weak emission but like SA the main path-way deactivation are the IC and ISC competing with the ESIPT in the E structure.

3. 2. Kinetics and Mechanism of the ESIPT with generation of the fluorescent (NH)* structure.

3. 2. 1. The general description of the phenomenon and methods of its study.

The Excited State Intramolecular Proton Transfer (ESIPT) is the primary step in the multistage complicated process of the photocoloration.

It has been established recently the ESIPT is the complex ultrafast reaction with the coupled electron –nuclear motion to occur in the subfemto or femtosecond time scale.

The experimental and quantum-chemical methods of the investigation of such reactions have been elaborated for the last a decade and half.

Therefore the kinetics and the intimate mechanism of the ESIPT have become a subject of the detailed experimental and theoretical investigations only not so long ago.

The excitation So-->$S_1^*(\pi\pi^*)$ of the E-structure is followed by the ESIPT OH->NH, and generation of the excited (NH)* form that can have the Zwitterion (Zw), Keto (K) or resonance Zw<-->K structures in dependence on the concrete Enole molecular structure, the environment, and the excitation energy.

The adiabatically formed (NH)*structure (see below) can usually be deactivated by emission S_1^*-hv->So with the Anomalous Stokes Shift (ASS flu), and as a rule, it is the necessary indication of the ESIPT in any structure with IHB /54/.

Table 3.21

Effect of the rings' substituents (crystals and solvents).

No	Structure	τ_f (s) (k × 10^{12})s^{-1}	Medium Temp. T_K	Method Excit./Prob. λ (nm)	Excitat. λ (nm)	Supposed mechanism	Notes		Reference
1	5ClSA	780 (1.28) 760 (1.32) 720 (1.39)	Crystal 293	SWff 480 Decay —"— ASS flu 560 Rise	385 410	Low barrier Tunneling Assisted by low-frequenting vibrations.	$\Delta E^{*)}$ 44	Slight temperature dependence. Deuterium effect.	/29/
2	4'CH$_3$5ClSA	2780 (0.36)	Crystal 293	SWff 480 decay	385 or 410		30		
3	4'CH$_3$SA	3570 (0.28)	Crystal 293	SWff 490 decay	385		23		
4	3-6ClSAN-Pyren	440 (2.27)	Crystal 293	SWff 490 decay	385		---		/69/
5	4C$_{12}$H$_{25}$OSA 4'COOHSA	360(2.78) 250(4.0)	Ethanol 293 Chloroform 293	Tabs 360 decay & rise	360	Low barrier (2.76kcal/mol). Probably tunnel. And relax. In NH Structure.	Deuterium effect was not studied.		/6/
6	SA-N-Phenantr.	<40 (>20)	ACN 293	SWff 480 decay	385	---	No Deuterium effect.		/10/
7	N-CH$_3$ (SALKo)	30 --- 11	---	Ab-initio (instant.appr) ---TDDFT	---	Low barrier (4-6 kcal/mol)	Large Deut. Eff.τ_H/τ_D=3.8 Large Deut.Eff. τ_H/τ_D= 2.3		/27/ /68/
8	3,5 Tertbut.SA 2'NPyr --- 3,5 Tertbut-SA 4'NPyr	<150 <1000 --- <150 <1000	CH Crystal --- CH Crystal	TABS 435,510,630 FDRS Pos Pos Neg TABS 440,515,650 FDRS	390 390	Similar to SA	In both cases the ESIPT is faster than the temporal resolution both in solvent (CH) and crystal.		

$^{*)}\Delta E$ kcal/mol=E(S$_1$*Enol)-E(S$_1^-$Keto). SWff—Short Wavelength fast fluorescence.
TABS —Transient Absorption Spectra
FDRS—Femtosec. Diffuse Reflectance Spectra

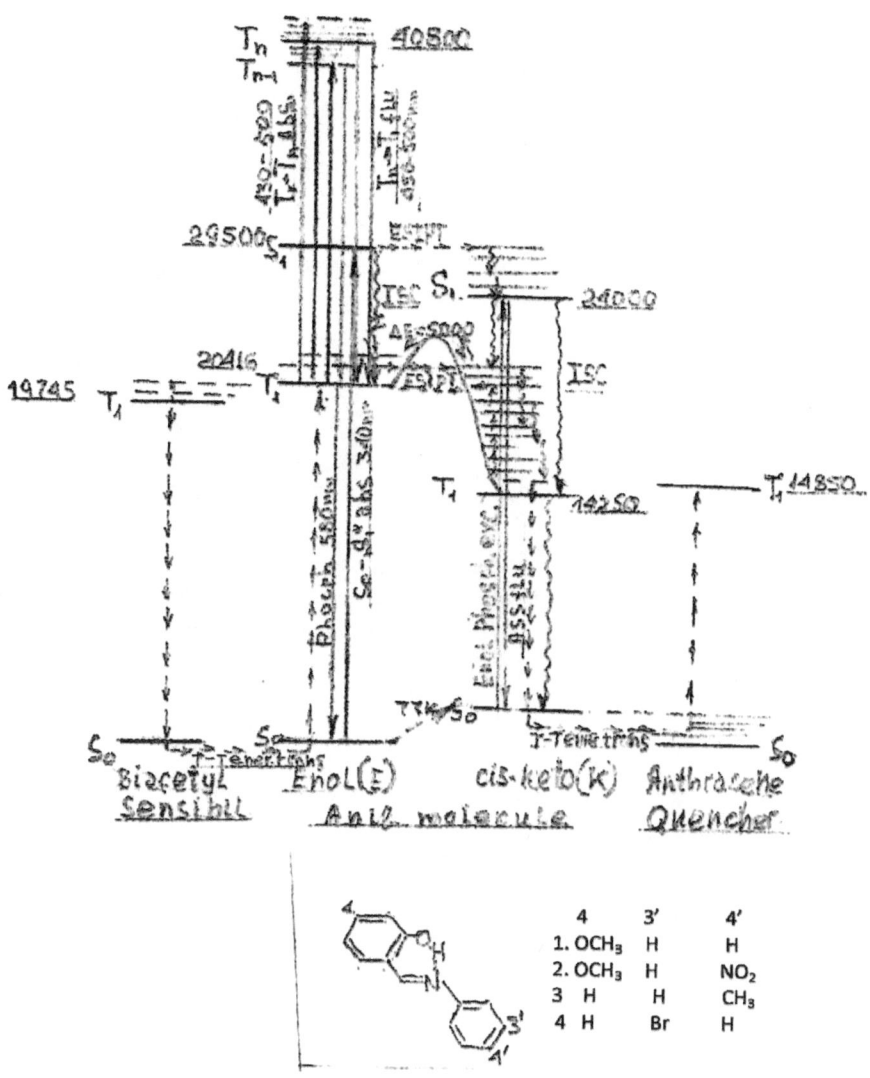

Scheme 3.9
The ESIPT in the Triplet (T_1) states of the Anil molecules (in MCH or CH).

Table 3.22

Fast-decay time constants' dependence on the hydrogen-bond length

Compound	Hydrogen-bond length (Å)	Time constant τ_f (fs)
I	2.534	440
II	2.537	710
III	2.584	780
IV	2.602	1000

I. 3,4,5,6 Cl N-pyrenyl .II.2-OH Naththyl-Ph-2-OHNathyl(4,5).III.5ClSA.IV.SA-Ph-SA(Sec.4.5).

==

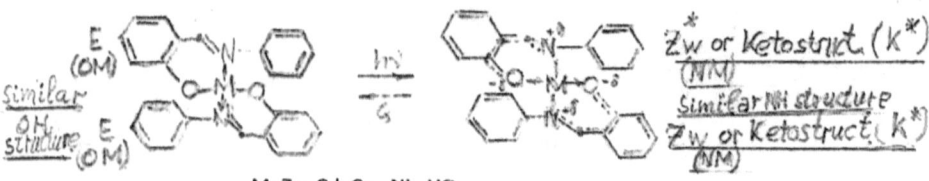

M: Zn, Cd, Cu, Ni, UO$_2$

Scheme 3.10

The photochromism of the Metal-Chelate complexes with SA caused by the generation of N$^+$-M (keto-zwitterion) twisted structure similar N$^+$-H "post TICT" photochromic structure for molecule SAs.

Table 3.23
The structural changes (bond length differences, ΔL Å) and the ESIPT parameters in anils with the different imine moieties.

Bond	SA /7/		SALKo/27/ [7)]		SALK$_1$ CH$_3$Ph /28/	
	ΔL$_{ETs}$ [1)]	ΔL$_{KTs}$ [1)]	ΔL$_{ETs}$	ΔL$_{KTs}$	ΔL$_{ETs}$	ΔL$_{KTs}$
O--H	-0.208	+0.693	-0.125	0.828	-0.119	+0.825
N--H	+0.911	-0.285	+0.318	-0.387	+0.288	-0.374
C$_2$-O	+0.44	-0.041	+0.018	-0.047	+0.018	-0.045
C$_7$-N	+0.004	-0.0002	-0.021	+0.015	-0.016	+0.018
C$_1$-C$_7$	-0.001	+0.0019	+0.007	-0.005	+0.005	-0.006
C$_1$-C$_2$	-0.030	+0.002	+0.006	+0.005	+0.002	+0.003
|ΔL$_{aver}$| [2)]	|0.133|	|0.172|	|0.083|	|0.214|	|0.075|	|0.212|
ΔH kcal/mol [3)]	5.7—7.2		1.62		1.27	
E$_{S1}'$-K$_{S1}'$ kcal/mol	7.2—11.4		18.4		15.15 -16.75	
τ$_{fs}$ [4)]	2200 (H)	49.6 (H) [5)]	30(H),115 (D)	11(H),25(D) [6)]	15(H)	

1)Here and further ΔL$_{ETs}$=L$_E$-L$_{Ts}$ and ΔL$_{ETs}$=L$_K$-L$_{Ts}$ where L$_E$, L$_K$,L$_{Ts}$—bond length in the E,K and Transition structures correspondingly.
2)\|ΔL$_{aver}$\|–The average absolute value of the bond length differ.3)The ESIPT potential barrier height in the excited state (see sch. 3.14)
4)τ$_{fs}$—time rate : for proton (H),deuterium(D), 5) From <1>,6) From <68>, 7) See also table 3.(struct. 7).

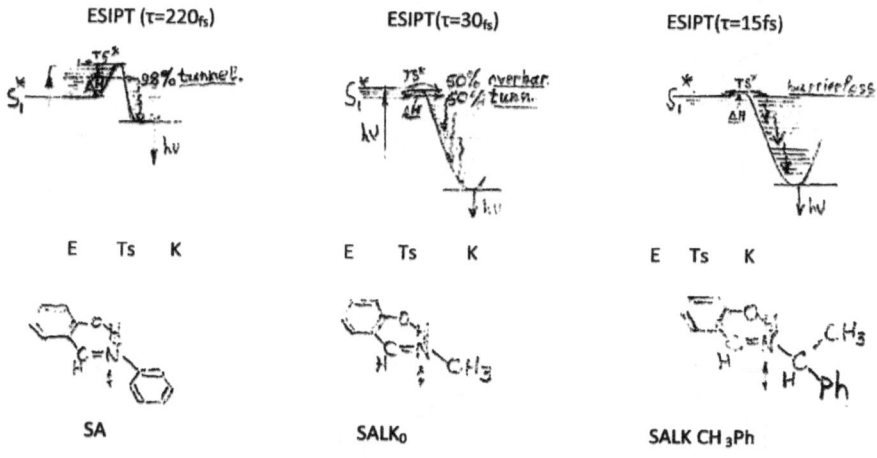

Scheme 3.11
Mechanism of the ESIPT on the base of the findings /7/,/27/, /28/.

Table 3.24 a

Dependence of the GSIPT (S_0) and ESIPT (S_1^*) reaction energy activation (ΔE) on the amount of the conjugated aldehyde rings in the structures with O-H...O bond (Quantum-chemical calculations) /78/)

Amount of the Aldehyde rings	ΔE (kcal/mole)	
	S_0	S_1^*
1	19.2	15.1
2	15.1	13.9
3	8.4	7.0
5	9.7	8.6

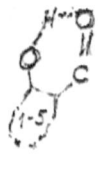

Scheme 3.12
The ring conjugated structures (see also ch.2).

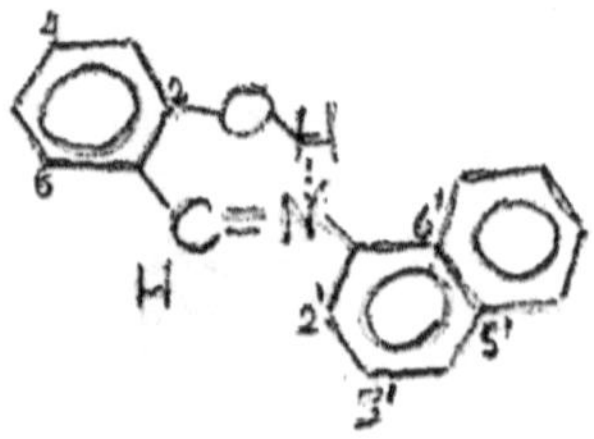

I(Ph,5',6' Np) 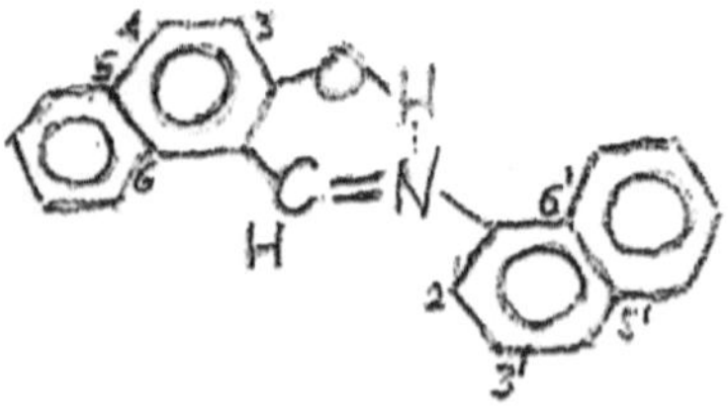   I(5,6 Nph,5' 6'Np)

Table 3.24b

The comparative data of the ESIPT time rate for the Naphthalidene derivatives[1].

Structure	τ (fs)	Media, room temperature	References
I(Ph,5'6' Nph)	40	ACN, no D eff.	/10/
I(5,6Nph,5',6'Ph)	<100 [2]	MCH, D eff. ?	/46/
I(Ph,Ph) SA	50	ACN, no D eff.	/10/

1) See also table 3.21 . 2) Time resolution ~100fs

Table 3.25
Some parameters of the (NH)* structures on the PES of S_1^* state of SA calculated by different authors [+]

Bond \ Type .length \ of (Å) \ state Angle (°)	1) $S_1(\pi\pi^*)$ C_s	1) $S_1(n\pi^*)$ C_1	2) $S_1(\pi\pi^*)$ C_s	2) $S_1(\pi\pi^*)$ TICT like
O-H	1.868	3.654	1.777 1.99 [3]	3.260
N-H	1.005	0.994	1.017 1.02 [3]	1.00
C_2-O	1.225	1.207	1.266	1.249
C_1-C_2	1.463	1.488	1.457	1.435
C_1-C_7	1.374	1.433	1.482	1.436
C_7-N	1.326	1.378	1.335	1.323
α	0	78.9 [1] 90.2 [3] 72.4 [6]	------	-90 [2,4]
β	0	30.9	0 0.10 [3]	0
μ (D)	3.6 [5]		9.7 [2]	14.4 [2]
$\Delta E(S_1^*_{(\pi\pi^*)}-S_1^*_{(n\pi^*)}) \approx 1055$ cm^{-1}		1)		

+) Methods of the calculation and the references : 1)Ab-initio CIS/6 + 31G* /7,28/; 2)Semiempirical PM3 /13,14/ 3)Semiempirical AM-SCF/6/. 4) Ab-initio HF/6-31GCl /35 / .5) Semiempirical INDO/S /80/ .6) Ab-initio B3LYP/6-31G*/9/.

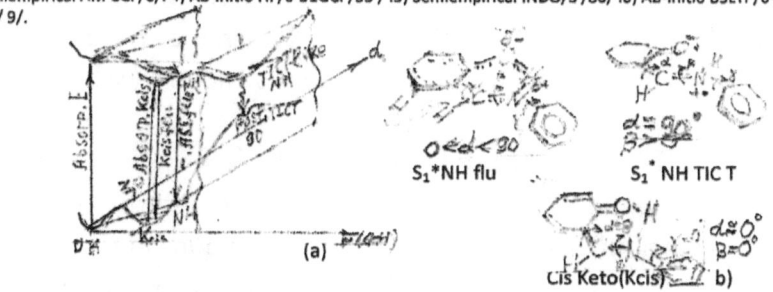

Scheme 3.13
The forming and the decay of the NH* fluorescent state (nature of ASS flu).

Therefore although the thermalized fluorescent (NH)*structure does not always arise immediately as a result of the ESIPT, its formation may compete with the reverse ESIPT reaction, and this process is usually considered as a final step of the direct ESIPT reaction(see below).

The principal experimental kinetic methods of the ESIPT study are based on the registration of the fast temporal dependence of the transient $S_1^* \to S_n^*$ absorption spectra of the Enol(E)(intensity drop) and (or) the NH one (delay of the appearing and intensity increase) and also of the delay of the appearance and evolution of the ASS fluorescence band after ultrashort (pico and femtosec scale) laser impulse excitation of the E structure (the pump-probe method).

The fluorescence can be registered both directly by the NH emission band and by delay of the transient absorption change induced by the rise of the NH emission, and probed in the wavelength region of the ASS fluorescence in the transient negative absorption spectra.

The dynamic quantum-chemical investigations of the ESIPT mechanism including the analysis of the vibrational modes' configuration have been conducted on the base of the high level ab-initio calculations with use of the method with the "instant on" approach /7/ and Time Depended Density Functional Theory(TDDFT) method/1/.

3. 2. 2. The ESIPT of the related structures.

The described above methods have been used for the investigation of the ESIPT with the IHBO-H...N in the series of the analogues of Hydroxyphenylazomethines with the rigid structure (Sch. 3. 7, Tab. 3. 17).

The ESIPT rates, τ(ESIPT) or k(ESIPT), have been obtained, and the conclusions have been made about intimate reaction mechanism and the contributions of the various nuclear motions in the ESIPT mechanism. The produced findings allow to make some general important conclusions/55-67/.

- i. The obtained values of the ESIPT rates lie within the subfemto-femtosecond time scale. Such an ultrafast rate, $k(ESIPT) > 10^{11-12}$ requires an understanding of interplay between the structure, dynamics and reactivity.
- ii. As long as the time scale of the ESIPT is comparable to the period of the low-frequency molecular vibrational mode one can

expect a better understanding of the process by learning about the vibrational dynamics that drives and follows the ESIPT. Really, all the findings point out the participation of the low frequency modes so called "Transfer Promoting Modes "(TPM) in the ESIPT reaction although the exact identification of the TPM and a contribution of the definite TPM are still not well established.

iii. The reproducibility of the results is unsatisfactory. The obtained values of the ESIPT rate for the same structure by various authors can be differed by the order of magnitude. Such a situation can be caused by the difference of the methods for the probe registration (ASS fluorescence or transient absorption), by various conditions of the experiments-different media(gas, solvent polarity, and especially by the difference of the excitation energy. Indeed, in the low polar solvents (CH, TCE) the lowenergetic excitation($\lambda \approx 370$, 347nm) results in the considerable higher ESIPT rate values(60-33fs) than those (500, 160fs) for the short-wavelength excitation (λ=310nm) (see Tab. 3. 17, data for HBTA and HBT) that can be obviously explained by the inclusion of the vibrational relaxation $S_1^{(*)} \rightarrow S_1^*$ of the NH structure in the intrinsic ESIPT reaction (see also below).

On the base of the analysis of the data of the ESIPT rates along with the role of the lowfrequency modes and the PES construction, a principal conclusion about dynamics and mechanism of ESIPT reaction has been made for the different molecular structures under various conditions.

The new development in the time –resolved spectroscopy and in the dynamic methods of the quantum –chemical calculations allows the observation and investigation of the highly localized wave packet motions during and following the ESIPT process.

For the ultrafast ESIPT reaction the two basic possibilities can only be realized on the PES-the barrierless (or almost barrierless)path-way or tunneling. The first case occurs for the structures HBTA, HMO, HBT and HBSe/55-57, 60-64, 67/, and the second one can take place for

HBO and HPPO in the nonpolar solvent and gas phase correspondingly /58, 59, 65, 66/.

However the principal role in the both cases is played by not an explicit proton motion but a changes of the O—N distance due to the low-frequency skeletal vibrations(TPM) attributed to the in-plane rotation (bending) of Hydroxyphenyl and Azole (or Triazole) subunits which modulates O—N distance and proton transfer barrier.

Thus in the first case with the closing in the reaction centers O and N the barrier almost vanishes and the E*->(NH)*reaction must be proceed practically instantly (τ<10fs).

Therefore the ESIPT time (30-200fs) obtained by the appearance of the (NH)*structure (ASS or transient absorption) is caused mainly by a skeletal motion inertia and no deuterium effect has been observed.

In the second case, only the initial step of the ESIPT is connected with contraction of the O-N distance owing to the TPM. The latter can be both in-plane bending and deformation modes /58, 59/ and out-of-plane bending and torsion modes/66/ so that O-H distance can be remain practically constant and proton moves together with atom O.

The following step of N-H bond formation occurs as a result of a strong increase of the tunneling probability due to the shortening of the N-O distance and strong lowering of the energy barrier for proton movement along the reaction coordinate O-H.. N. In this case the high ESIPT rate (60-220fsec) is caused by the coupling with the tunnel proton motion and marked deuterium effect is observed /66/.

In the final step of the both cases the large amplitude vibrations connected with variations of the O-N distance favor the strong closing in the N atom and proton and formation of the N-H bond switching wave packet to the electron motion including the electronic configuration change typical for (NH)* structure under the definite conditions (see above).

In both cases after (NH)*state appearance the new skeletal vibrational modes being involved in the process of deactivation and formation of thermalized (NH)*structure are responsible for the intensive ASS fluorescence. Such modes inhibit the reverse proton transfer and provide irreversibility of the ESIPT.

Thus the whole ESIPT process is really multidimensional with respect to both the nuclear and electronic motions intimately linked in the indivisible highly localized wave packet created by the electronic excitation of the OH structure followed by generation of the fluorescent (thermalized)NH* one.

3. 2. 3. The study of the ESIPT of Salycilideneanilines.

The findings obtained for SAs by the methods described above are produced in the Tab. 3. 18-3. 21. The values of the ESIPT rate (Tab. 3. 18, 3. 21) are the same order of magnitude as the related structures (Sec. 3. 2. 2) and lie within the subfemto-femtosecond time scale k(ESIPT)~10^{11}-10^{12} s^{-1}.

At the same time there are the two groups of the produced results. In the several recent studies the very high rate values have been obtained (ESIPT)<50fs, k(ESIPT)>20x$10^{12}s^{-1}$/1, 8-11, 68/ but in many cases the considerably lowerAQZ rate values have been obtained, τ>n10^2fs, k<5x10^{12} s^{-1} /4-7, 23, 29, 69/, and the large irreproducibility is observed within the second group (180 780)fs and more higher/29/). It requires a more detailed analysis of the utilized methods and conditions of the investigations in each case.

The solute-solvent interactions (viscosity, polarity bond, ect.) and uncertainty in acoplanarity of the different moieties influence the ESIPT rate in the solutions.

Therefore the investigation of the ESIPT rate in the crystal where the positions and conformation of molecules are fixed by intermolecular interactions in the crystalline lattice is expected by opinion of the authors /29, 69/to reveal the intrinsic dynamics of proton transfer along the Hydrogen bond coordinate.

Table 3.26
The dependence of the decay rate and the spectral characteristics of the S_1^* NH flu states on the solvent nature.

Solvent	ε (polarity Index)	η (cp) 293K	Transient absorption [2] τ(ps) (Excit λ= 380nm) Prob λ nm:			Fluorescence excitation λ^{exc}= 360nm			
			420	485	620 (negat.)	ν^{max}cm^{-1} [3] (λ^{max}nm)	Stokes shift[3] Δν(cm^{-1})	τ (ps)	Quantum Yield φx10^{-4}
Cyclohexane	2 (0.00)	0.90	3.9	3.5	4.0	18550 (539)	10650	5 [2] 26 [3]	0.7 [3]
Acetonitril	37.5 (6.20)	0.34	7.9 7.0 [1]	5.6 7.0 [1]	4.5 6.8 [1]	18000 [1] 18585 [2] (540-550)	11000	6.7 [1]	~1.0 [2]
Ethanol	24.3 (0.88)	1.20	6.0	5.4	8.3	19050 (524)	10450	11 [2] 20 [3]	2.1 [3]
Cyclohexanol	18.3 (2.45)	2.45	7.0	52.2	116.0	—	—	48 [2]	—
Butanol	17.1 (3.90)	2.98	24.5	30.0	8.0	18790 (532)	10620	27 [2] 56 [3]	2.3 [3]

The finding have been obtained from /8/ 1),/5/ 2) and /35/ 3).

Table 3.27
The dependence of the ASS flu lifetime (τ) and the quantum yield (Φ) on temperature and viscosity in octanol /35/.

T_K	η(cp)	τ(ps),(kx10^9)	Φ$_{rel}$ [2]
264	25	155(6.45)	1.000
278	18 [1]	98(10.20)	1.014
294	10	64(15.62)	1.000
308	7.0 [1]	52(18.88)	0.996
324	4.0	36(27.78)	0.981

[1] Obtained by the linear extrapolation. [2] Obtained from the dependence of the fast fraction on the temperature/35/.

Scheme 3.14
The two NH structures responsible for the different flu bands in the including complex Anil-CD(Anil is Salcilidene2-iminopyridine I(Ph,2Pyr).

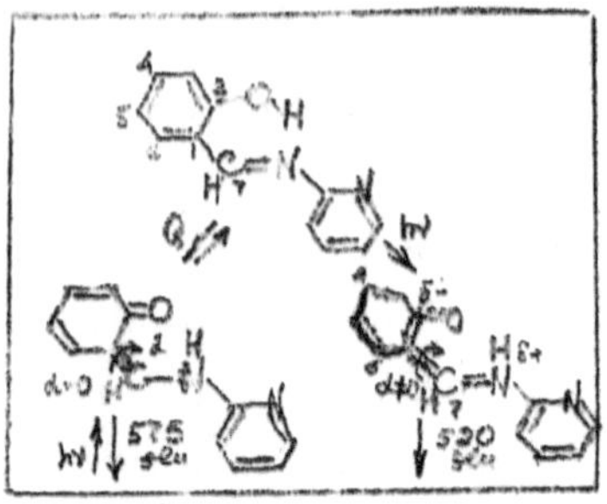

Table 3.28
Comparative fluorescent data of the model structure, L^{Ph}SA, and SA (IIP)*).

Structure	Absorption 293,77 K ν^{max}_{abs} cm^{-1}	Flurescence					
		ν^{max}_{fl} cm^{-1}		Stokes shift Δν cm^{-1}		Quant. Yield φ	
		293K	77K	293K	77K	293K	77K
L^{Ph} SA	31290	18580 19000	18800 19400	~12850	~12500	0.044	0.24
SA	30000	18850	19500	----	~11000	≤10^{-4}	~10^{-3}

*)IIP –Isopentane/isopropanol, 4/1(v/v)-Glassy solvent at 77K.

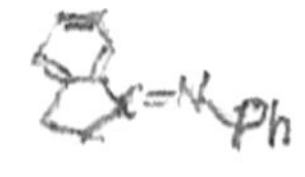

L^{Ph}SA Model structure

Table 3.29
The emission parameters for different fluorescent structures of molecule SA NH structure(IIP 77K).

Fluorescent excited State and its attribution	Excitation λ^{max} (nm)	Fluorescence two bands v_1^{max}, v_2^{max} (cm^{-1})	Splitting Δv_{1-2} (cm^{-1})	Relative Intensity I_{v_1}/I_{v_2}	Quantuum Yield (relative)	Estimated Angle α°
$S_1^* K_{cis}$ Thermochromic Solvatochromic	440	19500 18850	650	1.1	1.0	0
S_1^* NH flu ESIPT flu ASS flu	340	18350 19300	950	0.9	0.6	>0
S_1^* NH twist Photoclored strocture	480-530	1800 16750	1250	1.1	0.1	~70-90

Table 3.30
The ASS fluorescent characteristics of the thermochromic crystals *)

$T_K = 300$			$T_K = 4-70$			
λ^{max}_{flu} (nm) λ^{max}_{exc} (nm)	Band width ½Δv(cm^{-1})	Stokes shift	λ^{max}_{flu} (nm) λ^{max}_{exc} (nm)	Band width ½Δv(cm^{-1})	Stokes shift	τ(ns)
556 517	2016	1450	560,526 380	~1000,~1000	~7500	0.4

*)The parameters differ not strong for the structures 5ClSA,4'CH$_3$5ClSA,and 4'CH$_3$SA /69,89/ and the averaged values are given in the table. The fluorescence excitation spectra undergo the strong distortion as a result of the scattering in the crystal, and shifted towards to the red understating the Stokes shift strongly ,and that is not taken into account by authors.

In this connection the determination of the ESIPT rates and the study of the molecular structure, effectron transfer, and the ESIPT mechanism were conducted in the thermochromic crystals at the various temperature by use of the fast decay of the short-wavelength weak fluorescence /52, 53/.

The fast decay component of the fluorescence was observed according /69/, at wavelength just below or even inside of the Enol $S_0 \rightarrow S_1^*$ absorption band and was not found with the selective excitation in the thermochromic absorption band (the $S_0 \rightarrow S_1^*$ transition in the K structure).

Therefore according to /29, 69/ this fluorescence belongs to the S_1^* state of the E molecular structure in the crystals (Sch. 3. 8a). If to suppose that there is a little contribution of another path-ways of the E S_1^* deactivation (see however sec. 3. 1. 1.), the fast fluorescence decay rate-τ(decay), can be considered to be mainly determinate by the

ESIPT. The ESIPT rates obtained by such a method (Tab. 3. 21) have been utilized for the analysis of the ESIPT dynamics and its dependence on temperature and molecular structure(see also sec3. 2. 4).

However authors don't take into account the series of the important peculiarities of the structures and the properties of the thermochromic crystals (Ch. 2):

i. The E⇔K equilibrium is shifted strongly towards colored K structure with increase of the tem-perature (unlike the solvents!) especially at T>290K.
ii. The strong electrostatic interactions between the adjacent molecules closely arranged in the almost parallel planes. It creates the very favorable conditions for the exchange of the vibrational energy between the single molecule and the crystalline lattice.
iii. The strong steric interactions between the aldehyde structural fragments of the adjacent mo-lecules typical for the thermochromic crystals with the hindering of the deformational vibrations promoting the ESIPT and the exclusion of the ring twist ones providing the vibrational deactivation of the K S_1^* state and hindering of the reverse proton transfer in the excited state.

Due to (i) there is high probability for the excitation of the both E S_1^* state and KS_2^* one. The latter is excited mainly at T>200K. Therefore the ultrafast fluorescence can be the superposition of the two emission-from $S_1^*(E)$ and $S_2^*(K)$ states, but not from the $S_1^*(K)$ state in accordance with the experimental findings described above about the lack of the fast fluorescence under excitation in the thermochromic absorption band/29/. The fluorescence decay reflects the compound process of deactivation of the both states including the vibrational deactivation and ESIPT(Sch. 3. 8).

Thus for 5CLSA /29/the weak temperature dependence and the deuterium effect which is less by order of magnitude than calculated value for the tunnel mechanism /7/ can be explained by the large contribution of the vibrational relaxation rate comparing with the intrinsic ESIPT one.

However at the low temperature in the rigid matrices, where some deformation modes can be "frozen", the tunneling plays a marked role and τ-value reflects the ESIPT rate to a greater degree.

Obviously the both vibrational relaxations in the E and K structures and the EISPT precede a generation of thermalized fluorescent K S_1^* state, and therefore the fast fluorescence decay time must coincide with the slow ASS fluorescence rise time, τ(fast, decay)≈τ(slow, rise) that is agreed with the experimental findings for the solvent and the crystals(Tab. 3. 18/11/, Tab. 3. 21 /29/).

Table 3.31
Effect of the A ring substituents on the ASS fluorescence (Hexane 77K)(electronic factor).

N^0	$\lambda^{max}_{nm,}$	I_{rel}
1	520	1.00
9	454	0.35
10	515	0.52
11	500	0.85
12	520	0.47

*) Excitation 340-360nm (E-structure)

Table 3.32

Effect of B-ring substituents on ASS fluorescence (electronic) factor.

No Sfr uc- tur.	Cyclohexane /35/					Ethanol /35/						Octanol[4] /35/		Ethanol /33/				Crystal.film /30/			
	Abs λ_a^{max} nm	Fluorescence				Abs λ_a^{max} nm	Fluorescence					K_{TICT} ns^{-1} 4)	E_{TICT} kcal/mol 4)	Abs λ_a^{max} nm	Fluoresc.		Φ_{rel}	Abs λ_a^{max}	Fluoresc.	
		λ_f^{ma} nm	$\Delta\nu^{a-f}$ cm^{-1}	τ_{fl} ps	Φ_f x10^4		λ_f^{max} nm	$\Delta\nu^{a-f}$ cm^{-1}	τ_{fl} ps	Φ_{fl} x10^{-4}					λ_f^{max} nm	$\Delta\nu^{a-f}$ cm^{-1}		nm	λ_f^m nm	$\Delta\nu$ cm^{-1}
1	342	539	10690	26	0.7	338	525	10450	20	2.1	1.35	2.89	336	510	9800	1.00	347	---	---
2	357	552	9890	29	1.6	360	568	10170	19	1.4	1.49	2.78	354	540	10100	0.7	374	575	9365
3	352	542	9980	17	1.1	344	560	11190	34	1.6	1.34	2.87	---	---	---	---	---	---	---
4	346	529	9990	20	1.1	341	522	10170	28	2.7	0.97	3.10	---	---	---	---	---	---	---
5	353	526	9340	28	1.6	351	522	9340	41	4.9	0.80	3.22	349	540	10180	4.8	325	570	13225
6	384	528	7060	69	15.0	382	528	7250	33	12.4	0.23	4.01	380	545	8000	0.9	331	580	12970
7[1]	350	518	9265	---	14.0	---	---	---	---	---	---	---	---	---	---	---	---	---	---
8[2]	350	520	9340	25 20 3)	1.0	---	---	---	---	---	---	---	---	---	---	---	---	---	---

1) The findings from /32/. 2) The findings from /6/. 3) The same in chloroform. 4) See scheme for table

ESIPT Photochromism   153

Identification of the structures for the tables 3.31 and 3.32 (see Scheme)

No	1	2	3	4	5	6	7	8	9	10	11	12
R	H	H	H	H	H	H	H	$C_{12}H_{25}$	NMe_2	OH	$3-NO_2$	$5NO_2$
R'	H	NO_2	CN	Me	OMe	NMe_2	OH	COOH	H	H	H	H

The structure for the tables 3.31 and 3.32

Scheme for the table 3.32 (see 4))

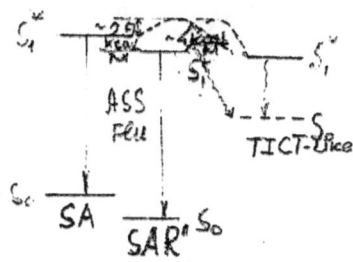

Table 3.33
The influence of the bulky ring substituents (steric factors.)
i-iv from /32/, v,vi from /36/

Sructure			T = 293 K a)				T=77K b)				
R	R'	No	λ^{max}_{flu} nm	λ^{ma}_{e} nm	Stok.shift $\Delta \nu$ cm^{-1}	Φ_{flu} 293 x 10^4	λ^{max}_{flu} nm	λ^{max}_{e} nm	Stok.shift $\Delta \nu$ cm^{-1}	Φ^{rel}_{flu77} (relative)	Φ_{77} Φ_{293} x 10^{-1}
H	H	i	518	350	9265	14.0	515 540	360	8360	1.0	6.9
Tr-But	H	ii	530	359	8988	7.0	520 540	370	8400	1.2	17
H	Tr-But	iii	518	354	8943	17.0	515 540	360	8360	0.8	5.1
Tr-But	Tr-But	iv	540	359	9337	12.0	520 545	360	8550	1.1	9.6
Tr-But	2'-Pyr	v	562 (5) c) 602(25)	370	10416	2.5	560(23) 608(250)	366	10870	--	--
Tr-But	4'-Pyr	vi	567(4.5) c) 610(14.5	361	11384	0.7	569(23) 630(25)	368	11300	--	--
SA			--	--	--	<1.0	513 530	342	9920	--	--

a)For i-iv in isopentane,T=293,for v,vi in cyclohexane,T=293K.b)For i-iv in isopentane,T=77K,for v,vi in the crystal film ,T=293
c) The main maxima are underlined, in the bracket for v,vi–the time constant (ps).

Table 3.34
The fluorescence of N-Alkyl structures (isopentane T=77K))/20//37/.

Structure	λ^{max}_{flu} nm	λ^{max}_{ex} nm	Stokes shift $\Delta\nu$ cm^{-1}	Φ^{rel}_{flu}	$\Phi^{flu}_{365}/\Phi^{flu}_{313}$	Φ^{flu}/Φ^{color} 365	Φ^{flu}/Φ^{color} 313
SALK$_0$	500 465 (465)	320 395	11250 3800	0.9	1.1	2.3	2.1
SALK$_{17}$	506 465(465)	320 400	11490 3500	1.5	1.3	0.9	0.7
L$_{Ph}$SALK$_{17}$	510(510) 468(465)	310 390	12650 4270	2.3	---	---	---
SA	512	340	9680	1.0	5.0	1.0	0.2

1) For convention signs of the structures see the scheme 3.3 . 2) The λ^{flu}_{max} at the room temperature. 3) The data for ASS flu. 4) The wavelength of the exsitation.

As a result of peculiarities of thermochromic crystal structure (ii), the characteristic time constants of the vibrational deactivation from the electronic excited states are increased strongly due to the effective exchange of the vibrational energy between the adjacent molecules and the crystalline lattice which structure is connected closely with the molecular one (see Sec. 3. 4).

At last the steric interactions (iii) hinder the bending vibrations of HO-Ph-C=N-structural fragment in the E structure excited state which promotes the ESIPT (see sec. 3. 2), and the torsional ring vibrations around C-C bond in the excited state of the K structure which provides its deactivation. As a result the steric hindrances can favor a decrease of the "brutto" ESIPT rate (k) determined by experiment. Thus since the peculiarities of the thermochromic crystals (i-iii) can favor the strong overestimation of the ESIPT time rate determined by the methods used /29, 69/ they are hardly appropriate to reveal the intrinsic dynamics of the ESIPT of the anil molecules in the crystals.

However the heightened ESIPT time rates can be also explained by the peculiarities of the ESIPT dynamics in the crystals. Therefore the information of the ESIPT dynamics in the solvents is needed.

The ESIPT rate study for SA molecule in various solvents at room temperature has been conducted by the transient absorption pump-probe method with the ultrashort time shortwavelength excitation within absorption band (λ^{max}=360nm) to the vibronic S_1^* state of E structure /4-6/.

The ESIPT rate has been determined by the rise of the delayed transient absorption at the probe wavelengths λ= 420 and 485nm attributed to $S_1^*(S_1^*)$-->Sn_1^*, $S_1^*(S_1^*)$->Sn_2^* transitions of the thermalized or "hot" (vibrational excited) S_1^* state of NHKeto) structure (Tab. 3. 18, Sch. 3. 8b).

The time of the vibrational relaxation to the thermalized (fluorescent) S_1^* state of the K structure (S_1^*->S_1^*)(τ=300-400fs) was estimated by the short-wavelength spectral shift and narrowing of the transient absorption band ($\lambda^{max}\approx$620nm) attributed to the S_1^*-->So emission of the K structure (ASS flu).

The values of the rate constants depend on the solvent nature (Tab. 3. 19). This dependence reflects a competition between specific interactions (Intermolecular H-bond), polarity, and viscosity of solvents.

A too little increase of τ in the protic solvents (compare 2 and 3, 5) can hardly been explained by fall of the ESIPT rate in the O-H...N complex with the protic solvent molecules (excluding ethanol). Decrease of the rate constant k s^{-1} in the polar solvents (compare 1 with other solvents) can't be connected with the formation of the polar K* structure /70/ as a result of the ESIPT but at the same time there is a marked tendency to the τ increase in the more viscose solvents (compare 1 or 2 with 5).

At the same time the rates in the various solutions are close to those of the vibrational deactivation in the S_1^* state of the K structure in the solvents ($\tau\approx$300-400fs) and the crystals (see above /29, 69/).

Thus the data above can testify to the considerable contribution of the Intramolecular Vibrational Redistribution (IVR) in the described experimental conditions (Sch. 3. 8a). At the same time the information about very fast ESIPT rate (ks^{-1}) can be derived by some indirect methods/9-11/ (see below, Tab. 3. 18).

Therefore in spite of distortion (overestimation) of the intrinsic ESIPT time rate involving IVR, the principal role in the estimation of the intrinsic ESIPT rate can be played by a correction of the method of the direct ESIPT time measurement. The correction consists in the assimilation of the instrumental response function (IRF) to the pump-probe crosscorrelation signal by measuring the two photon absorption signals in the different solvents where transient absorption intensity is displayed as a differential absorption at one time delay which is measured 4500 times and averaged /11/.

At the femtosecond time scale the solvent contribution and IRF have been considered to assess of efficient deconvolution of the time-depended concentration profile extracted from Multivariate Curve Resolution (MCR) method which has been developed earlier /41/.

On other hand the dependence of the IRF shape and maximum position on the difference between the probe and pump wavelength are taken into account /8/that excludes the overestimation of the rise and decay time at 420, 485 and 620nm as in /4-6/.

Furthermore the assignation of the broad spectrum in the region of 400-500nm observed just after excitation to the primary excited Enol species /4, 5/ is called in question /8/and therefore these bands can be unfit for the study of the radiationless deactivation in the E structure (Sec. 3. 1).

In the recent studies /8, 11/the appearance of the positive absorption bands(λmax$\approx$410-420nm) and the negative ones (λmax$\approx$620nm and 340nm) under excitation at λex=390nm/8, 11/ and λex=266nm/11/ in Acetonitril at room temperature is attributed to the transient absorption, stimulated emission and depopulation of the So E-state respectively as a result of the generation of the fluorescent S_1^*keto state due to the ESIPT that begins and proceeds within the temporal duration of the excitation impulse.

The application of the correlation methods with MCR described shortly above has allowed to estimate the ESIPT characteristic time within the limits 40-50 fs irrespective of the excitation wavelength (λ=266 or 390nm).

The ESIPT rate has been estimated also by some indirect methods (Tab. 3. 8).

A very weak short-wavelength fast fluorescence (SWFF) with λ^{max} $\approx$480nm that is overlapping with the strong ASS flu is attributed hypothetically to the E S_1^*state(see 3. 1) where it competes with the ESIPT and ultrafast radiationless deactivation within Enol structure.

Therefore the ESIPT rate could be estimated by the SWFF lifetime ($\tau \approx 50$fs)/8, 10/. However the similar fluorescence is also observed of the model structure MBA (see 3. 1)/11/without H-bond and therefore may hardly be a precursor of the ESIPT for SA. In any case the additional investigations are required.

The ESIPT rate has been estimated also by the Laser Induced Fluorescence (LIF) excitation spectra of the isolated SA molecules in the gas phase mixture of SA with Helium in the supersonic jet/9/. The rate value is estimated by a vibrational mode responsible for the ESIPT(about 3200-3400cm^{-1}). The ESIPT rate value obtained by such a way is not exceed $\tau \approx 50$-60fsec in the good agreement with the findings of /8, 10, 11/.

The method of the femtosecond time resolved Resonance Enhanced Multiphoton Ionization (REMPI) spectroscopy /23/ in gas phase mixture has been also used for the estimation of the

ESIPT rate. The characteristic time has been obtained by delay of the probe signal ($\tau < 750$fs).

That is not contradict to /9/ and another findings in the gas phase (see below). The overestimated upper limit of the time rate constant may be explained by the insufficient time resolution of the REMPI spectroscopy method.

However it is necessary to have in mind that the efficient ionization occurs mainly in the thermalized S_1^*NH state and therefore the delay of the probe signal can involve not only the intrinsic ESIPT process but also the "slow" vibrational deactivation from S_1^*Enol fluorescent excited state ($\lambda_{ex} \approx 320$-360nm)to the S_1^*NH structure ($\lambda_{abs}^{max} \approx 395$-400nm) which is considerably slower in the the gas phase than in the solvents /8, 11/.

Table 3.35
The conjugation effect on the fluorescent characteristics of anils

1) Compound		Methylcyclohexane /47/				Ethanol /53,42/				6) Crystal /30/			
	T_K	Exc. str.	λ^{max}_{abs} nm	λ^{max}_{flu} nm	Stok. shift Δv cm^{-1}	T_K	Ex Str	λ^{max}_{abs} nm	λ^{max}_{flu} nm	Stok.sh Δv cm^{-1}	T_K	Ex Str	λ^{max}_{abs} nm	λ^{max}_{flu} nm	Stok. Δv cm^{-1}
5-6 NA (R=H)	293	E	376	absent	----	293	E	360	530	12750	293	E	375	540	8150
		K	absent	----	----		K	456	540	3410		K	455	540	3460
	77	E	(350)	(540)[4]	10386	77	E	(362)	~540[4]	9106	77	E	----	----	----
		K	(462)[2]	484[3]	1100		K	(466)	482	712		K	----	----	----
SN	293	E	350	absent	----	293	E	360	absent	----					
		K	absent	----	----		K	absent	----	----					
	77	E	350[5]	535[5]	9880	77	E	(340)	528	9630					
		K	----	----	----		K	absent	----	----					
5-6 NN	293	E	384	absent	----	293	E	369	absent	----					
		K	----	----	----		K	455	absent	----					
	77	E	(375)	560[4]	8816	77	E	(368)	560[4]	9320					
		K	(450)	490[3]	540		K	(445)[3]	498[2]	2400					
PA	293	E	----	----	----	293	E	344	514	9330	293	E	340	530	10540
		K	----	----	----		K	435	514	4670		K	435	530	4120
SA	77	E	340	520	10180	77	E	336	510	9800	293	E	347	abs.	----
		K	(430)	515	3840		K	430	540	4700		K	450	abs	----

*) The modified scheme of the origin and the quenching of the fluorescence for the conjugated anil structures (on the base of /47/).

1) For conventional signs of the compounds see Scheme 3.4
2) It is shown and used for obtain of the Stokes shift the long-wavelength maximum of the two K structure absorption (excitation) vibration peaks.
3) The short-wavelength maximum is shown and used to obtain the Stokes shift.
4) The long-wavelength shoulder.
5) In 3MP, T=100K /45/.
6) The findings for the spectra at 77K are taken from /47/.

Table 3.36a
The substitutions' influence on the fluorescent properties of the ring conjugated anils.

1) Compound		Cyclohexane(CH) (room temp) /44/				Ethanol (room temperature) /42/				CH (room temp) /44/	
		Excited structure	λ^{max}_{abs} (nm)	λ^{max}_{flu} (nm)	Stokes shift (cm^{-1})	Excited struct.	λ^{max}_{abs} (nm)	λ^{max}_{flu} (nm)	Stokes shift(cm^{-1})	τ_{av} [2] (ns)	Φ^{aver}_{flu} [3]
5-6NA		E	376	509 sh	6950	E	360	540	9300	3.11[4]	2.2x10^{-5}
		K	469	482	990	K	456	540	3200	1.85[5]	
5-6NA 4' Me$_2$		E	390	514	6190	E	376	570	9100	1.80	3.2x10^{-4}
		K	497 sh	514	1740	K	457	570	4300		
5-6 NA 4' NO$_2$		E	397	500	5190	E	370	530	8200	1.88	1.2x10^{-4}
		K	484 sh	500	660	K	471	530	2400		
5-6NA 4'OMe		E				E	388	525	7600		
		K				K	463	525	2600		
5-6NA 4'COMe		E				E	377	500	6500		
		K				K	468	500	1400		
3-4 NA [6]		E	383	530sh	7240	E				4.23[4]	1.5x10^{-4}
		K	482	505	990	K				3.91	
3-4 NA 4'NMe2		E	405	518	5390	E					
		K	503	518	500	K					
3-4 NA 4' NO$_2$		E	403	520	5580	E					3.4x10^{-3}
		K	500	520	770	K				3.36	

1) See scheme 3.4. 2) Average of the value τ of the two exponents for the two bands. 3) The average for the two fluorescence bands.
4) In ethanol at 100 K. 5) In MC/T at 98 K. 6) Almost the same findings can be seen in MCH /46/.

Table 3.36 b
The effect of the structure and the ring conjugation in the imine moiety on the absorption and fluorescence of o-Hydroxynaphthaldehyde derivatives (in ethanol) /48/.

Compound [1] No	Keto (K) structure			Enol (E) structure		
	Absorp. λ^{max}_{abs} nm	Fluoresc. λ^{max}_{flu} nm	Stokes shift $\Delta\nu$ cm^{-1}	Absorp. λ^{max}_{abs} nm	ASS fluoresc. λ^{max}_{flu} nm	Stokes shift $\Delta\nu$ cm^{-1}
1	440,466	482,505	712	Not observed		
2	445,488	498,530	41	360	575	10390
3	465,495	502,542	302	360	585	10680
4	402,422	442,460	1070	350	~540 [2]	~10050
5	398,420	435,455	2200	350	530	9700
6	----			340	528,550	~10940
				350 [3]	535,565 [3]	9880 [3]
				350 [4]	550,560 [4]	~11230 [4]

1) For numeration of the compounds see scheme below. For compounds 1,2,6 see also scheme 3.4 and table 3.35.
2) The long-wavelength tail .3) In 3MP /45/. 4) In ACN ,τ=10 ps /10/

The structures 1-5 for table 3.36 b

R: Ph; 1-Npt; 2Ant; CH₂Ph; Bu
   1   2   3   4   5

                                                6

The successful attempt to estimate the ESIPT rate in the gas phase has been undertaken recently with use of the time resolved mass spectroscopy with the excitation (pump laser with λ =266nm) of the SA molecule in the gas mixture stream with helium followed by the multiphoton ionization and fragmentation by the laser probes (λ=792 or 399nm)/11/. The ions created are dispersed in mass by a time-of-flight spectrometer. The new fragmentation of the excited molecules and time evolution of the fragment peaks give the information about structural transformations of the excited molecules.

Such evolutions in the electronic excited (λ=266nm) MBA (without O-H...N bond and SA molecules are absolutely different. For SA, for example, mass peak corresponding to the loss of hydrogen in the ionic state shows the two time decay dynamics of 40 ± 10fs and 700 ± 100fs from which the first is absent of MBA and attributed to the ESIPT in very good agreement with the data of the LIF excitation spectra in the gas phase /9 and with those of the transient absorption spectra in the solvents/8, 11/.

The second component that can correspond to the relaxation of the S_1^* cis –keto structure will be discussed later. The theoretic analysis of the ESIPT dynamics of SA has been carried out for the first time /7/ by the ab-initio CIS/6-31G* method in the "instant on" approach. The results conform in principle to the experimental findings of the ultrafast ESIPT reaction with participation of the intramolecular vibrations.

According to /7/ the ESIPT is assisted by the harmonic low-frequency symmetric vibrational modes modulating the N-O distance (Tab. 3. 20) but is hindered by the high-frequency vibrational mode (ν=1830cm^{-1}) described as the asymmetric NHO stretching mode involving the movement of the proton.

However according to /7/ the existence of the potential barrier and the tunnel mechanism of the ESIPT reaction in the $S_1^*(\pi, \pi^*)$ planar excited state are not agreed with the experimental findings for the solvent and gas phase although can correspond to the some data for the crystal state (see below).

Indeed according to /7/ (Tab. 3. 18) the ESIPT rate is too low (τ=1200fs) with the calculated potential barrier (ΔE=7. 2kcal/mol) and can be increased to the experimental value /4/(τ=220fs) with the artificially corrected potential barrier height.

However this value also exceeds considerably the experimental data for time rate ($\tau \approx$40-50fs) obtained recently (Tab. 3. 18). At the same time the sufficiently large isotopic effect predicted by the calculations which is typical for the tunnel mechanism($\tau(D)/\tau(H)$=12. 3 and 17. 5) for the different potential barriers has not quite been found out by the experimental methods /10/ (Tab. 3. 18). Thus although authors /7/ believe "there is no need to postulate a "barrierless" mechanism for the explanation of the ultrafast ESIPT" , however the very high ESIPT rate in the solvents detected recently (Tab. 3. 18) can't be explained by the tunnel mechanism.

The recent theoretic study of the ESIPT dynamics has been carried out /1/ using the DFT and

TDFT calculations. The ESIPT process is considered in the completion with the fast relaxation channel of Internal Conversion (IC) via the structural transformations within Enol tautomer (see3. 1).

The ESIPT and IC are multidimensional processes, i. e. more than single coordinate is needed to described them. Therefore to indicate the vibrational modes which take part in the ESIPT the modeling is needed.

For this goal one can be assumed the motion of proton is involved essentially in the ESIPT with only a certain degree of motion of the donor (O) and acceptor(N)atom while the rest moieties remain approximately static.

Really according to TDFT calculations, reaction along the O-H coordinate on the PES of the S_1^* state from Enol*(OH)*structure to Keto*(NH)*one is barrierless process with the potential energy loss about 8kcal/mol.

In such a model the autocorrelation function (correlation amplitudes) with the vibration coordinates introduced above can be used for the formation of the wave packet moving away from its starting point connected with the FCS_1^*Enol state.

The analysis of the temporal changes of the correlation amplitudes (within the limits 1->0) allows to determinate the ESIPT rate corresponding to the model with O-H and N-H coordinates. In the present case the value of "t" at which the correlation amplitude goes down to 1/e is 49. 6 fs which can be accepted as estimate for the τ(ESIPT) in good agreement with the experimental data(Tab. 3. 18).

Thus the overultrafast ESIPT rate (τ<50fs) of SA in the bare molecule (and probably in solvent) are caused by the barrierless mechanism excluding the deuterium effect in accordance with the experimental data.

Such a mechanism can be provided by mainly both the O-H stretching vibration (proton motion) and the bending vibrations in the PhOCN structural fragment modulating the O-N distance.

The contribution of the latter is obligatory and can be varied within a large limits in dependence on the molecular structure and the media providing the realization of the two limits – overultrafast barrierless mechanism(with $\tau \approx 3\text{-}5 \times 10 fs$ without isotope effect) and ultrafast tunnel one (with $\tau \approx 1\text{-}3 \times 10^2 fs$) with the clear expressed deuterium effect).

Meanwhile there are Anil related molecular structures (see sec. 3. 2. 2)where the ultrafast ESIPT in the solvent can be interpreted as a multidimensional process assisted by the low-frequency vibrationalTPM modulating O-N distance without explicit proton motion (see also 8).

Furthermore as it has been shown earlier /72, 73/ and later /74/, the intramolecular complexes of SA with various metals (M:Zn, Cu, Cd, Ni, and UO_2)(Sch. 3. 10) in the solvents(T=77K) display under irradiation (λ=320-360nm) the spectral changes which are bound up with generation of the colored N-M(Keto or Zwitterion, like NH) structure stabilized by the rigid media.

Such structural transformations can be caused only by a modulation of N-O distance by the TPM bending vibra-tions without any motion or tunneling of the heavy metal atom. Apparently the analogues mechanism without proton motion is also possible for the ESIPT reaction in the solvents.

The situation can be considerably changed in the crystals/29/(Tab. 29/. Taking into consideration the criticisms concerning the results of the ESIPT rate measurements in the thermochromic crystals (see above) it is necessary, nevertheless, to have in mind that unlike liquid and gas phase the low-frequency inharmonic large-amplitude vibrations modulating N-O distance can be hindered due to intermolecular interactions in the crystal lattice.

The marked decrease of the ESIPT rate value in the crystal state (Tab. 3. 21), its correlation with the frequency of the O-H stretching mode(Tab. 3. 20), and the reduction with the H->D substitution /29, 69/ indicates the existence of the potential barrier and the considerable contribution of the tunnel mechanism in the ESIPT reaction especially at low temperature.

Apparently just that mechanism is described very well by the results of the theoretical investigations /7/ discussed above. Obviously the intermolecular interactions in the crystals, in particular, and perhaps the molecular structure may change markedly the ESIPT mechanism proceeding from the S_1^* state of E structure.

a) MCH (room temperature)

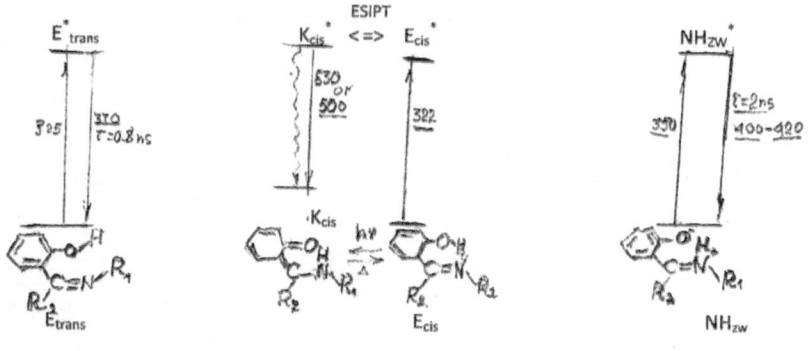

b) Crystal(room temperature), MCH(77K).

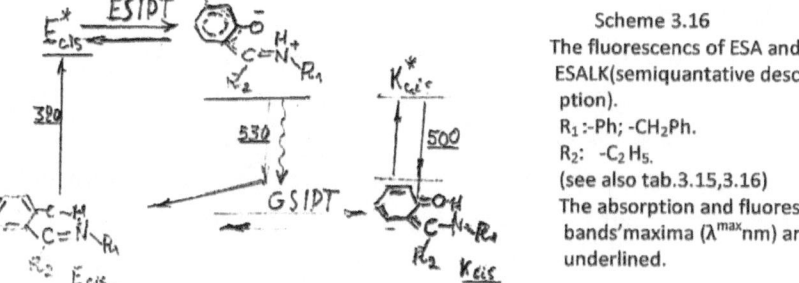

Scheme 3.16
The fluorescencs of ESA and ESALK(semiquantative desciption).
R_1 :-Ph; -CH_2Ph.
R_2: -C_2H_5.
(see also tab.3.15, 3.16)
The absorption and fluoresc. bands' maxima (λ^{max} nm) are underlined.

c) Ethanol(room temperature).

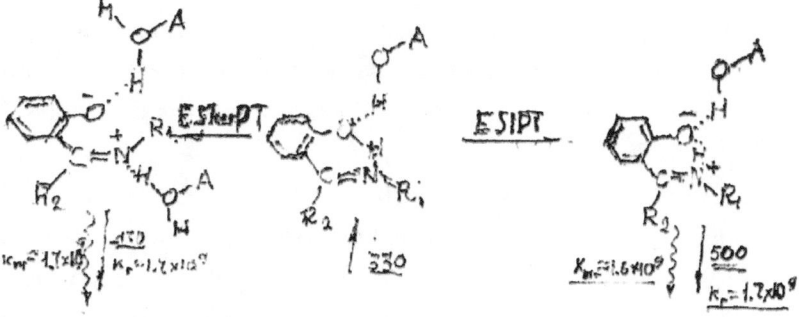

Such a mechanism with the reduced proton transfer rate in the thermochromic crystals (see 3. 24a) and probably in the model structure I(Ph], Ph) is realized owing to almost complete lack of the sterically inhibited IC competing in the S_1^* state which is caused by the aldehyde ring twist with the hydrogen bond O-H...N break followed by the formation of the twisted E structures(see above).

Thus like the related structures (Sec. 3. 2. 2.) the ESIPT in SA can proceed from $S_1^*(\pi\pi^*)$ state by the two alternative mechanisms in dependence on a media and the molecular structure: the barrierless path-way and the tunnel process with over-ultrafast (several tens fs) and ultrafast (several hundreds fs) rates correspondingly.

In consequence of the above discussed data the Intersystem crossing (ISC) $S_1^*\sim T_1(k\sim 10^{12}$ s$^{-1})$ is suppressed strongly in the competition with the ultrafast deactivation of the S_1^*Enol state due to the IC and the ESIPT with the rates $k\sim 10^{13}$-10^{11} ($\tau\approx$50-100fs). Thus the ESIPT (and photochromic reaction)via T_1 state under the direct So->$S_1^*$is ineffective.

Meanwhile the phosphorescence (T_1-hv->So) which allows to estimate the energy of the T_1 state has been observed first only recently (2001) for the series of anils (MCH at 77K)/3/(e. g. 1, 2 on Sch. 3. 9 and otherits attribution to the E or K tautomeric form is not definite enough(see below). However according to /3/ (Sch. 3. 9)the T_1 Enol state may be revealed $E(T_1)\approx 20320$cm^{-1} and populated directly by biacetyl sensitization due to the endothermic T-T energy transfer from biacetyl, $E(T)=19750$cm^{-1}, and can be displayed by the typical transient T_1->Tn absorption spectra with biacetyl excitation at 425nm under argon in benzene. It has reported also recently about analogous T_1->Tn absorption in cyclohexane solution without air at room temperature with the correspondding Tn-hv->T_1 fluorescence of the molecules 3, 4 (Sch. 3. 9)/75, 76/.

The another triplet state with the much less energy, $E(T_1, k)\approx 14250$cm^{-1}, has been also revealed after excitation of the Enol structure under the same conditions with use of anthracene triplet quencher(T_1, Q)$=14850$cm^{-1}, as a result of the endothermic T-T energy transfer from the cis-keto T_1 tautomer to the anthracene.

The energy difference of the observed triplet states indicates that the second triplet state is the NH(Kcis) tautomer produced by the ESIPT in the triplet state (Sch. 3. 9). However according to the results of the semiempirical SCFMO calculations /14, 77/ there is unlike S_1^* the sufficiently high potential barrier ($\Delta E \approx 5000 cm^{-1}$) on the triplet state PES path-way Enol-->Cis Keto and the ESIPT can proceed by both the over-barrier and the tunnel mechanisms.

In this connection there is the problem about attribution of the phosphorescent band to either

the Enol or Cis –Keto triplet states. Indeed the phosphorescence excitation band coincided with the absorption bands of the Cis –Keto structure and at the same time its spectral position ($\lambda max \approx 580nm (17240 cm^{-1})$ and onset $\lambda \approx 510nm (19600 cm^{-1})$ are very close to the triplet state energy of the Enol structure.

According to /3/ such a situation may be a result of the T_1 Enol state population owing to reverse proton transfer (NH->OH) from the T_1 cis-keto structure after $S_o \rightarrow S_1^*$ excitation of the of the cis –keto structure at 77K followed by ISC ($S_1^* \sim T_1$ cis-keto).

Thus sensitized reversible adiabatic triplet state ESIPT can be realized without a participation of the S_1^* state, and ISC $S_1^* \sim T_1$ (Enol) results probably in the generation of the colored structure.

One can believe the study of such sensitized photochromic reactions may be a novel prospective direction with the useful applications.

Table 3.37

The dependence of the relative quantum yields of the fluorescence and photocoloration on the excitation wavelength for molecule of SA (IIP,T=77K)/20/(see text).

λ^{ex} (nm)	Φ^{rel}_{fl}	Φ^{rel}_{col}	$\Phi^{rel}_{flu} / \Phi^{rel}_{col}$
365	1.0	1.0	1.0
313	0.56	2.8	0.2
254	0.48	12.5	0.04

Table 3.38

The dependence of the relative efficiencies of the ASS flu and the photocoloration on the concentration of the energy acceptor (singlet) for SA (IIP.T=77K) /20/(see text).

C, M/L	Φ^{rel}_{fl}	Φ^{rel}_{col}	$\Phi^{rel}_{fl} / \Phi^{rel}_{col}$
5×10^{-4}	1.0	1.0	1.0
2×10^{-3}	0.34	0.73	0.46
3×10^{-2}	0.18	0.50	0.36

Table 3.39

The dependence of the relative quantum yields of the ASS flu and the photoreaction on the wavelength of the excitation for $SALK_n$ (IIP,T=77K) /20/ (see text).

Compound	Φ^{rel}_{flu}	Φ^{rel}_{col}	$\Phi^{rel}_{flu} / \Phi^{rel}_{col}$	
			Exitation wavelength λ_{exit} nm	
			365	313
$SALK_{11}$	4.8	3.0	1.4	1.7
$SALK_{17}$	2.8	3.1	0.9	0.7
SA	1.0	1.0	1.0	0.2

Table 3.40

The dependence of the NH (keto) structure decay time constant (in S_1^* ($\pi\pi^*$) state) on the excitation wavelength of the E structure (SA in the gas phase, λ_{prob}=395nm /23/. (see scheme 3.19)

λ_{pump},nm	375	370	365	360	320
τ_{decay},ns	8.5 ±1.5	8.9±0.7	1.6±0.5	1.6±0.3	1.5±0.3

Scheme 3.17

The "hot" vibration state(s) as a common precursor of the ASS fluorescent and photochromic structures (non polar solvents)/5/.

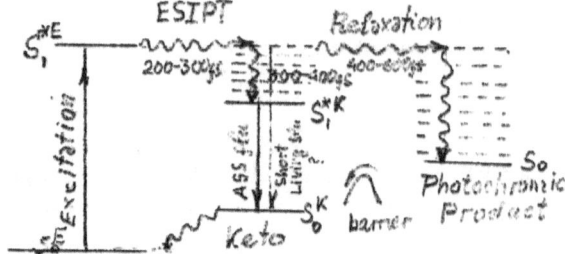

Scheme 3.18

The "hot" vibration states as a common precursor of ASS fluorescent and photocolored form in the gas phase according calculations /9/.

$$E \quad ---h\nu---> E^* - \underline{ESIPT} ->K^{\sim *cis}_{PL} ---->-K^{*cis}_{tw}(\pi\pi^*->n\,\pi^*)----->K^*_{trans}----->K_{trans} \text{ (Photocolor.product)}$$

| Short
| living
↓ ASS flu

| ASS
| flu
↓

ESIPT Photochromism

Scheme 3.19
The "hot" vibration states as a common precursor of the ASS fluorescent and the photocoloredgenerated via the twisted $(n\pi^*)$ state of the NH (K) structure(In the gas phase according /23/. (see also table 3.40)

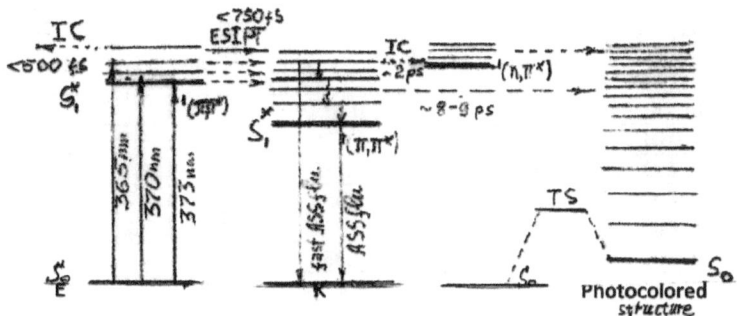

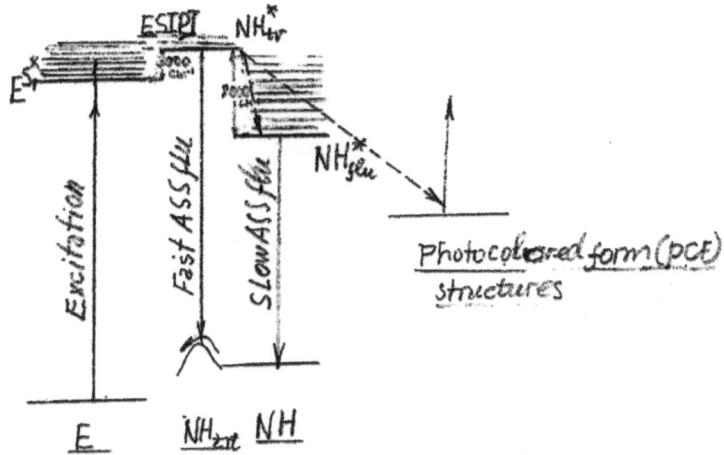

Scheme 3.20
The transient S_1^* excited NH structure as a common precursor of the ASS flu and the photo-colored form' (PCF) structure /14/ (see tab. 3.41).

Table 3.41a
The comparative data of the structure parameters of the NH transient (possibly precursor) (NH^*_{tr}), excited Enol ($E(S_1^*)$), and fluorescent state (NH^*_{flu}) structures in the S_1^* state /14/ (see scheme 3.20).

Structure (*) → ↓Parameters		NH^*_{tr}	$E(S_1^*)$	NH^*_{flu}
Bond Length Å	OH	1.197	0.979	1.777
	NH	1.331	1.726	1.017
	C_2O	1.287	1.328	1.266
	C_1C_2	1.467	1.456	1.437
	C_1C_7	1.460	1.421	1.482
	C_7N	1.338	1.350	1.335
Angle (degree)	α	~0	0	?
	β	0	0	0
Energy kcal/M	ΔH	118.9	110.3	95.1
		(0)	(-30005) [1]	(-8318) [1]

[1] The energy (cm^{-1}) relative to NH^*_{tr} structure.

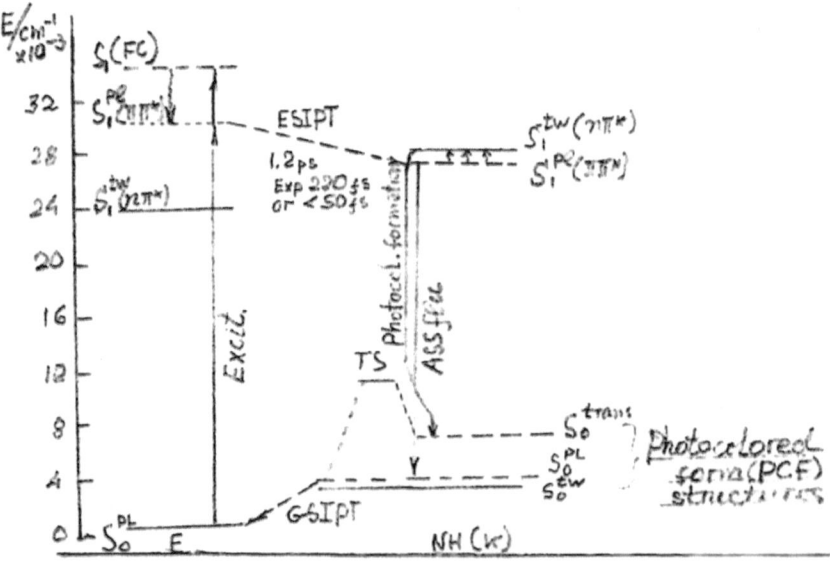

Scheme 3.21
The $S_1^*(\pi\pi^*)$ state of E structure as a common precursor of the ASS flu and a photocolored form (PCF) structure and some accompanying processes /7/.

Table 3.42
The absorption and the fluorescence of the SA photocolored structures (the findings in the solvents).

Absorption (Exitation) λ^{max} nm	Vibration splitting " Δv cm^{-1}	Flu.band λ^{max} nm	Vibration spliting Δv cm^{-1}	Stokes Shift Δv^{a-f} cm^{-1}	Solvent (T$_K$)	Referen.	Experim. method	Calcul. λ^{max} nm
476 555 [1]	2990	566 606	1150	350	IIP (77)	/12/	Steady State spec	----
474 541 [1]	2605	555 597	1250	466	IIP (77)	/20/	Steady State spec	—
470 500 [1]	1276	—	—	---	ACN (room)	/80/	Transient absorption	462
474 [2]	---	550 [2]	—	3000	ACN (room)	/8/	Time resolved $\tau^f_{decay} \geq 1$ns	290 [4] 445 [5]
475 488 [1]	516	563 605	1200	2730	ACN (room)	/90/	Time resolved Two step laser excitation	446
425 ~550 [2]	—	—	—	—	CH (room)	/4/	Transient absorption	412
485 [2]	—	—	—	—	Meth	/74/	Transient absortion	—
465 470 [3]	—	535 565	—	2804 3577	100K >140k 3MP	/45/	Transient abs. Time resolved fluorescence	—
480 500 [1]	833	—	—	—	ACN (room)	/11/	Transient abs. nanosecond spectroscopy	----
450 [6] 510 [1]	~4400	—	—	—	ACN (room)	/36/	Transient abs. nanosecond spectroscopy. Steady state absorption spectroscopy.	- ----
460 [7] 500 [1]	~4900	—	—	—	CH (room)			—

1)The low intensive maximum or sholder.2)The low intensive broad band 3)Te data for SN (scheme 3.23 and table 3.44) 4) and 5) From HF and CIS/7/and CASP2 /1/ calculation correspondingly.6) and 7) The data for 2P and 4P correspondingly

Table 3.43
Absorption bands of the photo (λ^{max}_{PC}) and thermo (λ^{max}_{TC})colored structures in the crystals. (λ_{ex}=360 nm, thin crystal films, room temp.)(see sch.3.22a).

Compound No	Photocolored structure band λ^{max}_{PC} nm	Thermochromic structure band λ^{max}_{TC} nm	Minimum Diference $\lambda^{max}_{(PC-TC)}$ nm	Notices	References
1 (SA)	489 510 sh 535sh	----	+33	The structures of SA and SAP are very similar.	/97/
2 (SAP)	—	456 490sh			/81/
3	480 520-530	~440	+40	—	/96/
4	480 520-530	407	+73	—	/39/
5	480 520-530	407	+73	—	/39/
6	~445	410	+35	----	/39/
7	~445 512 525	408	+47	----	/39/
8 (2P)	---	420		TC-steady state.	/36/
9 (4P)	480	---	+60	PC-trans. absorp	/36/

Scheme 3.22a
Photochromic and thermochromic anil structures in the crystal (see table 3.43)

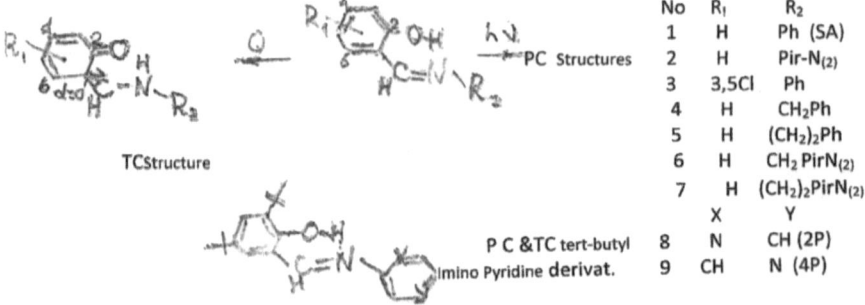

No	R_1	R_2
1	H	Ph (SA)
2	H	Pir-N$_{(2)}$
3	3,5Cl	Ph
4	H	CH$_2$Ph
5	H	(CH$_2$)$_2$Ph
6	H	CH$_2$ PirN$_{(2)}$
7	H	(CH$_2$)$_2$PirN$_{(2)}$

	X	Y
8	N	CH (2P)
9	CH	N (4P)

ESIPT Photochromism 171

Table 3.44
Conformation, generation and decay of the proposed Anils' PCF structures (Scheme 3.22a)

No	Proposed structure	Angles, degree [1]			Generation		Decay rate, τ (s)	Conditions and Method	Ref.	Notices
		α	β	γ	Rate τ(s)	Quant. yield Φ				
1	$NH_{tr\,cis}$ $NH_{tr,tr}$	12 180	180 0	15 0	170×10^{-9}	---	$>15 \times 10^{-9}$ 5×10^{-6}	ACN,BTN,room t-re Time res. flu	/2/	Two photon excitation (see /95/).
2	$NH_{tr,tr}$	180	0	0	---	---	$10^{-9}-10^{-4}$	ACN,3MP, room t-re.Transient ab	/80/	Large Zw-ion contribution in the NH structure.
3	$K_{tr\,tr}$ $K_{tr\,cis}$	180 ~90	0 ~180	0 0	---	0.1-0.3	8×10^{6} [2] $>10^{-10}$	>140 K 3MP, ≤100K Trans abs	/45/	The data for SN. Sch 3.22 b
4	$K_{tr\,tr}$	180	0	0	$~7 \times 10^{-12}$	0.1-0.3	$10^{-6}-10^{-9}$ $15\,10^{-9}$ [3]	ACN,room temp Trans.abs	/8/	-----
5	$NH_{tr,tr}$	180	0	0	$4-6 \times 10^{-13}$ [4]	---	$\geq 3 \times 10^{-5}$	CH,ACN,Alk,t. room,Trans.ab	/4-6/	Decay barrier is caused by trans-cis isomer.around $C_1:::C_7$ bond
6	$K_{cis,tr}$	0.5	4.0	0.8	Not observ.	---	---	Gas,Jet cooled temp.	/9/	The angles have been calculated by B3LYP/6-31G* -ab-initio.
7	$K_{tr,tr}$	180	0	0	1.5×10^{-9} [5]	---	---	GAS,REMPI	/23/	---
8	$K_{cis,tr}$	0	≠0	≠0	---	---	1.07×10^{4}	Cryst.30°C Steady state	/99, 100/	The data for 5Cl,4' tert-butyl SA
9	$K_{tr,tr}$	~185	1.4	11.6	$2 \times 10^{-13}-2 \times 10^{-9}$	6×10^{3}	8.5×10^{-12} [6]	Ab-initio calcul.	/7/	Back reaction -only GSIPT
10	$K_{tr\,tr}$	180	0	0	---	---	1.2×10^{9}	Vib.spCrisroom	/71/	Direct PCF observ.
11	$NH_{tr\,tr}$ $NH_{tr\,cis}$	180 180	0 180	0 ≠0	2×10^{-7} 9×10^{-6} [7]	---	$~15 \times 10^{-5}$	ACN d_3,room t-re Time resolv. IR	/95/	Direct PCF observation. Compare with /2/
12	$K_{tr\,tr}$	180	0	0	---	---	---	Cryst.X-ray ,Two photon experim.	/101/	Direct observation PCF 3,5 Tert-butil.SA
13	$K_{tr,tr}$	180	0	0	1.5×10^{-12}	0.34 0.56	$2.5\&0.7 \times 10^{-3}$ $3.8\&0.9 \times 10^{-3}$	$\lambda_{ex(nm)}$ 266 ACN 355	/11/	The two distinct PCF structure of the different stability have been observed !
14	$K_{tr,tr}$	180	0	0	$10^{-12}<\tau<10^{-9}$	---	186×10^{-6} 35×10^{-6}	CH ,room t-re	/36/	The data for the structure 8 (2P) See scheme 3.22
15	$K_{tr,tr}$	180	0	0	..."..."...	---	122×10^{-6} 19×10^{-6}	CH,room t-re	/36/	The data for the structure 9(4P) See scheme 3.22
16	$K_{tr,tr}$	180	0	0	25×10^{-9}	---	417×10^{6} 3×10^{4} (3%)	Crystal,room temperature	/36/	The data for the structure 9(4P)

1)The angles α and β are counted off the planar configuration of the cis atom O and NH positions respective to the $C_1:::C_7$ bond (α=0) and the planar configuration of the trans position of the ring A and the substituent R respective to the $C:::N$ bond (β=0) correspondingly.This corresponds to the conformation forming after the ESIPT immediately (Scheme 3.23) 2)This is the second order rate constant k_2.3) The data from " K. Kownacki Ph. D Thesis ,Institute of Phys.Chem. of Polish Acad. of Science,Warsaw 1991"4) For details see the table 3.45 and the scheme 3.17.5)The time of the IC from the planar S_1^* ($\pi\pi^*$) state to the nonplanar ICT state which is precursor of the PCF structures.6)The calculated time is bound up with PT. 7)The time between the beg.and completion of PCF accumulation.8)There are two conformers with diff.ang.9)Φ is obtained /11/.10)Rougth estimate,data/36/.

Table 3.45
The dependence of PCF structure generation time on the solvent nature for SA /6/.

N_o	Solvent	ε	η (cp)	τ_{gen} (fs)
1	Cyclohexane (CH)	2	0.90	400 ±100
2	Ethanol (Etl)	24.3	1.20	600 ± 50
3	Acetonitril (CAN)	37.5	0.34	380 ± 80
4	Butanol (Btl)	17.1	2.98	350 ±100
5	Cyclohexanol(CHL)	18.3	2.45	460 ± 75

Scheme 3.22 b
The supposed PCF structures with the flattened keto –moieties(α=0˚, β˚ =180)(see table 3.44).

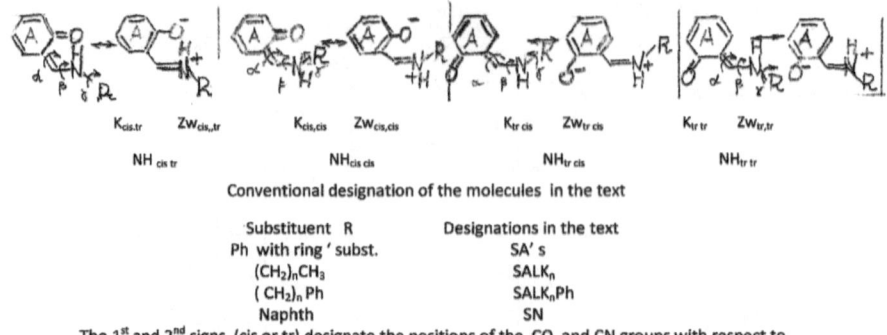

$K_{cis,tr}$ $ZW_{cis,tr}$ $K_{cis,cis}$ $ZW_{cis,cis}$ $K_{tr\,cis}$ $ZW_{tr\,cis}$ $K_{tr\,tr}$ $ZW_{tr,tr}$

$NH_{cis\,tr}$ $NH_{cis\,cis}$ $NH_{tr\,cis}$ $NH_{tr\,tr}$

Conventional designation of the molecules in the text

Substituent R	Designations in the text
Ph with ring ' subst.	SA' s
$(CH_2)_n CH_3$	$SALK_n$
$(CH_2)_n$ Ph	$SALK_n Ph$
Naphth	SN

The 1st and 2nd signs (cis or tr) designate the positions of the CO and CN groups with respect to $C_1=C_7$ bond (angle α) and A ring and R group with respect to C:::N bond (angle β) correspondingly.

3. 2. 4. The effect of the molecular structure on the ESIPT mechanism and dynamics.

3. 2. 4. 1. The influence of the ring substituents (solvents and thermochromic crystals)

For the best of our knowledge the systematic studies of the ring substituent influence on the rate and the mechanism of the ESIPT were not carried out but some data are presented in Tab. 3. 21.

In the solvents the marked increase of the potential barrier(to 2. 76 kcal/mol)and the ESIPT time rate(two or three times as much SA for the nonpolar and polar solvents respectively) is observed /6/with the insertion of the electron donor and withdrawing substituents in the aldehyde and imine rings respectively (Tab. 3. 21, line 5).

In this case the contribution of the interring Intramolecular Charge Transfer(ICT) in the S_o->S_1*transition is unfavorable for the ESIPT by the two main reason: the heightening(or arising)of the potential barrier along of the adiabatic path-way (OH*)->(NH)*reaction due to stabilization of the (OH)* structure relatively (NH)*one (especially in the polar solvents), and the unfavorable localization of the ICT transition (A ring->B ring) not affecting the reaction centers (O and N atoms).

The analogous but a pronounced effect is observed of the anil structure with the strong electron donor $4NMe_2$ substi-tuent in the aldehyde ring where the ESIPT (and hence photochromism) is absent completely /20/(see also Sec3. 1).

However the insertion in both aldehyde and imine rings of the bulky3, 5 tert-butyl substituents and 2' or 4' Nitrogen atom respectively (Tab. 3. 21, Struct. 8) does not influence the ESIPT rate in the solvents and even in crystal in spite of the difference in their molecular structure (different imine ring planarity)and in the properties (thermochromism and photochromism respectively).

The influence of the ring substituents on the ESIPT in the thermochromic crystals (but not in the photochromic those) can be discussed on the base of the findings of /29/. The general peculiarities of the ESIPT of the anil molecules in the crystal lattice unlike the solution can be connected with the intermolecular interactions in the thermochromic crystals(Ch. 2) that is only some modified by the molecular structure (and the ring substituents in particular)and consist of following:

i) An increased intrinsic time rate as a result of the potential barrier arising from a steric restrictions for the ESIPT assisting vibrations.

ii) An increased intrinsic time rate with the H->D exchange (isotope effect).

iii) A very weak or lack of the intrinsic ESIPT rate temperature dependence.
In the i)-iii) there are the signs of the ESIPT tunnel mechanism.

iv) The strong increase of the "observed "ESIPT time rate in the thermochromic crystals in comparison with the solvents as a result of the strong weakness of the interaction between the molecular vibrations and media in the crystalline lattice and solvents.

The manifestation of the above peculiarities is caused by the structure of the molecule in the crystalline lattice. The rise of the rate

(k) with the lowering and the contraction of the potential barrier (i) is provided by the shortening of the H bond length (r) in the series of the model compounds IV, III, II, I (Tab. 3. 2. 2) and bound up with the changes of the molecular structure in particular with the insertion of the halogen atom in the aldehyde ring (Tab. 3. 21, Struct. 1, 5).

According to /69/ the dependence k on r may be approximated by the exponential expression $k=pk_0 \exp[--(r-r_0)]$ where $k_0 =7.5 \times 10^{13} s^{-1}$ is the stretching frequency in the H-bonded O-H group, r_0 is the critical H-bond length without potential barrier for the ESIPT, and p is an arbitrary constant.

Although the calculated with help of the formula rate values are underestimated as compared with the experimental findings for the solvents (e. g. /5, 6/) they correspond to the rate values for the ESIPT with the tunnel mechanism.

Thus one may believe the time rate values (4-10)x100fs (Tab. 3. 22) correspond to the intrinsic tunnel ESIPT mechanism in the thermochromic crystals with the small contribution of the of the vibrational relaxation caused by the strong interaction of the molecular vibrations with the crystalline lattice (lattice assisted relaxation), and their variations reflect a dependence on the molecular structure (and on the ring substituents in particular).

However the dramatic increase of the observed ESIPT rate (Tab. 3. 21, Struct. 2, 3) when inserting of the 4'Me substituent without the marked electron donor or electron acceptor properties into the acoplanar imine ring can be explained mainly by a sharp fall of the rate of the intramolecular vibrational deactivation as a result of the strong weakening of the lattice-vibrational relaxations in the free space of the crystal packing that is formed due to the sterical interactions of the adjacent molecules with the bulky CH_3 groups (item iv).

At the same time the explanation of such a dramatic fall of the ESIPT rate only by a possible potential barrier rise in the series of the structures 1, 2, 3, only to decrease of the difference in the excited sate energies of the Enol S_1^* and the Keto S_1^* structures $\Delta E = E(Enol)S_1^* - E(Keto)S_1^*$ in this series (Tab. 3. 21, remarks line1-3) is seemed hardly probable.

Thus in the thermochromic crystal the ring substituents can influence (unlike solvents) the lattice assisted vibrational deactivation involved in the "observed" ESIPT rate with distortion of its real value.

3. 2. 4. 2. The effect of the imine moiety structure on the ESIPT mechanism.

The comparative theoretical investigations of the imine moiety structure influence the ESIPT mechanism have been conducted by the quantum-chemical ab-initial calculations (CIS/6-31G*) with utilization of the semiclassical instant-on approach and TDDFT method for SA (Tab. 3. 23 see also Sec. 3. 2) /1, 7/SALKo/27, 68/ and SALKCH Ph/28/(Tab. 3. 23, Sch. 3. 11).

According to the data/7, 27, 28/the calculated value(instant-on approach)of the ESIPT time rate (τ)decreases strongly in the series SA, SALKo, SALKCH$_3$Ph ($\tau \approx$220, 30, 15 fs respectively)due to the fall of the energy barrier height(ΔH=5. 7, 1. 6, 1. 2 Kcal/mol respectively).

The analysis of the data /7, 27, 28/ shows the decrease of the barrier height ΔH along the path-way of multidimensional ESIPT reaction is bound up with a closing in the S_1^* state of the Enol (E*) and the transition (TS*) structures in the above mentioned series of the compounds.

I. e. the mean absolute differences of the bond lengths for the E* and TS* structures decrease considerably for the series SA, SALKo, SALKCH$_3$Ph (0. 133, 0. 083 and 0. 075 A correspondingly) in accordance with the well known Hammond's postulate (see Sch. 3. 11.). Such a considerable closing in the E* and TS*structures in the SALKs in comparison with SA can be explain by the more distinct charge redistribution with the localization of the So->$S_1^*(\pi\pi^*)$ICT transition in the most simple chromophore of SALK.

The fall of the barrier height when changing of the imine moiety structure leads to the sharp decrease of the relative probability of realization of the tunnel /over-barrier path-ways-98/2, 50/50, and the almost barrierless reaction for SA, SAlKo and SALKCH3 correspondingly(Sch 3. 11).

In accordance with the above data deuterium effect value $\tau(D)/\tau(H)$ decrease also (for SA, SALKo-17. 5 and 3. 8 respectively, and of SALKCH$_3$ Ph it has not established at all).

The lowfrequency CH$_3$ torsional and C1 C7 N deformational ESIPT hindering vibrational modes (94 and 440cm^{-1} correspondingly) impede the ESIPT reaction in SALKo but do not play a marked role in the SAlKCH3Ph molecule, owing to the evident structural peculiarities. Thus lowering of the potential barrier, the corresponding increase of the ESIPT rate, and the absent of isotope eeffect of the latter as compared with SALKo can be caused by change of the imine moiety structure.

On the other hand the heightened exotermity of the E*->K* reaction and the considerable difference between the K* and TS* structures unlike the E* and TS* ones especially for the SALK compounds (Tab. 3. 23 and Sch. 3. 11) result in the high energy barrier along the path-way of the reverse ESIPT reaction, and the fast vibrational deactivation of the excited K* structure exclude the reverse ESIPT completely.

At the same time the superultrafast ESIPT of the SALK compounds suppresses completely the deactivation of the S_1* state bound up with the structural deformations in the E structure (see Sec. 3. 1).

The ESIPT dynamics of the SALKo has been recently analyzed also by means of the more accurate electronic structure calculations at the TDFT level for S_1*(π, π*) state /68/.

According to the results the ESIPT is exoergic and barrierless reaction which occurs from the unstable EnolS$_1$* state like that of SA. The wave packet is described by the three-dimensional set of the reduced vibronic coordinates, and involves the intrinsic proton (deuterium) motion from the reactant region towards the product that. The time evolution of the wave packet (multiconfiguration time-depended Hartry package) predicts a time rate of 11fs and 25fs ($\tau(D)/\tau(H) \approx 2. 3$) for the Hydrogen and Deuterium transfer respectively.

These time rates are more less than predicted by"instant-on" approach/27/ and have the less deuterium effect (Tab. 3. 21, Str. 7, and Tab. 3. 23).

The similar results have been obtained also for another dynamic models with the Oxygen atom both moving and motionless relatively phenyl ring. The difference between /110/and/41/ can be caused by the Frank-Condon excitation energy used in/68/ and barrierless nature of the ESIPT unlike/27/.

At the same time owing to the peculiarities of the applied methods which can underestimate (DFT/68/) the energy barrier for the ESIPT and overestimate that (CIS /27/) one can believe that the range of the actual ESIPT rate of anil is fixed properly by these methods.

The ultrafast ESIPT is common property of SA and SALK caused by the almost barrierless mechanism in both structures. However the strongly increased rate and the marked deuterium effect are the typical properties for SALK as compared with SA. It is evident that the difference in the ESIPT dynamics of SALK and SA is caused by the peculiarities of the multidimensional ESIPT nature in the molecules with the different structure of the imine moieties.

3. 2. 4. 3. The effect of the ring conjugation.

According to the findings discussed in detail earlier (Ch. 2, Sec 2. 4. 2) the E<=> K equilibrium in the Ground state is shifted strongly towards the K structure with the expansion of the ring π-system in the aldehyde moiety and change a little with the conjugation of the rings in the imine moiety.

Table 3.46
Comparative data of the longwavelength SA absorption band maxima (λ^{max}_{abs}) for Enol ,trans with respect to C=N(E_{tr}),planar Keto-cis (K_{cis}) and Keto -trans(K_{tr}) (with respect to $C_1=C_7$ bond)structures calculated by the different authors and methods.

λ^{max}_{abs} nm			$\Delta\lambda^{max}_{abs}$ nm	Method	Reference
E_{tr}	K_{cis}	K_{tr}	$K_{tr} - K_{cis}$		
333	454	446	-8	INDO/S	/2/
336	432	415	-15	CINDO/S	/12/
304	412	391	-21	AM-1	/6 /
247	292	291	-1	CIS/6-31G*	/7/
223	271	276	+5	CIS/6-31G*	/27/
223	370	372	+2	TD/B3LYP-31G*	/27/
351	448	412	-36	TDDFT	/1/
340	440 [a]	480, 520 [b]	+ 40--60	Exp.Abs. spect.	/37,96/

[a]Thermochromic band. [b]Photochromic band.

Scheme 3.23
The photoinduced transformations in the molecules of SN and related structures /45/.

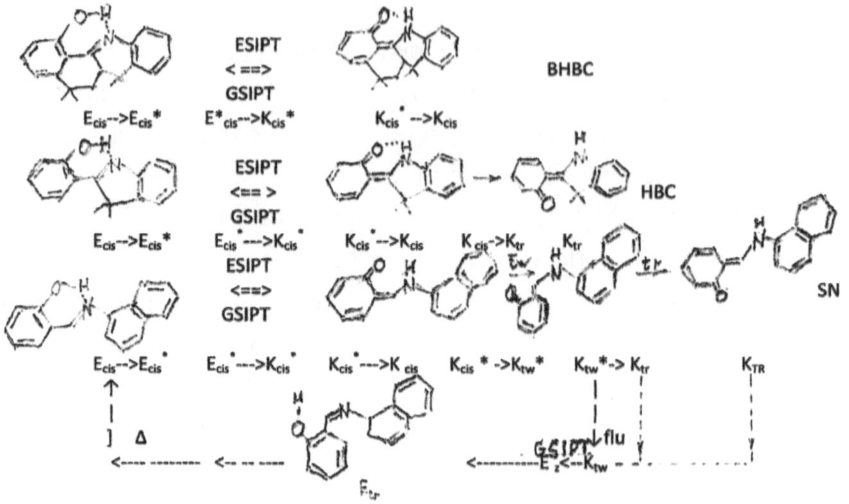

Such a situation is bond up with fall and rise of the potential barriers along the pathways of the E->K and K->E in the So state correspondingly with the aldehyde moiety ring conjugation.

Indeed the quantum-chemical calculations of the conjugated structures with O-H...O bond show/62/that potential barriers for the Proton transfer reaction in the S_1^* and So states fall with the increase of the number of conjugated rings till three in the aldehyde moiety but after that they do not change practically, and even increase a little(Tab. 3. 24a).

The analogous situation may be probably takes place for anils that influence strongly the ESIPT dynamics. Unfortunately the systematic investigations of the ESIPT dynamics and mechanism in dependence on the rings' conjugation in the anil molecules were not conducted.

The occurrence of the ESIPT has been shown for naphthalene ring in imine moiety, I(Ph, 5, 6Np), (Sch. 3. 12)by the typical for the ESIPT fluorescence with the Anomalous Stokes Shift (ASS) ($\Delta v \approx \approx 10000 cm^{-1}$)/45/.

The ESIPT rate has been determined recently /10/for this molecule by means of the decay time of the short-wavelength fast fluorescence

($\lambda\approx$480nm)(Tab. 3. 2. str6, Tab. 3. 24b). Determined time rate value($\tau\approx$40 fs) that is even some less than of SA($\tau\approx$50fs) and absent (like SA) of the deuterium effect show the identity of the ESIPT mechanisms of these structures.

The structure with Naphthalene rings in the both aldehyde and imine moieties, I(5, 6Np, 5'6'Np)

(Sch. 3. 12), has been studied /46/ in Methylcyclohexane (MCH) by the transient absorption spectroscopy (TABS)(Tab. 3. 24b) with time resolution 100fs in the spectral regions of the S_0 state depopulation(negative absorption band, $\lambda\approx$370-400nm) and of the S_1->S_n^* keto-absorption (positive absorption band with $\lambda max\approx$450nm). Both the positive and the negative bands rise "instantaneously" during the pulse excitation i. e. the ESIPT time rate τ<100fs, and may be estimated by a same order of magnitude as that for SA.

Thus it may be believed that the barrierless mechanism and dynamics are not change markedly with the one conjugated ring edition both in the aldehyde and the imine moieties and can be modified only by the participation of the different normal modes in the ESIPT reaction.

However the small amount of the investigated structures limited only by the Naphthalidene derivatives which have studied without utilization of the up-to-day methods of the calculations and the time high–resolved spectroscopy can't provide a sufficient possibilities for the general conclusions.

3. 2. 4. 4. The effect of the substituents in azomethine bridge.

The occurrence of the ESIPT has been established and its superficial study has been carried out of the C_7 substituted structural analogues of SA and SALK with the shortened H-bond(ESA and ESALK correspondingly)/49—53/(see also Ch. 2).

The peculiarities of the ESA and ESALK structures result in the existence of the equilibrium between two Enol isomers, Enol cis (Ecis) and Enol trans(Etrans))(See Sec. 3. 1. 2. 4 and Sch. 3. 5, 3. 6,

Tab. 3. 15, 3. 16), where IH bond O-H...N is realized or broken respectively.

Therefore the ESIPT can occur only in the former competing with the fast radiationless deactivation (IC and ISC) typical for the both isomer of the E structure (see Sec. 3. 1. 2. 4.).

The sole indication of the ESIPT utilized in the discussed works is the ASS fluorescence.

It should believed therefore in accordance with the data obtained (Tab. 3. 15, 3. 16, underlined findings for fluorescence) that the ESIPT can occur for both ESA and ESALK under the various conditions – in the nonpolar and protic solvents and crystals at the room temperature and 77K. At the same time the lack in some cases of the "observed "ASS fluorescence at the room temperature(Tab. 3. 15, 3. 16) can be a result of its quenching owing to diabatic (nonadiabatic) structural transformations like anils (see Sec. 3. 3), and cannot be evidence of the adiabatic ESIPT reaction absence as it was supposed /49-53/. For the direct evidence of the latter the methods of transient absorption or fluorescent methods of the ultrahigh time resolution are needed.

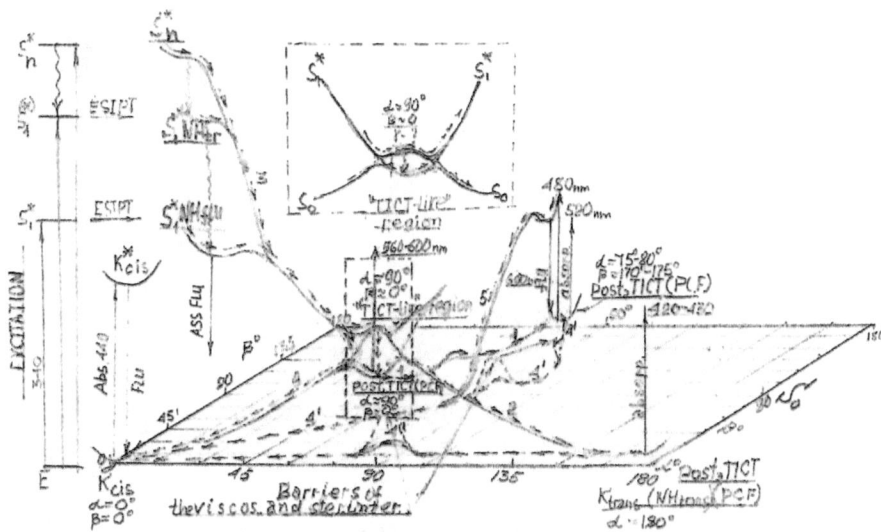

Scheme 3.24

The generation of the different post$_n$ TICT structures of the Photoinduced Colored Form (PCF) in the S$_0$ state from the TICT-like S$_1$*state region via post $_1$TICT(S$_0$ state) structure.
1.The path-way of the twisted post $_2$TICT structure generation (λ^{max}_{abs} ≈ 480,520 nm).2.The formation o the planar NH$_{trans}$ structure (λ^{max}_{abs} ≈ 420-440nm).3.The additional PCF generation path-way including Φ_{flu}/Φ_{color} value dependence on the wavelength of the excitation . 4,4',4''. The pathways of the dark bleaching .5.One of the path -ways of the photo-bleaching.
The qualitative scheme has given on a large and distorted scales for visuality.
(For the structures' identification see Scheme 3.25).

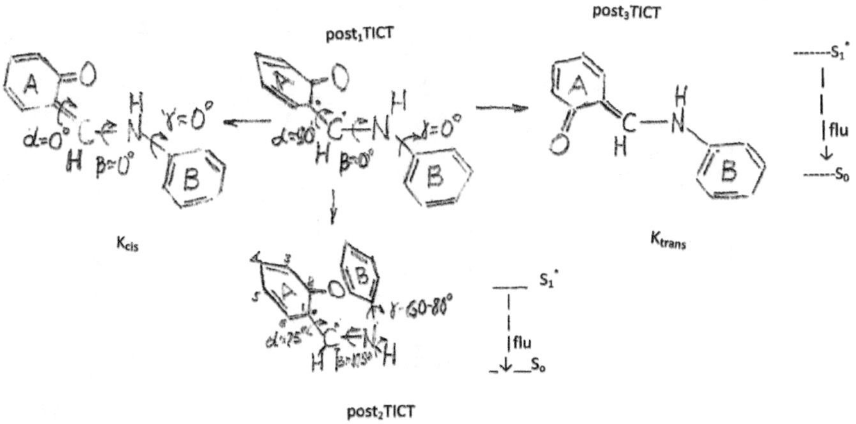

Scheme 3.25
The NH structures on the PES of the S$_0$ state associated with the photocolored form and its fluorescence
(see Scheme 3.24 and Tab 3.47,3.47').

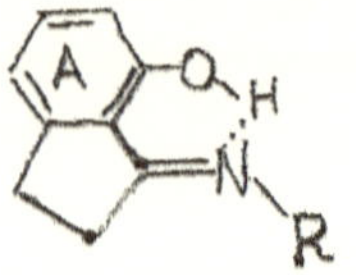

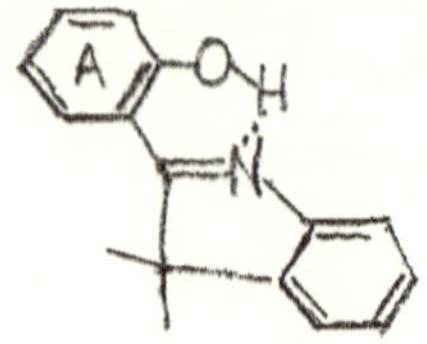

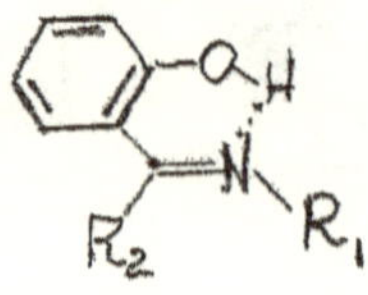

R: Ph,Alk
I

R₁,R₂:Ph.Alk
III

II

Scheme 3.26
The nonphotochromic model structures with the effective ASS fluorescence.

Table 3.47 a
The calculated parameters of the NH structure for SA on the S_0 state PES associated with PC

Type of struct.---->	K_{cts} (planar)		Post$_3$ TICT K_{tr} ,planar.	Post$_1$ TICT twisted	Post$_2$ TICT twisted
References --------> [3]	/7/	/13,14/	/7/	/13,14/	/13,14/
Bond & dihedral Angle ↓	Bond lengths (Å) and dihedral angles (degree) values				
C_1--C_7	1.374	1.386	1.359	1.414	1.412
C_1--C_2	1.464	1.466	1.479	1.472	1.475
C_7--N	1.326	1.371	1.342	1.380	1.375
C_2--O	1.225	1.297	1.211	1.231	1.231
O--H	1.872	1.826	4.676	3.160	3.00
N--H	1.008	1.017	0.995	1.00	1.00
α	0	0	180	~90	<90 (60-85) [2]
β	0	0	0	0	≤180
γ	0	0	0	≠0	≠0
μ(D)	----	3.3	---	3.5	3.8
$\Delta E(cm^{-1})$ [1] (λ^{max})(nm)	34280 (292)	22646 (442)	34361 (291)	18180 (550)	17540 (480-570)

1) The energy of the S_0->S_1 transition. 2) In the brackets the value of the angle α estimated by the findings of the semi-quantitative calculations (see text). 3) The methods of the calcu – lations –HF/6-31G*/7/ and PM3 /13,14/.

Table 3.47b

The bond length differences(Å) between the TICT-like and the post$_1$TICT structures

Bond → Structure ↓	C=O	C_1-C_2	C_1-C_7	C_7-N
TICT-like	1.249	1.435	1.436	1.323
Post$_1$ TICT	1.231	1.472	1.414	1.375
Difference TICT-PTICT	+0.018	−0.037	+0.022	−−0.052

*) See tables 3.35 and 3.47a..

Table 3.48

The dependence of the spectral characteristics of the most long-length absorption transition for the NH structure on the value of the dihedral angle α.

α degree	v^{max}_{abs} (cm^{-1})	λ^{max}_{abs} (nm)	Δλ/Δα *)	Oscillator Strength f×10^2
0	23148	432		22.6
30	22936	436	0.13	6.9
60	19608	510	2.46	0.7
80	16263	593	4.15	0.8
85	16207	617	4.80	0.8
90	15504	645	5.60 5.30	0.9
105	17700	565	2.50	0.45
115	18520	540	---	0.40
125	20830	480	2.40	0.30
150	23810	420	0.16	---
180	24100	415		43.1

*) The average value for each range of Δλ.

Table 3.49

The absorbance, fluorescence spectra, and the relative efficiencies of the generation of the photocolored species in the bulky (tert-butil) substituted salycilidene molecules (see also the text) (isopentane, 77K). Irradiation $\sum$Hg, t= 30min /32/.
(see also tab.3.9.)

Structure	Absorption λ^{max}_{abs} (nm)	Efficiency Φ^{rel}_{typ}	Fluorescence			Absorpion Φ^{max}_{abs}	Efficiency Φ^{rel}_{new}
			λ^{max}_{flu} (nm)	λ^{max}_{exc} (nm)	Stokes shift Δv (cm^{-1})		
			PCF structure, type I			PCF structure, type II	
i	480 510	1.0	535	470 510	2800	—	—
ii	480 500	2.7	560	480 500	3630	—	—
iii	480 515	1.1	540	480 510	2980	550 595	3.8
iv	485 510	7.2	555	470 510	3036	—	—
SA	480 520	—	567	480 520	3196	—	—

The structure for table 3.49

	R$_1$	R$_2$
i	H	H
ii	tert-But	H
iii	H	tert-But
iv	tert-But	tert-But

Table 3.50

o-Hydroxynaphtaldehyde derivatives I(5,6 Np,R). Influence of the imine moiety structure and solvent ..nature the absorption maxima (λ^{max}_{PCF}) and decay time of the photocolored structure (τ_{dec})/48/.

(see scheme 3.27 below).

Compound I (Np,R) No [1]	Benzene			Acetonitryl			Ethanol		
	λ^{max}_{stcf} (nm) [2]	λ^{max}_{PCF} (nm)	τ_{dec} (µs)	λ^{max}_{stcf} (nm) [2]	λ^{max}_{PCF} (nm)	τ_{dec} (µs)	λ^{max}_{stcf} (nm) [2]	λ^{max}_{PCF} (nm)	τ_{dec} (µs)
1	436 457	430	220	432 456	450	1400	434 457	480	190
2	450 417	460	2300	455 477	—	—	454 477	—	—
3	460 486	460	460	450 486	—	—	459 486	—	—
4	405 422	420	760	400 420	440	1800	402 418	440	13
5	400 418	420	490	400 420	440	1900	400 418	440	63
6 (SN)	--- ---	460 [4] 480	28	—	460	630	—	470	55
SA [3]	---	480	40	440	480	150	420-440	—	—

1) See scheme 3.27. 2) stcf-Solvato-thermochromic bands. 3) See Ch 2. 4) 3MP, T=160K /45/.

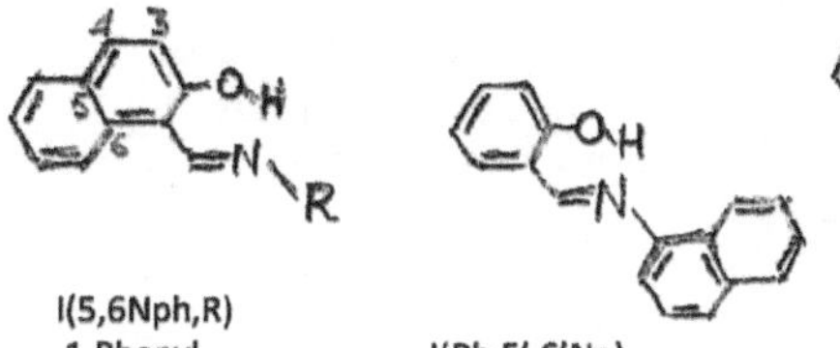

I(5,6Nph,R)
1. Phenyl.
2. 1-Naphthyl
3. 2-Anthryl
4. Benzyl
5. Butyl

I(Ph,5',6'Np)
6. (SN)

Scheme 3.27
Anyls of Naphthalydenealdehyde.

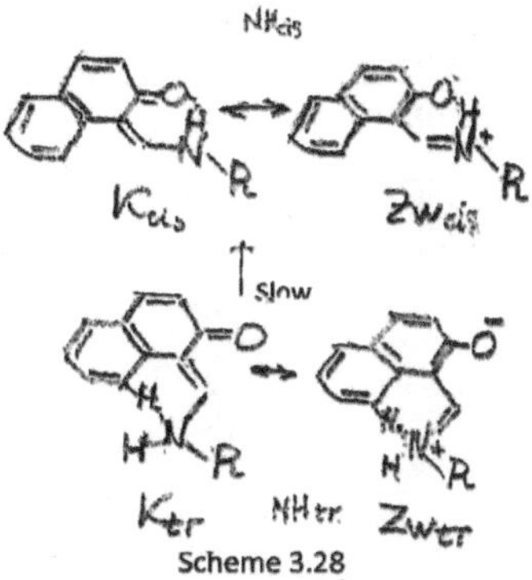

Scheme 3.28
NH structures of I(5,6 Np,R).

Table 3.51

The PCF(post$_n$TICT) structures' absorption bands (λ^{max}) and the thermal fading rate (k) dependences on the imine ring substituents in the 3,6 tert-butyl derivatives of anils in the crystals (30°C) /99,100/.

No	R	λ^{max}_{abs} (nm)	k_1 (s^{-1})	k_2 (s^{-1})
1	H	498	1.9x10^{-3}	1.0x10^{-5}
2	2'F	543	---	6.7x10^{-5}
3	2'Cl	520	2.2x10^{-3}	2.5x10^{-5}
4	2'OCH$_3$	528	1.0x10^{-3}	2.8x10^{-4}
5	3'F	501	---	2.2x10^{-5}
6	3'Cl	490	2.6x10^{-4}	2.7x10^{-6}
7	3'Br	490	---	4.6x10^{-6}
8	3'OCH$_3$	495	---	4.0x10^{-5}
9	4'F	524	---	1.6x10^{-2}
10	4'Cl	534	---	1.0x10^{-4}
11	4'Br	536	1.0x10^{-3}	1.8x10^{-4}
12	4'OCH$_3$	536	---	7.6x10^{-4}
13	4'CH$_3$	546	4.6x10^{-5}	2.3x10^{-5}
14	4'C(CH$_3$)$_2$	518	2.8x10^{-3}	3.8x10^{-4}
15	6CH$_3$	~500	1.7x10^{-2}	0
16	H	500	1.4x10^{-3} (7x10^{-4}) *)	6.0x10^{-5} (6.0x10^{-5}) *)

*)The rate constants for the deuterium substituted molecules are given in the brackets.

The scheme for the table 3.51.

1 - 14   15, 16

Unfortunately the methods with the insufficiently high time resolution (~10^{-9} s) used by the authors are not appropriate for the realization of the necessary dynamic investigations similar to those of SA molecules.

Furthermore the fluorescence with $\lambda \approx 370$nm /49/ belongs to the structure Etrans with the broken IHB and therefore can't be utilized for the measuring of the ESIPT rate.

On other hand as if very low value ESIPT rate (k~10^9 s^{-1}) estimated by decay of the fluorescence with λ≈470nm /52/ may hardly be bound up with the ESIPT rate because this fluorescence (Stokes shift Δν≈9000cm^{-1})is displayed only in ethanol and probably bound up with the Excited state Intermolecular proton transfer in the H-bonded complex ESALK-Ethanol(Tab. 3. 16).

Nevertheless the activation energy for the ESIPT reaction S_1*E- ->S_1*K (9. kcal/mol) and the relative stabilization S_1*K-S_1*E (-9. 51kcal/mol)have been estimated by the semiempiric method of the quantumchemical calculations (AM1).

Although the potential barrier height for the ESIPT reaction is usually overestimated by the semiempirical methods it could be believed that in spite of the shortened IHB, the activation energy for the ESIPT reaction of ESA and ESALK structures differ a little from the analogues value of SA calculated by ab-initio method, and can indicate to the tunnel mechanism similar to SA in the crystal.

However the findings that could allow to elucidate the ESIPT mechanism of the C_7 substituted structural analogs of SA and SALK are absent.

Thus the findings obtained can only indicate the occurrence of the ESIPT in the molecules ESA and ESALK. For elucidation of the C7 substituent effect on the ESIPT dynamic and mechanism it is necessary to utilize the adequate experimental and theoretical methods of the studies.

The findings of such studies can help in particular to obtain the important information about participation and the role of the certain vibrational modes in the ESIPT reaction.

3. 3. The fluorescence and the excited state structural transformations initiated by the ESIPT.

3. 3. 1. The nature of the fluorescent NH* state and the fluorescence with the Anomalous Stokes Shift.

The fluorescence with the Anomalous Stokes Shift ($\Delta\nu\approx10000 cm^{-1}$) (ASS flu) and with the low quantum yield (at room t-re $\Phi \approx 5\times10^{-3}$) is emitted from the thermalized $S_1^*(\pi\pi^*)$ state of the NH structure (S_1^*NH flu) as a result of the ultrafast ESIPT followed by the relaxation process $S_n^* \dashrightarrow S_1^*$ and $S_1^* \dashrightarrow S_0$ /1, 4, 6, 8, 10, 11, 36/ that can compete with the generation of the colored form (see below).

The S_1^*NHflu structure and its decay can be registered by the four spectral parameters independently —two bands of the positive transient absorption $S_1^* \rightarrow S_n^*$ ($\lambda \approx 420$ and 485 nm), negative transient band of the stimulated emission ($\lambda \approx 620$ nm), and the corresponding ASS emission band S_1^*-hν->So ($\lambda \approx 520$-540nm)(Tab. 3. 26).

The lack of the coincidence of the latter two bands occurs due to the overlap of the stimulated emission with the transient absorption bands.

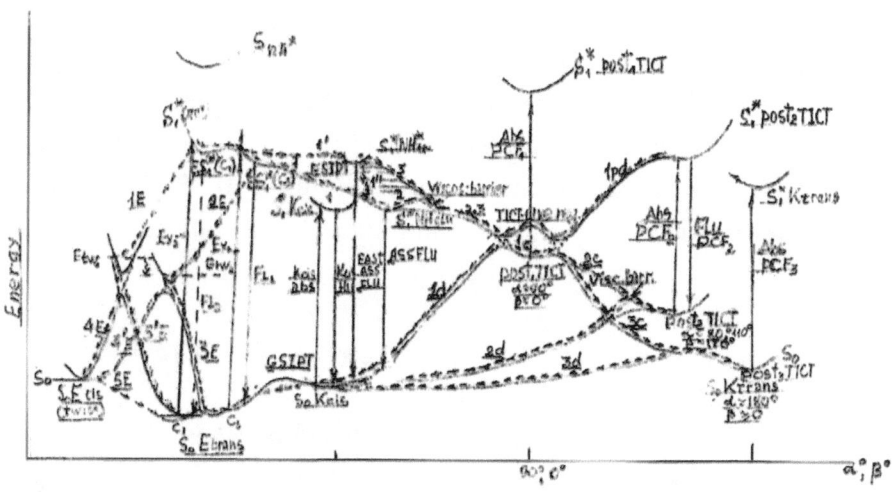

Scheme 3.29
The general scheme of the photoinduced processes of SAs (see also the schemes 3.2 a,b; 3.13 a,b; 3.17; 3.24; 3.26).

For the elucidation of the nature and the geometry of the S_1^* excited NH* structure responsible for the ASS flu the analysis of the data of the recent theoretic studies /1, 7/ is required.

According to /7/(ab-initio Hartree –Fock and CIS with the instant-on approach)the equilibrium geometry of the S_1^*NH flu state can be formed immediately after the ESIPT owing to closing in the lowest $S_1^*(\pi\pi^*)$ and $S^*(n\pi^*)$ states in the NH structure.

Obviously the real NHflu structure cannot be planar because of the effect of proximity($\Delta v=1055 cm^{-1}$) between the considerably nonplanar $n\pi^*$ or $\pi n^*(\pi)$/7, 27/(C_1 symmetry) and planar $S_1^*(\pi\pi^*)$(Cs symmetry) states (i. e. $\alpha \approx 80°$ and $0°$ correspondingly) /7, 9/ (Tab. 3. 25) and has to have the intermediate parameters (e. g.-dihedral angle α can be estimated as $70°$ /9/ but it is probably an overestimated value.

According to /1/ (ab-initio, TDDFT method) the planar $S_1^*(\pi\pi^*)$ state of the NH(Kcis) unstable with respect to the rotation (angle β) leading to the trans-cis isomerization about C=N bond and has only a single minimum on the PES along the torsion coordinate corresponding

90° twisting about C_1--C_7 bond with the energy difference between S_1^* and So states about 1006cm^{-1}.

The lack of the metastable (equilibrium) structure on the S_1^*PES of the NH form responsible for the ASSflu (S_1^*NHflu) contradicts to the experimental findings. However in spite of the incorrect

TDDFT data the region within the limits $\alpha=0-90°$ is the most probable for existence of the minimum with the nonplanar structure twisted around $C_1:::C_7$ bond (angle α).

Thus the calculated data obtained by the different up-to-date methods point to the marked nonplanar geometry($\alpha \neq 0°$) of the S_1^*NH structure that is agreed with the earlier hypothesis/79/.

This state is formed as a result of the ultrafast ESIPT and the motion of the wave packet with the low frequency large amplitude inharmonic torsional vibrational modes (involving rotation around C_1-C_7 bond to the(NH)structure with the considerable weakened H-bond and a marked charge separation ($\mu \approx 9.7D$). The contribution of such vibrations in the wave packet depend strongly on the excitation wavelength and the energy difference between (OH)* and (NH)* structures in the S_1^* state.

Therefore the S_1^* (NH)*structure formation can be proceed as either a barrierless reaction (short-wavelength excitation) or by the tunnel mechanism(long-wavelength excitation) in the dependence on the wavelength (energy) of the excitation and on the media nature (solvent of different viscosity, crystal state)(see below).

The thermalized S_1^* NHflu structure with $\alpha \neq 0°$ can be stabilized by the competition between the electrostatic (charge separation), the weak H-bonding, and the sterical (C-H)-(H-Ph) interac-tions (see Sch. 3. 13).

The ASSflu band position and shape can depend on the dihedral angle.

Indeed the parameters of the thermo-solvatochromic S_1^*-hv->So and the ASS fluorescence bands (and the colored structure also) are differed by the position and shape.

The broad ASS flu band in the free molecule /9/(jet-cooled emission) and in the condensed media /2, 5, 6, 8, 12, 29, 35/ has a fine vibrational structure with the two maxima well resolved in the rigid solvents at 77K ($\Delta\nu\approx950\,cm^{-1}$) shifted towards short wavelength with a small difference in the splitting in comparison with the flat keto-structure (Kcis) (Tab. 3. 29)/20/.

In the thermochromic crystals/29, 69/ (Tab. 3. 30)at high temperature (T=200-3000K) the broad emission band ($\Delta\nu\frac{1}{2}\approx2000\,cm^{-1}$) with the small Stokes shift ($\Delta\nu\approx1500\,cm^{-1}$) is observed.

This emission is superposition of the thermochromic (Kcis) and the ASS flu bands. At low temperature (T= 40-70K) the prevalent ASSflu band with the large Stokes shift ($\Delta\nu\approx7500\,cm^{-1}$) shifted towards short wavelength ($\lambda max\approx530\,nm$) is displayed well.

In the crystalline photochromic inclusion complexes of SIP(Sch. 3. 14)with β and γ Cyclodextrines (CD) /81, 82/the two emission bands are observed. The first band is the ASSflu ($\lambda^{max}\approx520\,nm$ (bound up with the structure distorted about $C_1:::C_7$ bond($\alpha\neq0°$), and the second one($\lambda^{max}\approx575\,nm$) is attributed to the planar cis –keto(Kcis) structure.

These results conform to the scheme suggested in /79/for SA in dibenzyl and stibene host crystals at low temperature.

Thus in the general case the ASSflu is located in the more short wavelength spectral region in comparison with the emission of the flat "thermochromic 'Kcis structure.

However the relative position of the different nature states and the corresponding structural transformations are very sensitive functions of the molecular structure(e. g. /27, 28/), and therefore the accident co-incidence of the emission bands of the S_1^*NH flu and S_1^*Kcis structures can be realized for some compounds/83/.

In any case the fluorescent Kcis S_1^*state bound up with the planar Kcis structure cannot be si-tuated on the path-way of the photocolored structure formation, and the excited Kcis* structure does not lead to the generation of the photocolored structure (photochromism).

The kinetic findings about dependence of the S_1^*NHflu state decay on the solvent nature and the temperature obtained by registration of the different parameters conform well.

The decay time increases with growth of viscosity of the various solvents (Tab. 3. 26) at the same temperature /5/ and the different viscosities (temperatures) of the same solvent(Tab. 3. 27)/35/.

This dependence is masked slightly by the influence of the solvent polarity and by the specific interacttions that can act in the opposite directions owing to high polarity of the quenching structure (see below) and (or) the protic nature of some solvents.

At the same time the position of ASSflu band and the Stokes shift value change a little and irregularly with the variation of the solvents that testifies to small structure modifications of the fluorescent state under such conditions.

The very low quantum yield ($\sim 10^{-3}$) increases markedly with the rise of the solvent viscosity (Tab. 3. 26, 3. 27)/5, 35/and strongly in the rigid media (77K) (Tab. 3. 28, the second line)/12, 20/.

In the model structure with the forbidden twist of the ring A($\alpha > 0°$, I([Ph, R)/12, 20, 84/ the ASSflu is displayed at the room temperature in the liquid solvents, unlike SA, has a considerable value of the quantum yield at 77K, and shifted towards the long wavelengths(Tab. 3. 28).

The three principle processes can be a basis of the very low efficiency of SA ASS flu:

i) a competition between the population of the fluorescent state and the generation of the photocolored species from a common transient state (however, see below),
ii) the high efficiency of the radiationless transitions due to the nonplanar geometry of the fluorescent structure, and
iii) the quenching adiabatic structural transformations("horizontal" deactivation).

The given above data including viscosity effect (Tab. 3. 26-3. 28) show the very important role of the third process.

Really the data of the semiempirical /12-14/and the recent ab-initio /28, 35/ calculations point out clearly the formation of the ICT state of high polarity with the strong charge separation (Tab. 3. 27) by the adiabatic A-ring twist($\alpha \approx 90°$)in the S_1^*state of the NH-structure (so called $\pi n^*(\pi)$)state by interpretation of /28/).

This state can be considered as the "quenching TICT-like "twisted structure /85-89/ that is generated from the S_1^*NHflu state above the very low potential barrier($\Delta E<5$kcal/mol) (Sch. 3. 13).

The low activation energy is provided by a nonplanar ("pre-TICT" /88/)structure of the initial fluorescent state.

The effective quenching is caused by the utmost proximity in the PESs of the $S_1^*$and So states ($\Delta E<5000$cm^{-1}) in the region of the "TICT-like "structure at the expense mainly of the very high potential barrier for the twist ($\alpha=90°$)in the So state.

Such a proximity of the So and S_1^* states has to ensure the effective radiationless deactivation S_1^*->So but however cannot favor an origin of any effisicient long wave length emission supposed of /35. 45/.

The definition "Twisted Intramolecular Charge Transfer" (TICT)-like state /89/means the structure with the Charge separation owing to the Intramolecular Charge Transfer (ICT) and the twist around the C_1--C_7 bond by the analogy with the well-known term of the TICT state /85-88/.

Thus the formation of the nonplanar structures in the S_1^*state plays the decisive role in the both generation and decay of the fluorescent state responsible for the ASSflu.

The both processes are closely bound since the formation of the acoplanar "pretwisted "fluorescent structure facilitates strongly a generation of the twisted structure responsible mainly for the quenching of the S_1^*NH flu state i. e. ASSflu(Sch. 3. 13).

3. 3. 2. The molecular structure influence the ASS fluorescence

3. 3. 2. 1. The effect of the ring substituents.

There is a point in the consideration of the effects of the electronic and sterical factors separately.

The electronic influence of the ring substituents the characteristic of the ASSflu has been studied by the steady-state /20, 30, 33, 35/and time-resolved /35/ methods in the solvents/20, 33, 35/and the crystal films/30/.

To the best of our knowledge the findings about A ring substituent effect are very scanty.

The table 3. 31 gives some examples with the very crude estimation of the relative fluorescence intensities (Irel)at 77K /20/ which reflects, fist of all, the strong shift of the E⇔K equilibrium towards K structure in the Ground state when including both the donor and acceptor substituents.

It is obviously, under such conditions, the decrease of the E structure concentration has to lead to the fall of the ASSflu intensity as compared with SA(1).

The spectral position of the ASS flu band changes a little with the strong electron donor and acceptor substituents due to the break of the conjugation in the A ring π-electron system of the NH structure.

However the compound 9 is an exception owing to the sharp change of the $So \rightarrow S_1^*(\pi\pi^*)$ transition nature connected with the ICT from the NMe group and the appearance of the ICT(not ESIPT) fluorescence($\lambda max \approx 454nm$)of the twisted Enol structure(see also Sec. 3. 1. 2. 1).

The influence of the electron factors when introducing of the substituents into the B ring has been studied much more carefully by various methods in/30, 33, 35/and also in /6, 32/(Tab. 3. 32).

The data produced show the ASS value drops with the inclusion of both the acceptor($\Delta\Delta va$-f ≈ 670-$750 cm^{-1}$ and especially donor ($\Delta\Delta va$-f≈ 1300-$3000 cm^{-1}$) substiuents at the expense mainly of the fluorescence band ($\lambda fmax$) blue shift in both the nonpolar and polar solvents.

Such a shift of the fluorescence band are caused by the different changes of the stabilization energy when introducing of the substituents.

In the case of the fluorescence band blue shift, the So state is obviously stabilized stronger than S_1^* one (Scheme under the Tab3. 32). Meanwhile the stabilization of the S_1^*NH state increases the energy barrier and reduces the rate of the adiabatic reaction with the quenching "TICT-like "state generation(Sch. 3. 32). As a result the strong increase of the lifetime($\tau(f)$) and the quantum yield($\Phi(f)$) of the ASS flu are observed (Tab. 3. 32)/35/.

At the same time the strong interactions between the adjacent molecules in the crystalline packing of the fluorescent (thermochromic) crystals(Ch. 2) stipulate the considerable increase of the Stokes shifts (unlike solvents) as a result of the opposite shifts of the absorption and fluorescence bands as comparison with the solvents (Tab. 3. 32)/30/ owing to the insertion of the donor substituents. To elucidate the peculiarities the special investigations have to be conducted.

The steric factor influence the ASSflu has been studied specially by the insertion of the bulky tert-butyl substituents into the A ring and B one with the electron donor 4'OH group(Tab. 3. 33)

/32/(see also Sec. 3. 1. 2. 1).

The variations of the fluorescent characteristics reflect the bulky(tert-butyl) substituent influence the change of the conformation and the path-ways of the generation and decay of the fluorescent state.

At the room temperature in the low viscous solvents the fluorescence band is red shifted and the quantum efficiency is lower for the structures with the bulky substituents in the A ring(ii, iv) as compared with the rest structures(i, iii)(Tab3. 33).

It could be explained by stabilization of the more flattened NH fluorescent structure (Sch. 3. 14) in (ii, iv) as compared with (I, iii) along the path-way of the adiabatic formation of the nonplanar fluorescent NH*state similar to SA molecule (see 3. 3. 1).

The more efficient radiationless deactivation and the lower fluorescence efficiency are provided by the C-H vibrations in the tert-butyl groups and the proximity of the S_1*and So states in such structures.

At the 77K the interactions with the rigid solvent play the key role especially for the molecule with the bulky substituents in the A ring (ii, iv) which prevent the A ring twisting in the way of the quenching TICT-like structure formation. As a result, the increase of the fluorescence quantum yields, $\Phi(77)/\Phi(293)$, is much more for the structures with the bulky substituents in the A ring than for other ones when lowering of the temperature from the room that (liquid solvent) to 77K (rigid solvent)(Tab. 3. 33). Owing to that the fluorescence quantum yield for ii, iv becomes higher than for I, iii, unlike the liquid solvents.

The effect of the bulky substituent in the A ring of the two compounds with the B pyridine rings on the kinetics of the S_1^*NHflu structure decay in the solvent (CH) and the crystal has been studied recently /36/(Tab. 3. 33, structures v, vi).

It is known the v and vi are thermochromic and photochromic in the crystals owing to planar and nonplanar arrangement of the N-pyridine ring correspondingly because of the steric interactions in the latter (see also below).

The longwavelength shift of the fluorescence bands for v, vi and the heightened Stokes shift in comparison with other structures (ii-iv) are bound up with the effect of the pyridine ring electronic effect, and can be explained by the more substantial conformational change with generation of the more nonplanar S_1^*NHflu structure responsible for the more longwavelength emission with very low quantum yields (Φ v, vi << Φii-iv) at the room temperature.

There are two excited NH* structures of the both v and vi molecules in the solvents which are distinctly differ in the energy (i. e. λ(flu) max) and in the kinetics of the decay (Tab. 3. 33). The shorter and longer decay times are bound up with the higher and lower state energy respectively.

The high energy state corresponds to the more planar transient structure that is the precursor for S_1^*NHflu one responsible for the principle fluorescence band with the longer lifetime.

There are analogous two bands in the ASS fluorescence spectra also of other anile molecules both with and without the bulk substituents that are insufficiently resolved in the liquid solvents but differ distinctly of the rigid media (including glass solvents)(seeTab. 3. 33). Therefore it may be believed that the transient short-lived structure is typical for the all anil molecules, and manifested clearly by the kinetic for the molecules with the bulky substituents in the A ring.

Table 3.52
Photochromisity of the substituted SAs and their 3,5-tert-butylsalicylidene derivatives (crystals) /110/

R'	Photochromisity R	
	H	3,5-di-^{t}Bu
H	+	+
2'F	--	+
3'F	--	+
4'F	--	+
2'Cl	+	+
3'Cl	--	+
4'Cl	--	+
3'Br	--	+
4'Br	+	+
2'OMe	+	+
3'OMe	--	+
4'OMe	--	+
4'Me	--	+
4'^{i}Bu	+	+

"+" and "—" –Photochromic and non-Photocromic crystals correspondingly

Structures for the tables 3.52-3.54

Table 3.53
Photochromisity of the substituted SAs and their DCA clatrate crystals /110/ [*]

R'	Photocromisity	
	Pure crystal	DCA clathrate
H	+	+
4'F	--	+
4'CL	--	+
4'Br	+	+
4'OMe	--	+

[*] For designations and structures see table 3.52

Table 3.54
Photochromisity of the substituted SAs and their 2'6'-disubstituted alkyl derivatives /110/. *)

. R	Photochromisity R"					
	H	2',6' Me	2',6' Et	2',6' Pr	2',4',6'Me	2',4',6' ^{t}Bu
H	+	----------	----------	+	--	----------
3,5Cl	--	--	+	+	----------	--
3,5Br	--	--	+	+	----------	----------
3,5I	--	--	+	+	----------	----------
3,5^{i}Bu	+	+	----------	+	----------	----------
5Cl	--	----------	----------	+	----------	----------
5Br	--	+	----------	+	----------	----------

*) For designation and structures see table 3.53.," ---------- " means the absent of the data.

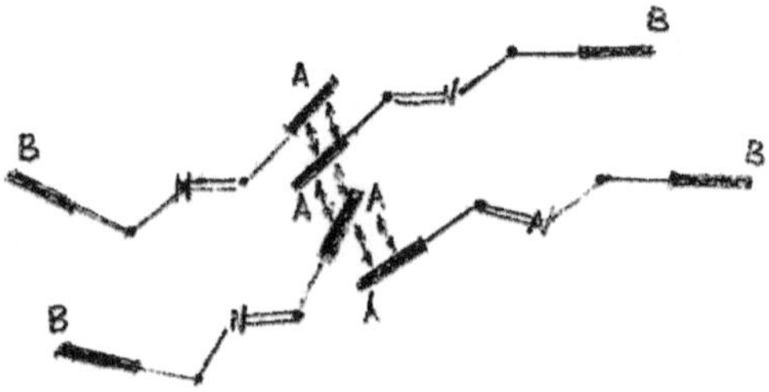

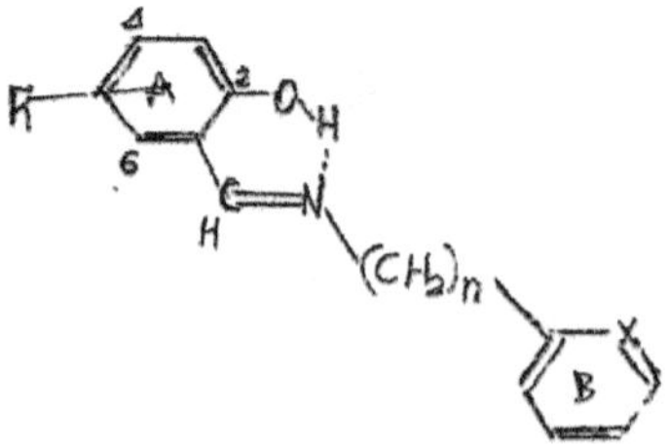

R	n	X
H	1	H
H	2	H
4OMe	1	H
5OMe	2	H
5Br	2	H
H	2	N

Scheme 3.30

The approximate qualitative scheme of the crystalline packing and examples of the molecular structures having both the thermo and photochromic properties in the crystals. (See also Scheme 2.13 and table 2.10 (Ch 2)).

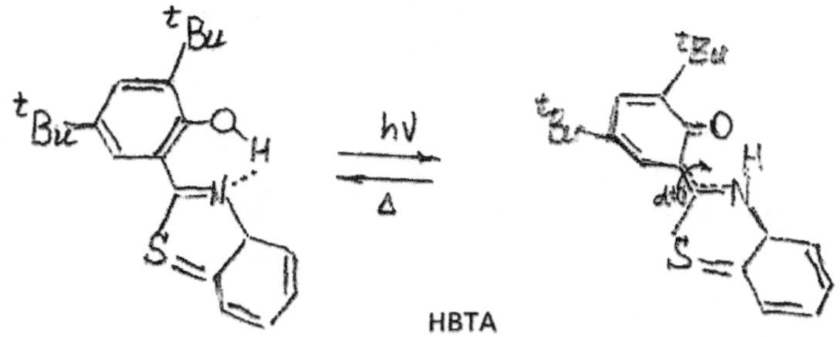

ESIPT Photochromism

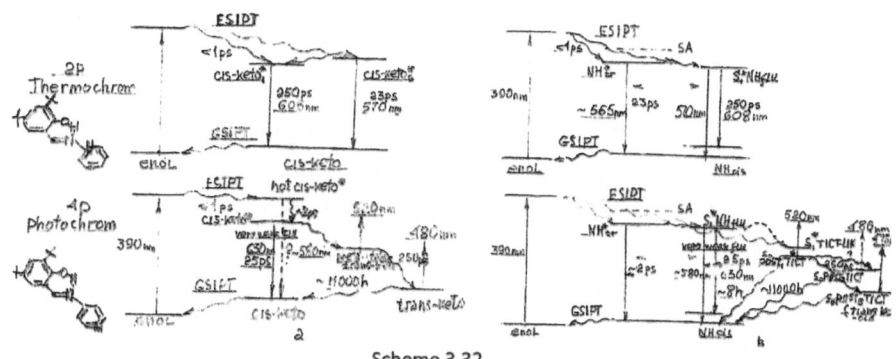

Scheme 3.31
The photochromism with the generation of only the post₁TICT structure in the crystal.

Scheme 3.32
Photoinduced processes of the thermochromic (2p) and photochromic (4p) crystalline Anils with the bulky (3,5 tert- butyl) substituents in the A ring on the base of findings /36/ and with alternative descriptions according both to /36/(a) and to suggested scheme 3.29 (b) (see text).

Table 3.55

The spectral and kinetic parameters /36/ with their attribution to the colored form structures in the crystals of the compounds 2p and 4p according to the 3.29 and 3.32b in comparison with SA.

Compound	Solvent		Crystal		Attribution of the structures .in the crystal at room temperature and type of photochr. (See ch 5)
	Absorp. band, λ^{max}_{nm}	Decay time, $\tau_{\mu s}$	Absorp. band, λ^{max}_{nm}	Decay time, τ_s	
2p	450 [1)]	186(62%)	–	–	Thermochrom
	500$_{sh}$	35(38%)			(L-photochrom⁻)
4p	450	122(45%)	475	4×10^7(99.9%)	post$_3$ TICT⁻ (LT-pho-
	480$_{sh}$	19(55%)	480 [2)]	3×10^4(0.1%)	post$_2$ TICT⁺ toch-
			520	25×10^{-11}	post$_1$ TICT⁻ rome)
SA	480	0.003 [3)]	480	[4)]	post$_2$TICT+ (LT-photo-
	520		520	$\sim 10^3$	+trace of post$_1$TICT) chr.)

1)/36/^CH. 2)The bands of the post$_2$TICT and post$_3$ TICT overlap but difference by life-time.3)From /4-6/. 4) From /71/.

The latter shows also the kinetic peculiarities in the crystalline state (see below).

Thus the decisive role of the aldehyde moiety deformation, especially A ring twist, in the formation of the anile fluorescence properties is supported well by the inclusion of the bulky substituents in this ring.

3. 3. 2. 2. The imine moiety structure influence the ASS fluorescence.

The main structural transformations bound up with the ASSflu (and also with photocoloration) occur in the aldehyde structural fragment.

However the replacement of –N-Ar (SA) by –N-Alk (SALKn) or by –N-(CH)-Ar (SALKn-Ar) radicals is the exceptionally important structural modification which affects photophysic and photochemical processes in the anile molecules /1, 4, 20, 26-28, 90/. (For the more detailed conventional signs of the structures see the Scheme 3. 3).

The ASSflu is practically absent in the liquid solvents (room temperature) of the SALKn molecule but is clearly observed in the rigid solvents(77K) (like SA) and in the crystal of some SALKnCH Ph molecules at room temperature /40/.

The fluorescence depends a little on the N-R structure. Meanwhile the presence of the fluorescence by itself casts doubt on the relative position of the $n\pi^*$ and $\pi\pi^*$ states of the NH structures of such molecules suggested on the base of the quantum-chemical calculations of /27, 28/ unlike of/41/.

The most important peculiarities of SALK in comparison with SA are the localization of the first electronic transition on the aldehyde moiety exclusively, and strong increase of electron density on the C=N group in the Ground state.

Owing to the first peculiarity the strong dependence of the quantum yields of the ASSflu(Φflu), photocoloration(Φcol)and their ratio (Φ flu/ Φcol) on the excitation wavelength is practically absent unlike SA (Tab. 3. 34)/20/.

Above mentioned data may be connected with the problem of the common precursor of the fluorescent and photocolored structures that will be discussed in the next section.

The second peculiarity leads to the strong shift of the E$\Leftrightarrow$K equilibrium in the So state towards K structure that fluoresces both at 293 and 77K with the $\lambda^{max}\approx$465nm(λ_{ex}=400nm)unlike the ESIPT NH* structure which emits the band with λ(flu)=512-520nm at 77K only.

However for the rigid structure (SALK$_1$, [Ph) with the fastened A ring the ASS flu is observed in the both liquid and rigid solvents (Tab. 3. 34)like SA([Ph) structure.

Unlike the longwavelength absorption band shifted towards the "blue" as compared with SA that (λ^{max}=340nm(SA)->λ^{max}=320nm(SALK)), the ASSflu band almost does not shifted since S$_1$*-> So transitions in the NH structures are localized in the identical structural(keto) fragments in the both molecules, and therefore the Stokes shift of SALK increases considerably (Tab. 3. 34).

Thus the SALKn molecules unlike SA ones display the two fluorescence bands of the different nature –the ASS flu band like SA emitted from the nonplanar, highly polar ($\mu\approx$8. 8D)S$_1$*NHflu structure and quenched by the "TICT-like" state, and the K*cis flu band emitted after excitation of the planar nonpolar SoKcis structure.

These bands are made out clearly by their spectral positions and temperature dependence of their intensity.

3. 3. 2. 3. The effect of the ring conjugation on the ASS fluorescence.

The study of the ring conjugation influence the spectral and kinetic parameters of the S_1^*NHflu state in the solvents and in the crystals has been carried out with help of the steady-state absorption and fluorescence spectroscopy /10, 33, 42, 44, 47, 48/(Tab. 3. 35, 3. 36, 3. 36') and also by the time-resolved nanosecond fluorescent spectroscopy/44/and sub-nanospectroscopy /10/ (Tab. 3. 36, and 3. 36').

Although the findings of the different studies are reproduced not so well however they allow to draw some clear conclusions. sssss

The considerable red shifts of the E structure longwavelength absorption (see also Sec. 3. 1. 2. 3) and the ASS flu bands as compared with the SA ones in the different solvents and crystals are observed only with the limited expansion of the π-ring system in the aldehyde moiety (up to the two ring) (5, 6NA; 5, 6NN; 3, 4NA) and after that (e. g. PA) the shifts can even decreased.

Unlike SA the very weak fluorescence is observed even at room temperature in the liquid solvents and crystals.

The common typical peculiarity of the structures with the expanded π-ring system in the aldehyde moiety is the strong shift of the E<=>K equilibrium in the So state towards the fluorescent K-structure under any conditions(Tab. 3. 35).

Therefore the measurement of the ASS flu parameters is impeded considerably. At the same time it should be noted that the K-structure fluorescence (λ^{ex}=400-500nm) and the ASSflu (λ^{ex}=340-380nm) differ considerably in the most cases in the band positions, the Stokes shift and the lifetime /44/ (Tab. 3. 36/.

These data are an evidence of the different nature of the fluorescent K*-structure and NH*that responsible for the ASSflu in the most of those compounds although it can be very hard to distinguish between the two fluorescence for some cases. The data produced above are described satisfactory by the general scheme (Sch. 3. 15) that somewhat differs from that proposed for nonpolar sol-vents/47/.

The two features are the basis of this modified scheme:

i) The high content and the lack of the K-structure in the Ground and excited S_1^* states correspondingly. The latter is caused by the high activation energy for the H-atom transfer unlike the very low energy barrier for the ESIPT in the S_1^* state of ICT nature (see above Sec. 3. 2. 4. 3).
Such a S_1^* state of the NH form is bound up with the charge redistribution from the ArOH group to the C=N one (unlike S_1^*K state of the K-structure) and leads to the ASSflu that differs from K-structure emission;
ii) The ASSflu quenching, at room temperature is provided by the efficient generation of the quenching "TICT-like" state which is suppressed by the interaction with the rigid (viscous) media (e. g. solvent at 77K).

The investigation of the influence of the substituents in the structures with the expanded ring conjugation in the aldehyde moiety has been conducted for compounds with the 4'-substituted

N-Ph ring in the nonpolar/44/and polar /42/solvents at the room temperature (Tab. 3. 36).

The Stokes shift is diminished slightly by the insertion of both the electron donor and acceptor substituents by the expense mainly of the absorption long-wavelength band red shift.

The quantum yield of the ASSflu is considerably higher for 4'substituted than for unsubstituted ones irrespective of the substituent nature. For almost all 4' substituted Naphthalene derivatives the spectral coincidence of the ASSflu and Kflu bands is observed unlike the phenyl substituted molecules of this kind (SAs).

The observed not too strong influence of the 4'substituents reflects the structural effects bound up with the change acoplanarity of the B ring and the electronic nature of the substituentson the characteristics of the ICT transition localized on the aldehyde moiety.

The detail investigation of the imine moiety structure effect including conjugation in B ring has been taken/48/by the steady-state fluorescent spectroscopy of the 5, 6 NA derivatives in the series (1-5) and SN(6) (Tab. 3. 36 band the structures there).

The findings presented have been taken from the fluorescence emission and excitation spectra in ethanol at 77K that show the most typical picture for the compound investigated.

The all compounds (with exception 1 and 6)have the two different emission bands. The first, intensive, band with two vibrational maxima, small Stokes shift and the mirror symmetry to the excitation band, belongs to the K structure.

The K excitation and fluorescence bands are shifted to the blue with the decrease of the π-electron system size in the imine moiety and absent of the compound 6 like SA due to almost complete E⇔K equilibrium shift towards Estructure. The second, very weak and broad, band with the anomalous large Stokes shift belongs to the NH structure generated by the ESIPT. This fluorescence is almost absent of 1 owing to the complete shift of the E⇔K equilibrium in the Ground state towards the K structure.

On the contrary, in the structure 6 the ASSflu arises only by excitation of the E form. Such a fluorescence is represented by the short ($\lambda^{max} \approx 530nm$) and long($\lambda^{max} \approx 560nm$) wavelength bands (Tab. 3. 36a)which, as it has been supposed /45/, belongs to the planar K*cis and strongly twisted (α, β >>0)K*tw excited keto structures correspondingly. However such an interpretation is seemed unlikely (see previous and next sections).

Thus the dependence of ASS and K_{flu} on the imine moiety structure is connected mainly with the E<=>K equilibrium shift in the Ground state, and also reflects the key role of ICT nature transition localized on the keto moiety (A)of the NH structures.

The position of the ring addition influences relative energy of E, K and NH structures, and the potential barriers for the TICT-like structure formation in the S_1^*state.

Therefore the ring addition to the 3, 4 position (3, 4NA) leads to the greater shift of the bands and to some increase of the ASS. The quenching of the ASSflu is more efficiently with the ring addition to the 5, 6 position(5, 6NA) that leads to the sharp fall of the fluorescence quantum yield and lifetime.

The substituent influence the ASS parameters –the increase of the quantum yield and lifetimes is expressed stronger for the structures 3, 4NA than for the 5, 6 those. The all above problems bound up with the expanded conjugation need the more detailed, especially quantum-chemical investigations.

3. 3. 2. 4. The influence of the substituents in the imine bridge.

It has been shown in the sections 4. 1. 2. 4., 3. 2. 4. 4, the most important distinction of the C_7 –substituted azomethine analogs, ESA and ESALK, from SA and SALK correspondingly is the high probability of the existence of the E_{trans} and E_{cis} forms (around C=N bond) with and without the H-bond in the Ground and the Excited states. It is obviously, only the first one can be responsible for the ESIPT and the ASS flu.

The fluorescence characteristics of the Enol structure for ESA and ESALK have been discussed earlier (Sec, 3. 2. 4. 4. See Sch. 3. 5, 3. 6) and presented in the Tables 3. 15 and 3. 16 correspondingly /49-52/ and in the Scheme 3. 16. In the liquid nonpolar solvent (Sch. 3. 16a) the excitation of the E_{trans} structure (λ_{ex}=322nm) leads to the formation of the structure responsible for the ASS flu ($\lambda^{max}_{flu} \approx$500nm, ASS $\Delta v \approx$11000cm^{-1}) which is similar to the K*cis one and deactivated radiationlessly strongly.

Simultaneously the excitation of the Ecis and Zw (NHzw) structures gives the fluorescence bands ($\lambda \approx$370 and 400-420nm) with the not large Stokes shifts.

In the rigid nonpolar solvents(T=77K)and the crystal(room t-re) (Sch. 3. 16b) $E_{trans} \Leftrightarrow K_{cis}$ equilibrium in the Ground state is shifted towards the K structure.

The ASSflu arises from the S_1*NHflu structure. ($\lambda^{max}_{flu} \approx$490nm, ASS $\Delta v \approx$10300cm^{-1}) as a result of the ESIPT after excitation (λ=322nm)of the Ecis structure.

Under excitation of the Kcis form (λ=400nm) the fluorescence band ($\lambda^{max}\approx$490nm) similar to the ASSflu appears (like that at the room temperature). At last in ethanole, room temperature (Sch. 3. 16c) the specific interactions with the solvent can stipulate the two kinds of the excited state proton transfer in the OH-complexes of ESALK with ethanol: the intramolecular PT(ESIPT) with the ASS flu band (λ^{max}_{flu}=500nm, ASSΔv=10300cm^{-1}), and the intermolecular PT(ESIterPT) with ASSflu band ($\lambda^{max}_{flu}\approx$470nm, ASS$\Delta v\approx$9030cm^{-1}).

Thus the S_1*NHflu state generated by the ESIPT emits the ASSflu with the band which coincides for ESA (ESALK) with the fluorescence band of the Kcis structure under various conditions, and therefore the ESIPT leads to the S_1*NH structure that is similar S_1*Kcis one which corresponds to the nonpolar planar keto structure Kcis in the Ground state in both nonpolar or polar solvents and crystal state.

The unusual feature of this structure (unlike SA and SALK) is the marked intensity of the ASSflu in the liquid solvents irrespective of their viscosity in spite of the low fluorescence quantum yields (e. g. $\Phi f\approx$ 2x10^{-2} for ESALK in ethanol at room temperature/52/) that also is typical for the fluorescence of the Kcis structure.

It could be testify to the difference of the ASS flu generation and quenching mechanisms between the anil molecules and their C –R derivatives (ESA, ESALK).

The low Φf value can be explained only by the high rate value of the radiationless deactivation (IC+ISC, Knr$\approx$10 s^{-1} /52/) without participation of the quenching TICT-like state which formation in the Kcis* structure needs overcoming of the high potential barrier for the twist around C=C double bond unlike the SA and SALK zwitterion structure.

Such a mechanism can also play the key role in the reaction of the photocolored structure ge-neration in photochromic anils (see below).

3. 3. 3. The development of the ideas about the post-ESIPT relaxation including the problems of the ASS fluorescence and precursor state

Due to the exothermicity of the ESIPT reaction the high vibrational ("hot") levels S_1^* of the thermalized S_1^*NHflu state or the higher electronic low-stable Sn^* states can be populated immediatly after the ESIPT. Therefore one can supposed at least the two path-ways towards Photocolored (PC) product.

The first path-way is the immediate diabatic (nonadiabatic) PCF structure generation omitting S_1^*NHflu state in the competition with the internal conversion (IC)(Sn^*~~>S_1^*)(see above) or (and) S_1^*~~>S0 vibrational relaxation. In such a case the Sn^*state(flattened structure) or $Š1^*$ vibronic states is the common precursor of the ASS flu and PCF structures. (see below).

The second path-way is the IC Sn^*~~>S_1^* and S_1^*~~>S0 vibrational relaxation to the S_1^*NH flu state that is the immediate precursor of the PCF product which is formed diabaticly (nonadiabaticly) competing with ASSflu via formation of the twisted structure (see also below).

The problem of the existence of a "primary" common transient state (structure) on the first path-ways of the ASSflu and PCF structures have arose at the first time /91/(and also the later studies /20/ in the glass-like solvent (T=77K) under the steady-state excitation).

At first, it has been found that the ratio $\Phi^{flu}_{rel}/\Phi^{col}_{rel}$ falls with the increase of the excitation energy (decrease of λ_{ex})(Tab.3.37), where Φ^{col}_{rel} and Φ^{flu}_{rel} are the relative quantum yields of the photocoloration and the ASSflu determined by the increase of the absorbance rate (λ=480nm)(Φ^{col}_{rel}), and the fluorescence efficiency (λ=530nm) (Φ^{flu}_{rel}) correspondingly under the identical conditions of excitation with the different wavelength(λ_{ex}). That dependence can indicate an existence of the common precursor of the fluorescence and PCF structures.

The second indirect confirmation of such a precursor transient states can be the short-lived fluorescence in the short-wavelength region of the ASS flu band($\tau \leq $ 1ps). Due to the lack of the marked Deuterium effect on the vibration frequencies they could be assigned to the active torsional modes bound up with the twisting of the keto-ring A around C_1-C7 bond /92, 93/.

There are many experimental findings have been obtained in the series of the recent studies (from 1990) conducted by the high time resolved spectroscopy (pico and femtosecond scales) that could be considered as if they corroborate such a point of view.

So, it has been shown by the femtosecond transient absorption spectroscopy in the liquid non-polar and polar protic solvents /4-6/that the band with $\lambda^{max} \approx 420nm$ responsible for the $S_1^* \to Sn^*$ transitions, which attributed to the NH* structure, shifts towards the "blue" and is narrowing with the rate with the time rate about 400-600 fs immediately after excitation due to the solvent assisted intramolecular Vibrational Redistribution (IVR) bound up with relaxation of the vibrational "hot" molecules to the equilibrium fluorescent S_1^*NHflu structure. (However latter /11/that shift have not been found, and the band was attributed towards $S1^* \to Sn^*$ transition in the E structure.)

The high vibrational states can also be responsible for the fast short wavelength fluorescence and according to/5/ relaxated simultaneously to the both "cold" ASS flu state and the photochromic product (s)(Sch. 3. 17)/5/.

The IVR and PC formation can be bound up with the torsional vibrations and rotation around C_1-C7 and C=N bonds(see also/45/) although according to /5/there is no direct correlation between the kinetic parameters and the solvent nature.

The search and investigation of the transient states in the photochromic reaction in the gas phase of Anil have been carried out by the two absolutely different methods /9, 23/.

The findings of the dispersed double band ASS flu spectra of the jet –cooled SA molecules with separation $\Delta v \approx 700-1000 cm^{-1}$ and the results of the ab-initio calculations may be embodied into the scheme(Sch. 3. 18)/9/.

According to the scheme the ESIPT reaction may generate a cis-Keto-(Kpl*cis)conformer in the vibrational "hot" S_1^*state that can easily be converted to the twisted ($\alpha \approx 72°$) cis-Keto (Ktw*cis).

The both keto-forms which are responsible for the ASS flu including the short-wavelength fast component and the twisted structure are predicted to be a precursor of the Ktrans structure that is considered as the photochromic one.

Thus the hot Kpl*cis state can be the common transient precursor for the both ASS fluorescent and PC structures.

The data obtained by the time resolved (femtosecond) REMPI spectroscopy(see Sec. 3. 2. 3)/23/ point to the extremely sharp shortening of the $S_1^*(\pi\pi^*)$NH state decay time with the decrease of the E-structure excitation wavelengths in the range of λ=370-365nm with the probe wavelength in the range of the efficient ionization of the S_1^*NHflu state(λ=365nm)(Tab. 3. 40).

The decay time dependence on the excitation energy has been attributed by authors to the opening of the IC channel from the $S_1^*(\pi\pi^*)$ state to the $S^*(n\pi^*)$(or $S^*(\pi n^*)$one of the cis keto(Kcis) form(Sch. 3. 19).

According to the prediction of /7/the S_1^*NH$(n\pi^*)$ structure stabilized by the A-ring twist leads to the K tans structure in the Ground state which is regarded /23/as a PCF structure. Therefore the "hot" vibrational levels of the S_1^*NH(or Kcis) structure can be a common precursor of the both fluorescent and PCF structures.

In the thermochromic crystals /29, 69/the "fast" fluorescence decay is observed in the higher energy region of the ASS flu band while the fluorescence decay in the lower energy region is relatively slow (just as in the solvents /93/).

At the 77K the decay time of the "fast" fluorescence (e. g. of 5ClSA, 3Me5ClSA) and the rise time of the "slow" fluorescence coincide practically (τ(decay)=0. 76±0. 05ps; τ(rise)=0. 72±0. 05ps). That coordination strongly suggests the fast decaying state is the precursor to the slow decaying (S_1^*NHflu) that. However the nature of such a precursor state and if it is the precursor of the PC product remain unknown.

Meanwhile one should note that fast short-wavelength fluorescence is not observed when excited in the all region of the flat Kcis (thermochromic) structure absorption band. And from the above mentioned viewpoint such a structure has no vibrational state responsible for the PC structure formation and does not responsible for the ASS flu.

Thus the date produced above are favorable to the idea of the hot vibrational states of the $Š_1^*$NHflu state differed from S_1^*Kcis one can be a precursor of the PC product.

At the same time one can believe that the role of transient precursor state can be played by the weakly stabilized flattened transient structure (S_1^*NHtr) responsible for the fast fluorescence rather than the vibrational $Š_1^*$NHflu state.

Indeed the detailed study of the S_1^*PES topography by the semiempiric PM3 calculations with the utilization of the gradient method /14/ allows to find such a structure stabilized by the low potential barrier ($\Delta E \approx 5$ kcal/mol) which could be a precursor for the S_1^*NHflu and PC structures (Tab. 3. 41a, Sch. 3. 20). It is placed higher by ~3000 and 8000cm^{-1} of the S_1^*Enol and S_1^*NH states correspondingly and can be responsible also for the fall of the ASS flu and photocoloration quantum yields' ratio(Φflu rel/ Φcol, rel)with the decrease of the excitation wavelength.

The variation of the stability of the S_1^*NHtr structure can also provide the dependence of the coloration and the ASSflu efficiencies on the molecular structure and media.

At last the quite different mechanism based on the results of the ab-initio calculations has been suggested /7/(Sch. 3. 21).

The FC $S_1^*(\pi\pi^*)$ planar Enol state is generated by excitation and deactivated to the unstable thermalized $S_1^*(\pi\pi^*)$ state which turns by the tunnel ESIPT mechanism(Sec. 3. 2) immediately into the $S_1^*(\pi\pi^*)$ state of the planar Kcis tautomer without any NH transient electronic or vibrational states. That S_1^*Kcis $(\pi\pi^*)$ state is responsible for the ASS flu and can undergo the change of orbital character to the $S^*(n\pi^*)$ state by utilize of the excess of the vibrational energy. At such a twisted conformation ($\alpha \approx 80°$) of the Keto-tautomer is very well "prepared" to adopt the planar Keto-trans (Ktrans) configuration regarded by authors as a photocolored structure.

Thus the above theoretical data suggest that the common precursor of the fluorescent and PC-structures could be identified with the low stable planar $S_1^*(\pi\pi^*)$ state of the Enol tautomer. Practically the same situation can be realized for the SALKn and SALKnPh structures /27, 28/, but the mechanism of /7, 27, 28/ does not explain the experimental findings of the relative efficiency dependence on the excitation wavelengths and of the fast ASSflu/20, 91/ discussed above.

Thus according to the above discussion, the ideas about the primary common precursor of the fluorescent and PC structures may correspond to the following states:

i) The vibrational excited ("hot") states of the fluorescent NHflu structure, (after ESIPT).
ii) The low stable flattened (S_1^*NHtrans) structure formed immediately after the ESIPT, and iii) The low stable planar $S_1^*(\pi\pi^*)$ Enol(OH) structure which fluorescence competes with the ESIPT.

However according to the most of the experimental data (see e. g. /20, 91/) just the post-ESIPT state (i, ii) has to play the key role in the ratio of the efficiency of ASS flu and PC generation.

On the other hand the important criticisms has been declared against the experimental evidences for the common transient precursor /8/ which can be supplemented with the new details:

1. The fast short wavelength component of the ASSflu /92-94/ is not be the direct evidence of the PC precursor existence since it is bound up only with the dynamics of the NH structure but not with the PC structure formation.

Really, one can be added that the decay time of the fast fluorescence, $\tau(decay) \approx 12ps/93/$, doesn't correlate with the delay time of the appearance of the PC structure fluorescence, $\tau(delay) \approx 170ns/2/!/$, and observed by the REMPI spectroscopy sharp shortening of the $S_1^*(\pi\pi^*)$ NH vibration states' decay time (see above)/23/ can't be caused by the formation of the PC structure because the latter is not generated in gas phase /9/.

2. The fall of ratio Φflu, rel/Φcol, rel with the decrease of the excitation wavelength (λ^{ex}) (Tab. 3. 37)/20, 91/ is observed within the very wide spectral range involving not only the S_1^* vibrational states but also those of the higher electronic energy(e. g. S_3^*) of E structure (Tab. 3. 37)

In that case of the diabatic (nonadiabatic) path-way of the immediate So state PC population (Sn*~~>So) (like methyl salicilate /66/) omitting S_1^*NHflu state), allows to explain the experimental Φflu, rel/ Φcol, rel dependence on the λ^{ex} with the idea about common precursor as a state (transient structure) connected with the ICT from the imine ring on the C=N group (see above) only in SA molecule.

That mechanism includes also a probability of the population of the short-lived state which is not bound up immediately with the PC structure generation but can be responsible for shifted ASS flu in SA molecule.

Really, the Φ ratio value dependence on excitation energy (λ^{ex}) typical for SA (Tab. 3. 37) has not been found for SALKn (Tab3. 39) in spite of the close likeness of their fluorescent and photochemical properties in the rigid matrices /20/.

Hence the observed wavelength dependence reflects the difference between SA and SALKn in the imine moiety structure and does not play evidently the key role in the mechanism forming the fluorescent and photochromic properties of Anils.

Thus in accordance with the data discussed above the additional primary common precursor for the ASS flu and PC structure can be situated on the PES of Sn* state of SA molecule before the area on the SA S_1^*PES of the pre –TICTor of the conical intersection of S_1^* and S_0 states in dependence on the mutual disposition of the S_1^* and S_0 PES.

In such a situation in the molecule of SA the both precursor on the path-way including the S_1^*NHflu state as a immediate precursor for the PC structure, seems to be the most probable.

There are at least two impressive experimental facts in favor of such a mechanism.

i) The longlived S_0 transient attributed to the trans-keto PC structure observed after excitation of the E form in the solutions /2/ and the low –temperature matrices /79/ has not been detected in the supersonic jet-cooled gas phase ("isolated state") /9/ i. e. under conditions unfavorable to the population and deactivation of the S_1*NHflu state by a medium assisted vibrational relaxation ($\check{S}_n^*$ ~ S_1^*NHflu *NHflu and S_1^*NHflu~>So correspondingly).

One can believe under such conditions the radiative deactivation (ASSflu) of the low populated S_1^*NHflu state competes successfully with the radiationless deactivation suppressing just by that way the PC structure generation (see below).

ii) The effective but different quenching is observed of the both ASSflu and PC structure formation by the S_1^*energy acceptor (diethylaminonantraquinonethilazole) with the S_1^* energy (17000cm^{-1}) that is lower than the S_1^*NHflu (18500cm^{-1}) (Tab. 3. 38) /20/.

The difference between the ASSflu and PCF quenching efficiencies is the argument of just the S_1^*NHflu state participation in the quenching process unlike the similar quenching efficiencies expected when participating of the S_1^*Enol state only.

The fact of the quenching of PCF structure generation by itself is the strong evidence that the S_1^*NHflu state is the precursor of PCF structure. The note should be taken that the results of the quantum-chemical ab-initio calculations also show distinctly PCF generation has to go via S_1^*NHflu (S_1^*pl($\pi\pi^*$)NH) state /7/ (see above, Sch. 3. 21).

Nevertheless the lower efficiency of the PCF product quenching in comparison with the ASS flu shows there is also the another path-way of the PC product formation that does not go via S_1^*NH flu state.

Meanwhile, the very clear kinetic findings concerning the post ESIPT relaxation process including the generation and the decay of the S_1^*NH flu structure have been obtained by the improved method of the time resolved fluorescence and the transient absorption spectroscopy in the pico-and femto-second time scales. According to / 10, 11, 36/(Tab. 3. 41b) the kinetic findings of the post ESIPT processes for the anils' structures under the definite fixed conditions (medium, temperature, excitation wavelength) are described mainly by the set of the three time constant $\tau_1 < \tau_2 < \tau_3$ corresponding to the different spectral domains.

The population (extra-population) of the S_1^*NH flu state takes place as an immediate result of the ESIPT with impulse duration and with rise time constant $\tau_1 \approx 0.04$-0.15ps for the different anil structures under various conditions, and proceeds obviously without marked contribution of the vibronic relaxation time.

Such a generation of the S_1^*NHflu state is typical for the low energy (longwavelength) excitation (λ=385-400nm) of the unsubstituted structure with the strong solvent assisted vibronic redistribution in the liquid solvents.

The strong rise of the population time as a result of the increase of the vibrational relaxation time in the different cases has been considered above.

Furthermore the marked contribution of the vibrational relaxation and its dependence on the excitation energy can be observed even for the SA molecule in the liquid solvent by the careful study with the absorption femtosecond spectroscopy: the time of the extra-population of the NH*flu state (rise time τ_1=0.4 and 0.8 ps that observed /11/ for the excitation by λex=390 and 266nm respectively can be explained by the generation of the S_1^*NHflu state according to/11/ via "hot" excited vibrational states or more probable, via the high transient state Ntr*(see below). The analogous and more distinct kinetic peculiarities are observed also for the compounds 2p and 4p in the crystals (rise time τ_1 in Tab. 3. 41b)/36/.

The fast decay ($\tau_3 \approx 4$-5ps) of the S_1^*NH flu state of SA can be caused by the both the "vertical" deactivation (IC, ISC and ASS flu) and "horizontal" adiabatic process towards the twisted quenching state (see below).

The considerable increase of the time constant τ2 is observed with the insertion of the bulky substituents in the aldehyde ring (structure 2p, 4p)especially for the thermochromic crystal(2p) with the inhibited aldehyde ring(A)twisting (Tab. 3. 14b), and also the quantum efficiency of the ASS flu increases sharply in the rigid media and of the model molecules I(Ph], Ph or Alk with the fastened aldehyde ring (see e. g. /12/)(see below).

Therefore the "horizontal" deactivation seems the most preferable for the efficient decay of the S_1*NHflu state. The typical feature of the ASS decay kinetics is the shorttime component(τ2) (Tab. 3. 41b)that is displayed within the limits of the spectral band of the ASS flu (emission spectra)or the negative band of the transient absorption spectra.

This time component is bound up with the shortwave-length band region, and can be attributed to the very low-stable transient "second" keto-structure NHtr*of the higher energy that is populated and deactivated irrespective of the S_1*NHflu state.

This state is obtained for SA(τ2 ≈ 0. 5ps)only at the short-wavelength excitation 266nm by the transient absorption spectroscopy method/1/ whereas it can be recorded even at the much more longwave excitation, λ=385nm, by only more sensitive time resolved fluorescent spectroscopy method for SA and SN (τ2=0. 3-0. 5ps)/10/and hence it has a higher probability of the population at the higher energy.

Such a state is considerably stabilized by the bulky substituents in the aldehyde ring even at the more low energy (with excitation at λ=390nm)in the solvent(τ2=3-5ps for the structures 2p, 4p), and especially in the thermochromic crystals(τ2= 23-25 ps for the same structures) where A-ring rotation is strongly inhibited.

Hence the stabilization of the S_1*NH*tr structure can also been bound up with aldehyde ring twist around the C_1-C7 bond.

According to the earlier calculations /14/just that transient state is 3000 cm^{-1} higher the S_1*enol structure of SA (see above), and can be efficiently populated only with the impulse excitation by λex=266nm(but not λex =390nm)followed by the marked population of the S_1^*NH flu state with the ASS fluorescence.

However the considerably more weak steady-state excitation with λex=266nm results mainly in the short-wavelength($\lambda \approx 450\text{-}480nm$) fast fluorescence with the very weak population of the S_1^*NHflu state (see fig. 2/11/).

It is easy to see for some anil molecules, the decay time of the NHtr* structure is very close to the additional rise time of the S_1^*NHflu state ($\tau_2 \approx \tau_1$)that is the direct evidence for the precursor nature of the former just as in the crystal 5ClSA and 3Me5ClSA (see above) /29, 69/.

Thus such a general picture is formed on the base of the vast results discussed above.

The nonplanar S_1^*NH flu structure responsible for the ASS fluorescence is generated mainly ultrafast ($\tau < 100fs$)as a result of the ESIPT ($\tau \leq 50fs$) in the planar Enol structure followed by a twist of the A ring along the PES towards the S_1^*NH flu structure without a significant contribution of the vibrational relaxation

The slower population of the equilibrium S_1^*state($\tau_1 \approx 10^2 n$ fs) can occur after the ESIPT via relaxation from the hot vibrational states and redistribution of the vibrational modes in the S_1^*NHflu structure.

At last the additional slow population from the precursor state $\tau_1 \approx (0.1\text{-}10)n$ ps(Tab. 3. 41b) is possible. This is very short lived instable almost planar transient ($S_1^*NHtr^*$)state with the high energy which stabilization relatively the S_1^* Enol state depend strongly on the molecular structure.

The $S_1^*NH^*tr$ state can be populated from the (S_1^*) or (and) S_1^*Enol states of the C_1 and Cs symmetry correspondingly immediately after the ESIPT. This structure is responsible for the fast short-wavelength ASSflu of arylimines (SAs) and connected probably with ICT from the imine ring(see above), and therefore is absent of alkilimines (SALKs).

Although such a transient state can be precursor of the S_1^*NH flu structure it cannot be obviously an immediate precursor on the pathway of the PCF generation(NH*tr τ(decay) /10, 11, 36, 93/<<PCF τ(rise)/2/), (see also/23/ and above).

The low-stable S_1^*NHflu structure, responsible for the ASSflu, decays fast ($\tau 3 \approx$ (1-10)n ps) towards the quenching twisted Zwitterionic(Zw) structure, owing to the adiabatic rotation of the A-ring along the dihedral angle (α) coordinate (around C_1-C7)bond, and may be considered with a high probability as a direct precursor for the twisted S_1^*NHTICT-like state followed by the ground state "post-TICT" or "post-CI" colored structures (see Sec. 3. 4. 2).

However according to the experimental findings /20, 91/(Tab. 3. 37 and 3. 38)the S_1^*NHflu structure of SA (unlike SALK) cannot be only precursor of the PCF structures.

It is obviously, the detailed studies are required in this direction.

3. 4. The nature of the Photocolored product.

3. 4. 1. The present ideas about structure and characteristics of photocolored form.

The most typical feature of the Photocolored form(PCF) is the absorption spectra band in the visible spectral region that is caused by the So-abs->S_1^* transition responsible for the fluorescence transition S_1^*-flu->So.

The absorption and fluorescence PCF bands in the solvents (Tab. 3. 42) have the approximate mirror symmetry and the double band vibronic structure with the split-ting depending a little on the solvent viscosity (temperature) especially for the fluorescence band characterizing a distribution of the Ground state vibrational modes.

Such vibrations can be connected with the stretching modes of the bonds typical for the K-structure, and correspond to the IR and RR spectra (sees, e. g. /71, 95/). The fluorescence bands of the Kcis, NHflu, and PCF structures are very similar. The similarity of the band shapes and regular alteration of the fluorescent parameters testify to the structure likeness, and the regular change of the conformation, probably, the increase of the angle in the above structure series (Tab. 3. 29).

There are two convincing experimental illustrations of the localization of the $S_0 \Leftrightarrow S_1^*$ transiti-ons on the Keto-moiety of the colored structure: the identity of the PCF absorption spectra of

SA and SALK molecules /20, 37/, and no participation of the N atom orbitals in the $S_0 \to S_1^*$ elec-tronic transition according to the findings of the Resonance Raman spectroscopy /71/.

The identical localization of the electronic transitions $S_0 \to S_1^*$ (21000-22000 cm^{-1}) in the sugges-ted nonplanar structures of SA and SALK is corroborated with the results of the quantum-chemical calculations /2/(see also below).

The considerable increase of the Stokes Shift in the liquid solvents in comparison with the rigid media shows the conformational alterations in the S_1^* excited state of the PCF inhibited by the

solvent viscosity. Thus the spectral data reveal some isolation of the structural fragments and flexibility of the PCF structure.

The position and shape of the PCF absorption bands in the crystals and solvents differ a little (Tab. 3. 42, 3. 43). At the same time the molecules that show unlike SA both photochromic(PC) and thermochromic (TC)properties in the crystal (e. g. /39/, /96/) can be a good indicators of the distinction between the PC and TC structures.

It results from the findings presented (Tab. 3. 43) that vabsorption bands of the PCF of the molecules 3—7 just as of SA are located in the consi-derably more longwavelength region than the TC ones of such molecules (just as for SAP) which have the planar Kcis structure (see also Sch. 3. 22a and Ch. 2).

The product of photochromic reaction (PCF) cannot be isolated from the colored crystal or the rigid medium due to its surface localization and fast decay at the ambient temperature, and so the direct identification of the PCF molecular structure is the very complicated problem (see be-low).

Therefore about the PCF molecular structure one can judge by the indirect experimental, structural and spectral, comparative findings including those for the model molecules, and the data of the quantum-chemical calculations.

Such a PCF structure with the longwavelength absorption or reflection spectral bands ($\lambda \approx 480$, 520nm)unlike the TC structure those($\lambda \approx 420\text{-}460$nm)(Tab. 3. 43)has to be sufficiently stable along the path-way of the reenolization (bleaching reaction) by the Ground State Proton Transfer (GSIPT)reaction.

The idea of the planar trans-keto (with respect to C_1--C bond) structure (Ktr) which meets un-like Kcis the condition of the stability, has been suggested in one of the pioneer work (see i. e. /97/).

From that time to the beginning of 90th the overwhelming majority of the authors believed that the photocolored product had the planar trans-keto (or the corresponding hybrid Zwitterionic (Zw)NH) structure) ($\alpha \approx 180$ °, $\beta \approx 0$°, y =0°)(e. g. see /22, 81, 98/).

The structure absorption, fluorescence spectra, and kinetic characteristics of generation and decay of the PCF have been studied from the early nineteenths until the present in the rather great number of the investigations by the up-to-date spectroscopic, kinetic methods, and by the quantum-chemical calculations of the various modern levels.

The most of the typical findings obtained are produced and discussed below(Tab. 3. 44, Sch3. 22b)

The almost planar NHtr, cis and planar NHtr, tr structures have been suggested/2/ on the base of the results of the MMX molecular mechanics method and INDO/S calculations for the determination if the So-->S_1* transition energy($v \approx 21\text{-}22000\text{cm}^{-1}$), which is localized mostly on the Keto moiety. The fluorescent PCF structure (Tab. 3. 42, 3. 44) is observed by only sequential two step laser excitation (λ_1=355nm and λ =495nm) with a delay time about 170ns.

Thus this PCF structure is generated after the ESIPT by the diabatic conformational transformations (S_1*NH flu-->SoPCF) with the rate constant Kcol$\approx 6 \times 10$ s^{-1} (Tab. 3. 44), and NHtr, cis structure can be a precursor of the planar PCF of the NH tr, tr structure /2/.

The findings of the decay rate are very scattered and depend strongly on the solvent nature and concentration. The NHtr, tr structure with a large Zwitterionic (Zw)contribution is suggested on the base of the results of calculations by INDO/S and Molecular mechanics methods and transient absorption spectroscopy/80/. The increase of the PCF structure lifetime in the polar solvent (CAN) with respect to nonpolar one (3MP) by three order of magnitude is explained by the contribution of the Zw structure.

Thus one can conclude /2, 80/in the liquid solvents the only resonance K <->Zw structure of PCF is observed with the decay time rate in the 10-100 microseconds time scale which depends strongly on the solvent polarity.

On the base of the comparative kinetic data for SN and the related model rigid molecules (HBC and BHBC)(Sch. 3. 23)/45/ the generation and stabilization of the two different structure of the PCF have been established and attributed to the twisted Ktr, cis (Ktw α =90°)structures without any kind of the quantum-chemical studies. According to /45/ the both structures are generated consecutively from the fluorescent Kcis, tr*(Kcis*)state via the diabatic formation of the SoKtr, tr(Ktr) PCF structure (Kcis*->Ktw*->Ktr). As a result, the intensive longwavelength fluorescence (Tab. 4. 42) is ascribed to the Ktw* structure that is hardly probable for such a twisted one.

At the same time Ktw structure is observed in the steady-state experiments at T<100K but the Ktr formation prevails at T>140K.

The later reenolized by the second-order double proton transfer reaction in dry nonpolar solvent with the H-bond dimerization or by the proton catalyzed first-order reaction in the presence of alcohol with the strongly shortened (by two orders of magnitu-de) lifetime. This flat Ktr, tr structure is responsible for photochromism (Tab. 3. 42) and formed after the ESIPT from Kcis* state apparently diabatically (Kcis*S_1*->Ktr(So)) with efficiency of 0. 1-0. 3, and decay with lifetime τ>>0. 1ns.

In the series of the studies/4-6/ the planar NHtr, tr polar resonance structure is supposed on the base of the experimental(transient absorption and time resolved fluorescence spectroscopy)data and the quantum-chemical calculations (AM1), which results overestimate the value of the first electronic transition energy and indicate its localization on the Keto-moiety including C=O group.

The time rate value of the diabatic generation of the colored structure from the vibrational S_1*Kcis structure is in the 0. 1ps time scale and depends markedly on the solvent nature /6/ (Tab. 3. 44, 3. 45, and Sch. 3. 17).

Such a dependence can be explained by the competition between influences of polarity (ε) and viscosity (η). The increase of their values can lead to the rate fall and growth correspondingly when forming of the colored structure via the strong twisted high polar precursor state (see below). (Compare e. g. 1 and 3, 3 and 4, 2 and 5 (Tab. 3. 45)).

The influence of and are prevailed in the first and the third cases respectively and almost equalized in the second one. The back reaction of reenolization with the time rate $\tau>3ns$ includes the two stages –the trans-cis isomerization around the C--C bond and the reverse GSIPT reaction with the calculated barriers $\Delta E \approx 28$ and 21 kcal/mol correspondingly.

Such high barriers although are smaller than the activation energy for the trans-cis isomerizations around C=C bond ($\Delta E \approx 30\text{-}40$ kcal/mol) but exceed considerably the numerous experimental findings obtained in the earlier studies ($\Delta E < 7$ kcal /mol)/91, 102-104 / and don't meet the high rate of the dark back reaction (k< 3×10 s^{-1} i. e. $\tau> 3ns$) determined by authors.

In this connection the realization of the trans-cis isomerization with respect to $C_1=C$ bond as the stage of the discoloration is questionable. However the probability of the alternative intermolecular mechanism of the dark bleaching reaction /45/has not been discussed.

The expected stabilized planar Ktr, tr structure predicted by the calculation has not been however detected in the gas phase/9/due to the unfavorable conditions for the population of the precursor S_1*NH flu state(see also above).

Therefore the IC to the nonplanar ICT state from the S_1*NHflu one /7/ leading to their dramatic lifetime decrease (from 8. 9 to 1. 5 ps) is not bound up with the PCF generation but reflects the peculiarities of the IC deactivation within S_1*NH struc-ture itself.

According to another mechanism proposed /100/ the formation of the PCF structure following the ESIPT is connected with the hybridization change of the protonated N-atom from sp² to sp³ to stabilize the resultant transient precursor of the PCF.

Thus in accordance with /100/, the PCF has Kcis, tr structure with the planar keto moiety and the imine one twisted around C-N and N-R bonds (Tab. 3. 44, Sch. 3. 22a). Such a structure with the double twisted conformation (α, y) can stipulate a two-stage kinetics with the different rates (see Tab. 3. 51 below).

However such an influence of the electronic factors in the imine moiety on the PCF structure seems to be unlikely owing to localization of the $S_o \rightarrow S_1^*$ transition on the keto moiety of the PCF molecule (see above), and therefore the two-stage kinetics can be explained by another structural factors (see below).

Meanwhile the dramatic fall of the dark reaction rate in the crystal as compared with the solvents testifies to the key role of the considerable conformational transformations in this reaction as well as in the PCF generation which are inhibited by the intermolecular steric interactions in the crystalline lattice along the reaction path-way.

According to the results of the quantum-chemical calculations /7/ the PCF is identified as an al-most planar Ktr, tr structure which is generated directly by the ESIPT with time rate $\tau \approx 200fs-2ps$ from the unstable ES_1^* planar structure via nonplanar $S_1^*(\pi^* n^*(\pi))$ state ($\alpha \approx 80°$)(see also /28/)with the quantum yield of about 6×10^{-3} (see ref. in footnote/6/ of the Tab. 3. 44).

The PCF back reaction of the free molecules goes along the pathway of the Ground state via the potential barrier of 10 kcal/mol for the Ktr, tr-> Kcis, tr isomerization with respect to C_1--C7 bond and the GSIPT with the time rate of 8. 5ps.

The attempts of the direct registration and the structure identification of the PCF have been made by the spectral and the X-ray analysis methods/23, 71, 95/.

The comparative analysis of the IR and Resonance Raman(RR) spectra of the photochromic form of the ^{1}N, ^{1}N, H and D substituted molecules in the crystal /71/ allows to attribute the spectral bands to the C=O, C-N stretching and the quinoid ring vibration modes in the keto-structure.

At the same time the data of the time resolved IR spectroscopy in acetonitril/95/ show the marked ZW structure contribution in the photocolored form unlike crystal caused by the solvent polarity.

As it follows from the findings produced above, the longwavelength electronic transition of the PCF structure is localized on the keto-moiety including the keto-ring with C=O group, and the data of the vibrational spectroscopy/71, 95/ cannot contain any specific information that allows to distinguish the cis and trans isomers with respect to the $C_1=C7$ bond.

Moreover there are no resonance enhanced Raman bands sensitive to ^{1}N substitution in crystal. It indicates that the PCF So->S_1* transition does not involve the orbitals connected with Nitrogen atom/71/, and therefore localized mainly on the ring twisted around $C_1=C7$ bond and isolated from C-NH group.

Thus such findings can rather testssify against planar trans structure and be evidence of the twisted structure of the PCF (see below). As it follows from /95/ the data of the time resolved IR spectroscopy allow estimating the PCF rise and decay time values which conform to the kinetic data obtained by the another methods /Tab. 3. 44/.

The first crystallographic determination of the structural changes accompanying the reversible phototransformations of an organic photochromic crystal has been performed for the one of 3, 5 tert-butyl 3'nitrosalycylideneaniline with help of the two-photon laser excitation (λ=730nm) to provide the deep penetration of the exciting light into the crystal and accumulation of the colored product in the amount sufficient for the X-ray analysis /101/.

The molecular structure, different from the original enol that, is generated as a result of irradiation coexists with the latter in the crystal. In this molecular structure the Oxygen and Nitrogen are in the trans posi-tion with respect to the C_1--C bond, and the bond length have been changes markedly in accor-dance with the E->K structural transformation. The colored crystal returned to the original form after irradiation of the light with λ>530nm which is within the limits of the absorption band of the PCF.

By authors' opinion /101/ such transformation in the crystal can easily occur through so called "pedal" mechanism with the motion of the pair of benzene rings which is analogous to the pedal motion of a bicycle and does not demand a large free volume in the crystal lattice.

In virtue of the data produced the conclusions can be drawn as follows:

i. The idea of the planar NHtrans(with respect of the $C_1=C7$ bond, $\alpha =180°$)PCF structure with the various contribution of the Zwitterion form in dependence on the medium remains of the most preferable hypothesis during the all period from the pioneer works to the present. The such structures have been suggested mainly as a result of the experimental studies by the indirect methods along with the search of the sufficiently stable NH species in the So state by the quantum-chemical investigations.

Only a few studies have been conducted by the direct spectral /71, 95/and X-ray analysis /109/ methods of structure identification. However the detail analysis shows that the data/1, 5 / in spite of the authors' interpretation are not coordinated enough with the idea of the planar NH trans structure, and only the consideration of the some crystallographic data /101/ could be interpreted in favor of the just planar Ktrans structure of the PCF, however, for the molecule with the structure different considerably from SA one(see above).

ii. The photocoloration is provided by the absorption longwavelength transition So-->S_1* with $\lambda^{max} \approx 480-550$nm which is responsible also for the emission S_1*-->So with $\lambda max \approx 560-660$nm (Tab. 3. 42).

The absorption and fluorescence band vibrational structures, the solvent dependent Stokes shift, and the data about localization of the So⇔S_1* transitions together with the findings of the vibrational spectroscopy hardly can testify to the strongly stabilized NHtrans planar structure of the PCF.

Furthermore the calculated values of the band maxima corresponding to such a planar structure is strongly shifted to the "blue" according to the results of the calculations of the different levels as compared with the experimental findings (Tab. 3. 42, last column).

At the same time the photochromic bands in the crystal and the solvents are shifted strongly to the "red" as comparison with the thermochromic and solvatochromic ones in the crystals and the solvents respectively which are identified as absorption of the planar Kcis structure(see Tab. 3. 43 and Ch2).

Thus the most of the spectral data show the existence of the nonplanar flexible structure of the PCF in the So state, and its transformation by excitation in the S_1^* state.

iii. The values of the PCF generation rates (Tab. 3. 44) can be divided into the two groups differing strongly by the 3-4 orders of magnitude in dependence on the attribution of the observed transient absorption bands and (or) of the utilized method of the PCF registration.

The very fast process with the time rate $\tau \approx 10^{-12}$-10^{-13}s is identified with the PCF generation in the solvents when observing of the decay of the transient band bound up with the immediate precursor of the PCF/ 4, 6, 8/or with the theoretical estimation of its decay time/7/.

The much slower process connected with the PCF generation ($\tau \approx 10^{-9}$-10^{-6} s) is recorded by the direct observation of the PCF fluorescence appearance /2/ or by the registration of the IR bands belonging to the PCF structure /95/.

Since the precursor can be deactivated also by additional path-ways, one can believe that the latter (with the slower rate) correspond to the true time of the PCF generation.

Meantime the observation of the precursor can give the useful information about the significant role of the solvent nature in the PCF generation (Tab. 3. 45) that testifies to the marked change of the polarity and conformation of the molecular structure when forming of the PCF product.

Thus according to the produced results the PCF structure recorded by its fluorescence is being generated diabatically (nonadiabatically) from the S_1^*NH state with the yield of 10-30 %. Its accumulation starts at ~200ns and is completed in ~9μs after the ESIPT correspondingly.

So long time formation and its dependence on the solvent viscosity and the polarity can be evidence of the existence of the high-polar (Zwitterion(Zw))transient precursor in the So state that undergoes the marked structural transformations on the path-way towards the PCF structure (see e. g. /36/ and sec. 3. 4. 2.

> iv. The data of the time rate of the PCF structure decay are scattered due to their hardly contro-lled dependence on the solute concentration, molecular structure, and medium nature (see ab-ove, and Tab. 3. 44).

However one can believe that value of the PCF decay time is higher than that for the PCF generation and varies within a very large limits of the time scale 10^{-3}-10^{-9} s decreasing under conditions providing the intermolecular H-bond formation and Proton transfer in the solutions The rate limiting step along the path-way of the reverse dark reaction is the rotation around the C1:::C7 or C 7:::N bonds since the GSIPT in the K-structure has much lower time rate ($\tau \approx 10^{-12}$s/7/), meanwhile the activation barrier typical for the trans-cis isomerization with respect to the true double bond(C=C or C=N)in the planar NH structure is too high in comparison with the experimental findings (ΔE<7kcal/mol) for the reverse reaction.

The different decay times are attributed in all cases to the various trans isomers about C:::;C and C::N bonds with unknown conformations which can be realized in common /2, 8, 11, 80/ or sepa-rately /4, 6/ or /95/with a large /8, 105/or a small difference in the decay time in dependence on the molecular structure, solvent nature, excitation wavelength and other factors.

The time falls sharply when the reverse reenolization reaction NH-->OH occurs via Intermole-cular proton transfer in the Keto-trans dimmers or in the alcohol –trans-Keto complexes /45/ avoiding the step of the rotation around $C_1=C\ 7$ bond which is necessary for the Intramolecular (GSIPT) back reaction.

Meanwhile the dramatic rise of the decay time rate in the crystal/36, 71, 100/(Tab. 3. 44) also testifies to the key role of the conformational variations inhibited by the steric interactions in the crystalline packing especially of the structures with the bulky substituents /36/.

It is obviously that the spectral and kinetic data indicate clearly the flexibility of the PCF structu-re and the key role of the structural transformations along the path-ways of the photocolored and bleaching reactions.

However many of them cannot be explained by the hypothesis about NHtans planar structure of the PCF. And what is more, according to the spectral findings, the PCF structures of SA molecule and their derivatives have to have the absorption bands in the considerably more red region ($\lambda^{max} \approx 480, 520nm$) than the thermochromic planar Kcis struc-ture($\lambda^{max} \approx 420, 520nm$) (Ch. 2 and Tab. 3. 43).

Meanwhile the analysis shows that the planar Ktr structure does not meet such experimental conditions since the spectral data calculated by the methods of the various levels don't reveal the necessary positive difference between the λ^{max} of the absorption bands of the planar Ktrans and Kcis structures mainly at expense of the strongly underestimated calculated values of λ^{max} ($\Delta\lambda \approx 30-200nm$) for the trans structure (Tab. 3. 46).

That discrepancy of the relative values ($\Delta\lambda$) for the calculated and the experimental findings cannot be caused by the methodical errors of the calculations, and reflects the fact that the flat K trans structure has no absorption bands in the more red region than Kcis one.

Thus the idea of the NHtrans structure as a single one for the PCF of the SA molecule and its simple derivatives does not meet both the kinetic and the spectral data.

3. 4. 2. The hypothesis of the twisted PCF structures generated via the "TICT-like" state.

In the situation described above one could be expected that the PCF has a nonplanar structure in which $So \rightarrow S_1^*$ transition is located on the Keto moiety(see above) having the A ring twisted around $C_1:::C_7$ bond ($\alpha >> 0°$).

Indeed, the calculated estimations of the dependence of the absorption spectra on the NHform structure by the INDO/S methods have showssssn that the longwavelength absorption band suffers the strong "red" shift with the angle α change from 0° to 90° relatively to the planar structures with $\alpha = 0°$ (Kcis, $\lambda^{max} \approx 430nm$) and $\alpha = 180°$ (Ktrans, $\lambda^{max} \approx 415nm$) correspondingly with the maximum ($\lambda^{max} \approx 645nm$) at $\alpha = 90°$ (Tab. 3. 48) /12, 106/. Moreover for the planar Ktrans and Kcis structures λ^{max}(Ktrans) < λ^{max} (Kcis) that contradicts experimental findings.

At the same time from the data discussed above it follows that there is the most probable and effective path-way of the formation of such a nonplanar structure following the ESIPT.

Really on the one hand it has been shown(Sec. 3. 3) that the S_1^*NHflu state is the precursor of the PCF and its main path-way of decay goes via the twisted "TICT-like "state (see Sec. 3. 3. 1), on the other hand there are no any zones of the $S_1^*$and So PESs which are close enough(or even undergo conical crossing in the area of the S_1^*NHflu state along the main torsional coordinates but the angles (A-ring twist)(Sec. 3. 1, Tab. 3. 25, Sch. 3. 13)/13, 14, 35/ to provoke an effective diabatic reaction of the PCF generation in the So state from the S_1^*NHflu structure.

In such a situation one should suppose that the principal path-way of the PCF generation is that via the "TICT-like" structure which therefore is the immediate precursor of the PCF structures in the So state.

With the conical intersection (avoided crossing in the case of the angle α)The realization of the diabatic population of the PCF ground state structure is conformed directly by the observation of the two-step laser induced PCF fluorescence (λ^{ex}=495nm, $\lambda_{flu} \approx$600nm)with the delay time about 170ns after excitation of the E structure (λ^{ex}=355nm)/2/ which coincides with the time of PCF structure appearance by its direct observation by the method of the time resolved IR spectroscopy/95/.

Indeed the detailed investigations of the So state PES maximum in the "TICT-like "region by both semiempirical (PM3) /13, 14/ and ab-initio(HF and CIS/6-31G)/35/methods show the existence of a clear potential pit with the barrier about 5 kcal/mol(Sch. 3. 24) that restricts the change of the angle within a narrow vicinity of 90° without variations of another coordinates.

As it has been discussed above (Sec. 3. 24)such a potential pit can be formed as a result of the avoided crossing of the S_1^* and So PESs in the TICT-like region (Sch. 3. 24).

Such a peculiarity, typical obviously for these compounds, allows to form with the time rate about 170ns the nonclassic, twisted, so called "post-TICT", structures with the low stability which can be stabilized in addition by the rigid media (Sch. 3. 24).

Apparently, for the creation of such a PCF structure in the So state an A-ring twist must be provided by the formation of the "TICT-like "structure in the S_1^* state followed by the "post-TICT" structures in the So state (Sch. 3. 13, 3. 24).

Indeed, no photochromism of the rigid model structure I (Sch. 3. 28)under any conditions is ob-served/12, 84/. The unusual behavior of the molecule III (Sch. 3. 26)for which no photochromism is observed /99, 100/in spite of the strengthening of the H-bond/52, 53/can probably also been explained by the lack of the quenching "TICT-like" state in such structures (see Sec. 3. 3. 2. 4), and that phenomenon needs subsequent studies.

However unlike the "TICT-like" structure (S_1^*state), the appreciable stabilization of the low-polar twisted structure in the Ground state can require the additional stabilizing intramolecular steric interactions arising due to the trans-cis isomeri-zation around C-N bond. Indeed, no any stable photochromic form like that of SA is observed in the model rigid structure II/12, 45/(Sch. 3. 26).

At the same time, the $SALK_{11}$ and $SALK_1$ molecules with the long Alkyl substituents (R=$(CH_2)_{11}CH_3$ or $(CH_2)_{13}CH$)have the absorption band of the PCF very similar to that of SA/37/ unlike the SALKo molecule with the short R radical(e. g R=CH_3).

The strongly blue shifted photochromic transient absorption band of the latter is observed only for laser excitation /38/, but is absolutely not displayed under the steady-state conditions even in the rigid media at 77K /37/.

Thus the generation or the PCF structure with a twisted A-ring is promoted by the steric interactions between the A-ring and the imine moiety due to the trans-cis isomerization about the C:::N bond from the precursor $post_1$TICT structure of the NH form of SA (and SALK) Sch. 3. 24, 3. 25).

Therefore the typical PCF with the extended structural absorption band ($\lambda^{max} \approx 480$ and 520nm) is apparently a set of the nonclassic twisted cissoid structures stabilized by the steric interactions and the rigid media. Such structures have series of the specific properties(Sch. 3. 25, Tab. 3. 48).

The strong intramolecular steric interactions between the keto and imine structure fragments arise when molecular moving from the $post_1$TICT state towards the equilibrium $post_2$ TICT state
along the concert coordinate(Sch. 3. 24), involving a comparatively small change (decrease)of the α angle (flattening of the keto-fragment), and the sharp change (increase) of the angle β (trans-cis isomerization with respect to the C::N bond).

The strong steric interactions between the structural fragments arising in the equilibrium $post_2$ TICT state prevent the subsequent motion along the concert coordinate the decrease and the increase of the angles α and β respectively.

The similar phenomenon is revealed earlier for benzylideneaniline (BA)/107/and recently for SA/11/ along the path-way of the trans-cis isomerization about C=N bond in the E-structure.

The position of the longwavelength absorption band is determined by the value of the angle α in the equilibrium $post_2$TICT structure which stability depends on the structure and the size of the imine moiety.

Thus in SALKo with the short alkyl radical(see above) the $post_2$ TICT structure, with the angles $\alpha \approx 60°$-$70°$, and $\beta < 175°$, is very instable and has the photochromic absorption band in the more shortwavelength region ($\lambda^{max} \approx 454$nm, see/27/). However in SALKn with n=11-17 the $post_2$TICT structure is considerably stabilized by at the angles α =70-75° and β =175-180° that causes the absorption band positions in the region at λ=480-520nm in accordance with the calculated (CNDO/S /13, 14/, Tab. 3. 47) and the experimental (Tab. 3. 48) findings.

It is obvious the position and the shape of the broad absorption band have to be very sensitive to the steric interactions caused by the influence of the bulky substituents which can stipulate an appreciable red shift of the absorption band when even small increasing of the angle α within the region 80°-90°($\Delta\lambda/\Delta\alpha \approx 5$nm/deg, Tab. 3. 48).

Indeed, a new colored structure with a red shifted extended absorption band having the fine vibrational structure (λ^{max}=550, 595nm-$\Delta\nu \approx 1300$cm^{-1})(PCF-II structure, Tab. 3. 39)is generated with a marked efficiency in the molecules having the tert-butyl substituents in the imine ring B only (Tab. 3. 49, structure (iii)), side by side with the normal PCF-I structure typical for SA. For such a structure the value of the angle α corresponding to the position of new bands can be estimated within the limits 75-80°(Tab. 3. 48) /32/.

Insertion of the bulky substituents in the A ring only or in the both rings does not lead to the appearance of the new photocolored structures but results in to the strong increase (three-se-ven times) the efficiency of the normal PCF-I generation (Tab. 3. 49) that is not bound up with the transformations of the NH form but can be explained by the inhibition of the structural transformations in the Enol structure competing with the ESIPT (see Sec3. 1. 1).

The twisted $post_1$ TICT and post2TICT structures (Tab. 3. 47) differ one from the other considerably only in the dihedral angles α and especially β.

In the both post TICT structures the lengths in the A-ring, CO, and CN bonds are markedly closer to the Keto structure than to the Zwitterion TICT one(Tab. 3. 47b)that is in good agreement with the findings of the IR and RR spectra /71, 95/ (see above).

At the same time the C_1-C7 bond is stretched out strongly both in the TICT-like and post TICT structures that is typical for the twisted ones, and in these structures the orbitals of the N atom are not involved in the S_o-S_1^* absorption transition as it follows from the experimental data /71/.

On the other hand the small dipole moment (3-4D) (Tab. 3. 4) of these structures testifies to a negligible charge separation between the twisted keto and imine moieties unlike the TICT-like structures. The discussed above characteristics of the post TICT structures together with a marked contribution of the very close T_1 state ($E(T_1)<3000 cm^{-1}$) can indicate the biradicaloid spin distribution typical for the homolytic break of the C_1=C7 double bond unlike the high-polar "TICT-like" state with the large contribution of the Zwitterionic stru-cture O^-NH^+(Sch. 3. 13, 3. 25).

It has been shown above (Sec. 3. 4. 1.), the data obtained recently by the special methods of the vibrational spectroscopy /71, 95/ and especially by the X-ray analysis/101/ allow to suppose that for SA the generation of the "true" twisted PCF structure with the longwavelength absorption band ($\lambda max \approx 480$, 520nm)is accompanied also with the appearance of the another NHtrans (about C_1-C7 bond) planar structure which does not cause the "red" photocoloration of SA.

According to the results of the quantum-chemical calculations /12, 106/(see also Tab. 3. 48)such a photoproduct has the absorption band in the spectral region differing a little from that of the planar Kcis structure responsible for solvatochromism(SC) and thermochromism (TC). Such a planar structure can be generated along the competing path-way 2(Sch. 3. 24)without trans-cis isomerization with respect of C=N bond over potential barrier about 5 kcal/mol for SA.

However the barrier height and comparative efficiency of the two competing path-ways of the generation of the twisted post TICT and planar NHtrans structures depend on the molecular structure and media.

One of the examples is the primary formation of the flattened NHtrans structure in SALKo with the short (CH_3)radical unlike SALKn with the long one (($CH_2)_{11}$or$(CH_2)_{17}$-(CH_3))/37, 38/ has been given above.

Another example is the influence the PCF structure of the ring conjugation(Tab. 3. 50, Sch. 3. 28) /48/.

The PCF absorption band positions of the structures I(5, 6Np, R) are in the same region or even shifted towards "blue" in comparison with the SC or TC absorption bands (see the column "Benzene" in Tab. 3. 50).

The position of the PCF bands depend a little on the imine moiety structure and shifted markedly towards "red" in the polar solvents unlike the PCF absorption band of SA and SN.

Thus one can believe that the So-->S_1^* transition has the ICT nature and localized mainly in the "Keto"-moiety of the trans-planar NH-form with the large contribution of the Zwitterion structure.

This transition is responsible for the photocoloration in Naphthaldehyde deri-vatives I(Np, R) unlike SA. Thus both the shortening of the alkyl radical in the imine moiety and the extension of the conjugation region in the aldehyde moiety result in the change of the twis-ted PCF structure with the planar Zwitterion that which leads to the change of the So-->S_1^* transition nature and to the short-wavelength shift of the photochromic band towards the TC (SC) absorption region.

Nevertheless the twisted post$_1$ TICT structure generated by the TICT-like state is the precursor of both the twisted and planar PCF structures. Besides the main path-way that is bound up with the excitation of the S_1^* states of the E-struct-ure and involves the S_1^*NHflu structure as a PCF precursor, there is the additional path-way of the PCF generation, as it has been discussed above.

It is necessary to note that any path-way of the PCF generation connected with excitation of the higher electronic states $Sn^*(n>1)$ of the E structure can lead to the dependence of ASSflu and photocoloration efficiencies, and also their ratio (Φflu, Φcol and Φflu/ Φcol correspondingly) on the excitation wavelength (λex) (Tab. 3. 37), if the hypothesis of the "diabatic generation" of the post-TICT colored structures is true.

Indeed, due to the exceptional role of the TICT-like state (S_1^* state) in the generation of the twisted PCF structures, one can believe that the molecule coming after the ESIPT on any high Sn*state (n>1) of the NH structure(Sn*E ESIPT->Sn*NH) and especially on S_1^*NHtr state can be deactivated diabaticly directly into PES of the S_1^*state in the region of the TICT-like structure missing S_1^*NHflu state with the subsequent generation of the post-TICT Photocolored structures in the So state according to the scheme 3. 24(the path-ways 3->1 and 3->2).

The hypothesis of the twisted PCF structures (Sch. 3. 24) unlike the idea of only planar trans-keto one (Sec. 3. 41) allows to find the qualitative explanation of the multiexponential nature of the decay reactions and their dependence on structure, medium, and excitation energy.

According to the hypothesis, the PCF form is the set of the different conformations of the three so called "post TICT" main kinds (Sch. 3. 24, 3. 25).

The first, very low stable $post_1$TICT structure with the twisted nonclassical conformation corresponds to the shallow potential hole on the top of the Ground state potential barrier limited by the low potential barriers (ΔE< 5kcal/mol)restricting the change of the tetrahedral angle within the limits of a vicinity of 90° without any alterations of another coordinates.

This PCF $post_1$ TICT structure with the longwavelength absorption band λabs$\approx$550-600nm comes immediately from the S_1^*TICT-like state, and is the direct precursor of other colored post-TICT structures.

The second, utmost twisted post TICT fluorescent colored structure ($\alpha\approx$80°, $\beta\approx$170°, $Y\approx$0°) with the longwavelength absorption bands, λabs$\approx$480-550nm, is generated from the initial $post_1$TICT one along the pathway 1 above the very low potential barrier (<5kcal/mol) with the time rate $\tau\approx$170ns.

In such a structure (see also above)the stability of the cissoid conformation with a twisted A-ring is provided by steric interactions of the A-ring with the imine moiety as a result of the rotation around C-N bond (towards the cis-isomer with angle y=180°). This is the main, most stable, colored structure of unsubstituted SA molecule for the visible coloration in the rigid matrix and the crystal. The scattering and the distribution of the slightly different conformations of the post TICT structure with various angles α, β, and y can extremely sensitive to the molecular structure and to such external factors as a solvent viscosity (microviscosity), solute concentration, wall effect, and the rate of the temperature change.

Therefore the observed scattering of the stability and the decay rate(see Tab. 3. 44)can be explained by the slightly changed conformations under various experimental conditions.

The third conformational kind, post3TICT, is the almost planar trans-keto (Ktrans)structure $\alpha \approx 180°$, $\beta \approx 0°$) suggested usually as a single PCF structure (see sec. 3. 41). Such a more stable colored structure is responsible for the more short wavelength bands ($\lambda^{max} \approx 400\text{-}440$nm) which is hard to distinguish from the cis-keto (Kcis) those responsible for the TC and SC(see above and Ch. 2).

The post TICT PCF structure is typical for Azomethine molecules with the heightened electron density on the C0N group (for example, N-alkylimines with the short alkyl radical or Naphthylimines), i. e. with impeded rotation around C:::N bond. This colored structure is generated from the $post_1$ TICT that above the low potential barrier ($\sim$ 5kcal/mol)along the path-way 2 with the rotation around C_1--C 7 bond (α angle change from 90° till $\approx$ 180°) and almost without β angle change.

The dark bleaching reaction for all colored structures involves the two stages--the rate limitative transformations (path-ways 4, 4', 4") to the Kcis structure (Sch. 3. 21) and the reenolization by the GSIPT NH->OH ($\tau < 10^{-11}$ s) probably with the tunneling /7/, and the tautomeric equilibrium O-H $\Leftrightarrow$ NH shifted strongly towards OH structure(Ch. 2).

The colored $post_1$-3 TICT structures are displayed by the different ways in the rate limitative stage of the bleaching reaction under influence of the various substituents and media.

The reactions $post_1 TICT$ and $post_2 TICT \to Kcis$ involving the A-ring rotation only or with that around C:::N bond (path-ways 4, 4') go in the liquid solvents at the room temperature with the time rates $\tau < 10^{-8}$ and 10^{-7} s correspondingly.

At the same time there is the typical example of the media and substituent effects on the PCF decay kinetic of the bulky substituted structures /100/ which can be explained by the Scheme 3. 24.

The dramatic rise of the PCF structure lifetime is promoted by the steric hindrances in the crys-talline lattice /108-110/especially for 3, 5 tert-butyl derivatives ($\tau = 10^3 - 10^4$ s) /100/that allows to observe photocoloration and thermal fading of the PCF structures in the crystal at an ambient temperature by the steady-state methods.

The two stages of the fading reaction with the first order constants indicate the different PCF structures which decay with the strong different rates ($k_1 \gg k2$) (Tab. 3. 51).

According to /100/ the both observed rate constants decrease in the molecules with the both series of the substituents: $F > Cl > Br > OCH_3 > H > CH_3 > C(CH_3)$ and $C(CH_3)_2 > CH_3 > H$ i. e. with fall of the electronegativity and the size correspondingly.

The influence of the substituents may be therefore explained by the combined action of the steric or (and)electrostatic repulsion between the adjacent molecules and the electronic nature of the substituents which must be depend on the type of the structure. At it has been mentioned above the reenolization (GSIPT, $Kcis \to E$), like solvent, proceeds with the very high rate ($k \approx 10^{10} - 10^{11}$ s^{-1}) and is not a limitative stage on the path-way of the bleaching reaction.

The additional investigations of the PCF thermal fading reaction for the structures 1, 15, 16(Tab. 3. 51)//99/show that the kinetic isotope effect with deuterium is observed only for the fast decay reaction (k_1) but it is absent for the slow that ($k2$)(structure 16) and not observed at all of the structure 15 with the 6 methyl substituent.

The fast decay reaction rate decreases considerably when introducing the 6Me substituent in A-ring and does not change when inserting of the 3, 5 tert-butyl groups (compare 15with 16, and 1 with 16 correspondingly).

The above findings have been considered /99/ as an evidence of the existence of two consecutive stages – the fast (k_1) and slow (k_2) ones with the formation of the Kcis structure which is the precursor of the E-form (GSIPT).

However such an interpretation seems to be doubtful for the following reasons. The planar (Ktrans, Kcis) structures cannot be responsible for the longwavelength absorption band of the colored structure ($\lambda \approx 500$nm). The isotope effect cannot be explained by the GSIPT as it not the first fast stage and it is not a limitative stage of the fading reaction in the crystal. The increase of the fast decay rate (k_1) when including 6Me substituent is also difficult to explain with the two consecutive stages.

On the other hand the Scheme 3. 24 allows to explain the experimental results of /99/. The two reverse reactions proceeding parallel, with the considerably different rates are bound up with decay of the different PCF structures which absorption bands are overlapped partially.

The fast reaction k_1 is the decay of the twisted post, TICT structure stabilized by the tert-butyl groups in the crystalline lattice. This rate increases considerably in the structure 15 as a result of the destabilization of the twisted NH structures due to the 6CH-NH interactions.

The slow reaction (k_1) which is absent of the molecule 15 is bound up with decay of the more stable post TICT structure which generation is unlikely in the molecule 15 due to the repulsing interactions of the 6Me group substituent with the adjacent molecules.

The both path-ways of the reenolization proceed via planar NHcis (or Kcis) structure followed by the very fast GSIPT without any accumulation of the Kcis structure (& Kcis – &E)>>0).

The little kinetic deuterium effect of the structure 16 (rate decrease) can be caused only by the large amplitude anharmonic bending vibrations of the structural fragment A ring –C—NH forming NH...O bond, but not by the GSIPT.

The typical example of the post3 TICT colored structure is that of imines of Naphthaldehydes,

I(5-6 Np, R). In these compounds (Sch. 3. 27, Tab. 3. 50)/48/ the planar trans-Keto structure responsible for the shortwavelength absorption band ($\lambda \approx 430\text{-}360$nm)is stabilized by the IHB CH...N (Sch. 3. 28)and decays by the exponential law with the time rate that higher by order of magni-tude than of SA and SN.

Such a peculiarity is typical for all compounds of this series with various imine radicals with the exception of the structure with N-Naphthyl radical which is stabilized additionally (Tab. 3. 50).

In terms of the Scheme 3. 24 the strong decrease of the SN PCF structure decay time in the nonpolar solvent (3MP)/45/(Tab. 3. 44) with the second order time rate constant (k_2) with the increase of the concentration or in the presence of alcohol(see above) is the realization of the new path-way of the reenolization without the isomerization about C 1::::-C7 and C 7:::N bonds i. e. with avoiding of the path-ways 4' or 4".

In the crystals and the rigid matrices at the room and low temperatures where the PCF struc-tures are strongly stabilized by the intermolecular steric or electrostatic interactions, the reverse reaction of the reenolization can be conducted by irradiation of the light with the wavelengths within the limits of the PCF absorption band ($\lambda \approx 480\text{-}500$nm)(see e. g. / 20, 40, 91/). In that case the complete discoloration can be achieved easily even in rigid solvents and crystals at 77K.

To the best of our knowledge the detail experimental and theoretic studies of the photoreverse reaction (photobleaching) are absent. However one can supposed that $So \text{-->} S_1^*$ excitation of the post TICT or the NHtrans structures has to promote both the fluorescence and the diabatic reaction. The large energy excess connected with such a reaction can provide the efficient back process even over the very high potential barriers caused by the rigidity of the media or the steric interactions between the neigbouring molecules in the crystal lattice(path-way 5).

Thus the photodecay of the PCF, like the photocoloration, involves the diabatic stage from the excited PCF structure to the planar Kcis that in the Ground state.

3. 5. The generalized scheme of photochromism and side photoinduced processes of anils.

In the previous sections (3. 1-3. 4) all the stages of the photochromic process and the peculiarities of the structures taking part in each step have been discussed in detail including the side reactions and processes starting with the excitation of the E-structure and ending with the generation of the PCF products followed by the reverse reaction of the discoloration closing the photochromic reaction cycle.

In the present section the separated stages of the photoinduced processes are included and united in the general semi-quantitative scheme (Sch. 3. 29).

On the ground of the analysis of the data conducted in the sections 3. 1-3. 4, the most probable structures and processes meeting the experimental and calculated findings have been selected for the description of each stage in the general scheme.

Within the frames of the general scheme the influence of the structural factors which is discus-sed in detail for each stage is not considered systematically.

According to the Scheme 3. 29 (see also Sec. 3. 1. 1, Sch. 3. 2. b) the excitation (Ex_1, $Ex2$) of the Enol (ESo trans) isomer results in the two lowest weakly fluorescent S_1^* states which differ in the energy and the symmetry- $(S_1)^*(E\,2, C_1)$ and $S_1^*(E_1, Cs)(E2 > E_1$, C_1 and Cs-nonplanar ($y\neq 0$) and planar ($y=0$) conformations respectively.

The fast structural transformations with the deactivation of the both excited states of the E structure competing with the ESIPT occur with the ultrahigh ($\tau\approx 30\text{-}40\text{fs}$) and the high ($\tau\approx 1\text{-}2\text{ps}$) time rates from the $(S_1)^*$ and S_1^* states correspondingly (pathways 1E and 2E).

Such transformations lead to the generation of the two twisted ($\alpha\neq 0$, $\beta\neq 0$, $y\neq 0$) transient structures, Etw_1 and $Etw2$, with the broken H-bond as a result of the conical intersection of the S_1^* and So PESs. These transient structures differing in the energies and conformations (tetrahedral angles α, β, and y) can undergo the structural transformations $Etw_1 > Etw2$ ($\tau\approx 12\text{ps}$)

and can be a precursor of the twisted SoEcis (path-way 4E, 4'E) and the initial planar SoEtrans structures ($\tau\approx 16ps$)(path-way 3E).

The equilibrium SoEcis <=> SoEtrans depends strongly on the viscosity, the temperature, and is shifted fast, completely ($\tau<1s$) towards SoEtrans structure(path-way 5E) in the liquid solvents at the room temperature.

However another path-way bound up with the closing in the S_1^* and So PESes and deactivation by the Intersystem Crossing $S_1^* \to T_1 \to So$ (see Sch. 3. 2a) just as the triplet state(T_1) sensitized Etrans->E cis isomerization can be realized /20/.

In any case the A-ring twist with the H-bond rupture is the initial step of the reversible trans-cis isomerization around C=N bond with formation of the twisted SoEcis and the stable initial S 0Etrans isomers. The efficiency of the photochromic reactions is decreased markedly by these competing transformations.

The over-ultrafast($\tau\approx 50fs$) ESIPT(path-way 1) can occur in the both $(S_1^*)(C_1)$ and $S_1^*(Cs)$ Enol states with the considerably different efficiencies competing with the described above deactivation of the S_1^* states (Sec. 3. 2. 3, Sch. 3. 2b).

The most efficient over-ultrafast ESIPT ($\tau\approx 40-50fs$) occurs with the excitation of the FC $S_1^*(Cs)$ (planar) state competing with both much slower "intraenol" deactivation ($\tau\approx 1-2ps$)(2E) and mainly with the weak Enol emission Fl_1 from the equilibrium $S_1^*(Cs)$ state which lifetime therefore is determined by the ESIPT rate. The less efficient ESIPT reaction (1) competing with the ultrafast "intra-enol" deactivation (1E) ($\tau\approx 40fs$) occurs also from the $S_1^*(C_1)$ state.

The over-ultrahigh ESIPT rate in the both states in the solvents is bound up with the barrierless mechanism when moving of the vibrational wave packet which evolves along the proton trans-fer coordinate (however without the marked explicit proton motion) in a purely repulsive potential on the S_1^* state PES towards the S_1^*NH flu structure(so named "ballistic motion") and disappears in time scale of a few vibrational periods with a minimal effect of a vibrational energy and isotope substitution (OH->OD) in accordance with the experimental findings.

The appreciable instantaneous(almost consert) A-ring twisting ($\alpha = 0°$ --> $\alpha \neq 0°$) as a result of (C-H)(H-C) sterical interactions competing with weak N-H...O bond is also provided by such a mechanism.

The one more additional path-way of the S_1*NHflu structure generation from the both states with a much slower rate involves the two stages –the barrierless ultrafast ESIPT (analogues de-scribed above) with formation of the low stable flattened transient S_1*NHtr state (pathways 1+1')responsible for the fast shortwavelength ASSflu and a much slower (τ <500fs)solvent assisted vibrational relaxation (path-way1") to the thermalized S_1*NHflu state. The path-way 1' is much more probable for the higher (S_1)*(C_1) Enol state. The S_1*NHtr state cannot be a direct precursor of the fluorescent PCF post TICT structure in the So state but can lie on the path-way 3 of the S_1*TICT-like state generation missing the S_1*NHflu state.

The latter, slightly twisted and polar structure is responsible for the double minima band ASS fluorescence(ASS flu)with the efficiency increasing sharply when raising of viscosity (rigidity)of the medium.

The ASSflu differs considerably from the more short-wavelength Kcis fluorescence (Kcis flu) arising with excitationof the planar Ground-state Kcis structure even at the room temperature from the excited K*cis state that does not lie on the path-way of the PCF generation.

The ASS flu is quenched mainly as a result of the adiabatic structural transformation from the S_1*NH flu state towards the TICT-like region over a very low (due to the pre-TICT, twisted structure of the fluorescent state)potential barrier which increases sharply in the rigid media (pathway 2)(see also Sch. 3. 13).

The "TICT-like" region of the S_1* PES is formed as a result of the utmost proximity interaction, or the avoided crossing of the S_1*and So state PESs along the angle coordinate in the vicinity of

α = 90° at the expense mainly of the very high potential barrier for the twist in the So state at 90°.

The TICT-like region involves on the PES S 0 the shallow potential hole divided by the very low potential barrier in the S_1* state.

There is the very small hole on the top of the potential barrier in the So state limited by the low barriers restricting the motion along the coordinate from a vicinity $\alpha \approx 90°$ (post$_1$TICT structure).

The TICT-like region is the only immediate precursor in the S_1^* state of the colored structures and can be populated by both via the S_1^*NH flu state (path-way 2) and the very instable transient NH*tr structure arising directly from the ESIPT(path-way 3).

Since the population of the latter increases considerably with the excitation of the higher $(S_1)^*(C_1)$ enol state, the increase of the ratio Pcol/Pflu with the excitation energy (decrease λ^{ex}) can be caused apparently by the localization of the higher electronic states.

According to the scheme the dependence of the photocoloration efficiency on the structure, media, and excitation is determined to a considerable extent by the TICT-like state as an immediate precursor of the photocolored structures.

There are three main kinds of the photocolored structures (PCF structures) which come from the TICT-like state, so called "post TICT" structures (see also Sch. 3. 24).

The most instable initial, strong twisted ($\alpha \approx 80-90°$) aforesaid post$_1$TICT colored structure with the absorption in the longwavelength region ($\lambda \approx 550-600$nm) is generated directly from the TICT-like region (path-way 1c), the second, strong twisted ($\alpha \approx 80°$), also with the longwavelength absorption bands ($\lambda \approx 480-520$nm) and the fluorescent post2 TICT structure stabilized in the cisoid about C-N bond ($\beta \approx 170-180°$) conformation is generated from the post$_1$TICT structure over potential barrier about 5kcal/mol with the time rate about 170ns (path-way 2c) and represents the main photocolored structure in the PCF.

The third post3 TICT, transoid, form about C_1--C 7 and C7--N bonds, almost planar ($\alpha \approx 180°$, $\beta \approx 0°$), structure with the more shortwavelength absorption band ($\lambda \approx 420-440$nm), similar to the thermochromic one in the crystal, is generated also from the post$_1$TICT structure (path-way 3c). It is the least probable colored structure among the PCF for the unsubstituted SA.

However for the structures with the heightened electron density on the C=N moiety (N-Alkyl, or C-Naphthyl, Antril derivatives) or with the impeded A-ring twisting such a photocolored conformation becomes much more preferable, and even the single one.

However in spite of the likeness of the absorption bands the photochromic and thermochromic forms have the different, trans and cis (around C_1--C7 bond), structure respectively (see above).

The each of the reverse dark bleaching reactions involves the two stages –the structural transformations from the initial colored structure towards the final cis-keto one(path-ways 1d, 2d, 3d) and the fast GSIPT reaction NH->OH ($\tau \approx 10^{-11}$s) with the formation of the initial Enol (SoEtrans) structure. The first stage is the rate limitative one.

Thus the observed multiexponential kinetics of the bleaching reaction in the liquid solvents with impulse excitation is caused by the decay of at least the two, $post_1$TICT and post TICT $_2$, colored structures (path-ways 1d, 2d).

The typical peculiarity of the reverse reactions of the PCF structures is the sharp rate dependence on the viscosity and the rigidity of the medium that is the forcible argument in favor of the instable twisted conformations of the colored structures.

Just because the PCF structures are strongly stabilized in the rigid matrix or in the crystal lattice and may exist a very long time in such a medium even at room temperature.

Under such conditions the discoloration can be promoted by the irradiation (excitation)of the PCF structures with the light within the limits of their absorption bands (photobleaching). This photoreaction proceeds like photocoloration diabaticly (path-way 1pd+1d+GSIPT) providing a complete discoloration.

Thus according to the presented data and the general scheme, the photochromic properties of anils in the solvents can be characterized by the following typical features.

    i. The lack of any irreversible reactions involved in the photochromic processes.

ii. The ultrafast rate of the primary reactional step (ESIPT).
iii. The fast generation of the Photocolored form (PCF).
iv. The very low stability of the basic PCF structures in the liquid solvents at ambient temperature.
v. The unique sharp PCF stability increase with rise of the medium viscosity and rigidity(especial-ly for anils with the bulky ring substituents) providng with the high opportunity a very long time conservation of the PCF structures.
vi. The marked efficiency of the photobleaching (discoloration) reaction in the rigid media even at low temperature.
vii. The efficient ASS fluorescence (ASSflu) can be available in the rigid media.
viii. The existence of the several processes of the degradation of the excitation energy, decreas-ing strongly the efficiency of the photocoloration:(a)The S_1^* fast energy deactivation in the E-structure competing with the ESIPT, (b)The strong solvent assisted vibrational deactivation compiting with formation of the TICT-like state, (c)The intensive energy degradation in the TICT-like region competing with the generation of the PCF structures.

These peculiarities of the photoinduced processes of Anils can be very important for the discus-sion of their advantages and shortcomings for applications of the Anil photochromism(Ch. 6).

3. 6. Photochromism of Anils in the crystals.

3. 6. 1. General information.

The especial role of the anil photochromism in the crystal is determined by the several reasons.

The first of all, the "photochromic anils "have been discovered visually and received their specific name as a crystalline compounds, and their photochromism was considered from the beginning and long time after as a property of the crystalline state exclusively (Ch. 1).

The second one, the crystalline anils show the photochromism, unlike the most well-known organic photochromes(e. g. spirocyclic compounds)which usually have no phtochromism in crystals.

Meanwhile the crystalline photochromes like anils, are the best ones for the applications due to their high technologicality.

The third, very important, reason is the unique posiibility to study the crystalline structures with the prediction their photochromic properties on the base of the close connection between the molecular mechanisms of the photochromism and the structure of the crystals.

There are the beautiful recent reviews /81, 98, 110/with the detail information about results of the both pioneers and recent studies of the crystalline anils and the up-to-day methods of their preparations and utilization.

Meanwhile some more precise information about the pioneer works concerned the photochromism of crystalline anils including additional references with data published in the initial period of the development of the ideas about the structure–property connection have been produced in the chapter 1.

In this section the general information about the previously known photochromic properties of

crystalline anils, the recently proposed methods of the prediction and the preparation of the photochromic crystals and up-to-day information about their properties briefly discussed on the basis of the reviews/20, 81, 110/and on the base of the produced above data (Sch. 3. 29).

It has been shown /20, 22, 81/(see Ch. 1)the crystalline anils undergo the typical variations in the visible region of the absorption (reflectance)spectra under irradiation (photochromism) or heating (thermochromism) in dependence on the crystal structure, that are manifested by the appearance of the bands with $\lambda^{max}\approx 480$, 520nm or 440-460nm for photochromism or thermochromism correspondingly.

According to the results of the more early studies the thermochromic and photochromic crystals have the mutually exclusive structures/22/

(see also Ch. 2). In thermochromic crystals the molecules are essentially planar($\gamma=0°$) and disposed in the parallel planes exhibiting the intermolecular interactions (close-packed structures with the short interplane distance (3-5 A) (see Sec 2. 13A, e. g. /98/).

In the photochromic crystals the planar salicylidene moiety and the imine that are acoplanar ($\gamma \neq 0°$) and the molecules are arranged in the unparallel planes. In the later case there is a free volume sufficient for the structural transformations connected with the generation of the PCF(Sch. 2. 13D).

The efficiency of the PCF generation and decay rate fall with the temperature decrease owing to the existence of the intramolecular potential barriers and the increase of the intermolecular interactions in the crystalline packing. Such a temperature dependence leads to the existence of the "working" temperature interval out of which the photochromic properties are not observed visually.

The ASS flu ($\Delta v \approx 10000 cm^{-1}$) similar to that of the thermochromic crystals arises below the lowest limit of the temperature interval.

Under irradiation in the absorption band of the colored form even at T<77K the reverse bleaching reaction is observed.

The PCF of the crystalline SA shows the excellent reversibility in the cycles of the photocoloration –photobleaching /111/.

3. 6. 2. The methods of preparation of the photochromic crystals.

In the last two decades the strong efforts have been exerted to produce new types of the organic photohromic crystals and to control their properties.

The most important conditions for manifestation of the crystalline photochromism of anils was considered to be provide reaction volume for the photoinduced isomerization in constrained media.

According to this conception the three effective methods for the selective preparation of the photochromic Schiff bases crystals have been reported.

This strategy are applicable to a wide range of the potential photochromic compounds /110/.

3. 6. 2. 1. Bulky group substitution method.

This method is based on the utilization of the volume surrounding of the bulky groups introducedinto aromatic rings /100, 108, 109/.

The bulky groups can act as a space opener in the crystal lattice and allow partial movement of the structural fragments.

For example introduction of the 3, 5 tert-butyl groups into the ring A turns nonphotochromic crystals into photochromic those (Tab. 3. 5. 2).

The same effect can be achieved by introduction of the 3, 5 di-tertpentyl groups/110/.

3. 6. 2. 2. Clathrate crystal method.

The second method /110, 112/ is based on the utilization of the crystal cavities constructed by host for the photoisomerization of the guest species.

The organic host used first for this purpose was deoxycholic acid (DCA). The clathrate crystals of DCA with the SA derivatives(SA@DCA) can be easily isolated by co-crystallization from alcoholic solution.

A number of SA@DCA compounds in the crystals show the photochromism unlike corresponding SAs in the pure crystal states (Tab. 3. 53).

The isolated existence of the guest species favorable for the monomolecular isomerization is thus established in the DCA clathrate crystals.

It has been shown also/81, 82, 113/ that the formation of the inclusion complexes of thermochromic 5-ClSA and SA 2-aminopyridine(SAP) with cyclodextrines (SA@CD, SAP@CD) results in the generation of the photochromic crystals. Such inclusion complexes have been discussed above (Ch. 2) and will be discussed also later for bis-Schiff bases (Ch. 4).

It is necessary to notice that the materials obtained by the clathrate methods can have properties that very similar to the nanoscopic those and can be considered as photochromic nanomaterials.

3. 6. 2. 3. Neighboring alkyl-group substitution method.

The third method is based on the introduction of the alkyl groups at the proper positions to direct the conformational preference for photochromic reaction.

The introduction of the two alkyl groups in the 2, 6 positions of the aniline ring is the most effective way (Tab. 3. 54) while the introduction of only one alkyl group is not enough. /114/.

Each aniline ring is significally twisted out of the SA plane and each molecule avoids tight packing forces /21, 44, 115, 116/. However the insertion of the 2, 6 isopropyl groups (iPr) is the most effective(Tab. 3. 54).

The results of the detailed X-ray structural analysis of the 3, 5 diCl derivatives /96, 100/show that such an essentially different behavior of the crystalline anils with various alkile substituents in the aniline ring (Tab. 3. 54) can be explained by the steric interactions between the salicylidene and aniline moieties of the adjacent molecules in the crystalline packing (for 2', 6' diMe substituted) or of the one molecule (for 2', 6' di tertbutyl substituted) which in the both cases prevent from the structural changes in the initial step of the PCF generation.

In contrast, the iPr groups adopt the structure which provides the conformational changes connected with the PCF generation, and size and shape of the iPr group are appropriate to weakened intermolecular interactions and to create a space for the transformations resulting in the PCF generation.

These facts indicate any conventional judgment for the photochromic anils that adopts a nonplanar structure is insufficient for a complete interpretation of the photochromic character.

Furthermore the presence of the crystal structure with the expanded free spaces is the necessary but an insufficient condition for the crystalline photochromism.

The detail analysis of the intermolecular and intra molecular steric and specific interactions and also electronic factors is needful for the prediction of the photochromic properties for the crystalline state of anils.

Some feeble attempts have been undertook to obtain photochromic crystalline anils with the additional hydroxyl groups.

It has been shown by X-ray structural analysis that in the crystal of 2'OH SA /110, 117/ the two types of the nonsymmetrical units with Enol and Keto structures are held together by the intermolecular hydrogen bonding, and in the crystal of 3'OHSA /110, 118/the 3'OH group forms a secondary H-bond between salicylidene moieties of the adjacent molecules which may prevent the ESIPT responsible for photochromism.

Photochromism is not observed also in crystalline 4'OHSA /21, 110, 116, 119/which shows only thermochromism (see Ch. 2)but not photochromism.

The failure in obtaining of the photochromic crystals of the azomethine bridge substituted SA (7 Alkyl or 7 Phenyl substituted)/100/ may be explained by the electronic and steric factors preventing from the formation of the TICT-like structure which is a precursor of the PCF structures.

3. 6. 3. The generalized view to the photo and thermochromism in the crystals.

The detailed analysis conducted in the Ch2 (Sec. 2. 6. 1) leads to the conclusion that the thermochromism is manifested as a collective phenomenon connected with the GSIPT reaction in the molecules which are in the crystalline lattice, engendered by the electronic interactions between mainly the aldehyde moieties of the adjusting molecules in the crystalline packing.

Therefore the GSIPT (OH->NH) efficiency and hence the formation of the thermocolored NH (Keto)structure depends on both the intramolecular (interfragmental) electron density distribution and the intermolecular interactions (Ch. 2).

At the same time photochromism is stipulated by the twisting deformation of the Keto-moiety (A-ring twist) in the S_1^* excited state followed by the ESIPT which efficiency is caused by the intramolecular factors.

Thus the competition between the processes of thermo and photochromism is caused by the comparative efficiencies of the intermolecular and intramolecular factors in the crystals which can be favorable to simultaneous existence of the both phenomena /81/. Such conditions arise for some Anil structures (e. g. SALKnPh)(Sch. 3. 30) (see also Sch. 2. 13).

In that situation the crystalline packing, on the one hand, promotes the strongly decrease of the interaction between the salicylidene moieties of the adjacent molecules and does not preven from the conformational transformations responsible for the photochromism in each separate molecule but, on other hand, provides the interactions of the close salicylidene fragments of the adjacent molecules in which the GSIPT(i. e. Kcis structure formation) efficiency is heightened strongly due to the increase of the electron density on the C=N group as a result of the-C=N-B ring conjugation break when inserting the alkyl bridge group.

Thus, although the display of thermochromism and photochromism in the same crystalline structure is the hardly probable occurrence, however, this very important phenomenon has been discovered recently /39, 81, 114 /, and give the base for the real understanding of the photo and thermochromism in the crystals.

The structural mechanism of the photochromism in the crystal can be described by the general scheme 3. 29 modified by the structure of the crystalline lattice which in the final analysis is also determined by the structure of the crystallized molecules.

However there are the general peculiarities caused by the strong steric and (or) electronic interactions of the adjacent molecules in the crystalline packing.

Decrease of the rate of the A-ring twist in the $S_1^*(\pi\pi^*)$ state of the E-structure(path-ways 1E, 2E) which competes with the ESIPT (pathway 1, 1') must increase the efficiency of the photochromic reaction.

However the change of the ESIPT mechanism from the barrierless (in the solvents) to the tunnel (in the crystals) that due to the exclusion in the latter of the ESIPT promoting vibration modes (see Sec. 3. 23) can lead to some fall of the ESIPT rate (from tens till hundred fs) and to decrease of the PCF generation efficiency.

Therefore the modifications of the first stages connected with the Enol structure result in the change of the PCF generation efficiency in the two opposite directions and cannot influence markedly the efficiency of photocoloration of the photochromic crystals.

The path-way of the TICT-like state population and the diabatic generation of the post$_1$TICT structure (path-ways 2, 3k)via S_1*NHtr and S_1*NHflu states in the photochromic crystals are connected with the A-ring twist only and require usually a less free volumes.

Therefore the rates of the 2, 3, 1c processes in the solvent and the crystals are similar for the most photochromic anils.

In particular the dependence of the competing processes of the ASS emission and the photocoloration on the temperature and the excitation energy (λex) are determined, like of solvents, by the intramolecular (and viscosity) pre-TICT-like state barrier and the population of the transient S_1*NHtr state correspondingly.

However the final stages of the PCF generation in the Ground state (2c, 3c)which are bound up with the considerable conformational changes can be proceed with the strongly different rates owing to the influence of the structural factors, the excitation energy and especially of the intermolecular interactions (see Tab. 3. 44).

In spite of that, the sufficiently high rates of the PCF generation are provided by the large energy excess in the diabatic stage of the photoreaction like the solvents.

Therefore some modifications of the direct photoreaction mechanism of SAs in crystals don't result in a marked change of the photocoloration efficiency as compared with solvents or amorphous rigid media.

At the same time the key role in the coloration of the crystals is played by accumulation of the PCF owing to the dramatic fall of the bleaching reaction rate that gives an opportunity of the visual observation of the photoromism at ambient temperature.

The dark bleaching processes proceed like solutions by the path-ways 1d-3d with the rate, depended on the kind of the colored structure and the difference in inhibition of decay for the various post-TICT structures that can influence the kinetics of the dark bleaching reaction.

For instance for the post$_1$TICT structure in the crystalline state the longwavelength absorption band ($\lambda^{max}\approx$500nm) differs a little but the decay rate can be more higher as compared with the post2 TICT one. In such a case the double exponent bleaching is provided by the two decay reactions (Tab. 3. 51) (see however below /36/).

The most high time rate constant is expected for the comparative stable post3 TICT, nearly planar trans-Keto structure that has absorption band in the more shortwavelength region ($\lambda\approx$440--460nm).

However the new possibility for explanation of the comparative fast its decay in crystals can be provided with the so called "pedal" mechanism of the trans-cis isomerization/101/ by which the strong steric hindrances in the crystalline packing can be avoided.

The very useful information for the understanding of the mechanisms of the photoinduced processes in the thermochromic and photochromic crystals involving various stages in the generation and decay of the PCF structures can be derived from the study of anil molecules with the bulky (tert-butyl)substituents/36/.

The experimental finding /36/can be explained on the base of the two different schemes which produced by /36/ (Sch. 3. 32a) and derived from the scheme 3. 29 (Sch. 3. 32b).

Both in the thermochromic (2p) and photochromic (4p) crystals the ESIPT with the same (tunneling) mechanisms ($\tau<$ 1ps)leads to the generation of the structures suggested which natures are different for each of schemes.

In the scheme 3. 32(2p(a)) the two, cis –keto$_1$* and cis-keto *excited structures, are responsible for the two different fluorescence bands ($\lambda^{max}\approx$570nm, τ=23ps and $\lambda^{max}\approx$608nm, τ=250ps), and therefore unlike SA derivatives the fine structure of the ASSflu band is explained by the overlapping of the two different radiative transitions S_1*-hν->So in the different species. However the similar double maxima band fine structure of the fluorescent and corresponding absorption bands is usually observed in crystals and solvents for the various NH forms (ASSflu, Kcis, and Ktrans planar structures, and the nonplanar colored ones)and attributed to the vibronic structure of electronic transitions between the same states ($S_0 <=h\nu=> S_1^*$) in the NH form of any nature being its typical sign.

Therefore in the scheme (b) the ASS flu band is attributed to one and the same electronic transition with the vibrational pattern ($\Delta\nu \approx 1130 cm^{-1}$), and the fluorescence lifetime $\tau=250ps$ for the most intensive spectral component.

The second, short-time, constant ($\tau \approx 23ps$) in the double-exponential fluorescence decay can be caused by the overlapping (in the shortwavelength region) emission from the NHtr* transient state.

Such a flattened structure is analogous to that on the scheme 3. 29 but markedly stabilized by the interactions of the bulky substituents with the adjacent molecules in the parallel planes of the crystalline lattice in the thermochromic crystals.

The comparatively high decay time rate constant ($\tau=23ps$) and population of the NHtr* transient state with the longwavelength excitation ($\lambda \approx 390nm$) are provided by the same interactions which inhibit the photochromic transformations in the thermochromic crystals.

According to the scheme, 4p (a), in the photochromic 4p crystals, the ESIPT leads to the "hot" vibronic state of the S_1^*(Cis –keto) one that which is the precursor of the PCF.

Thus according to /36/, unlike another photochromic anils (Sch. 3. 29), in the tert-butyl substituted those the transient structure NHtr* (or Cis keto *) is absent.

However the experimental fin-dings /36/ concerning the very weak fluorescence and the negative band in the transient absor-ption spectra can be interpreted as an evidence of the extremely low stability and population of the NHtr* transient state.

Therefore in the scheme 4p(b) in accordance with the scheme 3. 29 the very weak ASSflu of 4b, just as of 2b, is attributed to the low efficient and short-lived emission of the S_1^*NHflu state which competes with the effective transition to the S_1^*TICT-like region that is the precursor of the colored structure in the So state.

The extremely short-lived ($\tau \approx 2ps$) and low-intensive emission from the transient NHtr* state is overlap with the ASSflu providing the double-exponential emission decay. The sharp fall of the stability (and decay times) of 4pNHtr* in comparison with 2pNHtr* is promoted by inclusion of the additional effectice deactivation channel towards the TICT-like state along with the fluorescence and the S_1*NHflu state formation.

Thus the experimental data/36/are the forcible evidence of the existence of the common (but not immediate) precursor of the ASSflu state and the PCF structures.

The generation of the suggested colored planar trans-Keto structure in the scheme 4p(a) proceeds directly from the cis-Keto* state via the meta-stable "trans-Keto" structure of unknown con-formation with the time rate constant $\tau=250$ps.

The reverse bleaching reaction goes with the two stages —the trans-cis keto isomerization with the two very long time rate constants ($\tau_1 \approx 460$ days(!), and $\tau 2 \approx 8$ hours(0. 1%)) and the very fast GSIPT ($\tau \approx 10^{-11}$s) probably by the tunneling.

The spectral and kinetic parameters of the colored NH structures in connection with their conformations have not been discussed /36/.

According to the idea of the section 3. 4 and scheme3. 29, the absorption band ($\lambda^{max} \approx 440-460$nm)of the flattened, comparatively stable, transoid ($\alpha \leq 180°$, $\beta \approx 0°$)colored post3 TICT structure is shifted towards "red" when twisting around C_1:::C7 bond ($0° < \alpha \leq 90°$ $\leq \alpha \leq 180°$), and can be shifted to the $\lambda^{max} \approx 480-600$nm in the vicinity of $\alpha = 90°$.

The almost same broad absorption bands of the colored forms of 2p and 4p in the solvents with $\lambda^{max} \approx 450$, 500nm ==(Tab. 3. 55)/36/ with the double-exponent decay can be attributed to the flattened post3 TICT and twisted post2 TICT colored structures which are generated with high efficiencies from the unstable $post_1$TICT state with the very short time rates unlike SA in which only single short-lived post 2 TICT structure is formed.

However in the crystalline photochrome 4p(Sch. 3. 32b)under influence of the sterical hindrances, unlike solvents, almost only one post3 TICT structure is formed from the metastable $post_1$TICT structure with the time rate constant $\tau \approx 250$ps.

Owing to the sterical interactions in the crystal lattice the transoid nonplanar structure(α <180°, $\beta \approx 0°$) with the "red" shifted absorption band relatively that in the solvent ($\lambda^{max} \approx 475$nm, $\Delta\lambda \approx 15$-25nm, Tab. 3. 55) and decay with the huge time rate constant $\tau_1 \approx 10^7$ s is stabilized. Unlike solvents only the trace (0. 1%) of the twisted colored post2TICT structure ($\alpha \approx 80°$, $\beta \approx 180°$, $\lambda^{max} \approx 480$, 520nm) which is much less stabilized in the crystal lattice and manifested in the decay kinetic with much less time rate constant ($\tau\ 2 \approx 3 \times 10^{-4}$s) which however much more than that of crystalline SA ($\tau \approx 10^3$ s and τ(rise)≈ 100ps) where the twisted post2TICT structure is the main colored form unlike 4p (see above).

Such a PCF structure of the tert-butyl substituted SA in the crystal is corroborated by the data of direct determination of the PCF molecular structure with the X-ray structural analysis /86/ which is not agreed with the idea about the twisted NH structure of the colored form for the most (including unsubstituted) anils.

Thus according to the above presented data the mechanism of the photoinduced processes in the crystalline anils does not significantly change but is manifested much more distinctly with the inserting of the bulky substituents as a result of the different stabilization of the various transient and final colored structure owing to the steric hindrances with the media. Such kinetic changes lead to the variations of the colored structures and spectral properties of the PCF (see also Sec. 3. 4. 2).

With the insertion of the bulky substituents in the A-ring the opportunity of the more strong stabilization of the post$_1$TICT structure in the crystal state is realized in the molecule where such a structure is the only one responsible for the photocoloration. The typical example for the latter is 2-(2-hydroxyphenyl)benzthiazole (HBTA) with the 3, 5 tetrabutyl groups /110/ (Sch. 3. 31) showing the photochromism in the crystal.

The experimental data discussed above can also be evidence of the fruitfulness of the TICT-like state origin and the post-TICT nature hypothesis of the PCF for the understanding of the mecha-nism of the photoinduced processes in photochromic anils.

No any modifications of the photochromic properties of crystalline anils bound up with the irreversible side reactions have been found. Thus the efficient photocoloration (unlike solvents) along with the multiple recurrence of the photocoloration-bleaching (or photobleaching) reaction cycles create the very favorable conditions for the applications of the photochromic materials on the base of crystalline anils (see Ch. 6).

Conclusion

The ESIPT as a primary step of the photochromic reaction, has been investigated of SA and its simplest derivatives by the modern methods of the ultrahigh time resolved spectroscopy and by the dynamic quantum-chemical TDDFT calculations.

According to the general accepted notions based on the recent findings, the ultrafast ESIPT proceeds from the S_1^* Enol states with the different mechanism, barrierless or tunnel, in the solvents and the crystals correspondingly.

However there is a marked discrepancy between the experimental and the calculated data that is caused obviously by the limitation in the time resolution of the experimental devices or(and) probably in the systematic error of the calculation methods.

The new adopted view resulting from the recent data consists in the strong fall of the ESIPT efficiency due to its competition with the ultrafast H-bond break in the $EnolS_1^*$ state as a result of the Hydroxyl ring rotation influenced by the structure, medium, and the excitation wavelength, and followed by the trans-cis isomerization about C=N-bond in the ground state.

The common views on the post-ESIPT transformations in the S_1^* excited state are based on the very much numbers of the recent kinetic data including the both ultrafast and more slow populations of the S_1^*flu state responsible for the ASS fluorescence, generation, and decay of the transient high energy S_1^*NHtrans structure, vibronic deactivation, and ASS fluorescence quenching which have been discussed in detail.

The influences the post ESIPT photoinduced processes of the structure, media nature, and excitation energy have been also studied and discussed for the very wide circle of compounds under various conditions. The general scheme of the ESIPT photochromism of Anils in the crystal and solvents does not include, according to the experimental findings, any irreversible reaction that provides the unique fatigue resistance and stability of the photochromic properties.

Nevertheless there are some uncoordinated interpretations of the separate details of those processes which are connected ultimately with the problems of the path-ways of the generation and the structure of the S_1^*NHflu state and the PCF structures.

The traditional notions about S_1^*cis-keto structure of the fluorescent S_1^*NHflu state and the planar trans-keto structure of the PCF are connected with the idea about the diabatic formation of the latter from the S_1^*cis-keto flu state via the transient twisted ground state Kcis structure.

However in our view, one can believe that the notion about the nonplanar, Zwitterion S_1^*NH flu structure, and the "post-TICT" structures of the colored form originating from the precursor fluorescence quenching S_1^*NH TICT-like state, meets the recent experimental and calculated findings in the best way.

Thus one could be suggested the following most prospective directions of the investigations.

i. The more precise determination of the ESIPT rates under various conditions with the development of the experimental and theoretical methods for the better agreement between the experimental and calculated data and the more reliable conclusion of the ESIPT mechanism under different conditions.
ii. The elucidation of the conditions for the suppression of the H-bond break in the S_1^*Enol state by way of modification of the structure, medium nature, and the excitation wavelength for the increasing of the ESIPT efficiency.

iii. The more precise definition and the agreement of the various views about the structure and mechanisms of the formation of the primary NH* transient structures, fluorescent state, intermediate precursor, and the PCF structure generation in order to heighten its efficiency.

iv. The agreement of various conceptions about different structures (Ktrans or NHpostTICT) of the PCF for prediction of their stability (and decay time) in dependence on media nature and molecular structure.

References

1. J. Ortiz-Sanchez, R. Gelabert, M. Moreno, and J. Lluch, J. Chem. Phys. 129, 214308 (2008).
2. K. Kownacki, A. Mordzinski, R. Wilbrandt, and A. Grabowska, Chem. Phys. Lett. 227, 270(1994).
3. a)T. Suzuki, Y. Kaneko, and T. Arai, Chem. Lett. 756(2000). b)T. Suzuki, and T. Arai, Chem. Lett. 124(2001).
4. S. Mitra, and N. Tamai, Chem. Phys. Lett. 282, 391(1998).
5. S. Mitra, and N. Tamai, Phys. Chem. Chem. Phys., 5, 4647(2003).
6. S. Mitra, and N. Tamai, Chem. Phys. 246, 463(1999).
7. M. Zgierski, and A. Grabowska, J. Chem. Phys. 112, (14) (2000).
8. M. Ziolek, J. Kubicki, A. Maciejewski, R. Naskrecki, and A. Grabowska, Phys. Chem. Chem. Phys. 6(19) 4682(2004).
9. N. Otsubo, C. Okabo, H. Mori, K. Sakota, K. Animoto, T. Kawato, and H. Sekiya, J. Photochem. Photo-biol. A. Chem. 154, 33(2002).
10. W. Rodriguez-Cordoba, J. Zugazagoitia, E. Collado-Fregoso, and J. Peon, J. Phys. Chem. A. 111, 6241 (2007).
11. M. Sliwa, N. Mouton, C. Ruckebuch, L. Poisson, A. Idrissi, S. Alloise, L. Potier, J. Dubois, O. Pozat, and G. Buntinx, Photochem. Photobiol. Sci. 9, 661(2010).

12. M. Knyazhansky, A. Metelitsa, A. Bushkov, and S. Aldoshin, J. Photochem. Phtobiol. A. Chem. 97, 121 (1996).
13. M. Kletskii, A. Millov, A. Metelitsa, and M. Knyazhansky, J. Photochem. Photobiol. A. Chem. 110, 267(1997).
14. M. Kletskii, A. Millov, and M. Knyazhansky, J. Obsch. Khim. 68(10), 1700(1998) (in Russian).
15. M. Knyazhansky, M. Strjukov, and V. Minkin, Optica i Spectrosk. 35(5), 879 (1972) (in Russian).
16. V. Minkin, V. Simkin, L. Olekhnovich, and M. Knyazhansky, Theor. Exper. Khim. 10(5), 668(1974)(in Russian).
17. B. Tinland, Tetrahedron 26, 4795(1970).
18. M. Kamiya, and Y. Arahori, Chem. Pharm. Bull. Jpn. 18(11), 2190 (1970).
19. N. Bakhshiev, M. Knyazhansky, V. Minkin, O. Osipov, and G. Saydov, Usp. Khim. 38(9)1644(1969)(in Russian).
20. M. Knyazhansky, and A. Metelitsa, "Photoinduced processes in molecules of Azomethines and their structural analogs", Publ. House of Rostov Univ., Rostov-on-Don, 207pp(1992) (in Russian).
21. K. Ogawa, Y. Kasahara, Y. Ohtani, and J. Harada, J. Am. Chem. Soc. 120, 7107(1998).
22. E. Hadjoudis, in "Photochromism. Molecules and systems", H. Durr and H. Bounas-Laurent. Eds. Elsevier, Amsterdam, p. 685 (1990).
23. C. Okabe, T. Nakabayashi, I. Inokuchi, N. Nishi, and H. Sekiya, J. Chem. Phys. 121 (19), 9436 (2004).
24. G. Wettermark, in "Photochemistry of the Carbon-Nitrogen double bond" Patai Ed., London, N. Y. Sidney, Toronto, Interscience, p. 565 (1970).
25. O. Osipov, Yu. Zhdanov, M. Knyazhansky, V. Minkin, A. Garnovsky, and I. Sadekov, Zh. Fiz. Khim. 41(3)641(1967) (in Russian).
26. O. Osipov, Yu. Zhdanov, M. Knyazhansky, V. Minkin and I. Sadekov, in"Azometiny", Pub. House of Rostov State University, Rostov-on-Don, p43(1967) (in Russian).

27. M. Zgierski, and A. Grabowska, J. Chem. Phys. 113(18)7845 (2000).
28. M. Zgierski, and A. Grabowska, J. Chem. Phys. 115, (18)835 (2001).
29. T. Sekikawa, T. Kobayashi, and I. Inabe, J. Phys. Chem. A. 101, 644(1977).
30. S. Alarcon, A. Olivieri, A. Nordon, and R. Harris, J. Chem. Soc. Perkin Trans. 2, 2293(1996).
31. G. Saydov, and N. Bakhshiev, Dokl. ANSSSR, 175, 5 (1967) (in Russian).
32. M. Knyazhansky, A. Metelitsa, S. Bezugliy, Unpublished results (2006).
33. S. Alarcon, A. Olivieri, G. Labadie, R. Gravero, and M. Gonzalez-Sierra, J. Phys. Org. Chem., 8, 713(1995).
34. R. Morales, G. Jara, and V. Vargas, Spectroscopy lett., 34, (1) 1 (2001).
35. V. Vargas, J. Phys. Chem. A 108, 281(2004).
36. M. Sliwa, N. Mouton, C. Ruckebusch,, S. Aloise, O. Poizat, G. Buntinx, R. Metivier, K. Nakatani, H. Masuhara, and T. Asahi, J. Phys. Chem. C, 113, 11959, (2009).
37. M. Knyazhansky, A. Metelitsa, M. Kletskii, A. Millov, and S. Bezugliy, J. Mol. Struct. 526, 65, (2000).
38. A. Grabowska, K. Kownacki, and L. Kaczmarek, Acta. Phys. Polon. A. 88, 1081(1995).
39. E. Lambi, D. Gegiou, and E. Hadjoudis, J. Photochem. Photobiol. A. Chem. 86, 241(1995).
40. J. Zhao, B. Zhao, J. Liu, T. Li, and W. Xu, J. Chem. Res. (S)416(2000).
41. C. Ruckebusch, M. Sliwa, J. Rehault, P. Naumov, J. Huvvene, and G. Buntinx, Anal. Chim. Acta, 642, 228 (2009).
42. S. Alarcon, A. Olivieri, G. Labadie, R. Gravero, and M. Gonzalez-Sierra, Tetrahedron 51, 4619(1995).
43. L. Antonov, W. Fabian, D. Nedelecheva, and F. Kamounah, J. Chem. Soc. PerkinTrans. 2, 1173(2000).

44. H. Joshi, F. Kamaunah, C. Gooijer, G. VanderZwan, and L. Antonov, J. Photochem. Photobiol. A. Chem152, 183(2002).
45. J. Stephan, A. Mordzinski, C. Rios-Rodriguez and K. Grellman, Chem. Phys. Lett. 229, 541(1994).
46. P. Fita, E. Luzina, T. Dziembowska, D. Kopec, P. Piatkowski, Cz. Radzewicz, and A. Grabowska, Chem. Phys. Lett. 416, 305(2005).
47. A. Oshima, A. Momotake, and T. Arai. J. Photochem. Photobiol. A. Chem. 162, 47(2004).
48. A. Oshima, A. Momotake, and T. Arai, Bull. Chem. Soc. Jpn. 79, (2)305 (2006).
49. A. Koll, A. Filarowskii, D. Fitzmaurice, E. Waghorne, A. Mandall, and S. Mukherjee, Spectochim. Acta, Part A58, 197(2002).
50. A. Mandall, D. Fitzmaurice, E. Waghorne, A. Koll, A. Filarowskii, D. Guha, and S. Mukherjee, J. Photochem. Photobiol. A. Chem. 153, 67(2002).
51. A. Mandall, A. Koll, A. Filarowskii, D. Majumder, and S. Mukherjee, Spectrochim. Acta, p. A55, 2861(1999).
52. D. Guha, A. Mandal, A. Koll, A. Filarowskii, and S. Mukherjee, Spectrochim. Acta, p. A56, 2669, (2000).
53. A. Mandal, D. Fitzmaurice, E. Waghorne, A. Koll, A. Filarowskii, S. Quinn, and S. Mukherjee, Spec-trochim. Acta, p. A60, 805(2004).
54. A. Weller, Electrochemie, 56, 667(1952).
55. C. Chudoba, E. Riedle, M. Pfeifer, and T. Elsaesser, Chem. Phys. Lett. 263, 622 (1996).
56. W. Frey, and T. Elsaesser, Chem. Phys. 189, 565 (1992).
57. C. Chudoba, S. Lutgen, T. Jentzsch, E. Riedle, M. Woerner, and T. Elsaesser, Chem. Phys. lett., 240, 35 (1995).
58. Th. Arthen-Engeland, T. Bultmann, N. Ernsting, M. Rodrigues, and W. Thiel, Chem. Phys. 163, 43 (1992).
59. M. A. Rios, and M. C. Rios, J. Phys. Chem. 99, 12456(1995).
60. D. Zhong, A. Douhal, and A. Zewal, Proceed. Acad. Sci. Usa, 97, 14056(2000).

61. O. Vendrell, M. Moreno, J. Lluch, and Sh. Hammes-Schiffer, J. Phys. Chem. B 108, 6616(2004).
62. W. Frey, F. Laermer, and T. Elsaesser, J. Phys. Chem. A, 95, 10391(1991).
63. S. Lochbrunner, A. Wurzer, and E. Riedle, J. Phys. Chem. A, 107, 10580 (2003).
64. R. Vivie-Riedle, V. Waele, L. Kurtz, and E. Riedle, J. Phys. Chem. A, 107, 10591(2003).
65. A. Douhal, F. Lohmani, A. Zehnacker-Rentien, and F. Amat-Guerri, J. Phys. Chem. A, 98, 12198 (1994).
66. A. Douhal, F. Lahmani, and A. Zewal, Chem. Phys. 207, 477(1996).
67. M. Sanz, and A. Douhal, Chem. Phys. Lett. 401, 435(2005).
68. J. Orbit-Sanchez, R. Gelabert, and J. Lluch, J. Phys. Chem. A, 110, 4649(2006).
69. T. Sekikawa, T. Kobayashi, and T. Inabe, J. Phys. Chem. A, 101, 10645(1997).
70. M. Kletskii, A. Millov, M. Knyazhansky, and A. Metelitsa, Zh. Obsch. Khim. 69(8), 1335(1999)(in Rus-sian).
71. J. Turbeville, and P. Dutta, J. Phys. Chem. 94, 4060(1990).
72. M. Knyazhansky, P. Gilyanovskii, V. Kogan, O. Osipov, O. Schipakina, and V. Litvinov, Opt. Spektrosc. 35(6) 1083(1973)(in Russian).
73. M. Knyazhansky, P. Gylyanovskii, O. Schipakina, O. Osipov, and V. Kogan, Zh. Prikl. Spektr., 21(1), 183 (1974) (in Russian).
74. I. Khudyakov, N. Turro, and I. Yakushenko, J. Photochem. Photobiol. A. Chem. 63, 25(1992).
75. I. Karaaslan, and F. Bayrakceken, Spectrochim. Acta A, 68, 1379(2007).
76. F. Bayrakceken, Spectrochim. Acta A, 68, 1416 (2007).
77. W-H. Fang, Y. Zhang, and X-Z. You, J. Mol. Struct. THEOCHEM., 334(1)81(1995).
78. R. Herzfeld, and P. Nagy, Curr. Org. Chem. 5, 373(2001).
79. D. Higelin, and H. Sixl, Chem. Phys. 71, 391(1983).

80. K. Kownacki, L. Kazmarek, and A. Grabowska, Chem. Phys. Lett. 210 (5, 6) 373 (1993).
81. E. Hadjoudis, and E. Mavridis, Chem. Soc. Rev., 33, (9)579(2004).
82. G. Pistolis, D. Gegiou, and E. Hadjoudis, J. Photochem. Photobiol. A. Chem. 93, 179 (1996).
83. E. Ziolek, A. Kubicki, R. Naskrecki, and A. Grabowska, J. Chem. Phys. 124, 124518(2006).
84. M. Knyazhansky, S. Aldoshin, A. Metelitsa, A. Bushkov, and O. Filipenko, Khim. Fiz. 10(7)964 (1991)(in Russian).
85. Z. Grabowski, K. Ratkewicz, A. Siemarczuk, D. Cowley, and W. Bauman, Nouv. J. Chim. 3, 443, (1979).
86. Z. Grabowski, and J. Dobkowski, Pure Appl. Chem. 55, 245 (1983).
87. W. Rettig, and W. Majenz, J. Photochem. Photobiol. A. Chem. 65, 95(1992).
88. W. Rettig in "Topics in Carrent Chem." Ed. Springer-Verlag, Berlin, Heidelberg. 169, 254(1994).
89. M. Knyazhansky, Polish. J. Chem. 82, 795(2008).
90. F. Milia, E. Hadjoudis, and J. Seliger, J. Mol. Struct. 177, 191(1988).
91. T. Rosenfeld, M. Ottolenghi, and A. Meyer, Mol. Photochem. 5, 39(1973).
92. R. Becker, C. Lenoble, and A. Zein, J. Phys. Chem. 91(13), 3509(1987).
93. P. Barbara, P. Rentzepis, and L. Brus, J. Am. Chem. Soc., 102, 2786 (1980).
94. R. Nakagaki, T. Kobayashi, J. Nakamura, and S. Nagakura, Bull. Chem. Soc. Jpn., 50, 1909(1977).
95. T. Yuzawa, H. Takahashi, and H-o. Hamaguchi, Chem. Phys. Lett. 202(3, 4), 221 (1993).
96. H. Fukuda, K. Animoto, H. Koyama, and T. Kawato, Org. Biol. Chem. 1(9)1578(2003).
97. M. Cohen, and G. Schmidt, J. Phys. Chem. 66, 2442 (1962).
98. E. Hadjoudis, Mol. Eng. 5, 301 (1995).

99. K. Animoto, H. Kanatomi, A. Nagakari, H. Fukuda, H. Koyama, and T. Kawato, Chem. Comm. 870 (2003).
100. T. Kawato, H. Koyama,, H. Kanatomi, H. Tagawa, and K. Iga, J. Photochem. Photobiol. A. Chem. 78, 71, (1994).
101. J. Harada, H. Uekusa, and Y. Ohashi, J. Am. Chem. Soc. 121, 5809(1999).
102. A. Simonova, R. Nurmukhametov, and A. Prokhoda, Dokl. ANSSSR, 230, (1)146(1976)(in Russian).
103. W. Richey, and R. Becker, J. Chem. Phys. 49(5)2092 (1968).
104. E. Hadjoudis, I. Moustakly-Mavridis, and I. Xexakis, Isr. Jour. Chem. 18. 202(1979).
105. J. Sitkowski, L. Stefaniak, T. Dziembowska, E. Grech, E. Jagodinska, and G. Webb, J. Mol. Struct. 38 (1-3), 177 (1996).
106. M. Knyazhansky, B. Simkin, and B. Golyanskii, Theor. Exper. Khim. 18(1), 108(1982) (in Russian).
107. M. Kobayashi, M. Yoshida, and H. Minato, J. Org. Chem. 41, 20, 3322 (1976).
108. T. Kawato, H. Koyama, H. Kanatomi, and M. Isshiki, J. Photochem. 28, 103(1985).
109. T. Kawato, H. Kanatomi, H. Koyama, and I. Igarashi, J. Photochem. 33, 199(1986).
110. K. Animoto, and T. Kawato, Photochem. Photobiol. C. Photochem. Rev. 6, 207(2005).
111. R. Andes, and D. Manikowski, Appl. Opt. 7, 1179(1968).
112. H. Koyama, T. Kawato, H. Kanatomi, H. Matsushita, and K. Yonetani, J. Chem. Soc., Chem. Comm. (5) 579(1994).
113. G. Pistolis, E. Hadjoudis, and I. Mavridis, Mol. Cryst. Liquid Cryst. A242, 483 and 489 (1994).
114. E. Hadjoudis, M. Vittorakis, and I. Moustakali-Mavridis, Tetrahedron 43, 1345 (1987).
115. V. Vargas, and L. Amigo, J. Chem. Soc. Perkin Trans. 2, 1142(2001).
116. K. Ogawa, and T. Fujiwara, Chem. Lett. (7)667(2000).

117. A. Mukherjee, R. De, I. Barerjee, C. Samanta, X. Chitra, and N. Nayak, Acta Cryst., C55, 407 (1999).
118. Z. Popovic, G. Pavlovic, D. Matkovic-Calogovic, V. Roje, and I. Leban, J. Mol. Struct. 615, 23 (2003).
119. K. Ogawa, and J. Harada, J. Mol. Struct. 647(1-3), 211(2003)

Chapter 4

The Bichromophoric Systems On The Base Of The Anil Structures.

Introduction and classification of the structures.

In the chapter the results concerning the complex bichromophoric molecular systems including the structural fragments of anils' molecules are considered. As it follows from the brief discussion conducted in the sec. 4. 2. 1. the interest to the study of such structures arises first as far back as the sixties –seventies mainly with the object of creating of the novel organic luminophores with the large Stokes shift of the fluorescence (see /1/).

However the principal and detailed studies have been carried out for the last two decades in connection with search of the new photo and thermochromic anils in the crystal state and with the purpose of the study of the photo and thermochromic reaction mechanisms involving the GSIPT and the ESIPT in various media.

In such systems the modification of the thermochromic, phototochromic, and fluorescent properties is realized owing to interaction of the two identical or strongly different structural fragments in each of which the opportunity of the ESIPT or(and) the GSIPT is assumed.

In this connection the discussion of the results are carried out following to the classification of the molecular systems that is based on the identity and the difference of the structural fragments forming such systems.

Therefore in the in the first (4. 2) section the photochromism of homochromophoric systems consisting of the two identical (o-hydroxyazomethines) (Sch. 4. 1) fragments (bissalicylideneanilines and their structural analogues), and the possible novel approaches to mechanisms of the photoinduced processes are discussed.

The series of the structures I, II, III, IV (Sch. 4. 1) is considered in the consecutive order with the supposed strengthening of the interaction between the chromophores HOPhCN responsible for the electronic excitation, and it should be expected their influence the photoinduced processes.

In the second sections(4. 3)the heterochromophore bimolecular systems with only one anil moiety while another fragment is ether photoinactive π-electron conjugated structures (V, VI, VII) or the different photochromic structure(VIII) are discussed(Sch. 4. 13).

There are the well known very skilled reviews /2-4/ devoted to the discussion of some crystalline bichromophoric systems.

Therefore the crystalline bichromophoric compounds are considered only briefly in each section with the purpose to show the main differences or similarities of the separate stages of the photoinduced processes and the photochromic properties for the bichromophoric systems in the crystals and solvents on the one hand and for the bichromophoric and monoanil molecules in the crystals on the other hand.

4. 1. Homobichromophoric molecular systems

4. 1. 1. The early studies of bieanils.

To the best of our knowledge the first results of the investigations of the absorption spectra and the fluorescent properties of the type I compounds have been published in the middle of sixties /5, 6/(see also/1/) and somewhat later /7/.

It is known also that the studies of the photochromic, thermochromic, and fluorescent properties of the compounds I, II have been started as far back as early seventies and eighties /8-11/.

In these works the investigations have been carried out by the steady-state absorption and fluorescent spectroscopy methods without utilization of any data of the calculations. The results have mainly qualitative and descriptive characteristics.

In the rigid matrices (thin PMMA films, methylcyclohexane-decaline at t=-180°C, paraphine at t=-75°C) the compounds I (besides $R_1=NO_2$) display the thermally reversible formation of the colored NH structure (absorption band with $\lambda max \approx 460nm$) under irradiation of the Enol (E) form ($\lambda_{ex}=365nm$) i. e. photochromism like SA. At the same time tautomeric equilibrium in the So state E⇔Kcis shifted towards nonphotoactive planar Kcis structure (with the absorption band with $\lambda^{max} \approx 450nm$) with cooling of the solvents to -75--180°C (i. e. the Negative thermochromism is observed unlike SA)/11/.

The more detailed spectral investigations of the I compounds have been carried out in various solvents at 293 and 77 K /8-10/(Tab. 4. 1).

The absorption spectra of $I(v^{max})$ in the liquid solvents at 293K especially for the compounds I_2, I_5 can be presented as an almost exact sum (ε^{max}) of the absorption bands of the two weakly interacting SA structural fragments responsible for localization of the first electronic transition.

The structural ASS fluorescence bands with v^{max}_{flu}, and the Sokes shift values (Δv ASS) of the I and SA molecules are identical practically.

Such a weak interaction between E moieties is secured by joining of N atoms of azometine groups to the meta (compound I (2, Ar)) or para (compond I(1, Ar)) positions of the same coplanar phenyl ring or to the different coplanar phenyl rings unseparated I(3, 4, Ar) or separated I(5, Ar) by methylene group.

It should be noted that the strength of the n(N)-π(Ph) interaction connected with the coupling of the chromophores' interactions can be determined by the differences of the So--S_1^* transition energy between I and SA molecules (Δv^{max}(I-SA).

The interactions are very small for the structures I(1, Ar)-I(5, Ar), don't exceed the vibrational quantum energy value(Δv(I-SA))$\approx$600-3550cm^{-1}), and provide the stronger stabilization of the S_1^*state as compared with the So one both in the E and K structures that is manifested by the long-wavelength absorption band shift for the compounds I as compared with SA (Sch. 4. 2).

Thus the least shifts are observed for the structures with the weakest interactions between the Enol and the Keto structures of the different moieties (I(2, Ar), I(5, Ar)).

The more weak interacttions of the Keto-moieties are caused by the peculiarities their electronic structures securing the clear-cut separation between the π-electron systems of the Keto fragments.

The peculiarities of the I(Ar) molecules are the shift of the E$\Leftrightarrow$K equilibrium in the So state towards K structure with the Kcis absorption band ($\lambda\approx$440-460nm) that appearances even in the nonpolar solvents at room temperature unlike SA, and the considerable decrease of the rate the PCF accumulation as compared with the separate SA molecule. The absorption bands of the omobichromophors(dianils) and model structures are discussed in the Ch. 4.

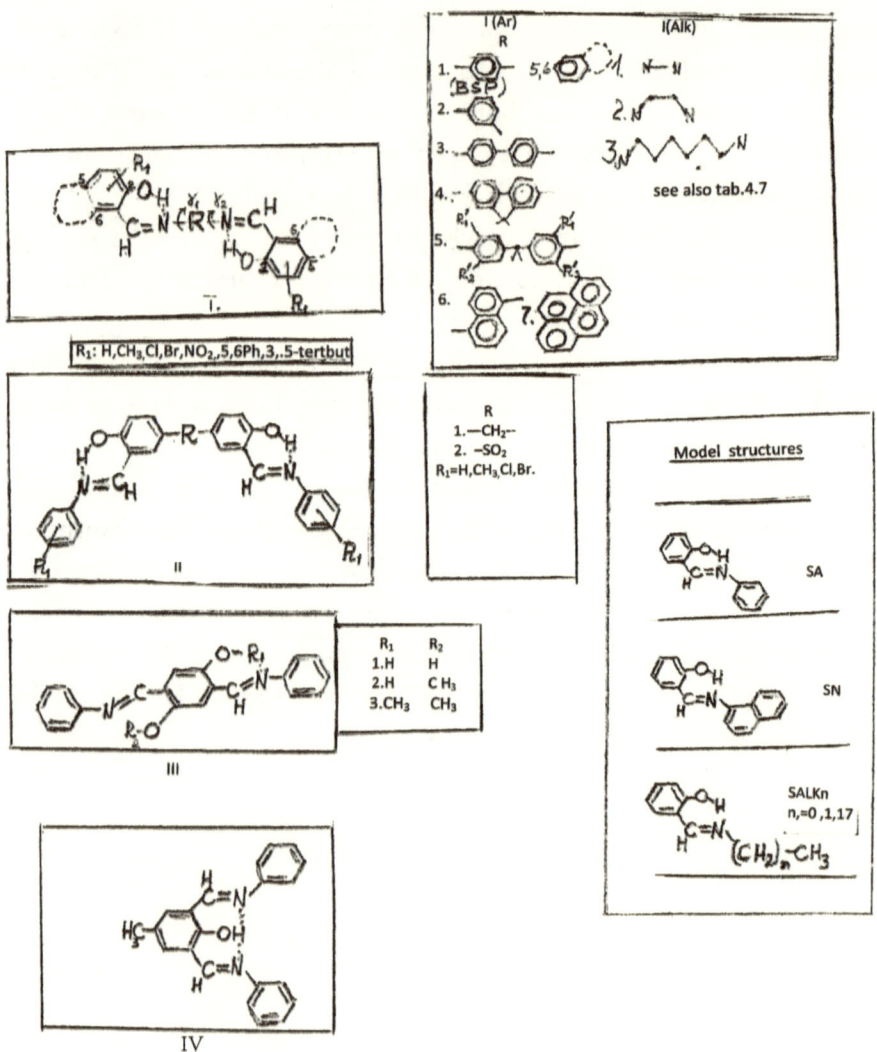

Scheme 4. 1
The Homobichromophors (dianils) and the model structures discussed.

Table 4.1
The comparative data of the absorption and fluorescence of the compounds I (IIP, room temperature) (see schemes 4.1 and 4.2).

Comp. I (Ar)	Absorption						Fluorescence	
	Enol			Keto				
	v^{max}_{abs} cm^{-1}	Δv^{max}_{abs} cm^{-1} (I-SA)	ε^{max}	v^{max}_{abs} cm^{-1}	Δv^{max}_{abs} cm^{-1} (I-SA)		N^{max}_{flu} cm^{-1}	Δv^{max}_{flu} (ASS flu) cm^{-1}
1	26596	-2990	----	21276	-724		----	----
2	28986	-600	28857	21276	-724		19131 / 17827	9855
3	27174	-2412	----	21052	-948		----	----
4	25974	-3612	----	20833	-1167		----	----
5	28986	-600	27654	21739	-261		9150 / 18050	9836
6 [1)]	27000	-1000 [2)]	----	----	----		19000 / 18000	9000
SA	29586	---	13428	22000 [3)]	----		19511 / 18406	10015

1)Butironitrile /12/.2)Comparatively with SANaphtylimine (SN). 3)IIP,T=77K.

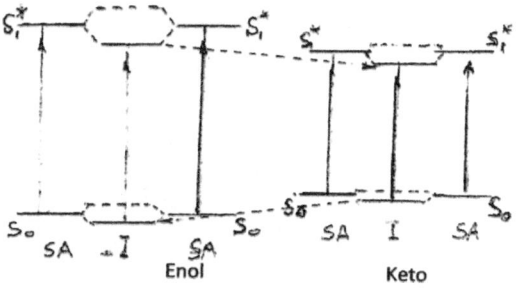

Scheme 4.2
Qualitative scheme of the two SA moieties which interaction in S_0 and S_1^* states of the structures I results in the long wavelength shift of the absorption bands I relatively SA(see table 4.1,see also values of ε_{max} of I in comparison with SA).

The PCF of molecules shifted($\lambda \approx$450-460nm) comparing with those of SA molecules, and it has therefore a more planar structure like solvato or thermochromic Kcis anil form.

Such peculiarities can be explained by the intermolecular interactions with formation of the H-bond dimers and by the steric hindrances preventing from the twist structure formation.

None of the compounds I(Ar) are photochromic in crystals but everything are thermochromic (coloration with the temperature increasing from 20° to 100-160°C) and have the ASS fluorescence at ambient temperatures.

Thus the results of the early studies of bisanils have already shown that the molecular systems of the I type consist of the two distinctly separated fragments(chromophores) by that the individual excitation of each can be secured.

Therefore the absorption, fluorescent and photochromic properties of the I compounds and SAs structures differ a little. However the weak interaction between the identical fragments and intermolecular interactions (between solute and solute-solvent molecules)can lead to the modification of the properties caused by the shift of the E⇔K equilibrium and by the changes of the Proton transfer rates in the So and S_1^* states.

It has been shown also that the compounds of the I and II types are of interest to obtain and to study of the new crystalline thermochromes and photochromes.

In conclusion one may suggest the results of the early studies have served as a base and the stimulus for the following development of the investigations.

4. 1. 2. The molecular systems with Aryl ring between imine moieties.

The data of the investigations of the type I(Ar) structures show that the separation of the identical fragments leads to both preservation of the likeness and appearance of the marked differences as compared with the initial anil molecules(SA).

For the investigation of the typical peculiarities of the complex molecular systems I(Ar) with the Aryl bridge it would have be more expediently to study the models I(Ar)with the most clearcut separated structural fragments joined to the meta position of the single phenyl ring, I(2, Ar), or separated by the several phenyl rings, I(3or4or5, Ar). However unfortunately such an aspect has been taken into account insufficiently when selecting the object to the study.

It has been supposed, however, in the compound I(1, Ar) in which the joining of the anil frag-ments to the para positions of the single phenyl ring may nevertheless secure their separation sufficiently. In this connection the compound I(1, Ar) is discussed in the detail in the beginning of consideration of the type I molecular systems.

4. 1. 2. 1 The structure and the Interfragmental ground state interaction of BSP molecule in solvents.

According to the data of the structural analysis /12-14/ the molecules I(1, Ar)(BisSalycildenePhe-

Nylenediamine, BSP)in the crystalline state are centrosymmetric and planar($y_1=y_2 =0°$)unlike SA ($y\approx49°$)/15/.

However in the liquid solutions/12, 16/the barriers for rotation around C-N single bond (angles y_1, y_2)are very low (for the isolated molecule the calculated value is 4kcal/mole) that is consistent with the NMR data and with the ground state experimental($\mu_g \approx 2. 8 0. 5D$)and calculated ($\mu_g=3. 0 D$)dipole moment values.

Thus according to above produced data the both "halves" of the molecule BSP in the solvent are almost freely rotating around the C-N single bond and the nonpolar rotamers contribute to the average value of the dipole moment.

The comparative analysis of the absorption and ASS fluorescence excitation spectra of I(1, Ar) molecule shows that the E-rotamers absorbing in the shortwavelength region ($\lambda<350nm$) don't take part in the origin of the ASS fluorescence and probably in the ESIPT/12, 17/.

At the same time the findings of the comparative analysis of the absorption spectra of BSP and SA molecules in different solvents /12, 16-21/(table 4. 2)allow to draw the conclusions:

(i). The value of the molar absorption coefficient (ε^{max})in the maximum of the E structures' longwavelength absorption band of BSP is not larger than doubled that for SA(ε(I, max)/ ε(SA, max$\leq$2)) see Tab 4. 2 column 3)however a little difference-2(SA, max) can be caused mainly by the marked contribution of the highly intensive Kcis absorption bands (So--S_1^* or So—S_2^*) in the E structure absorption band when shifting of E$\Leftrightarrow$K equilibrium towards K in the compounds I(1, Ar)as compared with SA (see below).

Thus the E-absorption band has additive nature. Such an additive character of the absorption band reflects the independence of the separate SA chromophore excitation. Really according to the data of the theoretical calculations /10, 22/ the excitation of I(1, Ar) is localized on one half of the molecule.

(ii). The Enol absorption band is shifted towards longwavelength (the negative shift) by less than $\Delta v^{max}_{abs}) \approx 3000 cm^{-1}$ as compared with that of SA (Tab. 4. 2, column 3). Such a shift is also connected with the weak interfragmental interaction which stabilizes the E structure mainly in the S_1^* state(Sch. 4. 2)and does not exceed of the vibration energy quantum value($<3000 cm^{-1}$)

(iii). The absorption band of the K structure unlike E absorption band is shifted towards short wavelengths (the positive shift) by $\Delta v(max, abs) \approx +500-1000\ cm^{-1}$ as compared with the absorption K-band of SA(Tab. 4. 2, column 5).

It points to the interaction between K structure moieties (K-K) or E and K structure ones (E-K) in compounds I (1, Ar) that is considerably weaker than E-Emoieties' interactions(see (ii)) and stabilizes the K structure mainly in the So state unlike E-E in-teraction (Sch. 4. 2)*).

As compared with the latter, the E-E and E-K interactions are caused obviously by the more clear separation of the π-systems of the chromophores due to the K structure peculiarities.

(iv). The marked shift of the equilibrium E⇔K in the So state towards K structure as compared with SA (Tab. 4. 2, column 6) can be explained partially by the difference in stabilization of the E and K structures in comparison with SA (see (ii), (iii)).

However the shift value depends strongly on the solvent nature (column 6), especially on the proton donating ability of the solvent.

---

* The data differ from those obtained in the early works (see Sec. 4. 1).

Due to intermolecular H-bond with alcohol molecule the stabilization energy for the cis Keto tautomer is nearly twice higher than that for the Enol structure and increases with the proton donating ability of the solvent /20, 21/.

Such an effect can be attributed to the strong H-bonding between the solvent and Oxygen atom of the carbonyl group /17/.

Thus the compounds I(1, Ar) in the So state consists from the two weakly interacted SA moieties connected by the almost freely rotated phenyl group in the liquid solvents. The SA moieties, in which the E⇔K equilibrium is weakly shifted towards K structure can be excited almost indep-endently.

4. 1. 2. 2. The peculiarities of the photoinduced processes in the solvents.

A. The notion of a single fragment activity in the excited state.

The photoinduced processes of BSP can be described by the scheme 4. 2a /17, 20/ that is based on the notion of the activity of the only one of the independent SA molecular fragments with participation of the solvent (especially alcohol) molecules.

Table 4.2
Absorption and fluorescence of bisazomethine I(1,Ar) (BSP) as compared with SA (Last row) in the solvents.

Absorption						Fluorescence				5)	Ref.
Enol			Keto			v^{max}_{flu} cm	$\Delta v_{Eabs-flu}$ cm^{-1}	Φ_f x 10^{-3} 6)	τ_{ps} k$_r$ s^{-1}	Solv. (room temp)	
V^{max}_{abs} cm^{-1}	Δv^m_{abs} . (I-SA)	$\epsilon(I)/\epsilon(SA)$ [M^{-1}Lcm^{-1}]	v^{max}_{abs} cm^{-1}	Δv (cm^{-1}) (I-SA)	K/E						
26818	-2700	-----	22500	+500	0.9	17500 18860	9300	1,0	60 4) ----- 2x10^7	BN 173K	/12/ /16/
27250	-2550	1) 2.4	2) ----	2) ---	2) ----	18000 18600	~9250	----	----	ACN	/18/
27000	-2700	2) 2.6 (35000)	22900	+900	0.08	17640 18360	~9360	1.0	11±3 ----- ~10^9	ACN	/19/
27200	---	---	22200	---	0.000	17540 -----	9660		23	HX	
---	---	---	---	---	0.000	17860 ----	-----		30	CLP	
27000	---	---	22210	---	0.003	17540 ----	9460		10	ACN	/17/
27000	---	(34000)	22220	----	0.008	18020 ----	8980		32	Eth	
27900	---	(29000)	22730	----	0.24	18350	9550		35	TTFE	
28653	---	-----	22990	-----	0.72	18398	~10300		37-40	HFIP	/20/
SA 29800	----	7) (13460)	2) 22000	---	---	18266 18933	10870	7) 0.12 0.6 8)	8 3) ----- 1.5x10^9	ACN	/18/

1) Obtained from the ratio of the solvent concentration with the equal absorbencrs (fig 1 /18/). 2) See table 4.1. 3) Methyl cyclohexane at the room temperature (P.Burbara, P. Rentzepis, L.Brus, J. Am. Chem. Soc. 102, 2786 (1980). 4) T.Inabe, New J. Chem. 15, 129 (1980). 5) BN-Butironitrile, ACN- Acetonitrile, HX-n-Hexane, CLP-1-Cloropropamne, Eth-Ethanol, TFE-Trifluoroethanol. 6) Averaged value of the Stokes Shift 7) ACN /23/. 8) BN /12/.

The instantaneous (instrumental function limited <50fs) increase of the intensities of the absorption transient bands is typical for the S_1^*NHflu (K*cis) state. Thus the generation of the latter by the ESIPT OH-->NH proceeds with time τ<50fs.

The ultrafast ESIPT competes with the very fast structural transformations in the Enol structure including the Aldehyde ring twist with the H-bond break which leads to the formation of cisenol structure via the conical intersection of the $S_1^*(\pi\pi^*)$ state with the $S^*(n\pi^*)$ one or with the So state directly according to the differing calculated data by the different methods(see Ch. 3).

The cis-Enol structure transformations to the initial trans-Enol one proceeds with the very low rate($\tau>10$ms) comparable with the colored structure decay rate (see below). Such Enol structure transformations are similar to those of SA molecule (see Ch. 3).

The global kinetic analysis of the post-excitation(impulse duration $\tau=100$fs)spectral changes/20/ (HFIP solvent)results in the three-exponential functions with time components $\tau_1=400\pm150$fs, $\tau_2=2.0\pm0.5$ps, and $\tau_3=35\pm5$ps.

The time component $\tau_1\approx400$fs is associated in the strong protic solvents with the additional reaction of the S_1^*NHflu state generation through the solvent assisted ESIPT by the intermolecular H-bond.

The component $\tau_2\approx2$ps can be connected with the vibration deactivation S_1^*NHflu--->S_1^*NHflu after the ESIPT(S_1^*E->S_1^*NHflu) and, at last, $\tau_3\approx35$ps is assigned to the time rate of the S_1^*NHflu deactivation resulting in the colored form(trans-keto structure) generation /20/ or adiabatic formation of the TICT-like structure which is a precursor of the colored structures according to the notion of the twisted post-TICT nature of the latters (see Ch. 3 and /24/). Really the ASS fluorescence lifetime is practically coincide with τ.

Thus the nonradiative deactivation of the S_1^*NHflu state(or Kcis*one)consists of the two com-ponents/21/:the temperature dependent structural transformations leading in the final analysis to the PCF structures formation, and temperature independent internal conversion (S_1^*NHflu- ->So) that is dominant below 200K (see alsoTab. 4. 3 and below).

The parameters of these components are markedly different of BSP and of SA structures.

According to /17, 18, 20/ (see also Ch. 3)the PCF of BSP is a trans-Keto(Ktrans)(around $C_1=C_7$ bond structure generated only in the one of two SA moieties.

However in the solvents of various nature there are at last the two photocolored structures differed by the decay time rates and moreover have the opposite signs (rise and decay) in the polar and protic solvents (Tab. 4. 4b).

Thus the interconversion between the different colored post-TICT structures generated via the "TICT-like" one can occur (dashed lines in the Sch. 4. 3) like of SA (see Ch. 3).

Although the spectral distinctions between the colored structures of SA and BSP are negligible, the generation and decay time rates are markedly higher of the latter (Tab. 4. 4a), that is connected probably with the steric hindrances.

The detailed investigations of the decay time rates in dependence on concentration in the solvents of the different polarity and protic strength lead to the conclusion of the wide range decay time rate constants with the considerable role of the intermolecular interactions including formation of the H-bonded complexes with a very strongly proton-donating solvent molecules.

Thus there are three main mechanisms of the PCF structure decay in the solutions /20/:

i. Monomolecular mechanism via formation of the cis –Keto structure and back GSIPT NH->OH in all solvents.
ii. Double proton transfer in the complex of the two BSP molecules in nonprotic polar and non-polar solvents.
iii. In the protic environment the decay is catalized in the complex between one BSP and one solvent molecules and the complex formation lowers strongly the rate of the bleaching reaction in the polar and and protic solvents

ESIPT Photochromism

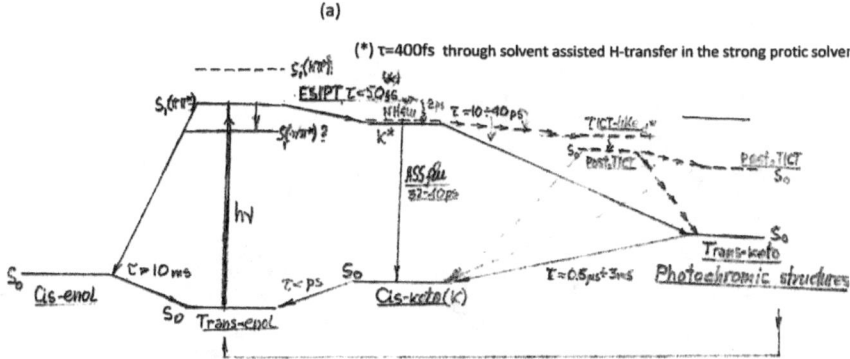

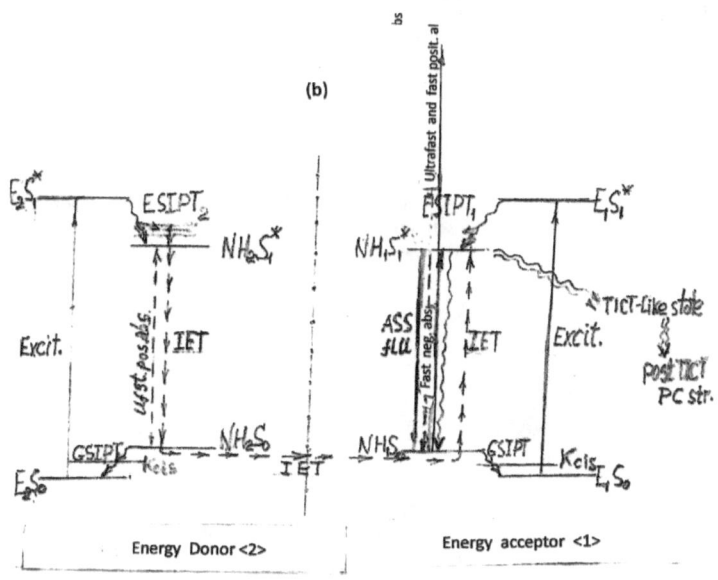

(see next page)

Scheme 4.3 (a), (b).
See next page for Scheme (c).

(c)

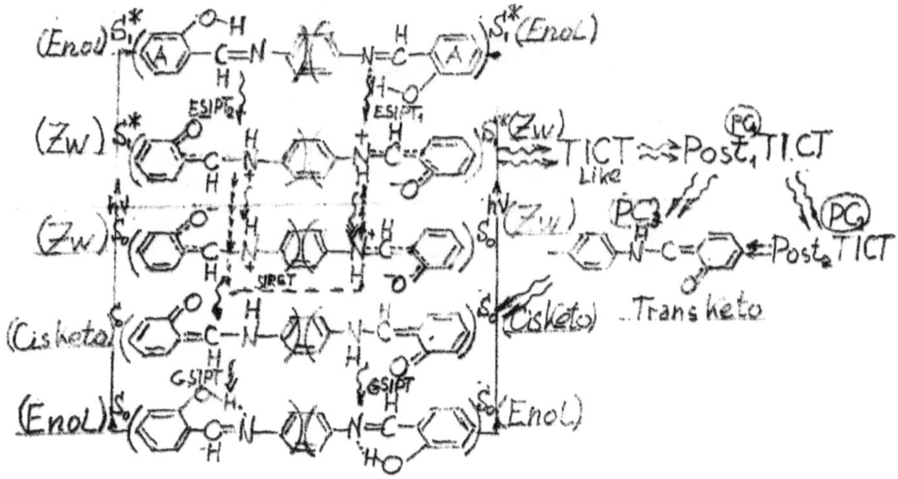

[Scheme 4.3 diagram]

Scheme 4.3
Schemes of the ESIPT photochromism in molecule BSP ,I (1,Ar),involving Intramolecular (interfragment) radiationless energy transfer(SIRET) a)Energetic scheme proposed in /20/ and alternative interpretation proposed by author(state position and transitions are shown by the dash lines).b)The supposed energetic scheme involving S_1^*-S_1^* (SIRET) between molecular fragments .c) The supposed structural scheme of the process with the SIRET.

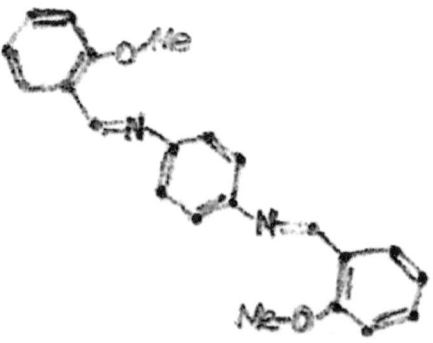

Scheme 4. 4
The model structure I(1, ArilOCH$_3$) without ESIPT and with the E structure of the both azomethine moieties.

Therefore efficient of coloration differs strongly in the different solvents, and location in cilica sieve(M-41)results in the appearance of the photochromic properties of the solid crystalline powders(unlike crystaline BSP) with increase in intensity (after prolonged irradiation by λ<400nm within hours) of the diffuse reflectance band (λ=550-700nm) of the photochromic structure(probably trans-Keto tautomer with a very long lifetime (several days in the dark) and high stability under irradiation with visible light (λ>500nm)/146/.

B. The supposition of the interfragmental interactions in the excited state.

The photoinduced processes of the compound I(1, Ar) have the very clearly expressed peculiarities which can be caused by interactions of the SA moieties in the excited state.

Realy as long as the two moieties are been excited independently, i. e. electron redistribution with excitation is localized on only one of them, one can supposed that an interaction in the S_1^* state between them may be caused by the Singlet-Singlet Intramolecular Radiationless Energy transfer (SIET).

There are the following principal potentialities of the SIET for the compounds with the two separated anil molecules:

(i). S_1^*(Enol)A<==>S_1^*(Enol)B-The energy exchange (migration) between the two S_1^*states of the identical fragments (A and B) with the equal S_1^*state energies which competes with the ESIPT.

(ii). S_1^*(Enol)A-->S_1^*(Keto)B-The energy transfer from the S_1^*excited Enol structure of the A-moiety to the Keto-structure ground state of the B-moiety which also can compete with the ESIPT.

(iii). (S_1^*NHflu)A-->(S_1^*Keto)B-The energy transfer after the ESIPT from the excited NH* structure of the A-fragment to the moiety B with the Keto structure in the Ground state that can compete with the ASSflu and the PCF generation but not with the ESIPT of the A-moiety.

(iv). $(S_1^*NHflu)A \rightarrow (S_1^*NHflu)B$—The exchange of the Energy between identical S_1^*NHflu states generated by the consecutive ESIPT reactions in A and B structural fragments.

The rest types of the suggested SIET processes concerning the participation of the Keto-structure moieties have a neglected small probability.

The SIET can be carried out with the most probability by the Forster dipole-dipole inductive resonance mechanism, and the following conditions are necessary for the effective realization of such a mechanism (see i. e. /26/).

(a) The excited singlet state energy of the acceptor (Ac) moiety must not be higher the donor (D) moiety singlet state energy: $E(S_1^*)(D) \geq E(S_1^*)(Ac)$.

(b) The separation between the SA moieties must be small enough because the efficiency of the SIET depends of the inverse sixth power of the distance between D and A.

(c) The high value of the probabilities of the electronic transitions for the emission (D) (the k_r is high) and for the absorption (Ac) (The oscillator strength, f, is high).

(d) The high value of the overlap of the energetic intervals of the radiation deactivation of the Donor and the excitation of the Acceptor (the overlap integral).

Table 4.3

Evolutions of the transient absorption spectra in the fast and ultrafast temporal intervals for I (I,Ar) and SA in CAN at room temperature (/17/,/19/,/23/.

Spectral range λ_{nm}	Temporal interval and lifetime (τ_{ps}) of the spectral changes *) SA	I(1,Ar)	Corresponding Transition (absorption)	Assignment of the Spectral evolutions.
375-390	----	--(0.05÷1.3) ↑ $\tau_1 \approx 0.5÷1.5$ $\tau_2 \approx 11÷13$	$E(S_0 \to S_1^*)$	Fast component is the S_0E state repopulation after pump pulse excit.
400-600	Absent +(0.5 ÷100) ↓	+(0.05÷0.50) ↑ +(0.5 ÷100) ↓	$NH(S_1^* \to S_n^*)$ $NH(S_1^* \to S_n^*)$	$S_1^*NH_{flu}$ ultrafast popul. $S_1^*NH_{flu}ASS$ flu depopul.
600-660	Absent - (0.5 ÷100) τ=6.8	+(0.05÷0.30) ↑ $\tau_i \approx 0.05;1.5;3.2$ --(0.5 ÷ 100) ↑ $\tau \approx 10.5$	$NH(S_0 \to S_1^*)$ $NH(S_1^* \to S_0)$	S_0 NH ultrafast popul. $S_1^*NH_{flu}$ depopulation (ASS flu decay).

*) " – " and "+" - The Negative and Positive absorption bands respectively.
↑and ↓ -The increase and the decrease of the transient absorption signals correspondingly.

Table 4.4
The spectral and kinetic parameters of the PCF parameters of BSP.

a)The comparative characteristic of the PCF structures of BSP and SA (CH_3CN, at room t –re) /18/.

b) Rate time dependence on the solv. nature

Compound	Absorption band,λ_{nm} [a]	Fluorescence band, λ_{nm}	Relative gener.eff.	Gener. time rate(ps)	Decay time rate(ms) [2]	Rate time, μs		
						Solvent	τ_1	τ_2
BSP	<u>475</u> 490	625	1.7	36 [3]	5.0±0.4 3.1 [3]	Hex	20(dec.)	100(dec.)
						ACN	600(dec)	100(rise)
SA	475 <u>490</u>	560-600	1.0	~7 >15 [4]	20 ±3	EtOH	250(dec)	130(rise)

1)Butyronitryle ar the room temperature .2)The maxima are underlined. 3)From /19/.4)From /23/..
Hex .-Norm hexane, ACN-Acetonitrile, EtOH-Ethanole

The efficiency of SIET can be varied within wide limits in the flexible structures in dependence on the fragments' separation (R), orientation factor (0<P<1) determining the mutual orientation of the D and Ac transition dipole moments and possible interactions between their π electron systems.

Therefore it is hardly take exactly into account and calculate the all above mentioned factors with consideration of the SIET in the molecules I(1, Ar) in the solvents but it is possible to estimate the contribution of each from their.

It is expected due to the condition (c) that the potentialities (i) and (ii) are not feasible because the S_1*(Enol) state is extremely weakly fluorescent and instable (see Ch. 3). The criticisms excluding the type (ii) SIET /27/are obviously correct but they do not unfortunately take into account another SIET types(see below).

The possibility (iii) is hardly probable due to the marked endoenergecity of such a process ES1*NHflu-NHketo =5000cm^{-1}..

The process (iv) is the most probable in such a situation. It cannot compete with the ESIPT and may have a nature of the energy migration between the two S_1*NH(1) A and S_1*NH(2)B structures.

However the efficiency and recurrence of its development depend on the energy transfer rates in both directions (+k(et) and –k(et)) and on the stability of the NH structures in both S_1*and So states.

On the base of the findings of the molecular structure analysis/28/ and from the numerous data of the inter and intra ET studies /29/ the Donor –Acceptor distance R and the critical one Ro can be estimated for I(1, Ar) as 15 and 60Å respectively.

With the above values of R an Ro, the high orientation factor and the ASS fluorescence lifetime$\tau \approx n \times 10^{-12}$s(see above)the rate value of SIET may reach Ket~10^{12}-10^{13}s^{-1}($\tau \approx 1$/nfs). The such an ultrafast SIET can occur with the efficiency (ET)$\approx$95-100% securing the complete deactivation of the S_1*NHflu state of the Donor moiety after the ESIPT.

The strong resonance conditions are broken fastly in theAcceptor moiety by a very fast of first step of adiabatic and vibration deactivation ofS1*NHflstate (10^{-11}-!0^{-12}s). The SoNH state corresponding S_1*NHflu state is also depopulated very fast (Tab. 4. 3). Therefore there are no any experimental indications of the reverse SIET.

Thus the irreversible SIET of the type(iv)is provided by the molecular structure of the I(1, Ar) compound and has to secure its differences from SA in the mechanisms and kinetics of the fluorescence and the photochromic reaction.

The schemes of the principal photoinduced processes with the ESIPT reactions and the SIET involving the both SA moieties are produced(Sch. 4. 3b, c). These processes occur by the two stages corresponding to the separate excitation of <1>(A) and <2>(B) moieties.

The processes begin after the $S_0 \rightarrow S_1^*$ excitation of only <1>SA moiety with Enol(E) structure. Like SA the ultrafast ESIPT OH->NH(τ<50fs or 100fs) competes with the radiationless deactivation as a result of the A-ring twist with H-bond break.

The population of the S_1NHflu state can occur either immediately after the ESIPT or via vibration levels(S_1^*NHflu~~>S_1^*NHflu) without participation of the flattened S_1^*NHtr structure also forming directly as a result of the ESIPT (see e. g. /19/).

The S_1^*NH$_1$ state is the precursor of the PCF which are generated via the "TICT-like" state of the <1>SA moiety as of SA. At the same time the S_1^*flu state is deactivated by the emission (ASSflu) and by the solvent assisted vibration relaxation to the low stable ($\tau_{NH1} \approx$ 1ps) ground state SoNH$_1$ structure S_1^*NH$_1$flu~~SoNH$_1$ with time rate $\tau \approx$ (10-11)ps.

The latter is different from the Cis-Keto structure and plays the decisive role in the second stage of the photoinduced process.

The second stage is beginning after E-structure excitation E So-hv->E $S_1^*$of the SA moiety <2>.

The high vibration levels (S_1^*NH) are generated by the ultrafast ESIPT (τ< 50fs) with the following vibration relaxation to the S_1^*NH state just of the <1> moiety. In this step the ultrafast SIET of the type (iv) can start competing with the deactivation processes.

The ASS emission and the generation of the PCF structures in the moiety <2> are suppressed completely by the ultrafast SIET(S_1^*NH flu--SIET-->S_1^*NH$_1$flu with a population of the S_1^*NH$_1$flu or Š1*NH1flu states(with the subsequent vibration relaxation).

Such a <1>S_1^*NH$_1$flu state repopulation is becoming possible within a narrow temporal interval between the decay of the ground state NH$_1$ structure arising due to the ESIPT$_1$ in the <1> moiety and the ESIPT with generation of the NH S_1^*flu state in the <2> moiety(10ps<τ<13ps).

The ultrafast population of the SoNH state in the <2> moiety and the corresponding additional population of the NH$_1S_1^*$state of <1>moiety due to the ultrafast SIET (NH S_1^*-->NH$_1S_1^*$) is displayed distinctly as a very fast arising of the positive absorption bands (0. 05-0. 5ps)(Tab. 4. 2 and "ufst. pos. abs" in the <1> and <2>moieties in Sch. 4. 3) which are absent, as it is expected, of the separate SA molecule.

The negative ($\lambda^{max} \approx 630$nm) following after the complete decay of the fast positive absorption band is connected with the stimulated emission band of the ASS fluorescence($S_1^*NH_1$flu –hv–>$SoNH_1$)with lifetime $\tau=10.5$ps like SA($\tau=7.0$ps).

Unfortunately in that spectral region there are also the positive transient absorption bands connected with the $S_1^*(\pi\pi^*)$state of the E structure /17/.

However the shapes and the kinetic parameters of these bands and the analogues characteristics of the "SIET" bands are differed markedly.

Really, the broad E-structure band of the model compound I(1, ArMe)(Sch. 4. 4) has the tic profile very similar to the instrumental function and monoexponential decay ($\tau \approx 100$ps)while the much more narrow positive bands of I(1, Ar) have the three values of the fitted decay parameters ($\tau \approx 50$fs, 1, 5ps, and 32ps)convoluted with the instrumental function/17/.

Thus the ultrafast evolutions in the spectral region of the positive absorption bands about 550 and 630nm of I(1, Ar)are associated with the SIET although can be overlapped with the spectral modifications caused by the deactivation of the excited E structure.

It results from the above the additional population of the $S_1^*NH_1$flu state has to be promoted by the SIET $S_1^*NH_2 \rightarrow S_1^*NH_1$ and the rise in the efficiency both the ASSflu and the PCF generation has to be provided by the SIET.

Really the efficiency of the PCF generation in I(1, Ar) is two times as much SA(Tab. 4. 4) and the ASS quantum efficiency is by order of magnitude higher that of SA (Tab. 4. 3). The significant increase of the latter may be caused by the exclusion of the vibrational relaxation with the population of the $S_1^*NH_1$flu state by the SIET.

Unfortunately the ground State depopulation (GSP) in the E structures of I(1, Ar)and SA(negative bands at $\lambda \approx 375$-390nm with two components of decay) (Tab. 4. 3) cannot be compared due to the lack of the realizable data for this spectral region of SA /23/.

Therefore it is difficult to bind the peculiarities of the GSD kinetics with the SIET of I(1, Ar).

Thus the I(1, Ar) molecule is the typical example of the two distinctly separated SA moieties between which the Intramolecular Radiationless Energy Transfer (SIET) can occur as a result of the consecutive ESIPT reactions in each moiety. Owing to the SIET the ASS fluorescence and the PCF generation with the enhanced efficiency are localized on only one moiety. Such a localization and the peculiarities of the photoinduced processes have not been interpreted previously.

4. 1. 2. 3. The photoinduced processes in the crystals.

The bichromophoric systems of the I(Ar) type have the thermochromic crystalline structures.

Their own distinctive feature is the close packed flattened adjacent molecules in the parallel (d≈4-5 Å) planes of the crystallisne packing /27, 30/ (see also Ch. 2).

The peculiarities (i.-iv.)caused by such a crystalline structure can be described by follows (Sch. 4. 7):

i: The strong steric and electrostatic interactions between the both aldehyde fragments (with A-rings) of the adjacent molecules involving the H-bonds O-H...N in the ground (So)and the exci-ted (S_1^*)states.
ii: The coplanarization of the bridge imine rings (B), their strong π-π interaction of the both frag-ments of the adjacent molecules and the lack of their rotation.
iii: The steric hindrances for the skeletal deformation and vibration modes in the crystal packing.

The dipole-dipole interactions of the fragments involving the H-bonds with a strong electrosta-tic components(i) and the hindrances of the deformation vibrations (iii) promoting the bar-rierless ESIPT OH->NH without intrinsic proton motion in the solvents(Ch. 3) favor the origin of the potential barrier in the ESIPT reaction.

On the other hand the absent of the temperature dependence, the marked deuterium effect, the H-bond length dependence, and other typical effects concerning the ESIPT rate testify to the tunnel mechanism of the ESIPT reaction (see Ch. Sec. 3. 2. 4. 1.)Thus the sharp fall of the ESIPT rate in the crystals compared with the solvents is caused by the essential change of the reaction mechanism. At the same time the inhibition of the A-ring twist (i) promoting the fast radiationless deactivation competing with the ESIPT in the solvents (Sec. 3. 1. 1.) leads to the arising of the fast E-structure emission ($v \approx 19000 cm^{-1}$, $\tau \approx 100n$, ps)competing with the ESIPT unlike the solvent (Sch. 4. 10). The flattening and interaction of the imine rings B of the adjacent molecules (ii) stipulate the increase of the efficiency of the Hydrogen atom transfer in the Ground state (GSIHT) with the formation of the planar colored Kcis structure to be one of the conditions for tyev thermochromism in the crystal.

All described above effects are peculiar not only for the bichromophoric molecules but to the monoazomethine ones also.

However an acoplanarity of the different azomethine moieties and various structural ragments inside each moiety, and especially the coplanarity and interactions of the bridge phenyl fragments with both different moieties of the bichromophoric molecules (ii) deprive of the autonomy of the anils' moieties π-electron systems, stipulate their interaction, and make their independent excitation impossible. Hence, such π-systems are excited as a whole ones, and the second anil moiety can be considered as a substituent in the imine fragment of the first one. Thus the $S_1^*(A) \rightarrow S_1^*(B)$ radiationless energy transfer between the separate anil moieties A and B in crystal is absent unlike solvents.

The general scheme of the photoinduced processes in the crystalline bichromophoic systems I(1, Ar) are based on the data /27/ and modified by us (Sch. 4. 5). It differs a little from the analogues scheme for the thermochromic monoanils as the ESIPT occurs only in the one of the SA moieties suppressing completely the tunnel ESIPT in another moiety.

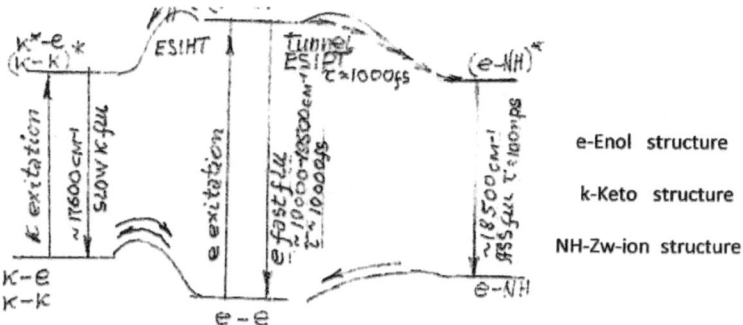

Scheme 4.5
The photoinduced processes in the crystalline I(1,Ar)

e-Enol structure
k-Keto structure
NH-Zw-ion structure

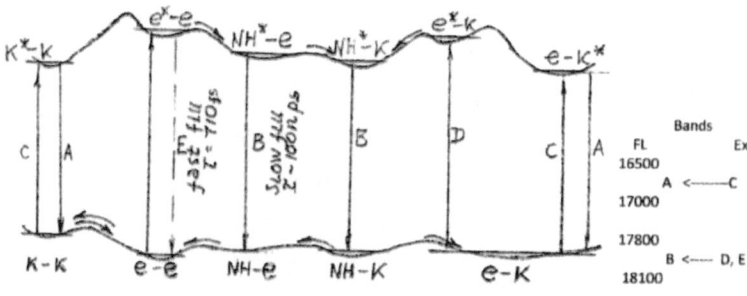

Scheme 4.6
The photoinduced processes in the crystalline I (1',Ar) and the nature of the various fluorescence bands.

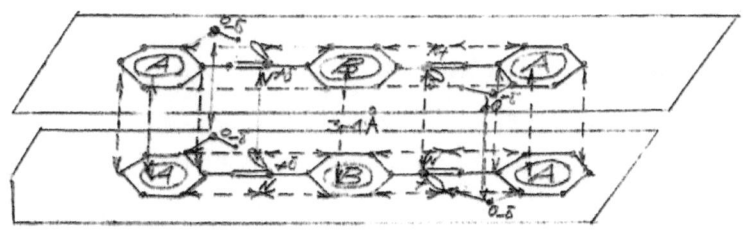

Scheme 4.7
The qualitative scheme of the mutual disposition, the steric, π-π(- - -), and dipole-dipole Interactions of the bischromophore molecules in the thermochromic crystals I (1,Ar).

The fluorescence excitation spectra in the crystals are distorted by the scattering and shifted towards "red" very strongly understating the Stokes Shift greatly. Unfortunately it is not taken into account by authors /27/(see also Tab. 3. 30).

Although such spectra don't allow of determining correctly the positions of the longwavelength absorption thermochromic bands, they give an opportunity to estimate the strong increase of their relative intensities with the temperature growth (14-290 K)i. e. the equilibrium shift in the So state towards the fluorescent Kcis structure in one or two moieties.

The excitation of the (e-e)form leads to the ESIPTwith the tunnel mechanism which competes with the fast fluorescen-ce of the Estructure, and the ESIPT rate is determined by the fluorescence decay time($\tau \approx 1000fs$).

The (e-NH*)structure is generated with such a rate and gives the ASS flu with lifetime $\tau \approx 100n$, ps and the rise time $\tau \approx 1000fs$. The NH*->NH fluorescence band is shifted towards "blue" relative to the K*cis –>Kcis fluorescence band ($\Delta\nu \approx 700$-$800 cm^{-1}$) and has the typical fine vibration structure. The analogous shift of the monoanil thermochromic structure (5CL-SA)-($\Delta\nu = 1000 cm^{-1}$).

The formation of the quenching Twisted S_1^* state in crystals is excluded due to the steric interacttions of the aldehyde rings of the adjacent molecules, and therefore the ASS fluorescence has a high intensity both at 77K and the room temperature.

Thus the mechanisms of the photoinduced processes of I(1, Ar)and SA molecules, differing stro-ngly in the solvents, changes considerably(including ESIPT), and become similar in the thermochromic crystals.

The inclusion of an additional conjugated ring in the aldehyde π-electron system, I(1', Ar) results (like monoanils)in the delocalization of the π-electron density and in the K-structure stabilization that makes for almost equal population of the Kcis and E-structures in the E⇔K equilibrium in the crystals at the room temperature. /27, 30/.

The scheme produced (Sch. 4. 6)is constructed on the base of the scheme and the spectral data of /27/, and modified with due regard for the recent findings of the ASS fluorescence nature. Such a scheme meets the experimental data /27/very good.

The identical anil moieties bound each with other cannot be excited independently. According to the scheme the excitation of the one of the

Enol (e) structures (e-e) leads to the ESIPT with production of the of the (NH*-e) form.

As a result of the strong electron withdrawing action the ESIPT does not occur in the second fragment.

In the first fragment the ESIPT competes with the lowintensive short-wavelength fluorescence with lifetime $\tau=710$ fs that corresponds to the ESIPT rate.

The tunnel ESIPT reaction like other crystalline anils occurs with the higher rate through a more low (thin) potential barrier due to the shorter H-bond (Ch. 3, Tab3. 22).

The ASS fluorescence includes the very close B bands (Sch. 4. 6) with the life time $\tau=100$ nps corresponding to the decay time of the two NH* forms (NH*-e and NH*-k) in the different bichromophoric structures (e-e) and (e-k) respectively, and the latter gives the fluorescence with the Stokes z shift that seems to be small.

The excitation of the K structure (the bound C) in the two different bichromophoric structures (k-k) and (e-k) gives the compound band A of the K structure (K*->K).

Thus the main difference between the crystalline bichromophores I (1, Ar) and I(1', Ar) are caused by the strong shift of the E⇔K equilibrium towards K structure in the latter.

The bisanils I(Ar) with the o-phenyl bridge as those with the p-Ph, I(1, Ar), are not photochromic /31/, planar and stacked in the parallel planes to form one-dimential columns/32/like the other thermochromic anils. In contrast bisanils I(Ar) with the meta-Phenil bridge, I(2,, Ar), have the more loose crystalline packing and show photochromism.

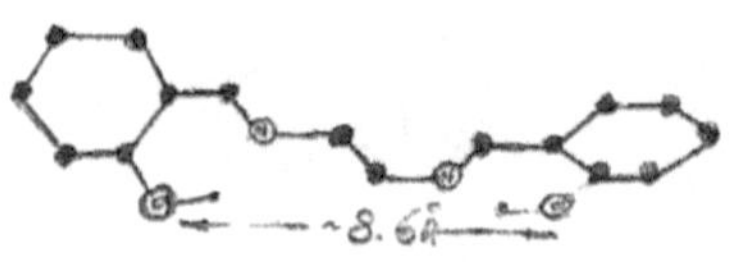

I (2,Alk) 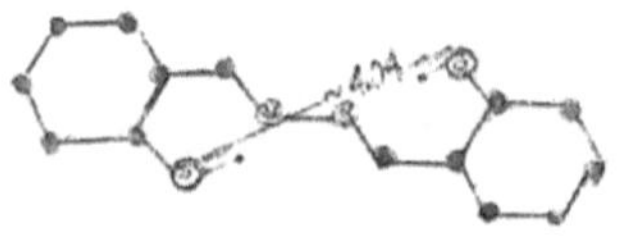   I(1,Alk)

I(3,Alk)

Scheme 4.8

Molecular structure analysis /37/ of the interaction between anil moieties of I(1,Alk),(2,Alk),and I (3,Alk) structures. The optimal structures calculated by AM1 method.

The bisanils with the broad π-electron systems bridge ring (I(7, Ar)/33/also form in the crystal state the parallel stacks typical for the thermochromic (not photochromic) crystals.

Unusual "vapor photochromism" of the crystalline 3, 5-di-tertbutyl derivatives, I(5, Ar, R_1=3, 5tertbutyl, R'_1=R 'or R'_1=OH, R' =H), are revealed and completely controlled by a contact with methanol vapor /34/.

Authors suppose that a necessary room for the photoinduced isomerization in the crystalline packing responsible for the PCF generation is maintained by the diffusible me-thanol molecules.

4. 1. 3. The molecular systems with the saturated bridge between the imine moieties.

4. 1. 3. 1. The structural and spectral peculiarities of the double enol form.

The findings of the studies of the type I(Alk) compounds in the solvents are produced in the Tab. 4. 5.

Unfortunately the only compound I(1, ALk)(Salicylideneazine (SAA))has been studied /23, 35/ by both the steady-state and high time resolved absorption and fluorescent spectroscopy. Meanwhile the compounds I(2, 3, Alk) have been analyzed by the time dependent absorption and fluorescent spectroscopy with the low time resolution /36, 37/.

Although the difficulties arise with the comparative estimation of these data, nevertheless such a comparison seems to be useful to the considerable degree. The structures I(1-3, Alk) differ in the distances between the identical chromophores each of those has mainly the Enol structure (SALK)n (n=0, 1, 2, 3....)(Sch. 4. 1, 4. 8, Tab. 4. 5). Thus just (SALK)n structure but not SA (like I(1, Ph)one (see above)must simulate the behavior of the separate chromophores.

The chromophores' interaction in the Ground state can be characterized by the red shift (ΔE) and the additive intensity (molar extinction, ε)of the first absorption transitions So-->S_1^* of the I(1-3) molecules as compared with $SALK_1$ structure 7(Tab. 4. 5, col. 2)(see also 3. 1).

The strong interaction of the E-structures of Salk chromophores, just as one been expected, is observed with the direct juncture of N atoms in I(1, Alk)structure($\Delta E \approx 3400 cm^{-1}$)and max I(1, Alk)/ maxSALK>3. 5 (i. e. >>2).

The anion structure of SAA in the solvents of the different polarities is also observed /35/. In the structures I(2, Alk), I(3Alk) the shifts of the absorption bands as compared with SALK are practically absent ($\Delta E < 300 cm^{-1}$)and the additivity of the two moieties' absorptions is well observed (maxI(2, 3, Alk)/ maxSALK$\approx$2) within the limits of the experimental errors(Tab. 4. 5, col. 2).

The spaces between the chromophores of I(2, Alk)and I(3, Alk) structures estimated by the distances between the OH-groups of the E-structure moieties /37/are very short(Tab. 4. 5, col. 12) and depend not only on the length but also on the Alkyl spacer structure (Sch. 4. 8/37/).

In addition the distance between the interacting chromophores solely without taking into account of their relative positions can not determine the degree of the interaction of the electronic transitions.

Therefore, as it follows from the above findings (col. 2 and 12, Tab. 4. 5) the weak π-π interaction even in the structure I(3, Alk) does not change the nature of the first electronic transition as compared with the SALK structure in spite of the short distance between the chromophores.

Thus it results from above that the SALK chromophores in I(2, 3 Alk) are being excited almost independently whereas in I(1, Alk) the π, n electron systems of the different SALK moieties are bound up with each other and cannot be excited independently.

Table 4.5
The Fluorescent and Photochromic characteristics of the bisazomethine structures with the imine moieties joined by the saturated spacer (in the solvents).

Compound	Initial form					Photocolored form					Chromophoric Distance (Å)		Reference
	Absorption λ^{max}_{abs} (nm) ϵ lxM^{-1}cm^{-1}	Fluorescence				Absorption λ^{max}_{abs} (nm)	Fluorescence		Kinetiks		Enole	Enole	
		λ^{max}_{flu} (nm)	Stokes Shift (cm^{-1})	Lifetime τ (ps)	Quant. eff.(Φ$_f$)		λ^{max}_{flu} (nm)	Stokes Shift (cm^{-1})	Quan. Effic. τ$_1$ Pop τ$_2$ Accum.	Back Reac (ms)	Enole	Keto	
I (1,Alk)	~400 354 23000	440 558	~1745 10300	<1 19+ nx10²	V. w. 0.6x10⁻³ Very weak	a) b) 490	c) ~580	3170	<0.2 τ$_1$ 19 ps	0.08	g) 4.04	--	a) 23
I(2,Alk)	320 14000	η 370 460	4220 ~9510	---	0.07 †0.13 0.20 h)	369	460	5360	--- τ$_2$ 7980 s	(60÷100) X10³	8.64	4.65	d) 37
I(3,Alk)	„ 316 10500	f) 370 460	4618 ~9900	---	0.06 †0.23 0.29 h)	362	460	5860	--- τ$_2$ 4980 s	(60÷100) X10³	3.88	5.05	d) 36 37
SALK$_n$	~400 312-320 ~6500	465 500 e)	~4130 11250	---	Weak <10⁻⁴	e) 450-460	e) 535	3530	<0.7 ---	<0.1	---	---	e) 38 39 40
SA	~400 336 10000	~430 540	~1740 11240	<2 7	Very Weak 1.2 x 10⁻⁴	a) b) 476	c) ~580	4770	<0.3 ---	3.5	---	---	a) 23

a) ACN ,room temperature,steady state excitation. b) Transient absorption spectra ,chlorophorm,room temperature. c) /18/ ,Two step excitation flurescence. d) Steady state excitation and time depended abs.,and fluorescence spectra,CHCl$_3$, ,room temperature and AM-1 quantum chemical calculations. e) Steady –state excitation,see also /38/ at 77K in isopentane f)Excitation by λ≈316-320 nm. g) Ab-initio,sto-321G /38/.h) Total fluorescent (enol flu+ +ASS) obtained /37/.

The most typical peculiarity especially pronounced for the E-structures of the dianil compounds I(2, 3 Alk) is the n(N)-->π(Ph) electron delocalization lack, and owing to that, the considerable strengthening of the H-bond O-H...N takes place. Therefore the mechanism of the very fast S_1^*state deactivation by the rapture of the H-bond with the subsequent "aldehyde"(A) ring twist typical for SA structure is excluded and sufficiently intensive fluorescence (Tab. 4. 5, col. 6) untypical for anil E-structure (Tab. 4. 5, col. 3) is observed /36, 37/.

The strong interaction between C=N groups in I(1, Alk)structure results in the shift of the E⇔K equilibrium towards the fluorescent K structure in both moieties obviously(λ^{max}_{abs}, ≈400nm, λ^{max}_{flu}≈420nm, Tab. 4. 5 col. 2 and 3 respectively). The E-structure fluorescence is not observe (see below, Sec. 4. 3. 3).

4. 1. 3. 2. The nature of the ASS fluorescence and the PCF structures in the solvents.

The ESIPT reaction takes place of I(1, Alk) (SAA) with the time constant shorter than 50fs /23/ or 80fs /35/ to produce S_1^*NHflu "hot" state followed by intramolecular vibration energy redistribution and vibration cooling S_1^*NHflu --->S_1NHflu state with the time rate from τ≈100fs to 2ps which produces the ASS fluorescence S_1^*NHflu-hv->So.

The adiabatic process of the structural transformations from S_1^*NHflu to the twisted S_1^*NHTICT-Like quenching structure is reflected by the viscosity-depended ASS emission decay (τ≈5-11ps).

The NH*twisted ("TICT-Like") structure is a precursor of the PCF (twisted, post TICT)structures with the longwavelength absorpion and with the fluorescence band like that of SA and SALK with the long Alkyl radical (Ch. 3) (Tab. 4. 5, Colon. 7, 8). Thus above described processes are typical for SA structure rather than for SALK one(see also Sec. 4. 2. 3. 3.).

On the other hand due to the E-structure peculiarities and strengthening of the H-bond of I(2, Alk)and I(3, Alk) structures the strongly pronounced H-atom (ESIHT) but not proton ((ESIPT) transfer in the excited singlet state occurs /41/.

Such a transfer can take place as a result of the adiabatic concert process dislocation between O and N atoms and the corresponding charge transfer OPh-e->C=N-with the assistance of the skeletal low frequency promoting vibration modes modulating the O-N distance. Therefore the low polar fluorescent K*cis structure is generated in the excited S_1^* state (instead Zw* one unlike SA) corresponding to the planar Kcis structure in the ground state.

The adiabatic reaction is interrupted (unlike SA) since the quenching TICT-like structure cannot be formed due to the high potential barrier for the A-ring twist around C=C double bond and to the absent of the necessary charge separation typical for the TICT-Like state.

As a result the intensive ASS fluorescence K*cis-hv->Kcis (λ^{max}_{flu}, ≈460nm, Φ_{flu} ~0.1-0.2, and ASS $\Delta v \approx 9500$-$10000 cm^{-1}$) with the E-structure excitation ($\lambda^{max}_{abs} \approx 316$-$320nm$) is observed in the liquid solvent at the room temperature unlike other anil molecules.

Thus for I(2 and 3, Alk) molecules the immediate precursor of the typical anil PCF structures can not be formed, and the twisted "post-TICT" red colored forms ($\lambda^{max}_{abx} \approx 480$-$500nm$) cannot be generated (see Ch. 3 and /24/).

Instead of such forms the planar low-polar fluorescent Kcis structure is generated adiabatically directly after the ESIHT and emission Kcis*-hv->Kcis with the isomerization into the planar photochromic Ktrans structure ($\lambda^{max}_{abs} \approx 360$-$370nm$) (Sch. 4. 8, Tab. 4. 5, col. 7). The latter is accumulated even in the liquid solvent owing to the heightened stability (Tab. 4. 5, col. 11).

Thus unlike SA, SALK and I(1, Alk) the PCF of I(2 and 3, Alk) have the flattened Keto-structure with the more short-wavelength absorption stabilized in the liquid solvents (see also Ch. 3).

4. 1. 3. 3. The role of the interfragmental interactions, and the general schemes of the photoinduced processes in the solvents.

The kind of the interaction between the anil moieties in the ground and excited states plays the key role in the mechanism of the Excited state deactivation of the I(1-3, Alk) structures.

The strong interaction between n and π anil electron systems in the So state of the Enol I(1, Alk) structure provides the delocalization of the electron density along the imine fragments likened to the electron structure of SA rather than to SALK that, which acts like the strong withdrawing substituent shifting the ICT transition towards the low energies ($\lambda^{max}_{abs} \approx 350$nm) as compared with SA ($\lambda^{max}_{abs} \approx 340$nm)(Tab. 4. 5, col. 2)/23, 35/.

Such an electronic interaction does not change apparently the mechanism and rate of the ESIPT (τ<50 or 80fs) which occurs however in the single moiety only /23, 35/ since right away after the ESIPT (O-H->H-N$_1^{+>}$---->-N$_2^{+}$--) in first moiety the positive charge on other N-atom is increased sharply also, so that the O-H...N bond in the second fragment is being weakened strongly (and may be it is breaking at all), and the ESIPT is getting a very low probability.

Thus the all photoinduced processes in the second moiety caused by the ESIPT including photochromism are annihilated by the ESIPT in the first structural fragment.

The mechanisms and the kinetic of the induced processes in the first moiety (Tab. 4. 5) involving the ESIPT, the formation of the low stable ZwS$_1$*NHflu structure (Tab. 4. 5, col. 3-6), and the generation of the low stable(τ≈1ms) fluorescent colored form (Tab. 4. 5 col. 7-11) with the longwavelength absorption($\lambda^{max}_{abs} \approx 490$nm) differ only a little from those of SA /23/ with the only exception (Tab. 4. 5, col. 11)of the hastened PCF structure decay (discoloration)owing to the strengthening of the NH...O bond in the flattened Kcis structure of I(1, Alk) and the following GSIPT with the formation of the initial E-structure.

For the molecules I(2, Alk) and I(3, Alk) the another situation is observed. According to the scheme 4. 9 the two identical azomethine moieties are excited independently. But in spite of the independent excitation, the strong inductive interaction of the electronic transitions results in the resonance energy migration between the two S_1^* states of the E-structures $ES_1^*<1> <-->ES_1^*<2>$.

The ESIPT competing with the energy migration occurs with the rate $\tau=n(ps)$ which is lesser than of SA ($\tau<50fs$) securing the existence of the Enol form fluorescence (ES_1^*-hν->ESo) with the marked quantum efficiency (Tab. 4. 5, col. 6).

That fluorescence being absent of I(1,, Ar), provides however the SIET of both (i) (energy migration, see above) and (ii) types in I(2, 3, Alk) unlike I(1, Ar) (see 4. 1. 2. 2).

The strongly exoenergetic process SIET of (ii) type ($ES_1^*<2 >-->KS_1^*<1>$) becomes much probable after the ESIHT and the SoKcis generation in one of fragments, and occurs with the very high rate.

Really for this process the all necessary conditions (a)-(d) (sec. 4. 1. 2. 2) have been kept, and the critical distance Ro<40 Å can be estimated taking into account the high values of the overlap integral, fluorescence quantum yield of the E-structure and the orientation factor.

With $\tau \approx 10^{-9}$ s the rate constant K (IET)(R)$\approx 1/\tau_0$ (Ro/R)6 can be estimated for I(3, Alk) (R$\approx$3. 88Å) and I(2, Alk) (R$\approx$8. 64Å) as ~10^{15} and $10^{13} s^{-1}$ correspondingly. (K(IET)3/K(IET)2>10^2 in dependence on the alkyl bridje length and the structure (Tab. 4. 4).

ESIPT Photochromism

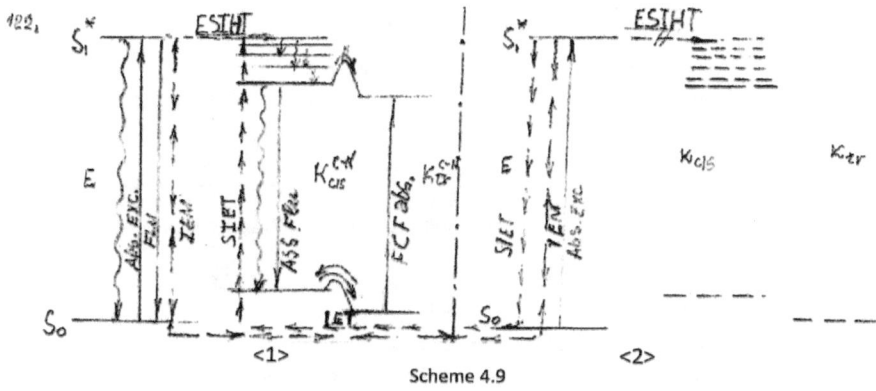

Scheme 4.9

a)
Energetic scheme of the photo-induced processes in the molecular systems I(2,Alk) and I(3,Alk) Including the Intramolecular energy migration and the intramolecular energy transfer with the generation of the PhotoColoratiom, PC, product. (see text).

(IEM) --> <-- -> <--Intramolecular Energy Migration between <1> and <2>.
(SIET)--> --> --> --> S-- S Intramolecular radiationless Energy Transfer ES_1^* <2>-->KS_1^* <1>.

b)
Fluorescent and photocolored structures

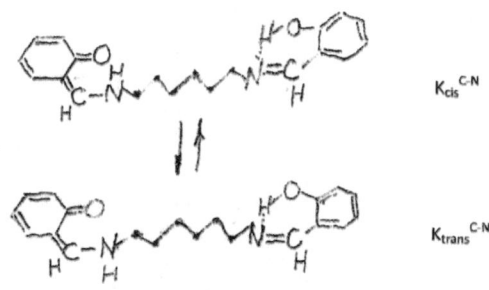

Scheme 4.10
Polymer structure on the base of molecule II(1)

Table 4.6
The comparative data for the solvent and crystal of the structures I(3,Alk) and SALK$_n$

Structure	Absorption, λ^{max}_{nm}				Fluorescence				PCF absorption λ^{max}_{nm}	
	Enol		Keto		Solvent		Crystal			
	Sovent	Crystal	Solvent	Ctystal	λ^{max}_{nm}	Stok.$\Delta\nu$ Shift cm^{-1}	λ^{max}_{nm}	Stok.$\Delta\nu$ Shift cm^{-1}	Solv.	Cryst
I (3,Alk)	316 [1)]	335	---	430 [2)]	370 460	4625 9906	480 and 510	~9030	490	435
SALKn	320 [1)]	317 [3)]	400	---	465 500	9745	---	---	460	420 ÷ 510

1) The variations of the parameters are very small for different "n" with the exception of PCF structures' absorption bands positions (see Ch.3).2)The thermochromic band (room temperature) that is absent at 77K.

Table 4.7
Influence of the substituents on the absorption and the fluorescence of crystalline I(3,Alk).
(For structures see below).

Substituted structures	Absorption band, λ^{max}_{nm}		Fluorescence		Photochromic abs.band, λ^{max}_{nm}
	Enol	Keto [1)]	λ^{max}_{nm}	Stokes shift [3)] $\Delta\nu$, cm^{-1}	
1	335	420	510 480 sh	9300 4200	435 [2)]
2	350	418	512 488 sh	8000 3428	----
3	336	450 375	550 430	4040 1410	450 [2)]
4	~350	435	511 sh 480	7740 5510	----

1) Thermochromic band. 2)Photochromic band at room temperature, at 77K photochromism is absent.
3) The Stokes shifts are estimated relatively E (upper) and K(lower)absorption bands with exception of 3(see text)
The structure for the table 4.7.

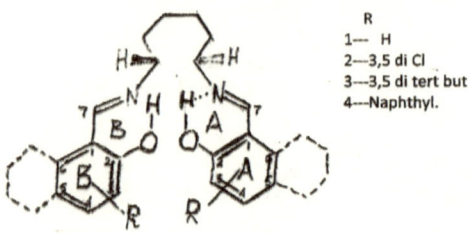

R
1--- H
2---3,5 di Cl
3---3,5 di tert but
4---Naphthyl.

Table 4.8

Absorption, Fluorescent, and Photochromic characteristics of the compound III(1-3). in the solvents (<45>--<47>).

Compounds	Absorption						Fluorescence				Photochromism		
	$S_0 \rightarrow S_1^*$				$S_0 \rightarrow S_2^*$								
	ν^{max} cm^{-1} (λ^{max} nm)	$\Delta\nu$ cm^{-1} relativ. SA	ε^{max} LM^{-1}cm^{-1}	$\varepsilon(III)/\varepsilon(SA)$	ν^{max} cm^{-1} (λ^{max} nm)	ε^{max} LM^{-1}cm^{-1}	ν^{max} cm^{-1} (λ^{max} nm)	$\Delta\nu^{max}$ cm^{-1}	τ ns [2]	Φ_{fl}	λ^{max} nm	Lifetime τ_{ms}	Φ_{col} [1]
III(1)	23600 (426)	6200	7140	0.53	29600 (338)	~19640	~16000 (625)	7600	0.2.	10^{-4}	~550	10^{-4}	0.07
III(2)	24400 (410)	5400	15900	1.18	30400 (329)	17940	~16400 (610)	8000	0.3-0.4	$1-2 \times 10^{-2}$	500 550 525	1-2 X10^{-2}	1.0
III(3)	25190 (397)	4610	27700	2.06	33300 ~300	20800	21100 (474)	3900	---	$<10^{-4}$	---	---	0
SA [3]	29800 (336)	---	1346t0	1.00	---	---	18266 18933 ('540)	~10870	0.7	0.1-0.6	480 500	2.0	---

1) Φ_{col} —Relative efficiency of the photocoloration. 2) The main component at $\lambda > 580$ nm. 3) From ref. /18/ and Ch. 3.

Table 4.9a
(see /47/)

The spectral-kinetic parameters and attribution of the spectral bands for the structure III(2) In the solvents (see scheme 4.11) (The rise-time of the all bands $\tau<100$ fs excluding $+A_2$ absorption of the PSF structures).

Designation	Spectrum type	Band region λ_{nm}^{max}	Decay time	Model struct.III(3)	Attribution of transition and strtuct.
$+A_1$	The strong pos.abs	550	300-400 ps	Absent	$S_1 \xrightarrow{abs} S_n$ NH_{cis} or K_{cis}
$-A_1$	The negative abs.	650-700	300-400 ps	Absent (see F)	$S_1 \xrightarrow{flu} S_0$ NH_{cis} or K_{cis}
F	The fluorescence	580-650	50-100ps 320-500ps 5-15ps, rise	Absent	$S_1 \xrightarrow{flu} S_0$ NH_{cis} or K_{cis} In the reg $\lambda<550$nm the main comp. $\tau<1$ps is assign.to ultrafast ESIPT
$-A_2$	The negative abs.	≤ 400	Till ms and s	Present	$S_0 \rightarrow S_1$ Depopulation of $S_0 E_{tr}$
$+A_2$	The positive abs.	500,550,675	400μs-1ms 40-80 μs	Absent	$S_0 \rightarrow S_1$ Population of the color. Form struct.- K_{trans} or post TICT

Table 4.9 b
The characteristics of the fluorescence of III$_2$ upon excitation by 400nm in the solvents /47/.

Solvents	Quantum yield $\Phi_f \times 10^{-2}$	Max.of flu. $v_f^{max} \times 10^{-2}$	Decay,τ_f ps	Stokes shift Δv cm^{-1}
n Hexane	1.3	17010	280	7380
1 Chrolopropane	1.9	16540	480	------
Acetonitrille	1.5	16280	480	8350
Methanol	1.3	16250	360	8380
Trifluoroethanole	2.0	16100	530	9216

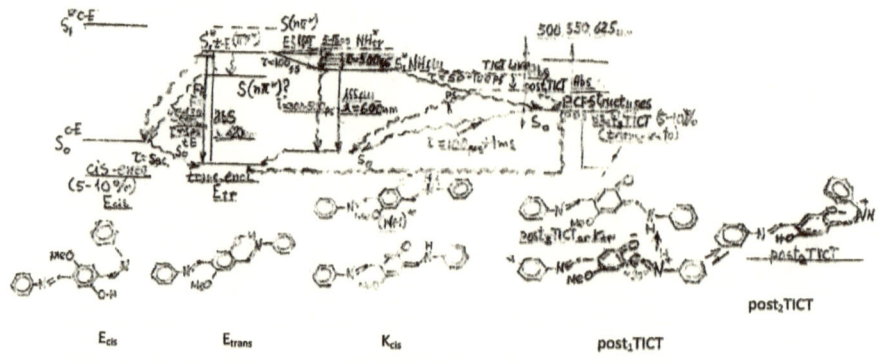

Scheme 4.11
Photoinduced processes in the structure III(2) (on the base of /47/).
The alternative recent data of the state and the transformations are shown by the broken lines.

Table 4.10
The absorption and fluorescent characteristic of the compound IV in the rigid media.

a) The thin polycrystalline film.

Absorption λ nm		Fluorescences [1]			
77K	Room t-re	77K		~293K	
		λnm	Δv cm^{-1}	λnm	Δvcm^{-1}
Absent	458	641 [2]	~11800 [3]	641	~11800 [3]

1) Excitation λ=365nm. 2) Intensity of flu increases strongly in comparison with SA. 3) The Stokes shift is done relatively 365 nm .(excitation).

b) The thin polycrystalline films of the including complex IV-βCD.

Absorption [1]				Fluorescence [1]					
77 K		~293 K		77 K				~293 K	
λ$^{max}_{nm}$		λ$^{max}_{nm}$		before		after		before	after
before	after	before	after	λ$^{max}_{nm}$	Δv$_{cm}^{-1}$	λ$^{max}_{nm}$	Δv$_{cm}^{-1}$	λ$^{max}_{nm}$ Δv$_{cm}^{-1}$	λ$^{max}_{nm}$ Δv$_{cm}^{-1}$
Absent	~491	463	463 [2]	640	~11700 [4]	640 [3]	~11700 [4]	640 ~11700 [4]	640 11700 [4]

1) irradiation with λ=365nm. 2) Without any change of the intensity after irradiation. 3) The flu. intensity increases twice after irradiation at 77K. 4) The Stokes is determined relatively to λ$_{ex.}$=365nm.

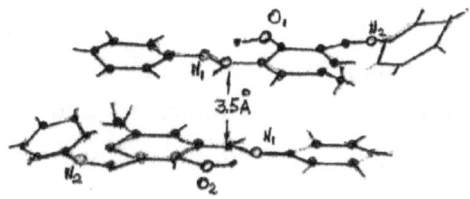

Scheme 4.12
Adjacent molecule IV. The close contacts ~3.5Å apart between molecules cease to exist in the inclusion complex with β-CD.

Thus all photoinduced processes in <2> moiety are being suppressed by the interfragmental SIET $ES_1^*<2>->ES_1^*<1>$.

Moreover as a result of the extra-population of the KS_1^* of <1> moiety in I(3, Alk), the efficiency of the PSF generation (Kcis structure) is being heightened with the increase of the SIET rate value about twice as much (Tab. 4. 5, col. 10).

Unfortunately the qualitative interpretation /37/ of the photochromism dependence on the molecular structure of I(2, Alk) and I(3, Alk) based on the coulomb interaction between the isolated anil fragments is not convincing and contradicts the known ideas about the cooperative mechanism of the Ground state proton transfer in the thermochromic crystals (see ch. 2).

However there are only indirect arguments in favor of the Interfragmental Energy Transfer and the supplementary investigations of its role in the photoinduced processes of I(Alk) structures.

4. 1. 3. 4. The peculiarities of the photoinduced processes in the crystals.

The comparative spectral data for the crystals and the solvents are produced in the Tab. 4. 6.

The significant shift towards the lower energy of the ICTnature $So->S_1^*$transition ($\Delta v>2000 cm^{-1}$) of the E-structure in the crystal and the absent of such a shift for the monochromophoric SALKn testifies to the interaction of the Enol moieties A and B of the different adjacent molecules in the crystalline packing that is impossible in that of SALKn.

Really the salicylidene moieties exhibit some close contacts(C C distance 3. 3-3. 4Å)with adjacent molecules and the closest distance involving the enolic oxygen atoms of the A and B moieties can achieve d=2. 67-3. 42Å.

Such an interaction stimulates also the E⇔Kcis equilibrium shift towards Kcis structure with appearance of the absorption band ($\lambda\approx430nm$)when increasing of with themperature (thermochromism) unlike solvents.

Due to the intensification of the photoinduced ICT(PhOH-e->N=C-)as a result of the $So->S_1^*$transition, the "protonity degree" of transfer and the transfer rate, and also the charge separation degree are increased, and C:::C bound order falls in the crystal as compared with the solutions.

Owing to above the competing short-wavelength E structure fluorescence is not observed in the crystal unlike the solvents, the potential barrier for the adiabatic formation of the quenching TICT-like state is lowered leading to the strong rise of the ASS fluorescence efficiency with the temperature decrease.

As it was to be expected the ASS flu band is band little shifted as compared with the Kcis structure fluorescence in the solvent and has the unclear fine vibration structure.

Thus the S_1^*NHflu(Zw) state has in the case the rather like S_1^*Kcis nevertheless different from that structure but is deactivated by emission to the So Kcis state.

In accordance with the findings /43/ the twist and even the rotation of the Ph ring around C_1-C_7 bond (structure below, Tab. 4. 7) in the crystalline packing is not prohibited.

Therefore the more flattened PCF with the blue shifted ($\lambda^{max} \approx 435$nm) relatively solution ($\lambda^{max} \approx 490$nm) absorption band can be generated by the two different path-ways: via the TICT like state followed by the post-TICT states and by deactivation (S_1^* ---> SoKcis) with the following cis-trans isomerization around C_1--C_7 bond.

It should be noted that due to absent of the E structure fluorescence the SIET with the participation of the ES_1^* state cannot be occur in the crystal and therefore the ESIPT and the following photoinduced processes including the ASSflu and PCF structures' generation can take place in the both moieties independently.

Influence of the molecular structure modification by substituents and conjugation in the Aldehyde moieties of the I(3, Alk) molecules in the crystal has been also studied /43/(Tab. 4. 7, 1-4).

All the molecules with exception of 4, have the E-form completely at 77K with the typical absorption band ($\lambda^{max} \approx 335$-$350$nm).

With the temperature increase from 77K to the room that the equilibrium E⇔K cis shifted towards the fluorescent Kcis structure with the more longwave-length absorption band($\lambda^{max} \approx 400$-$430$nm) (thermochromism).

According to the temperature dependence of the absorption spectra/43/the efficiency of the thermochromic reaction falls in the series (2)>(4)>(1)>>3. The efficiency is connected with the structure of the crystalline lattice and determinate by the π-π and electrostatic interactions between the salycilidene moieties of the adjacent molecules ($d \approx 3$-4Å).

The very low population of the K-structure and very weak thermochromism of I(3, Alk)(3) are caused by too long distance and unfavorable positions of the adjacent aldehyde moieties.

Therefore in that case the fluorescent aldehyde Kcis structure can be identified by the fluorescence excitation spectra only.

Unlike the rest crystalline I(3, Alk) compounds (1, 2, 4), the structure (3) is manifested by the two different, difficult for interpretation, fluorescent structures with the excitation in the bands ($\lambda_{ex}^{max} \approx 450$ and 375nm)attributed/43/to the two different K-structures.

The compound I(3, Alk) (3) as well as I (1, Alk) (1) shows the photochromism is most probably owing to the tert-butyl groups which secure the room sufficient for the formation of the strongly twisted PCF structures with the absorption bands $\lambda_{abs}^{max} \approx 450$nm and half-life of 4min at 25°C i. e. with higher stability than more flattened ($\lambda^{max} \approx 435$nm)PCF structure of (1)because of the bulky tert-butyl substituents.

Thus the interfragmental and intermolecular interactions in the crystal of the I(3, Alk) molecules influence the mechanism of the photoinduced processes involving the ESIPT, ASS fluorescence, and the PCF structure generation, as compared with the solvents on the one hand and with the crystalline monostructures on the other hand.

The modification of the molecular structure provides the changes of the crystalline packing, and can lead to the construction of the photochromic crystals.

4. 1. 4. Anil bichromophoric systems formed through the hydroxyl rings.

The second way of the anil bichromophoric system construction is bound up with the participation of the salycilidene rings. Such compounds which have been synthesized and studied with the purpose to obtain the novel photochromic crystalline anils can be arranged in the series II, III, IV (Sch. 4. 1) with the increase of interaction between the different anil chromophores.

Unfortunately the findings obtained for the structures II, III, IV are very difficult to compare with each other because of the differing methods and conditions of investigations.

4. 1. 4. 1. The structures with the separated anil moieties.

The information about the structure II with the most distinctly separated anil moieties is very scanty.

The compounds II(1, 2) have been studied by the steady-state method only /11/in the rigid matrixes (PMMA, methylcyclohexane –decalyne (t=180°C), paraffin (t=15°C). Under excitation of the E-structure of $II_1(\lambda \approx 365nm)$ the PCF structure($\lambda \approx 480nm$)is generated and decays after heating i. e. photochromism like SA is manifested.

In II_2 ($R=SO_2$)the phototochromism is not shown after the same conditions. In the crystals the compound II_1 and II_2 are thermochromic only. The polymer compound on the base of the structure II_1(Sch. 4. 10)is also thermochromic/44/.

Although the structures II are favorable to the independent excitation of the azomethine moieties however there are no any information about photophysical processes (including fluorescence) in such structures for discussion of the SIET.

At the same time the absent of photochromism in the crystals can be explained by a small room for isomerization step necessary to the generation of the PCF in the crystal packing/2/.

However it is not the necessary condition for the display of the thermochromism. (see Ch. 2).

4. 1. 4. 2. The double hydroxy structures with the common hydroxyphenyl ring

In the compound III(1-3)(Sch. 4. 1, Tab. 4. 8)/45-47/ the π-electron systems of the different moieties are not separated from each other, and the additivity of their absorption bands is absent unlike BSP.

The $So->S_1^*$ transitions are shifted strongly towards longwavelengths($\lambda \approx 400-420nm$, $\Delta\lambda \approx 80nm$)as compared with SA and belong to the Enol structures in accordance with the longwavelength absorption band of the model structure III (3).

Thus the long-wavelength shift of the $S_o \to S_1^*$ transition ($\Delta v \approx 4600\text{-}6200$ cm^{-1}) is caused by the interfragmental interaction of the anil moieties that is increased with the formation of the Itramolecular H-bonds which stabilize the planar geometry of the "Aldehydes" part of the molecule.

Such a transition can be attributed to the Intramolecular Charge Transfer(ICT) from the phenyl ring with the two electron donating OH groups to the electron withdrawing C=N groups in the different anil moieties.

Really the analysis of the spectra shows a very small shift of the absorption band maximum ($\Delta v \approx 600$ cm^{-1}) with the increase of the solvent polarity within the wide limits (from hexane to acetonitrile) that is typical for the ICT transitions with $|\mu(S_1^*)-\mu(S_o)| \approx 0$ (μ is dipole momentum) owing to a symmetric charge redistribution along the two anil moieties with excitation.

The intensity of the ICT transition (ε) drops considerably as compared with SA in the series III(3), III(2), and III(1) i. e with addition of the structural fragments with the Intramolecular H-bond.

Such a phenomenon is connected with the decrease of the charge redistribution favorable to ESIPT in the structural fragment Ph-OH...N=CH-with excitation leading to the sharp drop of the ESIPT rate in the structure III(1)(see below).

The detailed comparative kinetic and spectral investigations of III(1), III(2) and III(3) in the temp-oral range from 100fs to 1h with help of the time-resolved femto, pico, nano, and microsecond transient absorption and fluorescence along with the steady–state spectroscopy in the solvents of different nature (polarity, proticity) allows to elucidate the main mechanisms of the photoinduced processes connected with the fluorescent and the photochromic properties of the com-pound studied/47/.

The principal purpose of the comparative studies is attribution of the spectral bands to the definite structures in the excited (S_1^*) or ground (S_o) states on the base of comparison of the absorption and fluorescence spectra and their temporal changes both for the same structure and the related (model) ones.

The structural-kinetic scheme 4. 11 for III(2) in the solvents is drawn on the base of the data about the spectral-kinetic parameters and attribution of the main spectral bands (Tab. 4. 9a). The alternative interpretation of the data obtained later for SA structure (seeCh. 3) is shown in the scheme by the broken straight and wavy lines correspondingly.

According to the scheme the excitation So-->$S_1^*(\pi\pi^*)$of the E-trans structure occurs. The excited state of the $(n\pi^*)$type lies much more higher(but not lower /47/)in accordance with the data obtained later (see Ch. 3) for SA structure.

Therefore the twisting of the A-ring, the H-bond break, and the formation of the cis-Enol structure can proceed via the conical intersection of the $S_1^*(\pi\pi^*)$ and So states without participation of the $S(n, \pi^*)$ state unlike it proposed in /47/.

Such a transformation takes place with the ultrafast rate competing with the ESIPT in III(2), and also in III(3) competing with a very weak short-lived emission ($\lambda\approx$430-450nm) which is displayed also of III(3) in the shortwavelength region of the ASSflu band(see below).

The nonplanar Ecis structure is generated with the low efficiency (5-10%) and isomerizes into the initial E-trans structure with the utmost low rate of III(2) and III(1) structures that is reflected in the very low repopulation of the SoEtrans state with the corresponding decrease of-A absorption band intensity (Tab. 4. 9a).

The transformations of the Enol structure are similar to those of SA however manifested more distinctly as the azomethine moieties can be involved into the reactions in the S_1^* state.

Thus those structures are favorable for the study of the photoreactions competing with the ESIPT.

The main path-way of the S_1^*NHflu state generation is a result of the ultrafast ESIPT (S_1^*Etrans--ESIPT->S_1^*NH) (τ<100fs) that competes with the ultrafast transformation in the Etrans structu-re described above.

Thus the intensities of the absorption bands – A_2 and $+A_1$ which reflect SoEtrans and S_1^*NH flu states' depopulation and population correspondingly increase within the range of the instrumental function ($\tau \approx 100$fs), and the latter is present at III(2) only.

After that ultrafast process the vibration relaxation from the hot $\check{S}_1^*$NH state to the equilibrium S_1^*NH state takes place with $\tau \approx 500$fs which is reflected /47/in a red (≈ 500cm^{-1}) shift with the intensity increase of the $+A_1$ band along with the decrease of its "blue" and "red" wings.

The S_1^*NHflu state generated by the ESIPT is responsible for the ASS flu(band F_1,. Tab. 4. 9a), that is absent in the III(3) structure.

It displays a clearly expressed Zwitterionic (Zw)nature with the red shift of the fluorescent band $F(\nu^{max} \approx 16$-17000cm^{-1})with increase of the solvent polarity(Tab. 4. 9b).

In the general case such a Zw excited state which is responsible for the ASSflu does not coincide with the S_1^*Kcis state that is responsible for the fluorescence of another nature (S_1^*Kcis—$h\nu > $SoKcis)and unlike the former does not lie on the S_1^* state reaction coordinate of the photochromic reaction (see alsoCh. 3).

The more detail information about the S_1^*NHflu state population and decay can be obtained from the kinetic analysis of the F band(Tab. 4. 9a)/4. 7/. In the short-wavelength range (with $\lambda<500$nm)the main decay component which is beyond the temporal resolution of the spectrofluorimeter used ($\tau<1$ps) is not connected with the emission from the NHS$_1^*$flu state and can be assigned to the decay of the S_1^*Enol structure due to the ultrafast ESIPTand another trans-formation processes (fluorescence F(E), Sch. 4. 11).

The major decay component in the rest, long-wavelength, region of the band F, $\tau \approx 300$-500ps, (80-90% of amplitude, see also Tab. 4. 9b) is connected with the radiationless deactivation S_1^*-So. shorter decay component $\tau \approx 50$-100ps(10-20% of amplitude) determines the very important process of the PCF structure generation from the fluorescent state. There are two different notions of this process: the diabatic reaction of the PCF NH trans –keto structure formation /47/ and the adiabatic formation of the TICT-like structure (in the S_1^*state) followed by the generation of the post TICT PCF structures (like SA, see Ch. 3)/24/.

The much more shorter third component ($\tau \approx 5\text{-}15\text{ps}$) has been found as a rise time (negative amplitude). It can be connected with the picoseconds rise time observed in the transient absorption and stimulated emission bands (A_1 and-A_1 absorpion bands, Tab. 4. 9a).

That rise time can reflect the additional, slower population of the S_1*NHflu state via the higher metastable, transient (NH*tr) one of the planar structure which is also directly populated by the ESIPT (probably from the higher S*Enol state), and can be the second precursor of the PCF structures like SA.

However unlike SA the NH*tr state of the III(2) structure is manifested in the additional population of the S_1*NHflu state and is not in the short-lived shortwave-length fluorescence like SA(see Ch. 3).

Thus the additional low rate population of the S_1*NHflu state can be explained by the existence of the metastable transient S_1*NHtr state (see Sch. 4. 11) but not by the direct slow ESIPT from the nonplanar Enol structure /46, 47/.

According to the experimental data/47/there are the two or three PCF structures with the longwavelength absorption bands +A and with the different rates in the time scale from tens µs to1ms (Tab. 4. 9a).

Only a small fraction of the molecules III(2)are transformed to the PCF structures. According to /47/the PCF structures are the trans-Keto(around C=C and (or)C--N bonds)structures with the different polarities and the decay times.

From viewpoint of the postTICT notions (see Ch. 3 and /24/)there are a set of both planar (including Ktrans)and nonplanar NHstructures of different polarity and stability depending on the solvent nature with a common precursor of the "TICT-like" structure in the S_1*excited state (Sch. 4. 11).

The main differences in the mechanisms of the photoinduced processes of SA and BPHMe(III(2))

are caused by the unification of the two anil π-systems which leads to the strong red shift of the lowest electron transition in the E structure($\lambda^{max}_{abs} \approx 420\text{nm}$ vs330nm of SA).

In spite of the absent of the calculated data it can be believed that the influence of the two electron donor centers (OH and OCH)in the phenyl ring results in the strong increase of the electron density on the latter and leads to the stabilization of the Enol structure in comparison with SA that is expressed in the complete shift of E⇔K equilibrium towards E-tautomer in the ground state even in the polar solvents unlike SA.

The marked stabilization of the E structure in the S_1^* excited state results in the significant increase of the efficiency of the photoreaction inside of the E structure competing with the ESIPT in comparison with SA, and lifetime of the cis-Enol form is more 3 order of magnitude longer than photochromic structures.

The strongest differences are connected also with the S_1^*NHflu state which lifetime and fluorescence quantum yield by an order of magnitude higher of III (2) than of SAs /47/. Such distinctions can indicate to the different role of the adiabatic structural transformations in the deactivation of S_1^*NHflu state (see Sch. 4. 11).

The solvatochromic behavior of the ASS fluorescence is also different of III(2)and SA systems and characterized by the opposite, red(16000-17300cm^{-1})($\mu(e)>\mu(g)$)and blue(17500-18300cm^{-1} (($\mu(e)<\mu(g)$)), shifts correspondingly.

This is evidence of the different Zw-ion structures of the NH form for III(2) and SA molecules (Ch. 3)(see also, e. g. /24/and /38/). The additional population of the S_1^*NH state proceeds unlike SA via the transient metastable structure NHtr* that is generated immediately as a result of the ESIPT (see above and Sch. 4. 11).

The new photochromic absorption bands at $\lambda \approx 550$ and 675nm(Tab. 4. 9)as compared with SA($\lambda \approx$ 480, 500nm) appear of III(2).

Such considerable band shifts($\Delta v \approx 2000$ and 5200cm^{-1})can be attributed most likely to the generation of the new photochromic structures.

One can expect that the latter are the NH zwitterion(Zw) twisted post-TICT type structures (Sch. 4. 11) which a result of the steric interactions of the NH moiety with the bulky cisoid (about C—N bond) Enol one by an analogy with the SAs molecules with the bulky substituents(Ch. 3).

The typical spectral shifts in dependence on the solvent polarity (solvatochromism, Ch. 2)is pro-vided by such a PCF structure in accordance with the experimental data/47/.

The unexpected behavior of the "symmetric" III(1)(BPH) is manifested in the almost low efficiencies of the ASS fluorescence and PCF structures' generation which are nearly 2 and 1, 5 order of magnitude lower than of III(2)(BPHMe) correspondingly.

Such a behavior needs the special consideration.

It has been shown by the comparative analysis of the transient absorption and time resolved fluorescence spectra of III(1) from one hand and III(3) or III(2) from other hand in the solvents:

(i). In a picosecond time scale the expected intensive formation of the S_1*NHflu state (K*cis structure) in III(1)is not observed. Instead of that the transient spectra are strongly like those of the model structure III(3) in which the structural transformations in the Enol structure are reflected.

(ii). In the nano and microsecond time scales the absorption spectra of III(1) are represented almost completely by the two negative absorption bands ($\lambda \approx 320$ and 430nm)which correspond to the So cis and So trans structures just as of III(1). Only a very weak positive band of the photocolored structure ($\lambda max \approx 550nm$) is observed with decay time constant of the several hund-rends µs which increases with the solvent polarity (see Tab. 4. 8).

(iii). In the short-wavelength range($\lambda < 580nm$)of the ASS emission band there is a very shortlived component ($\tau \approx 10ps$, ACN)which correlates, like of III(2), with the structural transformations in the Enol structure involving the ESIPT, but with the time constant which exceeds considerably the corresponding time constant of the structure III(2).

The biexponential decay of the long-wavelength range of the ASS emission from the S_1*NHflu state of III(1) is characterized by the time constant $\tau_1 \approx 200$ and $\tau \approx 360$ ps which about two times shorter than corresponding ones of III(2).

It should be expected from the comparative analysis of the spectral and kinetic parameters of III(2) and III(1)(lower radiative rate constant, extinction coefficient, and considerably lower fluorescent quantum yield of the latter, Tab. 4. 8) that it is not more than 10% of the initially excited molecules are transformed to the S_1*NHflu structure.

Thus it follows from i-iii that the generation of the S_1*NH structure(and hence of the PCF strutu-res) is much less efficient in III(1) than in III(2)(Tab. 4. 8), and the main deactivation channel of III(1) operates within the Enol structure.

Such a situation can be caused by the sharp drop of the ESIPT rate of the "symmetric" III(1) structure owing to the following possible reasons:

(i). The strong stabilization of the Enol structure in the Ground state of the "unsymmetric" molecule III(2) with the two strong electron-donating substituents in the same phenyl ring(see above) is typical also for the "symmetrical" molecule with the two hydroxy-groups III(1).

Moreover in the latter (Tab. 4. 8) the excitation following the forbidden (ν max ≈ 7000 lxmol^{-1}cm^{-1}) transition with the ICT (So—ICT-->S_1*) unlike analogues intensive (ν max=16000 lxmole^{-1}cm^{-1}) transition of III(2) cannot lead to the charge redistribution O—e-->N in the S_1* Enol structure required for the ESIPT OH-->NH.

(ii). The exclusion of the active skeletal vibration assisting the ESIPT reaction with formation of the two symmetrical H-bonds as it has been suggested in /47/, and (or)

(iii). The effect of the enol transformations affecting the processes of deactivation with the for-mation of the two symmetrical H-bond as it also has been suggested in /47/.

Thus the strong thermodynamical and (or)kinetic stabilization of the Enol structure in the exci-ted state resulting in the sharp fall of the ESIPT reaction rate can be expected in III(1).

4. 1. 4. 3. The single hydroxy-structures with the common hydroxy-phenyl ring.

The structure IV has been studied only by the steady-state spectral methods in the crystal state and in the crystal of the including complexes with the cyclodextrine (CD)(IV-CD) with the purpose of the obtaining of the novel crystalline photochromic compounds /48/.

In the crystalline packing (Sch. 4. 12)the two adjacent molecules are placed very close ($\approx$3. 5Å) in the parallel planes and bounded up each other by the strong electrostatic and steric interact-tions. Such interactions are favorable to the thermochromism and prevent the structural transformations.

Really, according to the presented findings (Tab. 4. 10a)the molecule IV in the crystal state does not show photochromism but displays thermochromism within the temperature range 77-293K with the appearance of the absorption band ($\lambda max \approx 460nm$) unlike SA, and the fluorescence band ($\lambda max \approx 640nm$)with the ASS($\Delta v \approx 12000 cm^{-1}$) relatively to the absorption band.

In the combined crystalline packing of the molecule IV with CD there is the high probability of the formation of the including complex IV-CD.

The evidences of such a complexes' existence and information of their structures are absent. Nevertheless in such a structure the steric interactions between the adjacent molecules are considerably weakened, and the sufficient room for the realization of the aldehyde ring twisting necessary for the PCF structures' formation is created.

Therefore such crystals show photochromism: the absorption band with $\lambda max \approx 490nm$ appears under irradiation with the $\lambda \approx 365nm$ at 77K (Tab. 4. 10b)when the PCF is stabilized and accumulated with the amount sufficient for the registration by the steady-state methods.

At the same time the thermochromism is provided by the electrostatic interactions between the adjacent salycilidene moieties which remain sufficiently strong (see Ch2). The thermochromism in the temperature region of 77-293K, similar to that of the thermochromic anil crystals, is characterized by the appearance with heating of the absorption band with $\lambda max \approx 460nm$(Tab 4. 10b).

The thermochromic form is not photoactive, and photochromism is not shown by irradiation of the crystal in the absorption band of the thermochromic structure($\lambda max \approx 460nm$).

Thus the considerable differences between the bichromophoric IV and monoazomethine molecules in the crystals are not observed.

Unfortunately the lack of the spectral and kinetic findings in the solvents does not allow to conclude about the differences between the mechanisms of the photoinduced processes in the structure IV and usual anils.

4. 2. Heterochromophoric molecular systems with an anil moiety.

4. 2. 1. The bichromophors including the nonphotochromic structures.

4. 2. 1. 1. Pyridinium cation with salycylideneaniline.

A. Cation and Salycilideneaniline separated by the CH spacer.

It has been demonstrated above that the direct electrostatic and inductive interactions between the same anil chromophores can change the ESIPT mechanism or can lead to the Interfragmental S_1^*-S_1^* Radiationless Energy Transfer(SIET) which influence markedly the efficiency of the photoinduced processes of the anil molecules.

To study such processes the special chromophores coupled with the anil molecules has been selected /40, 49/(Sch. 4. 13). For instance, the pyridinium cation has the spectral-luminecent parameters and the photochemical characteristics satisfying the necessary conditions for the realization of the SIET in the structure V.

Pyridinium perchlorate (pyr)(Tab. 4. 11, see also /50/) has the typical absorption band, luminescent and photochemical characteristics (See Sch. 4. 11).

Under excitation at the room temperature in the region of the intensive broad absorption band (λmax$\approx$320nm, ϵ =35000 lx mole^{-1}cm^{-1}, Δv1/2$\approx$4400cm^{-1}) caused by the electron transition So-hν->S_1*of the ICT nature from the 2, 5 phenyl rings to the pyridinium ring, the ASS fluorescence (Fflu <0. 05, Δv(a-f)$\approx$12000cm^{-1}) and the effective reaction of the photocyclization (Fcyc$\approx$0. 01) occur from the S_1*state.

The both competing processes are the result of the flattening of the structure owing to the almost free rotation of the 2, 5 phenyl rings at the room temperature in the liquid solvents. In the rigid matrixes (in particular in the glassy solvents at 77K) the rotation is hindered that results in the sharp fall of the photocyclization efficiency (F cyc$\approx$0. 003) and in the lowering of the Stokes Shift (till Δv a-f$\approx$6000cm^{-1}) with the increase of the quantum yield (F$\approx$0. 26).

Meanwhile the very intensive structural phosphorescence band typical for Pyr appears at 77K as a result of the effective ISC in the acoplanar structure of the pyridinium cation($\tau\approx$4s) (Tab. 4. 11). The other moiety of the SA structure has also the very broad (Δv1/2$\approx$5000cm^{-1}) and more long-wavelength (v$\approx$29000cm^{-1})absorption band. The well known ASS fluorescence and photochromism are manifested especially clearly in the glassy solvents at 77K (see Ch. 3).

The autonomy of the structural fragments A(Pyr) and B(SA) in the molecule V is achieved by their separation with the bridge –CH – group. As one can be expected the absorption spectra of V are the superposition of the absorption bands of A and B although, owing to the very big differences in the intensities(A/ B $\approx$3), the absorption band B is overlapped almost completely by the A one. There are no any signs of the complex formation.

The A and B moieties are excited independently. With excitation of the structure V in the absorption band region of the A moiety mainly ($\lambda\approx$313nm) the fluorescence typical for the cation A (v$\approx$25000cm^{-1})

is almost absent even at 77K, the reaction of the photocyclization is not observed at the ambient temperature, and phosphorescence efficiency (T=77K) falls by the order of magnitude with no change of the lifetime(Tab. 4. 11). At the same time instead of the A(Pyr) fluorescence, the ASS flu, typical for SA caused by the ESIPT in the B moiety, arises.

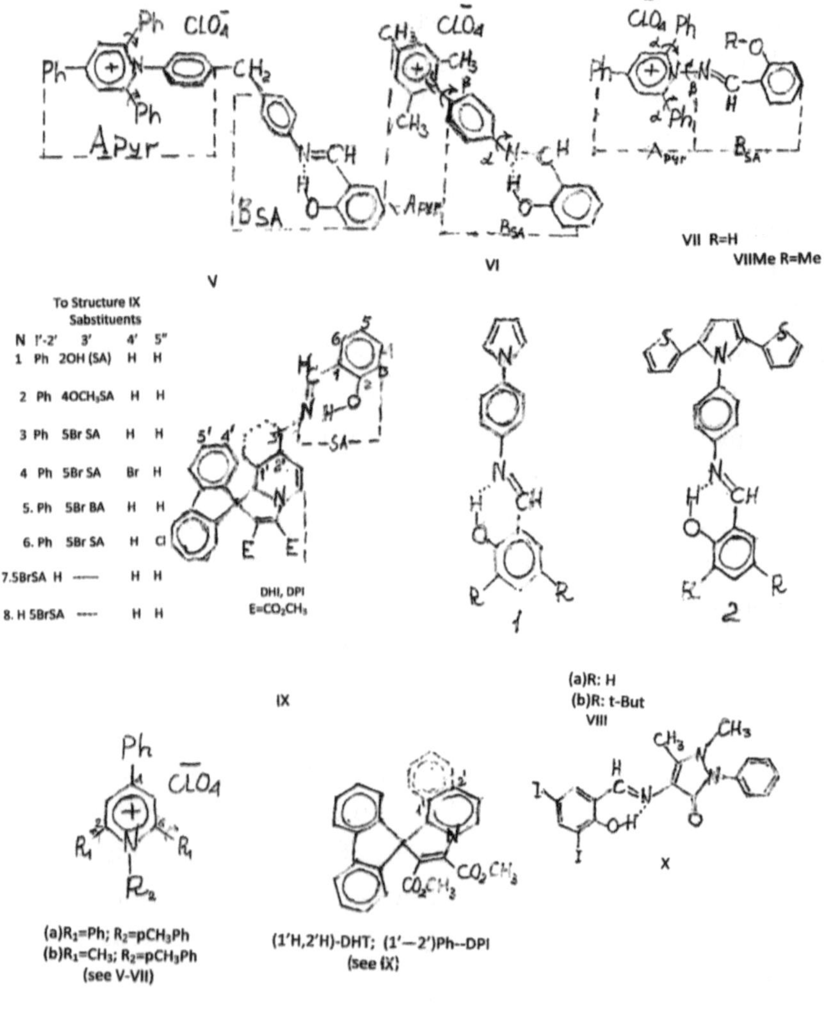

Scheme 4.13

The heterochromophoric molecular systems on the base of SA with the another structures, and the model compounds (see also scheme 4.1)

This fluorescence with the enormous Stokes Shift ($\Delta\nu\approx13000$cm^{-1}) has the intensity two times as much that of the equimolar SA solvent under the same conditions of excitation.

The excitation in the region of the A moiety absorption band at 77K leads also to the increase of the efficiency of the photochromic reaction half as much again as that of SA(Tab. 4. 11).

Table 4.11
The spectral photophysical and photochemical parameters of the bichromophoric molecules V and model structures (in ethanol and dichlorethane).

Struct-ure	Absorption 2)			Fluorescence 1)				Phosphorescence 1)			Photochemistry		
	ν^{max}_{abs} cm^{-1}	ε^{max} LM^{-1}cm^{-1}	$\Delta\nu^{1/2}$ cm^{-1}	ν^{max}_{flu} cm^{-1}	$\Delta\nu^{1/2}$ cm^{-1}	$\Delta\nu_{a\text{-}f}$ cm^{-1}	Φ_f	ν^{max}_{Ph} cm^{-1}	τ_{Ph} s	Φ_{Ph}	Photocolor.[1] λ^{max}_{ab}, nm	Φ_{SA} relative	$\Phi_{Photo\cdot cycliz.\,Pyr}$
Pyr	31200	35000	4440	24690	2890 2)	6510 2)	0.05	19880 3)	3.80	0.51	----	----	<10^{-2} 2
SA	29325	12000	4890	17860	2220	11465	0.25	17790 4)	----	----	480 520	1.0	----
V	31200	45000	6220	24690 17860	2900 2200	6510 11465	<10^{-4} 0..52	19880 3)	3.75	0.05	480 520	1.5	<10^{-4} 2

1)T=77K. 2)Room temp-re.3)The main max of the vibr. struct.4)Phosphor. Is absent.The energy of the T$_1$ state is calculated by CNDO/S method.

Table 4.12
The kinetic parameters of the photophysical processes in V and model structures (Sch.4.14)

Structure	Energy transfer		Excited states' deactivation rate, s^{-1}			
	Φ_{tr}	$K_{tr}\times 10^{-10}$s^{-1}	K_f	K_{ST}	K_{TS}	K_{Ph}
Pyr	---	---	5x10^8	1.4x10^8	0,13	0.26
SA	---	---	---	~10^{11}	---	---
V	0.89 0.67	2.3 [1] 1.7 [2]	5 x 10^8	~10^8	0.13	0.27

1)Calculates according Forster theory with taking in account the fluoresces' band A(Pyr) and absorption band B$_{SA}$ overlapping with d(A-B) =10Å. 2)Calculated by the phosphorescence quenching.

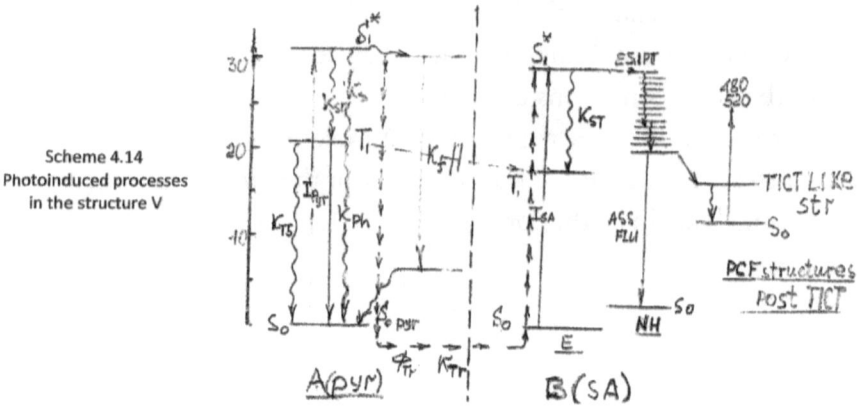

Scheme 4.14
Photoinduced processes in the structure V

The above data can be explained by the effective S-S Intramolecular radiationless energy transfer (SIET) from the A(Pyr) moiety with the high absorptivity to the B (SA) one.

According to the Scheme 4. 14 and to the Table 4. 12 the Energy of the $S_1^*(\pi\pi^*)$ state of the A fragment is transfered with the high rate (Ktr≈2. 3x10^1 s^{-1}) and efficiency (F≈0. 9).

Such an effective energy transfer proceeds owing to the favorable S_1^* relative positions of the Donor, A(Pyr), (E≈3. 9ev) and the Acceptor, B(SA), (E≈2. 3ev), to the short distance between them (≈10Å), and to the high value of overlap integral A(Pyr)flu –B(SA)abs.

Meanwhile the phosphorescence efficiency of the Pyr moiety (Fph) of the bichromophore molecule V is less by order of magnitude that of Cation Pyr, but phosphorescence lifetimes are the same (τ≈3. 75s, Tab. 4. 11).

This testifies to the lack of the interfragmental T-T energy transfer in spite of the favorable relative position of the D and A triplet states (27and 22 ev respectively) due to the almost complete orthogonality of A and B structural fragments with the negligible overlapping their π-orbitals which is necessary for the realization of the exchange –resonance mechanism of the T-T energy transfer/29/.

Thus the depopulation of the T_1(Pyr) state, and hence the strong fall of the phosphorescence efficiency, in the structural fragment of the bichromophore molecule V as compared with those of Pyridinium cation(Pyr) can be caused only by the S-S energy transfer in the molecule V.

It allows to utilize the alternative simple kinetic method for the estimation of the rate constant Ktr and the efficiency (Ftr) of the energy transfer by the comparison of the phosphorescence efficiencies of the structure V, F(ph, V), and Pyridinium cation, F(ph, Pyr) according to the scheme (Sch. 4. 14) with approximation Ks <<Kst:

$$Ktr=(Kst+Kf)(F(ph, Pyr)/ F(ph, V)-1) \approx 1.7 \times 10^1 \, s^{-1} \, ; \, Ftr = Ktr/Ktr+Kst+Kf \approx 0.67.$$

Thus the rate and efficiencies values of the energy transfer estimated by both methods are very close and very high.

The results obtained point to the important role that can be played by the IET in the complex molecular bichromophore systems which can increase the efficiencies of the photochromic rea-ction and the ASS fluorescence.

B. Cation and Salicylideneaniline are bounded directly.

The investigation of the bichromophoric systems VI (Sch. 4. 13) has been carried out in the crystal /40, 53/ with the intention to obtain the new crystalline photochromic systems on the base of anils.

According to the X-ray structural data the compound VI has acoplanar enol structure.

The dihedral angle between aldehyde and imine moieties $\alpha \approx 51°$, pyridinium and aniline rings are almost orthogonal ($\beta \approx 87°$, 2). The planar structure of the aldehyde fragment is stabilized by the H bond OH...N with the distance O.... N d=2. 6Å.

The pyridinium fragment in VI can be considered as a bulky, very strong electron withdrawing substituent. Such a substituent prevents the formation of the crystalline packing with the close disposed adjacent molecules in the parallel planes, and creates the sufficient room for the structural transformations.

At the same time, owing to the strong electron-acceptor nature, s it decreases considerably the electron density on the Nitrogen atom of the C=N group in the Ground state.

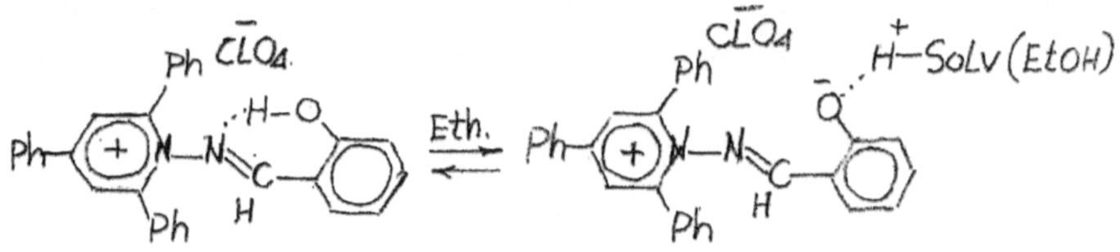

Scheme 4.15
The tautomeric equilibrium in the solution with IH-Bond break and H-bond complex formation

Table 4.13
The absorption and fluorescence data for the structure VII
that characterizes H-bond complex formation.

Solvent	Absorption		Fluorescence	
	λ^{max}_{abs} nm	I relative	λ^{max}_{flu} nm	$\Delta\nu_{a-f}$ cm^{-1}
Ethanol	315 1) (40000)	1.00	-----	-----
	420	0.01	480	~3000
Ethanol + Et$_3$N	310	0.92	-----	----
	420	0.23	480	~3000

1) In brackets ε (LM^{-1} cm^{-1}).

Such factors have to prevent the thermochromism and favor the photochromic properties in the crystals.

Really the typical photochromism which is very like that of SA is manifested by the compound VI in the crystal state.

Thus the utilization of the acoplanar bulky organic kation as substituent in the imine moiety of anils can be considered as one more method for the preparation of the new ptotochromic crystalline anils. The bichromophoric system VII (Sch. 4. 13) has been investigated in the crystal and the solvents /40, 49/.

Such a structure can be considered as an ionic anologue of SA molecule in which the positive charge is localized in the imine ring.

In the crystal the structure VII is strongly acoplanar, and the dihedral angles and α ≈ 70-80°, but the anil structural fragment is planar and keeps the IHB OH...N.

However in spite of the considerable mutual acoplanarity, the moieties A(Pyr) and B(anil) interect markedly in the ground state. This interaction is realized in weakening of the electron density on the Nitrogen atom of the C=N group as a result of the direct influence of the positive charged imine ring.

The IHB rapture and the formation of the H-bond complex with the solvent may be a consequence of that influence. In this case the equilibrium VII=VII(Zw) (Sch. 4. 15) is realized in ethanol, and manifested in the absorption and fluorescence.

The long wavelength absorption ($\lambda \approx 420$nm) and fluorescence($\lambda \approx 480$nm) bands appearing side by side with the intensive absorption band ($\lambda \approx 315$nm $\varepsilon = 40000$ LxM^{-1}cm^{-1}) of the structure VII (Tab. 4. 13) are absent of the model Methoxy derivative VII (Me) (Sch. 4. 13) and are connected with the typical charge transfer in the zwitterion structure VII(Zw).

Really, the equilibrium (Sch. 4. 15) is shifted towards VII(Zw) structure with the corresponding change of the absorption band intensities with the addition of Et N forming the H-bond complexes with VII(Tab. 4. 13) and with the decrease of the concentration of VII in the ethanol solution.

As it should be expected the compound VII is not photochromic.

4. 2. 1. 2. Pyrrole-salicylideneaniline bichromophoric system.

The compound VIII(1, 2) have been investigated with a purpose to obtain the new electrophotochromic polymers bearing the photochromic anils as a pendant functional groups on the base of the monomer structure VIII. (Sch. 4. 13).

The absorption spectra of VIII(1a, b), VIIi(2a, b) in dichloromethane consist of at last the two bands with the maxima at 330-350nm and 230-235nm which are obviously superposition of the Pyrrole and SA moieties' absorption bands.

However the necessary data concerning this probslem are absent. The X-ray crystallographic data obtained from the investigation of a single crystal of VIII(1, b) show that molecule in the crystal has the nonplanar structure in that the angles between the planes of the pyrrole ring and the aniline phenyl ring and between the aniline phenyl ring and salicylidene phenyl ring are 32. 6° and 33. 1° respectively. The salicylidene moiety has the enol structure with the strong IHB O-H...N.

The photochromic and thermochromic behavior of VIII has been studied only visually in the solvents and in the crystalline powders. As it should be expected no photochromism was observed in the liquid dichloromethane solution at ambient temperature. The crystalline powders of the compounds VIII(1a) and VIII(2b) show a photochromism under irradiation with 365nm light changing a color from the yellow to the red. The compounds VIII(1b) and VIII(2a) did not show any color change under the same conditions.

Table 4.14

The photochromism of the bicromophoric systems IX under the different conditions (see /52/)

Compa-und	Medium, Temperature,K	Before irradiat. λ^{max}_{abs} nm	Irradiation time,bands' λ^{max}_{abs} nm, and decreases of intensity(in brackets) after irradiation	Time decay(s)and(or) const.k(s^{-1})after flesh photolysis at the room temperature	Struct. moieties responsible for photochromism
1	Ethanol, room EPA, 77 MCH,room	350 350 400	350,550 350,500 400(0.8)15 min (0.5)60 min	0.08	DPT SA irreversible
2	Cryst .film, room Cryst, film, 77	345 360	345 (0.8) 480-500 360(0.7) 520	Dark reaction, 12 h	SA SA
3	Polymer, room Polymer, 77 PPA,77 Ethanol,room	400 400 420 ---	400(0.6) 400(0.8),530 420, 548 --- 550	Dark reac very slow Almost full restor-ation after dark reaction 0.42	Irreversible SA SA SA+DPI (overlap)
4	EPA ,room EPA, 77 Ethanol,room	400 400 ---	No irradiation 400,525 630	---- ---- 18700	---- SA DPI +SA
5	Polymer,room EPA,77 Ethanol,room	360 360 ---	No irradiation 360 600	--- --- 0.58	--- No change DPI
6	Cryst. film,room Cryst. film,77	400 400	400 (0.72) 400 (1.25),500	Almost no dark react Dark react.is very slow	Irreversible SA (very weak)
7	EPA ,room EPA,77 MCH,room Eth.room	435 435 300,440 ---	No irradiation 435,500 300,440(0.5)15min 550	----- Full restoration Full restoration 0.04	----- SA SA DHI
8	Polymer,room Polymer,77 MCH,room EPA ,77	400 430 336,400 330	400 (0.6) 15min 430(1.0), 517 336,440(0.6)15mn 330,500,630	~ 1,2 day Full restoration	Irreversible SA+DHI
DPI Or DHI	EPA,room EPA,77 Eth,room	360 360 -----	No irradiation 360. 620	---- --- 0.20	 No change DPI or DHI
SA	Ethanol,room EPA,77	340 340	340,~480 340,480,500	3×10^{-9} s	SA SA

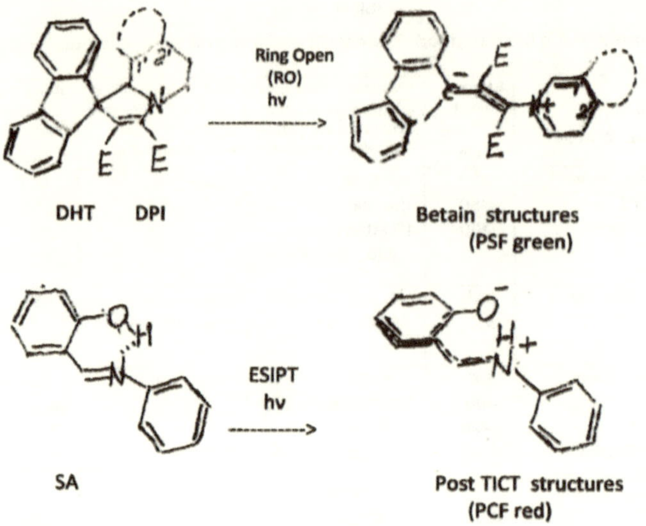

Scheme 4.16
The nature of the photocoloration of the molecular system IX
a) The structure of the PCF. b) The possible path-ways of photocoloration.

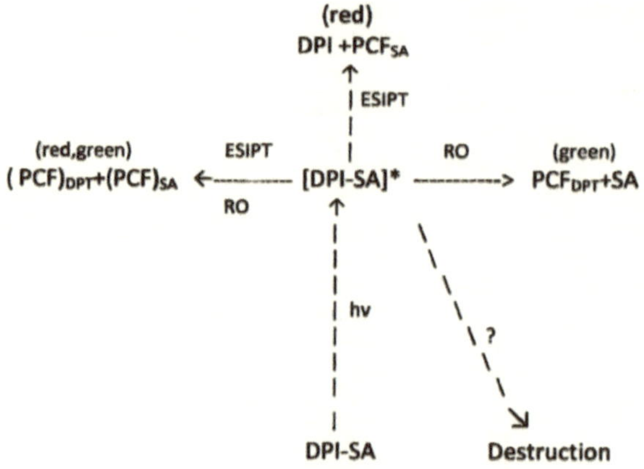

Both the photochromic and thermochromic properties are absent for all polymer films of VIII inspite of the existence of the photochromism of the corresponding monomers VIII(1a, 2b).

The detailed investigations of the photo and thermochromism of the structures VIII in solvents and crystals under various conditions are necessary.

4. 2. 1. 3. Schiff bases with the 4-aminoantipyrine structure.

The novel photochromic crystalline compounds on the base of the anil structure can recently been synthesized from 4-aminoantipyrine and 3, 5 dichlorosalicylidenealdehyde in the water solvent only(see/54/).

It is shown by the Density Functional Theory (DFT) calculation that from one to three water molecules participate in the water catalyzed mechanism to produced the E-isomer X (Sch. 4. 13).

The photochromism of crystalline X with the Enol structure is displayed, when the crystal was irradiated ($\lambda \approx 380nm$ at the room temperature), as the color change from white to green ($\lambda refl \approx 500$-$560nm$), and caused by the ESIPT (O-H...N—hν-->O...HN) responsible also for the ASS fluorescence ($\lambda max \approx 534nm$ in EtOH) in the NH("Keto") structure/54/.

The photochromic properties have been examined by the data of the Time-Dependent DFT (TDDFT) calculations without utilization of the experimental absorption spectra and X-ray strucral data /54/.

4. 2. 2. The bichromophoric systems on the base on anil and spirocyclic structures.

The bichromophoric systems IX(1-8) including the two photochromic structures-photochromic azomethine and photochromic spirocyclic structures, spiro-[1, 8a]-dihydroindolizines(DHI) or dihydropyrrolo-(1, 2a)-isoquinolines(DRI)(Sch. 4. 13) have been synthesized and studied in the various media, under different conditions of excitation at the ambient and liquid Nitrogen(77K) temperatures by the steady-state absorption spectroscopy mainly/52/.

In the general scheme (Sch. 4. 16b) the path-ways of the Photocolored Form (PCF)structures' generation can proceed (Sch. 4. 16a) via Ring Open(RO) in the moiety DHT or DPT, and the Excited State Proton

Transfer(ESIPT) in the SA moiety or in the both moieties combining the two reaction path-ways. In the latter case it is not known whether the concert mechanism or the step-by-step one takes place.

According to the finding produced (see also Tab. 4. 14) the definite pathway can occur in dependence not only on the structure but also on the nature of environment and on the conditions of the excitation.

In the liquid media at the room temperature with the impulse excitation the both (RO and ESIPT) mechanisms can be realized.

The RO mechanism leads to the formation of betaine structure with the slow back dark reaction (comp. IX(3, 5, 7)), and the ESIPT reaction gives the post-TICT structures with the fast back reaction (comp. IX (3, 4, 8)) along with the RO mechanism in the DPI moiety.

The latter cannot be realized in the rigid media (including crystals and polymer matrixes, especially at 77K), but the ESIPT reaction followed by the formation of the PCF(SA) structures is held easily sufficiently in the SA moiety under such conditions in the compounds IX (1-4, 7).

For some compounds the destruction is observed at the room temperature under steady-state excitation in the nonpolar solvents (comp. IX (1, 7, 8)), polymer matrix (comp. IX 3, 8), and the crystal state (comp. IX(6)).

At the liquid nitrogen temperature decomposition of the material is prevented, and the compounds are stable or photochromic in the polymer matrix and the crystals (comp. IX (1, 2, 6)).

In addition the compound IX(6) shows the photochromism and thermochromism at low temperatures.

The above findings show the very similar behavior of the compounds in the crystals and the polymer matrix, and that points to the weak influence photoinduced processes of the crystalline packing.

Thus the opportunity of the variation of the photochromic properties with the changes of the media and the conditions of excitation is the principal advantage of the bichromophores IX.

However the mutual suppression of the photochromic efficiency of each of two moieties and the effective photodestruction of the complex bichromophoric molecule especially at ambient temperature are very strong defect of such compounds.

Conclusion

The principal findings about the composed molecular systems including the anil structural fragment have been obtained in last two decades.

However the generalization of the date of such molecular systems as the separate class of the photochomic compounds has been carried out for the first time in this chapter on the base of the structural classification.

There are the two main groups of these systems: the Homochromophoric systems, and the Heterochromophoric ones.

The systems of the first group (bisanils) involve the two same anil moieties, in the second ones the anil moiety is bound up with the another, nonphotochromic or photochromic, structures.

The Homobichromophoric systems can be classified in accordance with the strength of the π-electron systems' interactions between the anil moieties in the ground state.

In the bisanils with the weakly interacting moieties the marked shift of the tautomeric equilibrium E$\Leftrightarrow$K towards K structure is observed even in the nonpolar solvents unlike SA. In spite of that, such systems are characterized by the high degree of the independence and the superposition in the absorption spectra of the anil moieties in the E structures.

In this situation both highly distinctive feature of the localization of the photochromic activity on only one anil moiety and the kinetic peculiarities of the phototransformations in comparison with the separate anil molecule can be explained satisfactory by the assumption of the ultrafast interfragmental radiationless S-S energy transfer (ISET).

The efficiency of interaction between the anil moieties' π-electron systems which causes the differences between the bisanil molecules depends strongly on the molecular structure of the anil moieties, the mode of their joining, and the media nature.

The contribution of the SIET in the mechanism of the photoinduced processes and the ESIPT efficiency fall with the strengthening of the interfragmental interactions. With the very strong interfragmental interactions the ESIPT and the photochromism can be completely suppressed in both anil fragments.

Unfortunately the above ideas bear in common the semi-quantitative character, and need the strict additional evidences on the base of the data of the detailed and accurate purposeful experimental and theoretical studies.

The Heterochromophoric systems on the base of the anil molecules coupled with the nonpho-tochromic pyridinium cations have also been considered in the sequence of the strengthening of the interaction between their π-electron systems in the series of the structures.

The contribution of the SEIT in the mechanism of the photoinduced processes and the efficiency of the photochromic reaction fall in this series of compounds, i. e. in the first system the high efficiency of the SEIT is combined with the effective photochromism, and in the last one the ESIPT (and the photochromic activity) is suppressed completely.

The new bichromophoric systems on the base of SA with various bulky lowpolar nonphotochromic structures have been also studied to search and to obtain the new crystalline and polymeric photochromes.

On the other hand the bichromophoric systems on the base of SAs and the photochromic spirocyclic structures show the photochromic properties typical for the both moieties in various conditions that is the principal advantage of such systems.

However there are the essential defects of those connected with the mutual suppression of the photochromic activity in each of two moieties, and with the effective photodestruction of the bichromophoric molecules especially in the solvents at ambient temperature.

It may be supposed that such defects are typical for any bichromophoric system, especially with the moieties of different molecular structures, and this problem needs the special investigation.

References

1. B. Krasovitskii, and B. Bolotin, "Organic luminophores", Moskow, Publish House "Khimia" (1984) (in Russian).
2. E. Hadjoudis, Mol. Eng. 5, 301 (1995).
3. E. Hadjoudis, and E. Mavridis, Chem. Soc. Rew. 33(9) 579 (2004).
4. K. Animoto, and T. Kawato, Photochem. Photobiol. C. Photochem. Rewievs 6, 207(2005).
5. B. Krasovitskii, V. Smeliakova, and R. Nurmukhametov, Optica i Spektrosk. 17, 558, (1964)(in Rus-sian).
6. B. Krasovitskii, N. Mal'tseva, and R. Nurmukhametov, Ukr. Khim. Zhurn. 31, 828, (1965)(in Russian).
7. N. Vasilenko, R. Nurmukhametov, and Ya. Pravednikov, Dokl. Akad. Nauk. SSSR 224(6), 1334, (1975) in Russian).
8. B. Krasovitskii, O. Asmaev, M. Knyazhansky, O. Osipov, N. Levchenko, V. Smeliakova, A. Nazarenko, N. Maltseva, and L. Afanasiadi, Zh. Fiz. Knim. 45 (6)1467(1971)(in Russian).
9. M. Knyazhansky, O. Asmaev, O. Osipov, and Krasovitskii, Zh. Fiz. Khim. 46, (1), 178(1972)(in Russian).
10. O. Asmaev, M. Knyazhansky, B. Krasovitskii, and O. Osipov, Zh. Fiz. Khem. 46, (3), 638(1972)(in Rus-sian).
11. M. Rawat, and J. Norula, Indian J. Chem. B26(3), 232(1987)).
12. K. Kownacki, L. Kaczmarek, and A. Grabowska, Chem. Phys. Lett. 210(5, 6), 373(1993).

13. N. Hoshino, T. Inabe, T. Mitani, and Y. Maruama, Bull. Chem. Soc. Jpn. 61. 4207(1988).
14. S. Takeda, H. Chihara, I. Inabe, I. Mitani, and Y. Maruyama, Chem. Phys. Lett. 189, 13(1992).
15. R. Destro, A. Gavezzotti, and M. Simonetta, Acta Cryst. B34, 2867(1978).
16. A. Grabowska, K. Kownacki, and L. Kaczmarek, J. Luminescence 60-61, 886(1994).
17. M. Ziolek, A. Kubicki, R. Naskrecki, and A. Grabowska, J. Chem. Phys. 124, 124518(2006).
18. K. Kownacki, A. Mordzinski, R. Wilbrandt, and A. Grabowska, Chem. Phys. Lett. 227, 270(1994).
19. M. Ziolek, J. Kubicki, A. Maciejewski, R. Naskrecki, and A. Grabowska, Chem. Phys. Lett. 369, 80(2003).
20. M. Ziolek, G. Burdzinski, and J. Karolczak, J. Phys. Chem. A. 113, 2854(2009).
21. K. Filipczak, J. Karolczak, and M. Ziolek, Photochem. Photobiol. Sci. 8, 1603 (2009).
22. Y. Zhang, and Z. lu, Mater. Chem. Phys. 57, 253 (1999).
23. M. Ziolek, J. Kubicki, A. Maciejewski, R. Naskrecki, and A. Grabowska, Phys. Chem. Chem. Phys. 6(19) 4682(2004).
24. M. Knyazhansky, Polish J. Chem. 82, 795(2008).
25. M. Ziolek, I. Sobczak, J. Incl. Phenom. Macrocycl. Chem. 63, 211(2009).
26. N. Turro, " Modern Molecular Photochemistry" , University Scince Book, (1991).
27. T. Sekikawa, T. Kobayashi, and T. Inabe, J. Phys. Chem. A. 101, 10645 (1997).
28. T. Inabe, N. Hoshibo, T. Mitani, and Y. Maruyama, Bull. Chem. Soc. Jpn. 62, 2245(1989).
29. V. Ermolaev, E. Bodunov, E. Sveshnikova, and T. Shakhverdov, "Radiationless transfer of the electronic excitation energy" , Publishing House "Science" , Leningrad, (1977) (in Russian).
30. T. Inabe, J. Lineau, T. Mitani, Y. Maruama, and S. Takeda, Bull. Chem. Soc. Jpn. 67, 612(1994).

31. T. Kawato, H. Kanatomi, H. Koyama, and T. Igarahi, J. Photochem. 33, 199(1986).
32. T. Inabe, N. Hoshino, T. Mitani, and Y. Maruama, Bull. Chem. Soc. Jpn. 64, 801(1991).
33. T. Inabe, New J. Chem. 15, 129(1991).
34. M. Taneda, H. Koyama, and T. Kawato, Chem. Lett. 36(3), 354(2000).
35. M. Ziolek, M. Gil, J. Organero, and A. Douhal, Phys. Chem. Chem. Phys. 12, 2107 (2010).
36. J. Zhao, B. Zhao, J. Liu, A. Ren, and J. Feng, Chem. Lett. 29(3), 268(2000).
37. J. Zhao, B. Zhao, J. Liu, W. Xu, and Z. Wang, Spectrochim. Acta. A57, 49(2001).
38. M. Knyazhansky, A. Metelitsa, M. Kletskii, A. Milov, and S. Besugliy, J. Mol. Struct. 526, 65(2000).
39. A. Grabowska, K. Kownacki, and L. Kaczmarek, Acta Phys. Polon. A88, 1081(1995).
40. M. Knyazhansky, and A. Metelitsa, "Photoinduced Processes in Molecules of Azomethines and their structural analogs" Publ. House of Rostov State Univ., Rostov-on-Don (1992) (in Russian)
41. J. Waluk, Polish. J. Chem. 82, 947(2008).
42. J. Zhao, B. Zhao, J. Liu, T. Li, and W. Xu, J. Chem. Res. (S)416(2000).
43. E. Hadjoudis, A. Rontoyianni, K. Ambroziak, T. Dziembiovska, and I. Mavridis, J. Photochem. Photobiol. A. Chem. 162, 52(2004).
44. J. Laverty, and Z. Gardelung, Polymer lett. I, 161(1961).
45. A. Grabowska, K. Kownacki, J. Karpiuk, S. Dobrin, anL. Kaczmarek, Chem. Phys. Lett., 267, 13
46. M. Zio-lek, J. Kubicki, A. Maciejewski, R. Naskrecki, W. Luniewski, and A. Grabowska, J. Photochem. Photobiol. A. Chem. 180, 101(2006).
47. M. Ziolek, G. Burdzinski, K. Filipczak, J. Karolczak, and A. Maciejewski, Phys. Chem. Chem. Phys. 10, 1304(2008).

48. E. Hadjoudis, T. Dziembowska, and Z. Rozwadowski, J. Photochem. Photobiol. A. Chem. 128, 97 (1999).
49. M. Knyazhansky, N. Makarova, Ya. Tymyanskii, and V. Kharlanov, Zhurn. Prikl. Spectr. 62(3)64 (1995)(in Russian).
50. M. Knyazhansky, Ya. Tymyanskii, V. Feygelman, and A. Katritskiy, Heteroicycles, 26(11)2963 (1987) and references there.
51. B. Tompson, K. Abbod, J. Reynolds, K. Nakatani, and R. Andebert, New J. Chem. 29(9)1128 (2005).
52. M. Holderbaum, H. Durr, and E. Hadjoudis, J. Photochem. Photobiol. A. Chem. 58, 37(1991).
53. S. Aldoshin, M. Knyazhansky, Ya. Tymyanskii, and L. Atovmian, Khim. Fiz. 8, 1015(1982)(in Russian).
54. A-G. Zong, D. Chen, M. Lei, and Sh. Liu, J. Theor. Comput. Chem. 7(5), 1071(2008).

Chapter 5

The Esipt Phpotochromism Of The Molecular Systems With The Nonanil Structures.

Introduction and classification of the structures.

The great number of the compounds different from Anils, involving the Intramolecular Hydrogen Bond(IHB) that show or can show the ESIPT photochromism has been found and studied in the last two decades or a little earlier. However to the best of our knowledge there are no any review or general discussion in that direction till now.

Therefore the attempt has been took to summarize and discussed the main results of those studies on the base of the suggested classification by the structures and the photochromic properties.

Since the ESIPT is the primary step of any ESIPT photochromic reaction it should be expediently for discussion to distribute the structures by the groups of the Proton (Hydrogen) transfer between the Proton donor and acceptor reaction centers including:

i. The different heteroatoms (O⇔N);
ii. The same heteroatoms (N⇔N, O⇔O);
iii. The C-atom and heteroatom (C⇔N, C⇔O);
iv. The double transfer involving the iii-type as a primary step followed by the i-type.

However the distinctions in the photochromic properties are not reflected by such a classification.

By the strict definition, the photochromism is the **phenomenon** of the visually (by a naked eye) observed reversible color (**chromus**) change under irradiation (**photo**) in the common ambience.

The irreversible processes connected with the photochromic "fatigue" don't consider here.

From the spectral and kinetic views it corresponds to the reversible (thermal or by irradiation) variations of the visible absorption spectra under irradiation, and to a long space of the time enough for the visual detection of the colored form respectively.

The above parameters are the functions of both the molecular structure and the media nature which have to secure the conditions for the realization of the photochromic properties.

However while the spectral parameters are determined by the differences between the initial and colored structures, the kinetic parameters are caused by both the molecular structure and the medium nature.

In the search of the potential photochromic structures it is expediently to follow above notions.

At the same time according to the experimental findings the structures considered below can be classified by the photochromic properties and distributed in the three types:

1. The photochromic molecules displaying the visible coloration in the liquid solutions under the steady-state irradiation can be named as the "True" photochromes (T-photochromes).
2. The "True" photochromism is displayed in the rigid matrixes (including crystals) but the fast changes of the absorption spectra in the visible spectral region, inaccessible for the naked eye can be observed in the liquid media (including solvents) under impulse irradiation.

    The later spectral manifestations can be named "Latent" (L) photochromism, and the molecules with such

Latent and True photochromic properties can be named LT-photochromes.

3. However the overwhelming majority of the molecules with the ESIPT displays the "Latent" (L) photochromism only both in the liquid and rigid (including crystals)media showing the fluorescence with the Anomalous Stokes Shift (ASSflu)/1-3/, and can be named as "Latent" photochromes. Only some of such structures which have a prospective to be modified for the LT or T groups /4-6/ have been considered in the Chapter.

The brief qualitative consideration of the connection between the molecular structure, intermolecular interactions, and photochromic properties for each group, and attribution of each structure type to the definite group are carried out in the general discussion in the conclusion of the chapter.

The photochromic molecules in the crystals with the suggested intermolecular ESPT mechanism (e. g. /7/) have not been discussed in the chapter.

5. 1. The ESIPT between the different heteroatoms (OH⇔NH).

5. 1. 1. Oximes of orthohydroxyaldehydes.

The experimental investigations of orthohydroxybenzylideneoximes, I(OH), (Sch. 5. 1)have been carried out by the steady state methods in the liquid and glassy(T=77K)solvent only/8/(Tab. 5. 1).

The oxime molecule I(OH)is structurally similar to that of Salicylidenealkylimine(SALKo)(Sch5. 1).

Like SAlKo the longwavelength transition ($\lambda^{max} \approx 300$nm) has the ICT (PhOH-e->C=N)nature.

The indirect evidence of that is the appearance of the intensive E-structure fluorescence(Stokes Shift $\Delta v \approx 3000 cm^{-1}$) of I(OMe) model molecule unlike I(H) one (Sch. 5. 1) because of the strong energy decrease of the $S_1^*(ICT)$ state and corresponding increase of the $S_1^*(ICT)-S(n\pi^*)$ energy gap ("proximity" effect) with the introduce of the electron donor (OMe) substituent.

Table 5.1
Photochromism and fluorescence of oxime molecules (solv.IIP, T=77K)
Comparative data relatively $SALK_0$ (see scheme 5.1).

Structure	Before irradiation			Photocolored form (PCF)			[4]
	Absorpion λ_{abs}^{max} nm	Fluorescence[3,5] $\lambda^{max}(\Phi_f)$ nm	Δv^{a-f} cm^{-1}	Absorp.[2] λ_{abs}^{max} nm	Fluorescence [2]		Φ_{rel}
					λ_{fl}^{max} nm	Δv^{a-f} cm^{-1}	
I_H	282	---	---	---	---	---	---
		---	---	---	---	---	---
I_{OCH_3}	303 (E)	330 (0,12)	3000	---	---	---	---
		---	---	---	---	---	---
I_{OH}	300(E) 360(K)	432,440 (0,07)	10185	400,432	475,500	1650	0.4
/1/ $SALK_0$	312(E) 380(K)	522,555 (0,002)	12900	385,420	573,590	6350	1.0

/1/See ch.3 and ref/8-10/./2/Two maximum is vibronic structure./3/The Stokes shift is relatively to the absorption band of the Enole (E)structure./4/The efficiency of the photocoloration./5/In the bracket- The relative efficiencies

Scheme 5.1
Photochromic transformations in the oxime molecule I(OH) and the model structures.

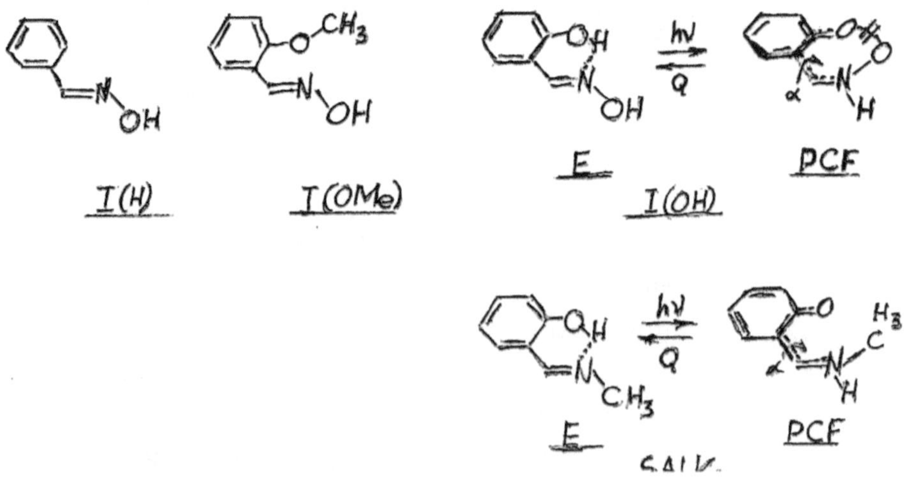

In the structure I(OH) the strong OH...N bond is formed similar to SALKo molecule. It results, especially, at 77K in the glassy solvent, in the shift of the ground state equilibrium E⇔K towards K structure with the appearance of the typical Keto-structure absorption band ($\lambda^{max} \approx 360$nm with the corresponding fluorescence band ($\lambda^{max} \approx 430\text{-}440$nm), and in the ESIPT with the ASS fluorescence band ($\Delta v_{a\text{-}f} \approx 10000 \text{cm}^{-1}$) suppressing completely the E-structure fluorescence.

By analogy with SALKo in the glassy solvent at 77K the irradiation ($\lambda=313$nm) results in the effective generation and stabilization of the fluorescent Photocolored (PC) structure (Tab. 5. 1) which is not observed in the liquid solvent at the ambient temperature with the steady-state irradiation.

Unlike SALKo under influence of the electronegative Oxygen atom joined to the C=N group instead of the Alkyl one, the ASS fluorescence is shifted towards "blue" with the decrease of the ASS but with the heightened quantum efficiency (Ff). On the contrary, the efficiency of the PCF generation falls, and its absorption bands are shifted towards red. In view of the latter a note should be taken that the considerably nonplanar trans (C—N) structure of the PCF may be stabilized by formation of the O...H-O bond (Sch. 5. 1) with the corresponding "red" shift of the absorption band as compared with SALKo structure (see Ch. 3).

It seems the study of the photochromism of the crystalline oximes is rather prospective.

Thus like SALK (and SA) molecule, I(OH) one is the LT photochromic structure with the T photochromism in the rigid media.

5. 1. 2. Photochromism of orthohydroxybenzylidenehydrazones.

The absorption, fluorescent, and photochemical properties of II(OH), III(OH), IV(OH) molecules and their model structures III(H), III(OMe), IV(OMe) (Sch. 5. 2) with the imine N(i) and amine N(a) atoms joined each other directly, have been studied by the methods of the absorption and fluorescent steady-state spectroscopy in the solvent only /8/.

The experimental data (at 77K) for the structures III, IV are presented as an example in the table (Tab. 5. 2). Like anils the first electronic transition in absorption has the ICT nature(from Ph-R to C=N group) of all the II-IV molecules and shifted towards "red" with inclusion of the donor substituent (OMe, OH) in the aldehyde ring.

The hybridization type of the n orbitals of the N(a) atom is a common decisive factor determining the difference of the excited state deactivation mechanism of the II(OH), III(OH), and IV(OH) molecules and the model structures(Sch. 5. 2).

In the case of the sp^3 hybridization (structures II_1, III) the interaction of the n and $p(\pi)$ orbitals of the N(i) and N(a) atoms respectively is very weak and such molecules have the fluorescent and photochemical properties similar to the SA and SALK structures. Really the structure III(OH) unlike III(H) and III(OMe) ones shows the strong ASS fluorescence and photochromism caused by the ESIPT followed by the structural transformations with the effective generation of the non-planar fluorescent PCF structure(Ch. 3).

On the other hand the N atom joined to the groups with the π-electron systems has mainly the sp^2 hybridized orbitals that results in the N(i)-N(a) orbitals interaction and secures according to the calculation (CINDO/2) the considerable decrease of the electron charge on the C=N group (-0. 40e and -0. 10e for SA and III(OH) in the So state correspondingly) and of the N(i) atom basicity in the So state.

The immediate effect of the later is the utmost weakening of the H-bond O-H...N, fall of the ESIPT efficiency and the formation of the ionic structure in the polar solvents owing to formation of the H-complexes with a solvent molecule.

Such structures are characterized by the very low efficiency (II(OH), 2, 3) or even lack of the photocoloration, and they cannot be prospective photochromes. Meanwhile independently on the structure for all molecules II-IV with and without o-hydroxyl group the specific reversible side photoreaction is observed.

That photoreaction occurs in the initial E-structure and is responsible for the additional fall of the integral intensity of the ASS fluorescence band under irradiation at 77K and the room temperature (Tab. 5. 2).

Table 5.2
The fluorescent and photochemical characteristic of hydrazone structures.
(isopentene, 77K, see sch.5.2)

STRUCTURES	Initial form					Photocolored form				Accumul.	Side photoreaction	Notes
	Absorption		Fluorescence			Absorption		Fluoresc.				
	λ^{max} (nm)	ε $M^{-1}l^2$	λ^{max} (nm)	Φ_f^{rel}	Stokes Shift (cm^{-1})	λ^{max} (nm)	ε $M^{-1}L^2$	λ^{max} (nm)	St.okes Shift (cm^{-1})	k_{ac} (s^{-1})	k_{ac} (s^{-1})	For the back dark Side react. $k=10^{-4} s^{-1}$ in the liquid solvents at room t-re
III(OH)	320	14000	/3/ 515	1.0	11840	440	<4000	571	5215	$9x10^{-3}$	$1.2x10^{-2}$	
III(Me)	301	17000	$<10^{-4}$	-	-	-	-	-	-	-	+	
III(H)	299	12000	$<10^{-4}$	-	-	-	-	-	-	-	+	Photocolor. and side reaction are absent at any conditions.
IV(OH)	345	19000	$<10^{-4}$	-	-	-	-	-	-	$<10^{-4}$	-	
IV(1,OMe)	332	22000	<1> 0.2	4700	-	-	-	-	-	-	-	
SA	340 420	11000	/3/ 520	—	11000	480	3000	560	3000	$5x10^{-3}$	-	

<1>Absolute quantum yield(liquid solvent, room temperature).<2>"+" or "-" is the present or absent of the side reaction respectively. The value k_{ac} is the accumulation rate constant<3>ASSfl.

Scheme 5.2
Photochemical transformation and model structures of hydrazone.

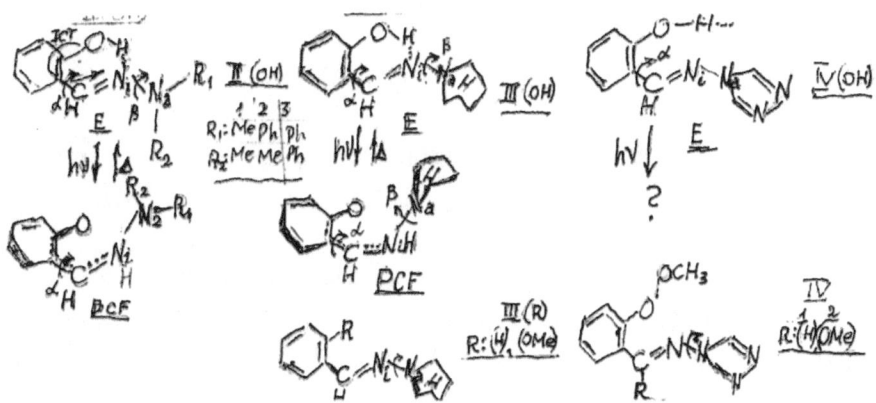

The kinetic data obtained for the structures with the OMe groups (III(OMe), IV(1, OMe)) show the side photoreaction with some indication to the photo trans –cis isomerization around C=N bond: the direct photo and back dark reactions have monomolecular nature, and the first one can be sensibilized by the T-T energy transfer (benzophenone, $E(T_1)=24200 cm^{-1}$).

However the more detailed analysis shows the reaction cannot be attributed to such an isomerization owing to the series of the traits: unlike the latter the efficiency of the direct photoreaction depends strongly on the solvent viscosity, and does not occur at 77K (i. e. for the compound IV(1, OMe), the back reaction takes place at the room temperature with the rate constant $k=10^{-1}s^{-1}$ which is much less of that for the cis-trans isomerization about $C=N$ bond. Moreover, it is observed for the corresponding ligands((III(OH), IV(OH))in the intracomplex compounds with metals in which the trans-cis isomeization is impossible because of the structural restrictions.

At the same time in the last case the reaction of the twist around the formally single N(i)-N(a) bond is quite possible.

Moreover according to the data calculated /11/ the final acoplanar photoproduct has to fluorescent model structure, IV(2, OMe), has the very similar absorption spectra.

Thus the efficiency of the PCF generation for Hydroxyhydrazones in the solvents is determined by the N (i)-N(a)-R fragment structure and its photoinduced transformations.

5. 1. 3. Orthohydroxyaryltriazines (HTrs).

5. 1. 3. 1. Photoinduced structural transformations, generation, and decay of the final metastable photoproduct of HTrs.

The light induced appearance of the comparatively longlived absorption band in the visible spectral region (about 450nm) is observed of the structures with the ESIPT, V, VI (Sch. 5. 3), in the solvents of the different nature.

For the first time the time resolved studies/12-14/of the V type structures have been carried out by the absorption and fluorescence spectroscopy methods.

The long-wavelength absorption band of V($\lambda max \approx 350nm$)has the nature of the ICT from the hydroxyphenyl moiety to the heteroaromatic ring. The extremely large Stokes Shift ($\Delta v \approx 10000$ cm^{-1}) of the ASS fluorescence of V (λ, max$\approx$530nm) is a direct evidence of the ESIPT occurrence.

It has been determined from the rise time of the ASS fluorescence that the ESIPT rate constant is greater than $2 \times 10^{11} s^{-1}$ ($\tau < 5$ps).

However the nature of the structures followed by the ESIPT which responsible for the long-lived transient absorption bands, and kinetics of their decay have not been established due to the overlapping of the $T_1 \to T_n$ absorption of the Enol or Keto structures with the transient absorption of the ground state metastable form.

Therefore a naphthalene analogue VI(1) and its related structure VI(2) have been studied for clarify a mechanism of the photoinduced processes of intramolecularly hydrogen bonded triazine derivatives in solutions /15, 16/.

In those structures the well spectral separated $T_1 \to T_n$ absorption band (λmax$\approx$520nm) which intensity does not depend on the medium viscosity could be clearly identified by the triplet sensitization experiments.

Therefore the more shortwavelength transient absorption bands (λmax$\approx$450nm and 470nm for VI(1) and VI(2) respectively) can be attributed to the absorption of the final photoinduced longlived structure in the ground state (Final Transient Structure-FTS).

The process of the FTS generation, with the ESIPT followed by the structural transformations and the ASS fluorescence, is produces by the schemes (Sch. 5. 4 and 5. 6) /15, 16/.

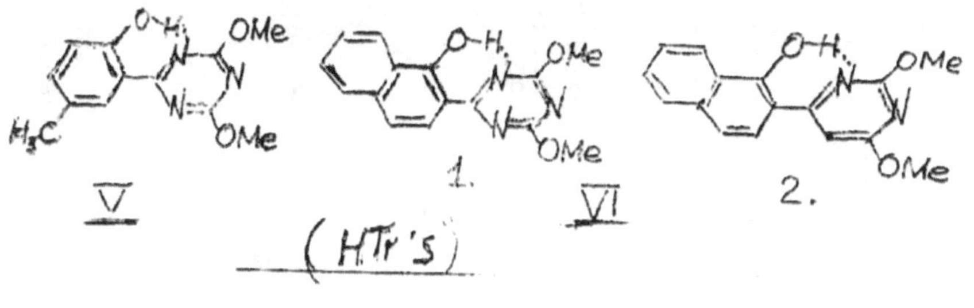

Scheme 5.3
Hydroxyaryl-s-triazine(V,VI(1)) and 1,3 pyrimidine(VI(2)) derivatives.

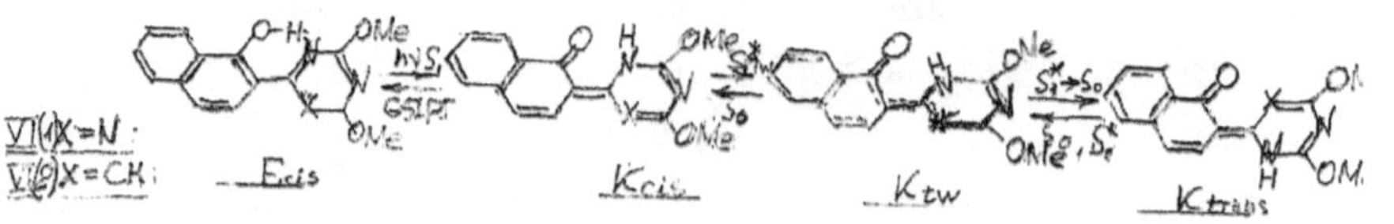

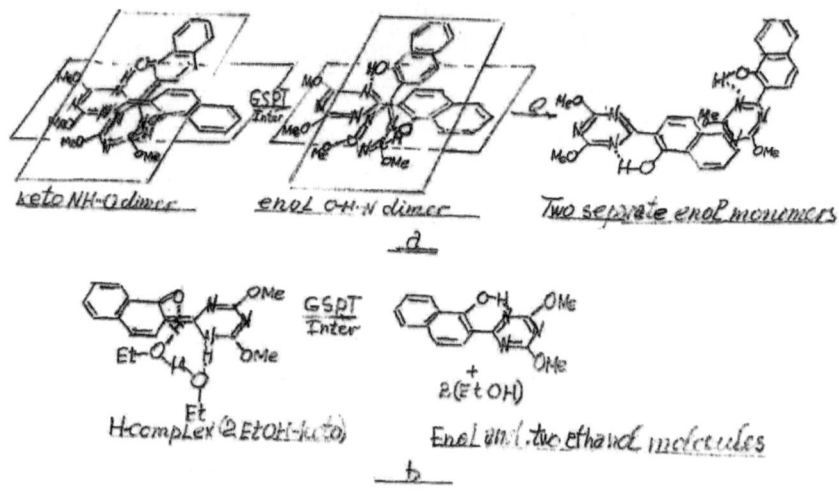

Scheme 5.4
The ESIPT and following suggested structural transformations with the transient generations and decay by the intramolecular mechanism in the ground state including GSIPT.

Scheme 5.5
Intermolecular mechanism of the reverse reaction.
a/The bimolecular (H-dimmer) mechanism .b/Ethanol-catalyzed decay reaction.

+ The schemes are suggested and drawn by author on the base of the data /15.16/.

Table 5.3
Spectral and kinetic characteristics of S-triasines (VI) (MCH 293K)<15,16>

Compound VI	Initial structure (E)				Photoinduced trans. structure			
	Absorption		Fluorescence		Absorption		Reverse reaction	
	λ^{max} nm	ε^{max} M-1 L-2	λ^{max} nm	$\Delta\nu$ cm^{-1}	λ^{max} nm	ε^{max} M^{-1}L^{-2}	$k_1(s^{-1})$ (τ) µs	k_2 (s^{-1})/ε (τ) µs
(1)	367	11000	480	6415	450	$\sim 10^4$	2.7x10^4 (37) 10^5 (10) (*)	5.8 x10^5 (1.7) —
(2)	356	12300	482	7300	470	2x10^3	4.7x10^4 (21)	—

(*)In acetonytril

Table 5.4
The kinetic characteristic dependence on the medium nature and the temperature (MCH or 3MP) for the structure (VI)

	Photogeneration of transient						Transient decay		
Viscos.293K <1>	Rel.effect. $\Delta A_l/\Delta A_0$	τ(Ass flu) (ps)	$T^{-1} \times 10^{-3}$ (K^{-1})	$\dfrac{\Delta A(T)}{\Delta A(293)}$	$\dfrac{\Phi_f(T)}{\Phi_f(77)}$	$\dfrac{\tau(ASSflt(T)}{ASSflu(77)}$	MCH+ (add.M)	$k_1(s^{-1})$ $(\tau_1)(\mu s)$	K_2/ϵ (s^{-1}) $(T_2)(\mu s)$
0.31	1	36	3.4	1.0	≥0	0	+EtOH ($<10^{-2}$)	3.1×10^4 (32.2)	---
0.96	0.65	96	4.0	0.8	0.02	0.02	(3.2×10^{-3})	---	6×10^5 (1.6)
2.31	046	154	6.0	0.2	0.3	0.28	(10^{-2})	---	5.1×10^6 (0.1)
5.6	038	181	7.0	---	0.65	0.62	+TEA (10^{-4})	---	4.4×10^5 (2.2)
11.0	031	190	8.4	---	0.82	0.84	(10^{-3})	---	3.2×10^6 (0,3)
19.0	017	830	13.0	---	1.0	1.00	(2×10^{-3}) <2>	---	4.7×10^6 (0.2)

<1>The variation of viscosity is obtained by the change of the volume ratio n-hexane/liquid paraffine. <2>In ACN

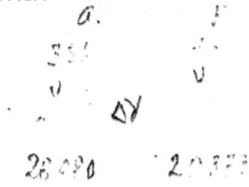

The long-wavelength transition ($\lambda max \approx 360$nm. Tab. 5. 3) typical for VI(1) and VI(2) is bound up with the ICT from the naphthalene moiety to the heteroaromatic ring, and the blue shift of this transition for the structure VI(2) testifies to the decrease of the electron affinity and planarity of the electron acceptor heteroaromatic ring with inclusion of the CH group instead atom N.

The ICT with $So \longrightarrow S_1^*$ excitation decreases markedly the dipole moment of the E-structure in the S_1^* state (Tab. 5. 5) and has to strongly facilitate the ESIPT reaction due to the increase of the acidity and basicity of the OH group and the Nitrogen atom correspondingly.

Therefore the excitation Ecis-hv->E*cis leads to the very effective ESIPT. The following ASS fluorescence with the quantum yield $\Phi(f)$, the rate constant $K(f)$cis, and the lifetime $\tau(f)$ (Sch. 5. 6) is a direct result of the ESIPT. Hence $\Phi(f) = \Phi(ESIPT) K(f) \tau(f)$, where $\Phi(ESIPT)$ is the quantum efficiency of the ESIPT.

It follows from the experimental findings $\Phi(f)/\tau(f)=K(f)$ $\Phi(ESIPT)=\text{const}(T)$(see Tab. 5. 4).

Since K(f) can be assumed to be insensitive to the temperature change, $\Phi(ESIPT)$ is found to be independent on temperature. Hence the ESIPT reaction has no potential barrier, and there are no competing radiative or radiationless processes with the ESIPT in the S_1^* state of the Ecis structure because of the ultrafast nature of the ESIPT.

This result does not contradict to that of the structure V(see above), and is similar to the findings of Anils. However unlike V and Anil molecules in the case of VI(1, 2) ones the Stokes shift of the ASS fluorescence is much smaller ($\Delta\nu(a-f)\approx 6000-7000 cm^{-1}$), Tab. 5. 3)at the expense of the absorption band longwavelength shift ($-1510 cm^{-1}$) and the opposite shift of the fluorescence band ($+1955 cm^{-1}$).

According to the scheme(Sch. 5. 7)(see also /15/)such shifts of the band positions can be attributed to a considerable decrease in the stabilization energy with the ESIPT reaction at the expense of the lowering of the S_1^* state of the ICT nature in the E structure and to a sharp decrease of destabilization of the ground state Kcis structure with substitution of the benzene moiety by a naphthalene ring.

It would be noticed the polarity of the S_1^*Kcis fluorescent structure of VI is very small ($\mu e\approx 1.6D$) (Tab. 5. 5)unlike the strongly polar S_1^*NHflu structure of SA ($\mu e\approx 10D$) responsible for the ASS fluorescence following the "net" proton transfer(ESIPT). In the case of VI the S_1^*Kcis flu structure responsible for the ASSflu is a result of the ESIPT in which with the high probability the surplus positive charge is compensated by the additional synchronous Itramolecular Charge Transfer (ICT) to the C=N group in the $S_1^*(\pi\pi^*)$ excited state (see also below).

Thus one could be supposed that the large difference of the Stokes shifts values between VI and SA molecules is connected with the different nature of the fluorescent states and the radiative transitions of the E, NH and K forms which react in a different manner to the variations of the molecular structure.

Like anils the S_1^*Kcisflu state responsible for the ASS fluorescence, is the precursor of the final photoinduced transient structure(PTS).

The PTS generation has to compete with the deactivation of the S_1*Kcisflu state stipulating of the variation of the fluorescence kinetic parameters with change of the conditions which influence the PTS generation efficiency.

Really in the nonpolar solvent (3MP)(Tab. 5. 4) the relative efficiency of the PTS generation ($\Delta A /\Delta Ao$ where ΔA is the absorbance at $\lambda=450$nm) falls strongly and fluorescence intensity (I_f) and lifetime (τ_f) increase sharply with the viscosity growth (T=const=293K) or with the temperature decrease /15/, and reach the constant value below 25K.

In addition the both fluorescence intensity and lifetime are found to increase in the polymer matrix in comparison with 3MP within temderature limits 293-150K.

Table 5.5
The dipole moments (D) in the ground and excited states (the MO calculations/16/)(Sch.5.6).

State--> Structure ↑	E_{cis}	E^*_{cis}	K_{cis}	K^*_{cis}	K_{tw}	K^*_{tw}	K_{tr}	K^*_{tr}
VI (1)	3.6	0.2	5.4	1.6	14.6	7.2	10.4	7.1
VI (2)	3.9	2.4	4.7	3.3	13.1	6.4	8.1	4.6

Table 5.6
The calculated and experimental values of the So-->S_1^* transition energy for VI(1) structure.

Structure--> Data & method ↑	E	K_{cis}	K_{trans}	K_{tw}
Calculated <1>				
$V cm^{-1}$	29570	24558	34324	9725
λnm	338	407	411	1028
Experimental		<2>		
$v^{max}_{cm}{}^{-1}$	27248	≈22833	22222	-----
λ^{max}_{nm}	367	≈438	450	
$\Delta v\ cm^{-1}$	+2322	1725	2102	
$\Delta\lambda_{nm}$	13	33	39	-----

<1>Method INDO/S-Cl.<2>Obtained from the ASSflu band with v^{max}=20833 cm^{-1}(480nm)+Stokes Shift ≈2000cm^{-1}.

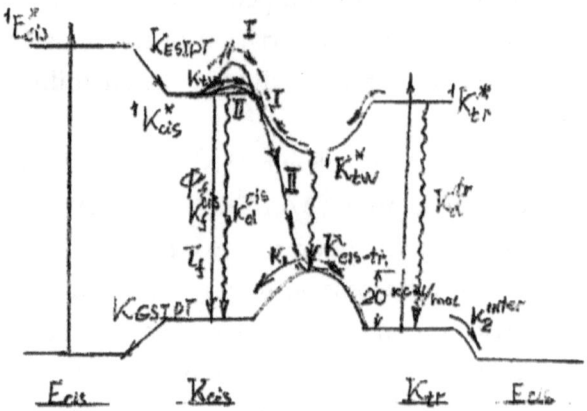

Scheme 5.6

General qualitative scheme of the photoinduced transformations of VI(1) and VI(2) with the ESIPT followed by cis–trans isomerization and by back reactions with the mono and intermolecular (includin ethanol-catalyzed) proton transfer in the Ground state (see schemes 5.4, 5.5 and ref./15,16/).

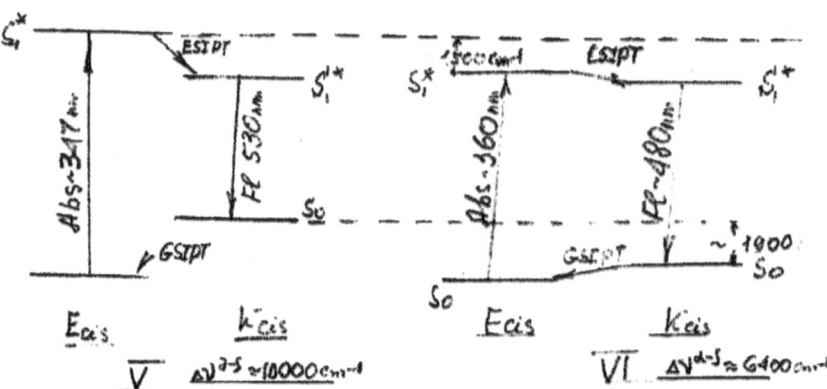

Scheme 5.7

The qualitative explanation of the strong ESIPT-ASS fluorescence Stokes Shift for the structure VI as compared with V ones.

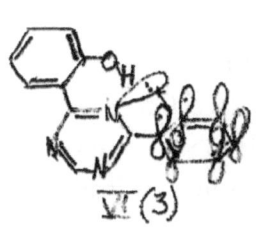
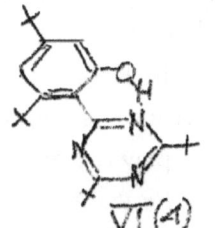

Scheme 5.8

The supposed triazine VI(4) photochromic structure.

The strong effects of viscosity and temperature on the PTS generation, I_f, and τ_f reflect the present of the strong viscosity and temperature dependent processes of the structural transformations that occure immediately at the beginning of the pathway of the deactivation $S_1^*Kcis \rightarrow So(FTS)$.

The structural mechanism and efficiency of the deactivation are determined directly by the structure of the PTS in the ground state.

The trans keto(Ktrans) structure is the most widespread idea of the PTS both for Anils (see Ch. 3) and their structural analogues with the C=N double bond (see, e. g. /17, 18/), and it is also highly probable for the molecules VI (1, 2)/15, 16/.

However such a notion is not only that in the case of anils(Ch. 3) and therefore needs the special basing of it. As far as the direct experimental study of the structure of the metastable photoproduct is extremely difficult, the main method of the PTS identification consists (like SA) of the comparison of the experimental and calculated findings for the spectral positions of the absorption bands in the visible spectral range (So-abs->S_1^* transition). The corresponding data for the all structures participating in the photoinduced processes are presented in the table(Tab. 5. 6)/15, 16/.

Unfortunately the absorption band of the Kcis is absent (the Ecis⇔Kcis equilibrium in the ground state is shifted completely towards Ecis structure, unlike anil molecules), and its position may be determined only approximately by use of the fluorescent band position and the typical Stokes Shift value ($\Delta v(a-f) \approx 2000 cm^{-1}$).

According to the data presented in the table (Tab. 5. 6) the calculated values of the transition energies are systematically shifted towards higher energies by ≈2000cm⁻¹ as compared with the experimental ones for the Ecis, Kcis structures, and therefore have to correspond well enough to the experimental position of the Ktrans absorption band.

Thus for the VI(1, 2)structures unlike anils (see Ch. 3)the flat trans-structure can obviously play the principal role in the composition of the metastable forms with the absorption in the visible spectral range (λ≈450, 480nm).

According to the scheme proposed on the basis of the data calculated by INDO/S-CI method the final PTS structure is generated in the ground state via its direct precursor, the twist structure, which is formed adiabatically from the fluorescent Kcis* structure and located on the reaction path-way to the PTS on the S_1^* state.

The similar post ESIPTmechanism of the PTS generation is typical for the series of the molecules with OH...N bond(see e. g. /19, 20/) including SA(see Ch. 3).

However there are the important peculiarities of the twisted structure of VI. Its comparative low polarity and especially the sharp increase of the later with the transition to the top of the potential barrier in the ground state (Tab. 5. 5) are not typical for the twisted structures.

It also not secures the effective quenching of the ASSflu in the unviscose media in spite of the strong proximity in the PES of the S_1^* and So states(Tab. 5. 6). Therefore there is the marked potential barrier located on the path-way Kcis*-->Ktrans caused by not only viscosity but also and the molecular structure.

Such a barrier can be connected with the impeded rotation about the high order C-C bond in the low polar Kcis*structure (Tab. 5. 5) generated by the Excited State Hydrogen (atom)Transfer (ESIHT) unlike the high polar S_1^*NH structure which is formed by ESIPT.

The temperature and viscosity (T=293K) influences the ASS fluorescence kinetic parameters (Φf, τf) (Tab. 5. 4) are connected with both the structural barrier and the viscosity one which is located on the S_1^*PES (Sch. 5. 6).

At the same time the temperature and viscosity dependences of PTS formation efficiency (ΔA/ΔAo and ΔA/ΔA(273))and those of the ASS fluorescence parameters are very similar, and caused apparently by one and same barrier.

Therefore the final step of the ring rotation with the formation of the trans Keto structure takes place without any potential barrier in the ground state, and stabilization of the twisted forms in the ground state is hardly possible (see also sec. 5. 1. 3. 2).

Unfortunately the height of the excited state potential barrier has not been determined by calculation and its estimation is difficult on the base of the findings /15, 16/.

However it may be believed that the barrier height for the twist around the almost double C—C bond is too high for the adiabatic path-way of the twist structure formation $^1Kcis^*$--adiab.-->$^1Ktw^*$-->Ktw-->Ktr (path-way I) and the observed dependences of the kinetic parameters are connected mainly with the diabatic path-way of the PTS generation($^1Kcis^*$--diab-->$^1Ktw^*$-->Ktr)(path-way II)(Sch. 5. 4, 5. 6).

Since the fluorescence of the Ktr structure is not observed, one may supposed the reverse photoreaction with the excitation Ktr—hv-->$^1Ktr^*$ can occur via the twisted structure suppressing completely the emission from the Ktr* state.

The last step of any monomolecular back reaction is the ultrafast ($k > 10^{10} s^{-1}$) GSIPT (Kcis->Ecis) which could not be detected by the utilized methods with a nanosecond time resolution, and it is not the limiting stage of the reaction.

Such a stage of the monomolecular reverse reaction is the trans-cis isomerization around C=C bond (Sch. 5. 4)with the calculated potential barrier about 20kcal/mol (Sch. 5. 6)/15/.

The reaction has the first order constant ((Tab. 5. 3)at the room temperature /15/where an increase by the order of magnitude in CAN is caused by the stabilization of the high polar transition state ($\approx$13-14D) (Tab. 5. 5).

The mechanism and the kinetics of the ground state reverse reaction are strongly changed by addition of the protic solvents (ETOH or TEA) to nonpolar one.

In such cases the back reaction is characterized by the second order rate constant k that is increased strongly with the rise of the concentration of the solution and that of the protic extra solvents(Tab. 5. 4)with small deuterium effect.

The various mechanisms can be realized in such back reaction. The intermolecular Ground State Proton Transfer(GSIterPT) in the Ktrans structure (Ktr-->Etr) followed by the Strans—cis isomerization Etr-->Ecis about C-C single bond results in the formation of the initial Ecis structure with the O-H...N bond.

The formation and the GSIterPT of the H-bond VI dimers (Sch. 5. 4a), the concerted biphoton transfer process in EtOH-VI H-Bond complexes (Sch. 5. 4b)for the ethanol catalyzed decay reaction, and the mechanism of the dynamic catalysis for TEA have been proposed /15/ (see also/21/).

The diffusion controlled rate constant of the above processes depends strongly on the viscosity, temperature, and concentration which determine the efficiency of the accumulation and decay of the FTS.

On the other hand, the inclusion of the bulky substituents into the rings can secure the decrease of the k_1 rate constant and the accumulation of the FTS in the rigid media(and crystals) with display of the visual coloration.

5. 1. 3. 2. The crucial role of the primary step character (ESIPT or ESIHT) in the photochromic reaction (Htrz vs SA).

The photoinduced processes of H-triazine (Htrz)VI and SA molecules with the primary step of the ESIPT or ESIHT character in spite of their apparent likeness result however in the considerably different properties of the ASS fluorescence and of the final stable photoproducts.

It may be supposed that comparative consideration of each stage in the consecutive order can help to understand more clearly the role of the primary steps of the different nature of Htrz and SA molecules and to propose on that basis the new Htrz structures with the effective photochromic properties.

The very low polarity ($\mu_e \approx 1.6D$) of the excited $S_1^*NH(Kcis^*)$ structure (Tab. 5. 5) generated by the Excited State Proton or Hydrogen

Transfer of the structure VI(HTrz) testifies to the later mechanism (ESIHT) unlike SA molecule where the S_1^*NH state has the highly polar Zwitterionic struct-ure ($\mu e \approx 10D$)(Ch. 3, Tab. 3. 25)which is result of the Proton Transfer mechanism(ESIPT). In Htrz structure unlike SA the Nitrogen atom is included into the rigid heteroaromatic ring.

Therefore the ESIPT-like or ESIHT-like characters of the reactions can be cause by the two main reasons connected with the structural peculiarities of SA and Htrz molecules.

The first reason is the delocalization of the $n(N)sp^2$ lone-pair of Nitrogen atom as a result of the $n(N)$—$\pi(Ph)$interaction in the SA structure vs its clear-cut localization in the Htrz molecule.

The second one is the exclusion of the promoting vibration deformational modes with the participation of the $C=N$ group included in the heteroring of the Htrz molecule.

Obviously the above peculiarities typical for Htrz, secure the synchronism of the proton and electron transfers which is necessary for the realization of the ESIHT and vice versa (see also/22/).

Although the experimental and theoretical studies are required for the elucidation of the struc-tural conditionality of the ESIPT or ESIHT characters, however at the same time it is obviously that their difference results in the very clear distinctions of the observed fluorescence and photochemical properties.

Really, unlike SA where the potential barrier for the twist around almost the single C-C bond in the S_1^*NHZw state is mainly caused by the viscosity (Ch. 3), the high barrier of twisting about the double $C=C$ bond of the nonpolar Kcis*flu structure doesn't allow to form adiabatically the que-nching TICT-like structure in the excited state (Ktw*).

Therefore the twist reaction occurs diabatically over the much lower potential barrier (Sch. 5. 6) to the highly polar, ground state twisted transient (Tab. 5. 5), competing with the emission.

Thus unlike SA the ASS fluorescence with the decreased considerably Stokes shift of the Htrz molecules is observed with the excitation of the E-structure (see above, sec 5. 2. 3. 1.) at ambient temperature, and attributed to the planar Kcis*structure which is generated only directly in the excited state.

This clearly observed experimental distinction in the fluorescent properties between SA and Htrz is the direct consequence of the difference between the ESIPT and ESIHT mechanisms.

In the So state the tautomeric equilibrium Ecis <=>Kcis is shifted completely towards Ecis structure unlike SA under the same experimental conditions, and the SoKcis-->S_1^*absorption band is ab sentIn the case of the ESIPT mechanism typical for SA the formation of the twisted colored "post-TICT" structures occurs via the polar "TICT-like" structure in the S_1^* state to the shallow but clearly marked potential pit($\Delta E \approx 5kcal/mol$)on the top of the potential barrier in the So state which restricts the subsequent free ring rotation in the So state(see Ch. 3).

However for the Htrz molecules with the ESIHT mechanism such a path-way with the stabilization of the twisted "post-TICT" structures (PCF structures, Ch. 3)in the So state is hardly possible, and instead of it the planar metastable Ktrans ground state structure is generated diabatically (seeSec. 5. 2. 2. 1.).

There is the sharp difference between the kinetic parameters of the decay reactions of the twisted structure of SA and the trans-planar one of Htrz molecule which is caused by the different nature and the height of the potential barrier connected with the molecular structures of the metastable forms.

The low viscosity dependence of the metastable structure decay rate of Htrz molecule in spite of its much lesser value in the liquid solvents than of SA(compare Tab. 5. 4 and 3. 44) cannot secure, unlike SA, the accumulation of the PCF in the rigid media enough for the visual manifestation of the T-photochromism.

Thus the Htrz molecules show unlike SA only L-photochromism even in the sufficiently rigid media. This clearly observed distinction in the properties of the photoinduced metastable forms also caused directly by the ESIPT and ESIHT mechanisms of their generation.

Thus both the succession of the photoinduced processes and the structures of the transient forms are determined completely by the ESIPT-like or ESIHT-like character of their primary step.

It is obviously to obtain the novel triazine structures which can show visually observed "true" photochromic properties it is necessary to increase strongly the dependence of their FTS structure decay rate on viscosity.

It may be attained by the two methods.

1. 1. To secure the twisted structure formation like SA with the increase of the ESIPT character of the primary step by way of the delocalization of the n(N) electrons as a result of the n(N)-π(Ar)-interaction with the introduce of the aryl substituents (see structure VI(3), Sch. 5. 8).
2. To impede the rotation about C—C-bond by means of the introduction of the bulky substituents in the both rings (Structure VI(4), Sch. 5. 8).

One may hope, the combination of the both methods will allow to obtain the desired effects with the high probability.

However it is necessary to remember also that the rate of the FTS generation falls strongly with the increase of the viscosity and in the rigid media (Sec. 5. 2. 2. 1.).

Therefore the obtaining of the photocolorated materials on the base of the triazine molecules is the rather difficult problem.

5. 1. 4. Photochromism of orthonitrobenzylidene derivative.

The photochromism of OrthoNitroBenzylidenes (ONB)(Structure VII, Sch. 5. 9) is described in the reviews /7, 23/.

For the first time the T-photochromism of the solid (polycrystalline powder) ONB has been found out in /24/.

The reflection spectra of the initial pale-yellow solid form contain the intensive broad band with the long-wavelength limit about $\lambda \approx 450$ nm.

The irradiation by UV light leads to the appearance of the red form with the broad intensive band ($\lambda^{max} \approx 500$nm).

The reverse reaction proceeds with the half-life about 52 min at 80°C. The temperature dependence of the back reaction rate gives the activation energy about 10kcal/mol. Authors /24/ have suggested the intermolecular mechanism involving the Hydrogen shift in an excited state due to the high concentration of the molecules in the crystalline packing.

However the similar band has been detected by the flash photolysis technique in the PMMA rigid glasses at the room temperature but at the low concentration(10^{-4} M) /23/. It points to the intramolecular mechanism proposed below in which the ESIPT is not a primary step of the photochromic reaction unlike SAs (Sch. 5. 9).

The excitation of the Atr structure (Atr->A*tr) leads to the adiabatic trans-cis isomerization around C-N single bond in the excited state (A*tr->A*cis) with the formation of the N-H...O-N-O H-bond followed by the GSIPT (Acis->Bcis).

The colored "red" structure B trans ($\lambda max \approx 500$nm) is generated as a result of the Ground state cis-trans isomerization about C=C bond (Bcis->Btrans) and the formation of the initial structure Atrans (Acis->Atrans) over the small potential barrier of the cis-trans isomerization around the almost single C-N bond.

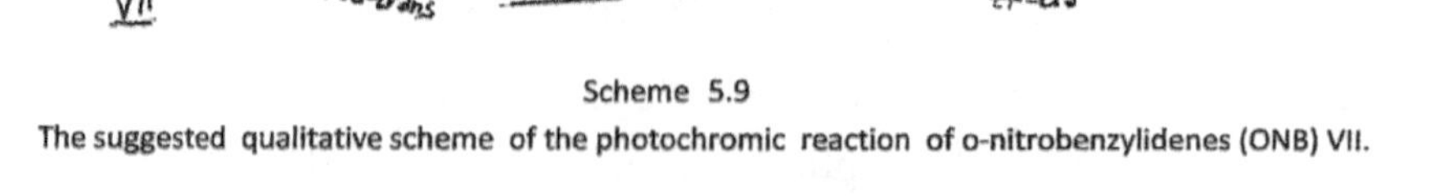

Scheme 5.9
The suggested qualitative scheme of the photochromic reaction of o-nitrobenzylidenes (ONB) VII.

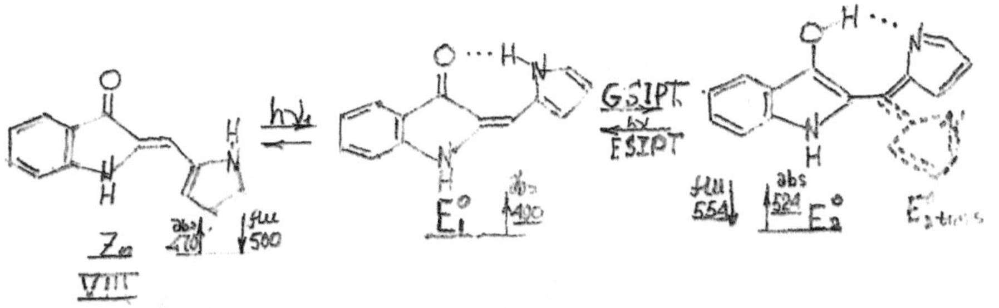

ESIPT Photochromism

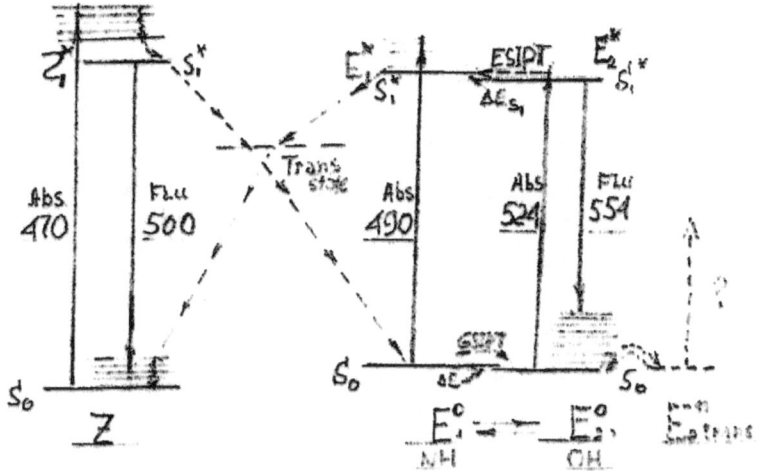

Scheme 5.10
The suggested structural schemes (with the additional structure (E_2^0) of the photochromic reaction of 2-(2-Pyrrolylmethylidene)indolin-3one (PM I)(VIII).

Scheme 5.11
The energetic scheme Photochromism of PM I (VIII) (see Scheme 510).

Table 5.7
Photochromism and fliorescence of compound VIII (PMI).

<1> Structure	Absorption λ^{max}_{nm}	Fluorescence λ^{max}_{nm}	Stokes shift $\Delta\nu$ (cm^{-1})	Quantum Yield Φ_f	Life time T(ps)	Solvent	Photo-coloration	Photo-Bleach.
Z^0	470	500	1276	0.023	210	benzene	0 03	0.003
E^0_1	490	----	----	----	----	ACN	0 03	0.02
E^0_2	524	554	1003	0.06	350	MeOH	0,03	0.05

<1>See scheme 5.10

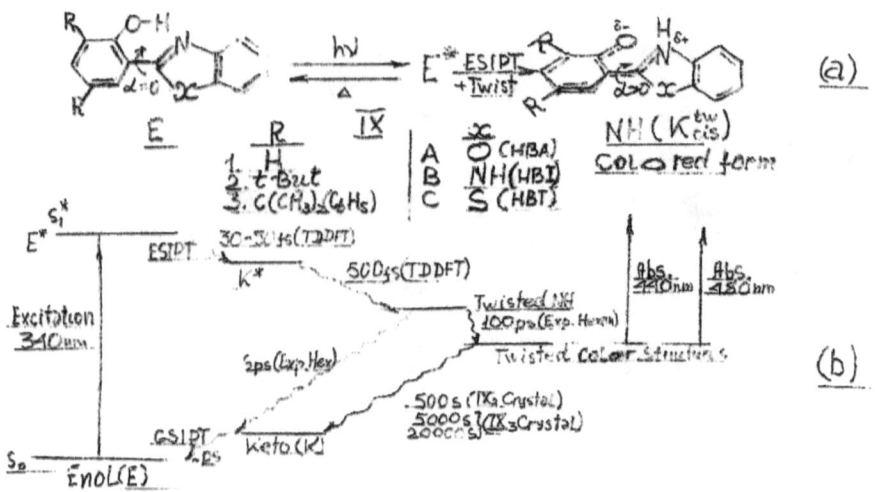

Scheme 511₁
Photochromism of 2-(2-hydroxyphenyl)benzothiazoles (HBT), IX ,C₂.
(a) The structural scheme of the photochromic reaction.
(b) The kinetically-energetic scheme on the base of the estimated data of the photoinduced processes including formation and decay of the photocolored structure of HBT in the solvent and the crystal (on the base of the findings /6,27/)
The method of calculation and (or) the condition of the experiment are shown in the brackets.

It should be note however that the much later X-ray crystallographic study /25/ did not show preorganized bonding between the postulated proton donor (NH) and acceptor(NO).

Therefore the additional study is required for the elucidation of the photochromism nature of VI structure.

5. 1. 5. Novel photochromic Dye based on the formation of the Hydrogen bond.

Unlike the structure discussed above the primary ESIPT (ESIHT) stage of the photochromic reaction of 2(2-PyrrolylMethylidene)Indoline-3one (PMI)/26/ (VIII Sch. 5. 10), the new colored forms are generated (in opinion of authors/26/)without ESIPT OH-->NH but as a result of the formation of the NH...O bond in the ground state of the final structure (Sch. 5. 10).

The T-photochromism of PMI(reddish-orange to greenish-yellow) is manifested under UV irradiation (λ^{max}=360nm) in solvents by the fall of the initial absorption band (λ^{max}≈470nm) intensity with the appearance and the intensity growth with the different rates of the two close and overlap-ping bands(λ^{max}≈490 and 524nm)(Tab. 5. 7).

The appearance and the rise of the two intensive bands (but not a single one (λ^{max}≈524nm) as it has been suggested of /26/) are clearly observed (/26/, fig 1 and 2), and point to the generation of the two different(not single) colored forms. The emission bands(λ^{max}≈500 and 554nm) with the different parameters(Tab. 5. 7) belong to the initial and final structures correspondingly.

The back bleaching reaction is observed un-der irradiation only (photobleaching).

The high quantum yield of the photocoloration is independent on the solvent polarity while the quantum yield of the photobleaching is lower by the two-three orders of magnitude and depends on the solvent nature(Tab. 5. 7). The photochromic reaction is multiple repeated at least in the nonpolar solvents.

The scheme has been drew up on the base of the findings presented /26/ with the additional colored structure (E°) proposed by us (Sch. 5. 11, Tab. 5. 7).

Excitation of the Z structure (Zo->Z*) (λ^{max}≈470nm) is followed by the diabatic process of the trans-cis isomerization about C=C bond (Z_1^*->$E_1°$) via the transient, probably triplet, state with the twisted structure.

The process occurs over the small potential barrier with the high quantum yield (Fr) and competes with the fluorescence (λ^{max}≈500nm) of the Zo form with the small Stokes shift, the marked quantum yield (Φ_f), and the lifetime (τ_f) in the hundreds picoseconds scale(Tab. 5. 7).

Along with the generation of the $E_1°$ structure (λ^{max}_{abs}, ≈490nm) the formation of the fluorescent E° (OH) structure(λ^{max}_{abs}≈524nm, λ^{max}_{flu} ≈554nm) takes place, and the tautomeric equilibrium ($E_1°$⇔E°) is established in the ground state as a result of the GSIPT. At the same time the formation of the Ground state trans-structure with brake of the O-H bond can be excluded.

The high quantum yield of the direct photoreaction and its independence on the solvent polarity(Tab. 5. 7) are provided by the absent practically of the potential barrier in the diabatic reaction along the path-way of the reverse photoreaction with irradiation at $\lambda=546$nm the structure E° is mainly excited(E°->E*), and E*structure undergoes the adiabatic ESIPT reaction OH-->NH to E_1^*.

The very low quantum yield of the back photoreaction is caused mainly by the low rate of the endoergic ESIPT reaction probably with tunneling through the potential barrier which height depends on the solvent polarity due to the different polarity of the E_1^* and E*structures(μE_1^*-> μE *).

The strong dependence of the back reaction quantum yield on the solvent polarity (Tab. 5. 7) is explained by such a reaction mechanism.

The final step of the reverse photoreaction involving the trans-cis isomerization around C=C bond (E_1^*->Z) proceeds diabatically with formation of the initial structure.

Thus the principal peculiarities of the photochromic reaction of VIII are described well by the schemes presented on the base of the findings of /26/ (Sch. 5. 10 and 5. 11).

The note should be taken that in the structure VIII unlike the other photochromic reactions discussed above(including this Chapter) the GSIPT is the final step of the direct reaction of the photocoloration while the ESIPT is the primary step of the photobleaching reaction.

5. 1. 6. 2-(2-Hydroxyphenyl)benzazoles, and photochromism of 2-(2-Hydroxyphenyl) benzthiazoles as a simulation of generation and structure of anil photocolored form.

2-(2-Hydroxyphenyl)benzazoles (HBA) are the structural analogs of salicylideneaniline molecules with the rigid structure in which only rotation around Ph:::C bond after the ESIPT in the NH structure can occur to form the photocolored one.

However HBA molecules with O or N atoms in the five-member heterocycle show usually no any indications to such a structure formation after irradiation following the ESIPT and the ASS fluorescence in the solvents or the crystals.

However 2-(2-Hydroxyphenyl)benzthiazole (HBT) (Structure IXc_1, Sch. 5. 11_1a)unlike other azoles shows photoinduced spectral changes connected with the L-photochromism in the liquid solvent manifested by the appearance of the shortlived longwavelength band (see e. g. /4, 5/).

Recently /27/the study of the structure IXc_1 has been carried out by the femtosecond pump probe absorption spectroscopy and the quantumchemical dynamic method (TDDFT).

According to the findings (Sch. 5. 11_1b) after excitation E->E* the ultrafast ESIPT(OH->NH)($\tau \approx$30-50 fs)in the planar (Enol)* structure leads to the K*cis structure followed by the internal conversion ($\tau \approx$500ps)with the formation of the twisted NH structure and the conical So/S_1*crossing. After the internal conversion the two competing path-ways take place.

One closes fast($\tau \approx$26ps) the proton transfer cycle by the GSIPT (OH->NH) via Keto structure and another ($\tau \approx$100ps)lead to the colored form attributed as usually to the trans-keto structure.

However the T-photochromism of solid IX_1 in the crystal is not displayed /6/, and hence the PCF structure is not generated in the crystalline packing.

Really the molecules IX_1 is planar and packed face to face with the short intermolecular distance in the crystal /28/and there is no reaction room to secure even a little distortion with the photoinduced isomerization after the ESIPT in the crystal lattice/29/.

Such a room in the crystalline packing is provided by inclusion of the bulky substituents in the structures (IX and IX)/6/. Indeed the T(visible) photochromism is revealed in the crystal of the latter molecules unlike IX_1.

The color of the crystals IX_1 and IX_2 is changed from yellow to the dipper color by the UV-light irradiation and returned to the initial color in the dark/6/.

The maximum changes in the optical density after irradiation are observed at λ^{max}_{abs} = 440 and 480nm for IX and IX correspondingly(Sch. 5. 11_1b).

The thermal decay reaction passes at 30°C slowly with time rates increasing with the size of the bulky substituents (Sch. 5. 11_1b) and for IX depending also on the crystalline lattice structure which changes with enclathration of a small amount of water molecules in the crystal.

Thus the bulky substituents is not only effective to maintain a room for the molecular deformations in the excited state but also(that is very important)act as a kinetic stabilizer of the PCF structures in the ground state by steric interactions between the adjacent molecules in the crystalline lattice.

The observed spectral and kinetic characteristics of the photocolored structure of $IX_{2,3}$, and SA in the crystal are very similar and must be caused by the same structural transformations.

From the most widespread viewpoint, the trans-Keto structure(about C=C double bond)is only single one which is responsible for the coloration of Anils in the solvent and the crystals(Ch. 3).

However in the crystals the usual mechanism of the cis-trans isomerization about double bond especially for molecules with the bulky substituents is unlikely because of insufficient free volume in the crystal packing for such a transformation.

Therefore for the explanation of the photochromism of SA in the crystals the mechanism with the "pedal" motion of two phenyl rings has been suggested/30/ (see also Ch. 3). The interatomic interactions of the adjacent molecules do not play an important role in such a "pedal" mechanism which can therefore provide photochromic reaction even in the tight crystal packing.

However in the molecules IX the cis-trans isomerization cannot occur via a "pedal" motion owing to their rigid cycle structures without a possibility of axial rotation of the rings in the crystal lattice cavity.

Therefore the PCF species of IX_2 and IX_3 (just as of SA) can be assigned only to a twisted NH-cis structure that is identical to the $post_1$TICT structure (according to the terminology of Sec. 3. 4. 2), and requires a rather small room in the crystal packing.

Thus the findings /6/can be considered as the direct evidence of the twisted structure of the main colored form of the Anil molecules which has a "TICT-like "precursor state with the high probability(see Ch. 3).

5. 2. The structures with the ESIPT (ESIHT) between identical heteroatom.

5. 2. 1. Oxygen-Oxygen (OH⇔OH) proton transfer.

5. 2. 1. 1. Nitrones of o-Hydroxyaldehydes, and their vinilogs.

The structures X_1, X_2 (Sch. 5. 12, 5. 13)with the N->O structural fragment involved in the O-H...OH-bond (instead of the O-H...N H-bond of photochromic anils)have been studied as prospective photochromic compounds by steady-state absorption and fluorescent spectroscopy with utilization of the model structures X (R, OMe), X(R, ion)(R:Ph, Me) in the solvents/8/.

The excitation of the of the initial structure X_1(Ph)or X_1(Me) in their longwavelength absorption bands ($\lambda \approx 340$-360nm)at the room temperature leads to the appearance of the double band fluorescence($\lambda^{max} \approx 440$ and 542nm), and promotes the irreversible changes in the absorption spectra (Tab. 5. 8).

According to the schemes (Sch. 5. 12a, b)the changes of the absorption spectra are explained by the irreversible diabatic photoreaction Nitron-(1)->Oxaziridine-(1')->Oxazirane(see e. g. /31/)which is accompanied with the fall of the initial Nitrone's absorption band(340-360nm) and appearance of the new short-wavelength one ($\lambda \approx 250$-320nm).

The quantum yield of that reaction falls sharply with the formation of the H-bond O-H...O in the compounds X_1(R), as compared with X_1(R, OMe)(Tab. 5. 8), and the reaction competes with the ESIPT O-H->H-O(reaction 2).

Table 5.8

Spectral and photochemical characteristics of nitrones $X_1(R)$ (ethanol, room t-re)

Structures of compounds	Absorption band λ^{max} nm	Fluorescence band λ^{max} nm	Stokes shift $\Delta\nu^{a-f}_{cm^{-1}}$	Photoreactions Irreversible reaction Abs. λ^{max} nm	Efficiency Φ_r Ethanol	Hexane	Photo-cro*) mism.
$X_1(Ph)$	358	443 / 542	5400 / 9450	≈325	0.04	0.03	+
$X_1(Me)$	345	---- / 520	------ / 9755	≈255	0.09	0.02	+
$X_1(Ph,OMe)$	314	-----	--------	≈230	0.20	0.22	--
$X_1(Me,OMe)$	314	-----	--------	≈210	0.17	0.17	--
$X_1(Ph,ion)$	435	----- / 549	4800	-----	-----	----	--
$X_1(Me,ion)$	391	- / ----- / 525	6530	-----	-----	----	--

*)"+"weak photocoloration (λ^{max}≈480-500nm) in glassy solvent(IIP)at 77K ."—" photocoloration is absent

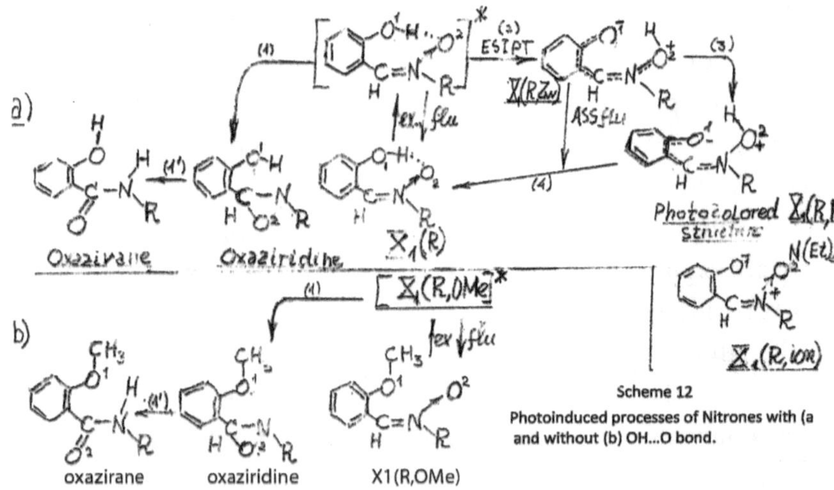

Scheme 12
Photoinduced processes of Nitrones with (a and without (b) OH...O bond.

As a result of the latter the ASS fluorescence band arises which spectral position and shape correspond well to those of the model structures $X_1(R, ion)$, and obviously the Zwitterionic structure $X_1(Zw)$ is responsible for the ASS fluorescence of both $X_1(Ph)$ and $X_1(Me)$ molecules.

As it has been shown/8/the lack of the fluorescence of the Nitrone molecules without o-OH group is caused by the complete suppression of the emission S_1^*-hv->So by competing fast initial step of the reaction 1(Sch. 5. 8).

Therefore inhibition of that reaction by IHB O-H...O with decrease of its quantum yield leads to the appearance of the Nitrone structure's emission with the normal Stokes shift, and Nitrone molecule X_1(Ph) has the double band fluorescence.

However the fluorescence of the structure X_1(Me) is suppressed by the competing ESIPT reaction which rate is increased as compared with that of X_1(Ph) due to the increase of the electron density on the N->O group with the exclusion of the n-π(Ph)interaction, and therefore the moleculeX_1(Me) has only the single band of the ASS fluorescence.

Thus since the ESIPT occurs of the both X_1 molecules it should be expected of the PCF generation and stabilization under conditions similar to those for photochromic anils.

Really at the heightened concentration (~10^{-2}M) in the glassy solvent (IIP at 77K) the accumulation of the photocolored product (path-way 3) and its decay (path-way 4) with increase of the temperature are distinctly observed like the photochromic azomethine (T-Photochromism).

With the including of the vinil radical into the o-hydroxyaldehyde moiety (X (Ph), X (Np))instead of the simple imine fragment, the cisoid structure Az is formed and stabilized by the strong bond O-H...O (Sch. 5. 13)/8, 32, 33/.

The irreversible photoreaction of oxaziridine formation typical for simple Nitrones is excluded completely in such structures.

At the same time under irradiation of the structure Az(So) in its intensive absorption band with $\lambda^{max}\approx$440-450nm($\varepsilon \approx$13000 L^{-1}M^{-1})in the solution at room temperature and 77K the unusual for Nitrones and Anils photo and thermo transformations take place in the molecules X (Ph) and X (Np).

As a result of the adiabatic ESIPT reaction, $Az(S_1^*)$->$Bz(S_1^*)$, the excited N-O-H structure $Bz(S_1^*)$, responsible for the ASS fluorescence($\lambda^{max}_{flu}\approx$630-640nm, the Stokes shift $\Delta\bar{\nu}\approx$10000cm^{-1} which intensity grows strongly at 77K, is formed.

The following, diabatic, stage of the trans-cis izomerization about $C_1=C_7$ bond, $Bz(S_1^*)$-diab.----->Bz'(So), occurs with the activation energy in the $S_1^* \Delta E^* \approx 0.3-0.8$ kcal/mol.

That isomerization competes with the ASS fluorescence, and since the fluorescent $Bz(S_1^*$ structure is the precursor of the colored ones in the So state (B (So) and C(So)), the increase of the potential barrier correlates with the drop of the photocoloration quantum yield Φcol) and growth of the fluorescence efficiency (Φ_f)(Tab. 5. 9).

The structural transformations in the So state lead finally to the formation of the cyclic chromen structure C(So). Under the steady-state irradiation the equilibrium in the So state Az(So)<=>Bz(So)⇔Bz'(So) B (So)⇔C(So) is strongly shifted towards the most stable C(So)and B (So)structures with the absorption bands with the λmax≈440 and 550nm correspondingly.

After stopping of irradiation the equilibrium is shifted almost completely as a result of the reverse dark bleaching reaction towards the most stable initial Az(So)structure with the activation energies typical for the trans–cis isomerization about C=C bond (Tab. 5. 9.

That energy barrier lowers strongly with the enlarging of the aldehyde ring π-system and also with the substitution of the methyl radical by the phenyl ring.

Table 5.9
Kinetic parameters of the photochromic reactions for vinilogs of aldonitrones X_2 (in toluene,293 K) (see scheme 5.13).

Structure R	Coloration eff. Φ_{col}	Fluorescence eff. Φ_{flu} (relative)	Color.activ.energy $E^*_{dec..}$ kcal/mole	Decay activ.ener. Edec. Kcal/mol
X_2(Ph CH$_3$)	0.42	0.25	0.57	17.4
X_2(Ph Ph)	0.45	0.25	0.37	13.8
X_2(Np CH$_3$)	0.11	1.00	0.74	9.9
X_2(Np Ph)	0.20	0.45	0.84	5.1

Scheme 5.13
Photoindused processes including Photochromism of Nitrons' vinilogs (X_2).

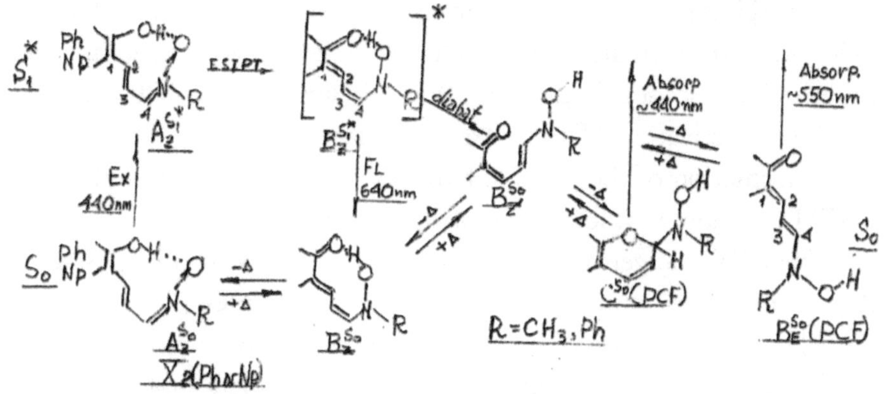

Thus the spectral and kinetic parameters of the photochromuic reaction of vinilogs of aldonitrones (T-photochromism) differ strongly from those of anils and are connected with the absolutely another mechanism of generation, and structure of the PCF in spite of the same primary stages (ESIPT).

5. 2. 1. 2. 2-hydroxychalcones as a "latent" (L) and prospective "true" (T) photochromic molecules.

It has been studied the two types of Hydroxychalcone structures which differ from each other by the position of the carbonyl group. The first type, 2'Hydroxychalcone (2'HC)(XI_1), has an IHB between the carbonyl oxygen and the hydroxyl-group in the both trans and cis E-structures (Sch. 5. 14)/34, 35/.

The position of the long-wavelength bands of the trans and cis isomers (XI_1a) are almost the same (λmax≈408-410nm) but the molar extinction coefficient of the cis structure is about four time as less.

There are no the mutual trans formations of the trans and cis structures in the ground state due to the very high barrier for the twist around C=C bond. No evidence of the existence of the E⇔K equilibrium in the both tans and cis forms.

The irradiation of XI_1a structure leads to the one-way Etrans=>Ecis isomerization (Sch. 5. 14).

On excitation of the trans-form (So-S_1^*) the ESIPT(ESIHT) occurs followed by the formation of the K*trans (or Zwitterion) structure with the ASS fluorescence (λmax, fl$\approx$510nm, Stokes shift $\Delta\nu\approx$7000cm^{-1}).

The deactivation of the K*trans structure (S_1^*-flu --->So) results in the formation of the instable Ktrans (So) structure followed by the GSIPT reaction with formation of the Etrans structure.

The adiabatic generation of the Kcis* and Ecis*in the S_1^* state of XI_1a Etrans form cannot occur by the energetic causes.

Thus the excitation of the XI_1a Etrans form does not result in the fomation of the cis-isomer or some kind of other metastable structure.

The direct excitation of the Ecis structure gives the Kcis structure in the T_1 state (K*cis(T)). The Kcis*(T)undergoes twisting around double C=C bond to give the perpendicularly twisted triplet state structure which gives the triplet trans (Ktrans*(T))tautomer with the very low energy.

The latter is deactivated to the very instable Ktrans ground state structure followed by the back GSIHT reaction with the formation of the initial Etrans structure.

Thus the excitation of the Ecis structure results in the formation of the Etrans one with the absorption band in the same spectral region.

The typical properties of the XI_1a compounds is the one-way transformation, the luck of the considerable spectral changes with the structural transformations of the stable structures, and the absence of the metastable transient one show the photochromism is scarcely realizable under any conditions.

At the same time the insertion of the donating (R=OMe)group in the phenyl ring /35/ results in the strong ASS fluorescence (λmax$\approx$600nm, Stokes shift $\Delta\nu\approx$10000cm^{-1})under excitation of the Enol form (λabs, max$\approx$370nm) owing to the ESIHT OH-------> OH and the adiabatic formation of the fluorescent NH*keto structure(Ktrans *)in the S_1^* state.

The quantum yield of the ASS fluorescence is increased sharply (by three order of magnitude)with the decrease of temperature from room that to 77K, and also by introduction of the bulky substituents in the phenyl ring.

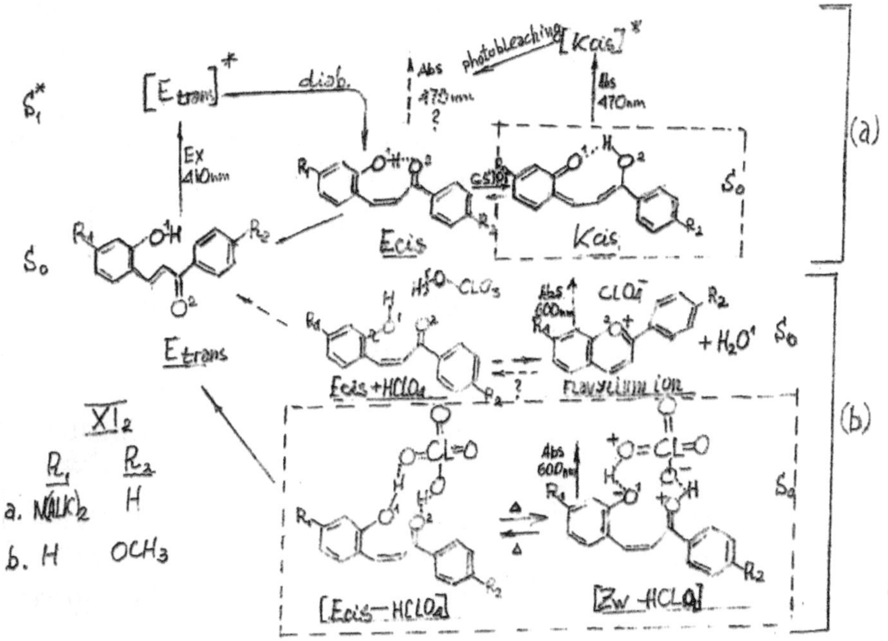

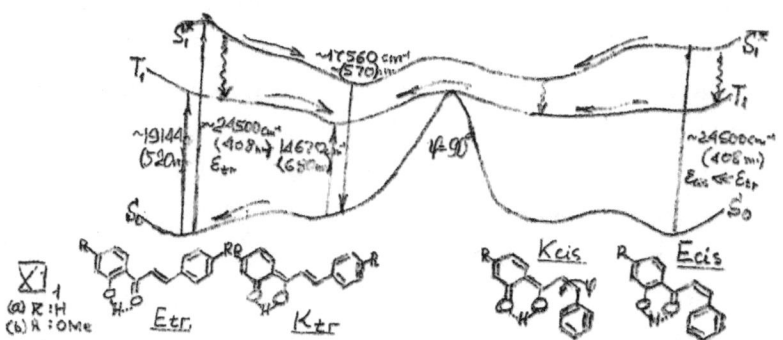

Scheme 5.15
The qualitative scheme of the photochromic reaction of Chalcones X_2 in the neutral aromatic solvent with (a) or without (b) addition of acid. Te structure framed by broken line are proposed in the present work (see text).

Scheme 5.14
Photoindused processes of the structure XI_1 with the ESIHT (ESIPT) in the T_1 state.

These modifications of the fluorescent properties are similar to those of Anils' spectra.

However no any changes in the absorption or fluorescence spectra under irradiation (including photochromism) are observed unlike Anils.

Meanwhile one can supposed the slight structural modification can result in the appearance of the photochromic properties.

Such a modification can be realized in the second type of Hydroxyhalcone structure in which the IHB between the carbonyl Oxygen and Hydroxy group can occure in the cis-form only (XI_2 a, b, see Sch. 5. 15).

The compounds XI_2 have been studied in the solvents of the different nature by the steady-state absorption spectroscopy /36, 37/. In the neutral solvent with excitation of the initial structure Etrans in the long-wavelength absorption band ($\lambda \approx 410\text{-}420 nm$) its intensity falls and new absorption band ($\lambda \approx 470 nm$) appears. By addition of perchlroric acid the irradiation leads to the appearance of the another, more long-wavelength, absorption band ($\lambda max \approx 600 nm$).

The all above described transformations of the absorption spectra are reversible at least by visible light irradiation.

The simple scheme (Sch. 5. 15a) is suggested /36/ involving the diabatic E*trans-->Ecis isomerization upon excitation (Etrans-hv->E*trans) with the formation of the photocolored structure (Ecis) in the So state with the absorption band ($\lambda max, abs \approx 470 nm$) stabilized by the $O^1\text{-}H...O^2$ bond.

The absorption band with $\lambda max \approx 600 nm$ is explained by the production of flavilium ion in presence of perchloric acid with chipping off the Hydroxy group with the formation of H_2O molecule (Sch. 5. 15b, unframed by the broken line).

However the two criticisms should be done. The first one: such a large red shift of the absorption band ($\Delta\lambda \approx 60 nm$) cannot be caused by the trans-cis isomerization with the O-H...O bond formation according to the findings for the analogous molecular structures (see above X_1, Etrans and Ecis structures), and such a shift should not to exceed 8-10nm.

At the same time the absorption band of the colored form of XI is very typical for the Keto-structure, Kcis, with the extended π-electron system which can be generated by the Proton(Hydrogen)transfer O^1H->O^2H(Ecis->Kcis)in the ground state.

It can be supposed that as a result of the diabatic reaction E*trans-(diab)->Ecis (Sch. 5. 15a) the tautomeric equilibrium Ecis<=GSIPT=>Kcis established in the ground state that is almost completely shifted towards more stable colored keto structure Kcis with the typical absorption band ($\lambda max \approx 470 nm$), and the Etrans structure is the transient precursor of the colored form (Sch. 5. 15a framed by the broken lines). The photobleaching reaction only, that is observed experimentally /37/, is the most probable in such a situation.

Thus according to the new scheme the GSIPT is the final step of the direct photochromic reaction while the ESIPT can be a primary step of the reverse photochromic reaction in the neutral aprotic solvents.

The second criticism: it is very doubtful that the alienation of the hydroxyl group with the formation of H_2O molecule for the generation of the heterocyclic cation (with the absorption band about $\lambda \approx 600 nm$)in the presence of perchloric acid can be a reversible reaction/19/.

It is more probable that the double H-bond complex[Ecis-acid] is formed (Sch. 5. 15b framed by the broken line)with the conversion to the [Zw cis –acid] complex with the longwavelength ICT-transition($\lambda max \approx 600 nm$)which is a result of the double intermolecular proton transfer $O_1H => HO_2$...

The reversibility of the photoreaction is secured well by such a mechanism of the ptotocoloration.

Thus the compounds XI_2 show T-photochromism in the solvents (unlike XI_1)which is more adequately described by the new scheme 5. 15.

5. 2. 2. The Nitrogen-Nitrogen (NH ⇔ NH) transfer.

5. 2. 2. 1. Phenoxazine derivatives.

The photochemical reactions of 1-p-toluenesulphonyl(TS) derivatives of tert-butyl phenoxazine XII(Sch. 5. 16a) and the model structures XII_1, XII_2 (Sch. 5. 16b) have been studied by the steady-state absorption and fluorescent spectroscopy methods with the utilization IR, PMR, and X-ray structural-analytic methods, and the semiempiric PM3 quantum-chemical calculations /38, 40/.

A. The initial structure.

The PMR signals of the aromatic and phenoxazine protons and especially the shift of the NH proton signal towards the weak fields (ΔH=7. 5ppm), the IR absorption band of N=N stretching vibration (ν=3460cm^{-1}), and the typically modified NH group vibration band testify to the ZAzostructure(Az(Z)) of molecule XII in which the phenoxazine moiety and azogroup are bound by the weakened H-bond $N_1H...N$.

The weakening of the H-bond can be caused by the acoplanarity of the structure XII ($\varnothing$ >0) due to the steric interactions of the azo and tert-butyl groups.

Such an acoplanarity is confirmed by the likeness of the absorption spectra of XII Az(Z)(Tab. 511) and the noncoplanar molecule of cis-azobenzene(λmax, abs≈280nm, ε = 5200 LM^{-1}cm^{-1}) unlike the planar trans-one(λabs, max≈320nm, ε=20000LM^{-1}cm^{-1})/11/.

The considerable charge separation between the azo and oxazine moieties(μ=9. 06D) can also be provided by their acoplanarity.

Such a charge separation just as a moieties' acoplanarity promotes the weakening of the H-bond and considerable decrease of the rate of the GSIPT reaction N_1H->N_2 with formation of the Hydrazo Zcis (HAz(Zcis)) structure which is less stable than Az(z) one by 15kcal/mol according to the quantum-chemical data.

Really the formation of the HAz(Zcis) structure are not observed in the absorption spectra with the variation both temperature and solvent polarity within the large limits.

The nature of the first electronic transition observed in the absorption spectra (So->$S_1^*(\pi\pi^*)$) is also caused by the charge separation. The calculated and experimental data of the energy values for that transition are agreed with each other very good. (Tab. 5. 10).

According to the calculated data the transition is mainly connected with the ICT from the phenoxazine moiety to the aza-group which promotes a sharp increase of the electric dipole moment ($\Delta\mu=\mu(e)-\mu(g)$, D) in the S_1^* state (Tab. 5. 10) and manifested in the considerable red shift of the absorption band with the solvent polarity increase(Tab. 5. 11).

Also according to the data of the quantum-chemical calculations there are at least the two low located $S_1^*(n, \pi^*)$ states and corresponding $T_1(n, \pi^*)$ and $T_1(ICT)$ ones which are not directly observed due to the extremely low transition intensity (Tab. 5. 10).

The strong radiationless deactivation caused by existence of such a set of the nonspectral states is typical for the azostructures and conformed by the lack of the fluorescence of the structure XII and of the model structures under even the very favorable conditions.

B. Photoinduced processes.

The ESIPT N_1H->N_2H occurs as a result of the strong increase of the electron density (basicity) on N_2 atom in the $S_1^*(ICT)$ state and has to compete with the very fast radiationless decay of that state.

Therefore the ESIPT of the structure XII like that of SA is the adiabatic barrierless ultrafast reaction which is the primary step of the following photoinduced processes.

According to the scheme(Sch. 5. 16a) the $N_2H(Zw\ Z)$ structure ($\mu(e)=20D$) is formed as a result of the ESIPT followed by the almost barrierless diabatic Z->E isomerizaton about $N_1=N_2$ bond with the formation of the Zw(E) structure in the ground state.

The typical for the ESIPT ASS fluorescence from Zw(Z)* state is completely suppressed by such a competing diabatic reaction, and are not manifested even under the most favorable conditions.

However although the direct evidence of the ESIPT (ASS fluorescence)for the structure XII is absent there are the several important arguments which demonstrate its decisive role as a primary step of the photocolored reaction of the structure XII.

Really the ASS is clearly observed for the nonphotochromic model compound XII_1 in which the ESIPT is apparently not followed by the adiabatic reaction with the PCF structure formation.

On other hand both the photocoloration and the ASS fluorescence are not observed of the model structure XII without H-bond and the ESIPT. The Zw(E) structure is an intermediate of the two competing path-ways of the reversible and irreversible reactions. Along the path-way of the reversible(photochromic)reaction the equilibrium between the two metastable colored structures-nonplanar zwitterionic HAz(Etrans) and low polar HAz(Zcis)-is established, and those structures are generated consecutively with the absorption bands of $\lambda(max, abs) \approx 600$ and 380nm correspondingly in the solvents(Tab. 5. 11)(see also below).

The twist around C:::N bond due to the steric interactions between Ts and tert-butyl groups acompanied by the charge redistribution and separation when forming the HAz structure from Zw(E) one secures the realization of the long-wavelength ICT transition ($\lambda max \approx 660nm$) in the HAz structure while the planar low polar HAz structure has absorption band in the much more shortwavelength region ($\lambda max \approx 380nm$)(see below).

The reverse dark reaction in the So state is connected with the E->Z isomerization of Hydrazostructure (HAz(Etrans)->HAz(Zcis)) controlled by the medium viscosity, and completed by the back GSIPT with the formation of the initial Azastructure ((HAz(Z)->Az(Z)).

Such a reversible photoreaction is observed with the absorption spectra in the liquid solvents only under the shorttime (2-3min) irradiation of the initial form. Upon the prolonged irradiaton under such conditions the spectral changes are caused only by the competing irreversible reaction.

ESIPT Photochromism 379

The products are generated along the irreversible path-way, with the participation of the dissolved oxygen, and could be isolated from the solution after prolonged irradiation.

They have been identified with help of the X-ray structural analysis, the absorption steady-state spectroscopy, and have been attributed to the high and low polar structures Ox_1 and Ox_2 with the absorption band maxima about λmax, abs≈340-360nm and 580-620nm correspondingly depending on the solvent nature(Tab. 5. 11).

These structures are bound up with each other by the tautomeric equilibrium $Ox_1 \Leftrightarrow Ox_2$ owing to the OH⇔NH Intramolecular proton transfer(Sch. 5. 16a).

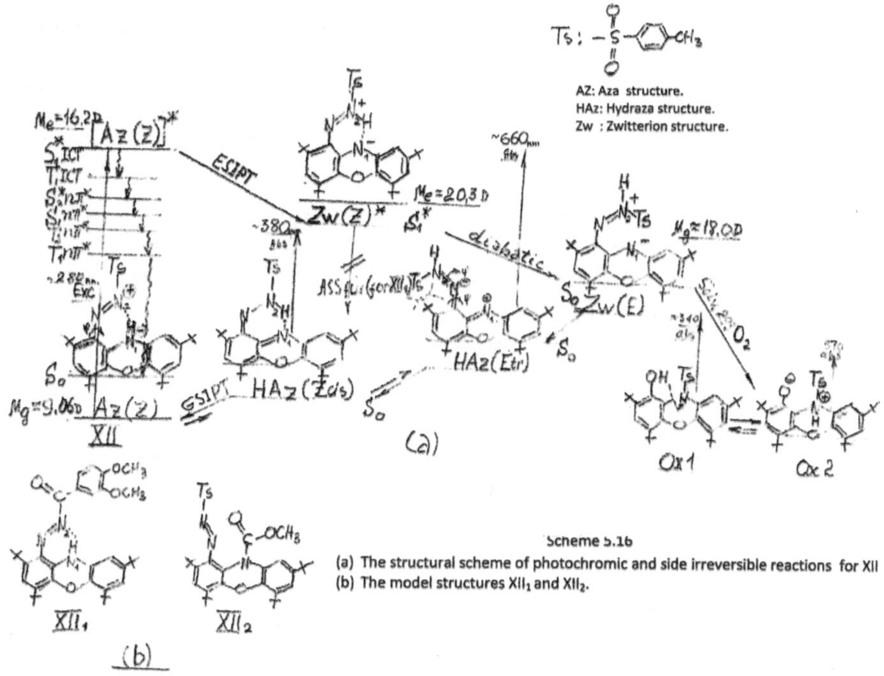

Scheme 5.16
(a) The structural scheme of photochromic and side irreversible reactions for XII
(b) The model structures XII$_1$ and XII$_2$.

Tables 5.10
Structural and spectral characteristics of the initial structure of Phenonxazine molecule XII

PMR N-H δH (ppm)	IR N=N (cm^{-1})	Absorption transition							
		Experiment <1>			Calculation <2.>			
		λ^{max}_{nm} ($\nu^{max}_{cm}{}^{-1}$)	f	ε(M^{-1}Lcm^{-1})	λ^{max}_{nm}($\nu^{max}_{cm}{}^{-1}$)	f<3>	Nature of S*	Δμ(D)
7.5	3460	278 (35971)	0.13	7750	270(39674)	0.21	S$_1$*ICT	+7.11
		Unobserved	<10^{-3}	Unobserv.	283(35347)	0.00	T$_1$ICT	+13.6
		Unobserved	<10^{-3}	Unobserv.	376(26550)	0.04	S$_2$* nπ*	-4.00
		-----	---	---	388(25767)	0.08	S$_1$*nπ*	+1.97
		-----	---	---	403(24793)	0.00	T$_2$ nπ*	-0.86
		-----	---	---	436(22925)	0.00	T$_1$nπ*	+8.93

<1>Hexane, room temperature. <2> By PM3 method.

Table 5.11
The longwavelength absorption of the initial and final structures in the photoreaction of XII

Structure	*)	Tempe-rapture T K	Isopentane		Ethanol		PMMA	
			λ^{max}(nm) ν(cm^{-1})	ε (M^{-1}Lcm^{-1})	λ^{max}(nm) ν(cm^{-1})	ε (M^{-1}Lcm^{-1})	λ^{max}(nm) ν(cm^{-1})	ε (M^{-1}Lcm^{-1})
Initial	Az(Z)	293	278 35970	7750	284 35211	9500	----	---
Reversible (photo-chromic)	HAz(Z)	293	---	---	---	---	370 27030	---
		77	380 26316	6670	380 26316	6000	----	---
	HAz(E)	293	---	---	---	---	620 16630	---
		77	662 15106	2300	660 15152	1500	620 16130	---
Irreversible	Ox 1	293	344 29070	---	363 27548	11702	360 27778	---
	Ox 2	293	577 17330	---	617 16207	5750	560 17857	---

*) See scheme 5.16

ESIPT Photochromism

Table 5.12
Absorption and fluorescence of heterocyclic arylethinenes

Structure XIII	Absorption				Fluorescence				[5]
	E_{trans}	E_{cis}	K_{cis}		E_{trans}		E_{cis}	K_{cis}	
	λ^{max}_{nm}	λ^{max}_{nm}	λ^{max}_{nm}	τ ps	λ^{max}_{nm}	$\Delta\nu^{a\text{-}f}_{cm^{-1}}$	λ^{max}_{nm}	λ^{max}_{nm}	$\Delta\nu^{a\text{-}f}_{cm^{-1}}$
1 [1]	344	360	530	70 85 [3] 180 [4]	430	6070	—	570	~10234
2 [2]	377	414	580	300	440	3800	—	600	~7490

[1]/ACN [2]/Benzen [3]/THF [4]/Transient absorption [5]/Excitation of E_{cis} structure.

Scheme 5.17
The structural and energetic schemes of the photoinduced processes (including photochromism) in heterocyclic arylethylenes

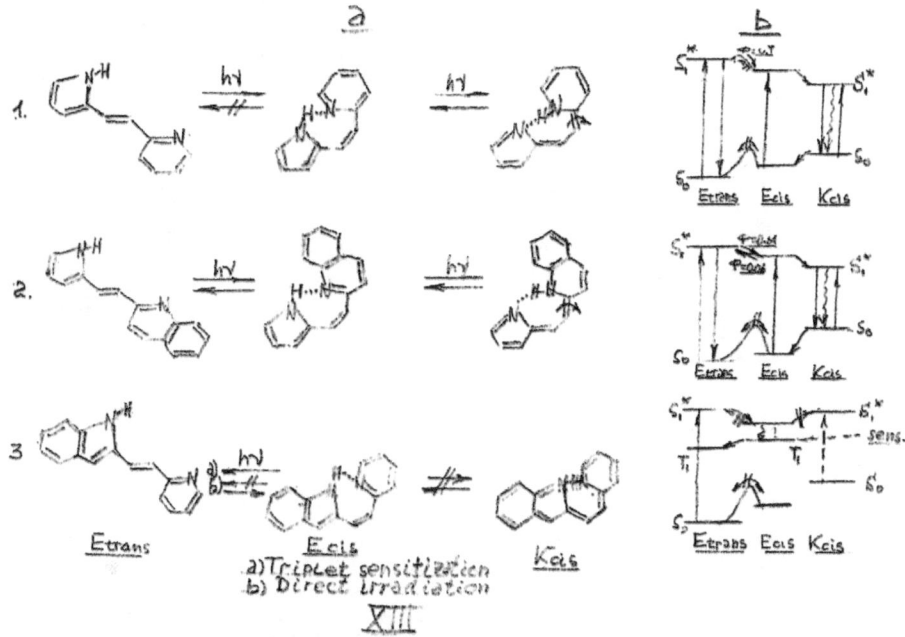

a) Triplet sensitization
b) Direct irradiation

XIII

Table 5.13
The spectral properties of spyranes of the pyrimidine (SPP) series (XIV), and of the model structures in the solvents (see scheme 5.18).

Structure → Spectra ↓ Solvent ↓	XIV Absorp λ^{max} nm	XIV Absorp ε $M^{-1}L$ cm^{-1}	A Absorp. λ^{max} nm	A Absorp. ε $M^{-1}l$ cm^{-1}	A Fluoresc. λ^{max} nm	A Fluoresc. Δv^{a-f} cm^{-1}	B Absorp. λ^{max} nm	B Absorp. ε $M^{-1}L$ CM^{-1}	A+B 2/ Absorp. λ^{max} nm	A+B 2/ Absorp. ε $M^{-1}L$ cm^{-1}	XIV_M Absorp. λ^{max} nm	XIV_M Absorp. ε $M^{-1}L$ cm^{-1}	Attribution and localization of transitions. (In brackets, Sch.5.18 a,b)
Heptane	234	55750	236	44250	---	---	---	---	233	55000	233	53500	$\pi\to\pi^* S_0\to S_3^*$ (a0)
	259 1/	5000	---	---	---	---	263	20750	261	23000	---	---	$\pi\to\pi^* S_0\to S_4^*$ (b)
	330	12500	333	11000	---	---	---	---	333	15500	---	---	$\pi\to\pi^* S_0\to S_3^*$ (a)
	344	12680	347	11750	376	2140	352	5000	345	16000	342	13700	$\pi\to\pi^*_2 S_0\to S_1^*$ (a+b)
	415	1760	---	---	---	---	---	---	---	---	424	3200	Ict(a→b) $L_N\to C=O S_0\text{-}S_1^*$ $C=O(nn^*)$
	---	---	---	---	---	---	395 1/	2500	395 1/	2500	---	---	3/ ICT(B) $I_{OH}\to\pi_{CO} S_0\text{-}S_1^*$ in B
Acetonitrile	232	47500	235	58500	---	---	---	---	235	68000	233	49300	$\pi\to\pi^* S_0\to S_3^*$ (a)
	263	8500	---	---	---	---	266	17500	266	19500	---	---	$\pi\to\pi^* S_0\to S_4^*$ (b)
	332	13000	336	16250	---	---	---	---	336	21250	---	---	$\pi\to\pi^* S_0\to S_3^*$ (a)
	345	13250	348	2200	386	2830	---	---	---	---	343	12200	$\pi\to\pi^* S_0\to S_2^*$ (a)
	---	---	---	---	---	---	376	3750	377	5000	---	---	$n\to\pi^* S_0\to S_2^*$ C=O in B
	406	1770	---	---	---	---	---	---	---	---	411	1500	ICT(a→b) $S_0\text{-}S_1^*$ $L_N\to C=O$
	---	---	---	---	---	---	442	3750	442	3750	---	---	ICT in B $S_0\text{-}S_1^*$ $L_{OH}\to\pi_{CO}$

1/ The holder 2/Two component solvent with equimolar concentrations of A and B. 3/Calculated for the structure B (the sch.5,18b)

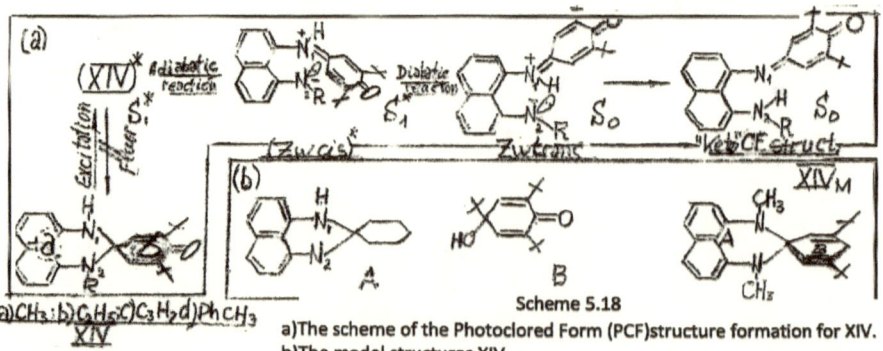

Scheme 5.18
a) The scheme of the Photoclored Form (PCF) structure formation for XIV.
b) The model structures XIV_M.

Thus the irreversible reaction involves the connection of Ts group with the Nitrogen atom of the phenoxazine moiety in the Zwitterion (ZwE) structure and removal of the aza-group with the oxidation of the phenoxazine moiety. The latter step is controlled by diffusion of Oxygen, and reaction is excluded in the rigid glasses at 77K.

The absorption bands of the structures Ox_1 and Ox2 are similar to the photochromic absorption bands of HAz(E) and HAz(Z) structures respectivelly, and shifted only a little towards the short-wavelengths (Tab. 5. 11).

Thus the structure Ox_1 and Ox2 are perfect counterparts for the interpretation of the absorption bands' nature of the photochromic products HAz(Z) and HAz(E) correspondingly.

Obviously in the both cases the electronic transitions $So->S_1^*$ are localized in the phenoxazine moiety and caused by the ICT from the electron donor to the electron acceptor substituents of the final structures of both the reversible and irreversible photoreactions.

The coloration is mainly caused by the Zwitterionic structure with the long-wavelength absorption band (λabs max$\approx$600nm).

Thus the effective T-photochromism of Phenoxazine derivatives in the solvents is accompanied by the degradation that can be suppressed in the deoxygenated and the rigid media.

The photoinduced transformations of Heterocyclic arylethylene XIII_1-3 (Sch. 5. 17a), including the trans-cis isomerization around C=C bond, controlled by the NH...N bond formation have been studied by the steady-state and the transient absorption and fluorescence measurements in the liquid solvents at an ambient temperature. /41-43/.

The compounds XIII_1-exist in the form of the mixture of the trans and cis isomers but there is no equilibrium between the isomer structures in the ground state.

The cis-structure is stabilized by the strong H-bond NH...N that is absent in the trans-structure (e. g. XIII) exhibiting H^1NMR signal at δ=15. 1 and 8. 7 ppm of the cis and trans structures correspondingly) (Sch. 5. 17a).

Moreover due to the strong intramolecular H-bond there is the only one-way of the trans-cis isomerization (XIII_1, 3) or the very low-efficient cis=>trans photoisomerization (XIII2) with the $So-S_1^*$ excitation of the trans or the cis-isomers (Sch. 5. 17b). At the same time the cis –>trans photoisomerization of the compound XIII takes place only with the triplet sensitization via low-lying triplet(T_1) state of the cis-isomer structure /42/. The long-wavelength absorption band of the cis-isomer(Ecis) is shifted towards "red" in comparison with the trans one (Etrans) ($\Delta\lambda$=20-40nm, Tab. 5. 12), and the latter have the marked fluorescence with the normal (but increased)Stokes shift competing with the trans->cis photoisomerization in the S_1^* state (Sch. 5. 17b, Tab. 5. 12).

The irradiation of the mixture of the trans and cis-isomers results in the accumulation of the latter owing to the one-way trans->cis photoisomerization.

The ESIPT takes place of the structures 1 and 2 only (but not of 3) and results in the appearance of the typical ASS fluorescence(Δv(a-f)$\approx$7500-10000cm^{-1}, Tab. 5. 12) from the Kcis*structure.

The colored structure(Kcis) with the transient absorption band in the long-wavelength spectral range (λmax$\approx$600nm)(Tab. 5. 12) is generated as a result of the Kcis*deactivation(Kcis*->Kcis).

These Kcis transient structures of XIII$_1$ and XIII exist in the liquid solvents in the nanosecond time scale (Tab. 5. 12), and can be considered as a mani-festation of the short-lived L(Latent), photochromism.

The similar L-photochromism is typical also for photochromic anils in the liquid solvents (Ch. 3) with the colored form decay time in the micro-submicroseconds scales in dependence on the solvent nature while the calculated GSIPT time rate lies in the subpicosecond scale (Tab. 3. 44).

Thus the latter is not a limiting stage of the reverse reaction that has to have like SA the preceding steps connected with the decay of the twisted forms which time depends on the media nature (viscosity, polarity) and the molecular structure.

Really the lifetime of the Kcis transient structure(τ) decreases markedly with the growth of the solvent polarity and the broadening of the ring π-electron system (Tab. 5. 12).

One can supposed the sharp rise of the colored structure decay time will take place like SA with the insertion of the bulky substituents in the rings and especially in the rigid media including crystal state.

Thus in view of above findings Heterocyclic arylethilenes with the NH...N hydrogen bond can be considered as very favorable structures for the rigid media and crystalline T-photochromic systems.

5. 2. 2. 2. Spirans of perymidine series.

The photochromic spirans of the perymidine series (SPS) have been synthesized for the first time and investigated preliminary by the PMR, IR, X-ray structural methods and by the UV-VIS steady-state spectroscopy in the works/44-47/.

The reversible Ring⇔Chain transformation in the solvents and the structure-influenced synchronous mechanism of the E⇔Z isomerization along with the spirocyle closing⇔opening, involving the Ground state Hydrogen transfer (GSIHT) have been proposed /48/.

The structural scheme of the phototransformations with the generation of the transient Zwitterion structures and production of the colored structure has been proposed/49, 50/.

The detail experimental and quantum-chemical investigations of the initial cyclic structure with the utilization of the model molecules have been conducted.

At last the detail investigation of the nature of excitation and of the reaction mechanism involving generation and decay of the colored structure has been carried out recently/51/by the steady-state spectroscopy and the quantum chemical methods with utilization of the model structures.

On the base of the results of the studies the general view of the photoinduced processes of the structures XIV(Sch. 5. 18a)with the utilization of the model structures A, B, XIV-M (Sch. 5. 18b)is produced below.

The initial form of SPS in the solvents in the ground state is the mixture of the two main structu-res in the tautomeric equilibrium of the spiro-structure(XIV)and the Keto Colored Form (KCF) one (XIV=CF) shifted almost completely (especially in nonpolar solvents) towards the photoactive spiro-structure XIV.

The latter consists of the two, heterocyclic (a) and quinonimine(b), structural fragments (Sch. 5. 18a) separated by the sp³-hybridizated carbon atom and located in the almost orthogonal planes (Sch. 5. 19(1)).

The localization and the nature of the electronic transitions are ensured by such a structure. Really, it results from the comparison of the absorption spectra of the model molecules A and B in the solvents and their equimolar binary solutions with those of the studied SPS molecule(Tab. 5. 13)that all transitions of the high energy (So->S_n^* n=2-5)are localized in the π-system of one of the separated fragments(a or b) and have the characteristic features for the π-π* transitions-high intensity and very small spectral shifts.

Unlike those the lowest So->S_1^* transition is the typical for the Intramolecular (Interfragmental)

Charge Transfer (ICT) that between the essentially acoplanar moieties (a and b).

Really, the So->S_1^* transition is characterized by the low intensity(ε<2000 M⁻¹L cm⁻¹)and "blue" shift in the high-polarity solvents (ITab. 5. 13).

The latter corresponds to the decrease of the dipole momentum in the S_1^* state that is typical for the transitions involving the lone(l)or nonbonding (n)electron pair.

The same transition is observed also of the model compound XIV-M and caused obviously by the ICT from the NR groups of the "a"-moiety towards the π-system of the "b"-moiety involving the C=O group(Sch. 5. 19(1)and (2)).

The n-π*transition localized on the C=Ogroup is overlapped partially with the latter(Sch. 5. 19(2).

At the same time the ICT of the another nature (from the OH group to the C=O one)is manifested of the model structure B and characterized by the "red" shift in the polar solvents(see Sch. 5. 19(2)). The supposed mechanism of the photoinduced processes for the structures XIV in the solvents is presented by the schemes (Sch. 5. 18a, 5. 20).

The excitation of SPS structures XIVa-d in the ICT nature transition ($\nu=24000cm^{-1}$) results in the strong weakening and fast rupture of the C-N spiro-bond in the S_1^* excited state with the adiabatic formation of the Cis –zwitterion structure $(ZwCis)^*(\mu(e)=15D)$ that completely suppresses the fluorescence from the S_1^* state of XIV structure even in the glassy solvents at 77K.

Really, such a reaction occurs under excitation also in the analogous ICT transition of the model structure XIV-M ($\lambda abs=424nm$ or $\nu=23580cm^{-1}$)(Tab. 5. 13) but does not take place of the model structure A in which the typical ICT transition is absent, and therefore the efficient fluorescence of that molecule is observed.

The formation and stabilization of the KCF in the So state occurs as a result of the Ground state Proton transfer (GSIPT) in the precursor Zwitterion trans (Zwtrans)structure ($\mu(g)\approx 10D$) that is generated by the diabatic reaction of the cis –trans isomerization $(Zw\ cis)^* \to (Zw\ cis)^\circ (S_1^* \to So)$.

The efficiency of that step and hence the efficiency of the KCF formation has to rise with the decrease of the energy gap between S_1^* and So state ($\Delta E<6000cm^{-1}$) i. e. with the growth of the asymmetric energy barrier (see Sch. 5. 20) in the So state($\Delta E\approx 20$-$25 kcal/mol$) caused by the steric interactions which is increased with the size of the substituent R in accordance with the observed findings(Tab. 5. 14)(see also below).

At the same time the stabilization of the Zwitterion transition state (decrease of the potential barrier)in the ground state and of the precursor Zw trans structure in the polar media must result in the observed sharp decrease of the KCF structure generation efficiency in the polar solvent (ACN)(Tab. 5. 14).

Such a photocolored structure is generated only as a result of the GSIPT with the formation of the metastable Keto cis-structure.

Really, the structure, the set of the electronic states, and the nature of the structural transformations are the same for the studied (XIV) and the model (XIV-M) molecules excluding the lack of the NH group and the GSIPT of the latter in which the colored structure is not generated. Thus the GSIPT is a final step and the decisive factor in the Photochromism of SPS molecules.

Table 5.14

Dependence of photochromic properties XIV on the molecular structure and solvent nature (ambient temperature).

Strucf. XIV (Schem. 5.18)	Octane (O CT))				Acetonitrile (ACN)				Polystyrene			
	λ^{max}_i nm <1>	Φ^i_r	τ_i s <1>	τ_i/τ_a	λ^{max}_i nm <1>	Φ^i_r	τ_i s	τ_{isolv}/τ_{ioc} <2>	λ^{ma}_i nm	Φ^i_r	τ_i s	τ_{isol}/τ_{ovT} <2>
a	622	024	670	1.0	602	0.057	365	1(0.54)	626	0.12	6430	1(9.6)
b	626	0.42	1680	2.5	602	0.055	910	2.5(0.54)	646	0.20	32100	5..0(19.1)
c	630	038	2944	4.4	607	0.053	2280	6.2(0.77)	633	0.13	33430	5.2(11.4)
d	611	0.26	12400	18.5	595	0.055	101600	278(8.2	615	0.08	130370	20.3(10.5

<1>i-a,b,c,d .< 2>The ratio of the corresponding lifetimes of the structure in the solvent and octane $\tau_{i\,solv}/\tau_{i\,OCT}$ is shown.

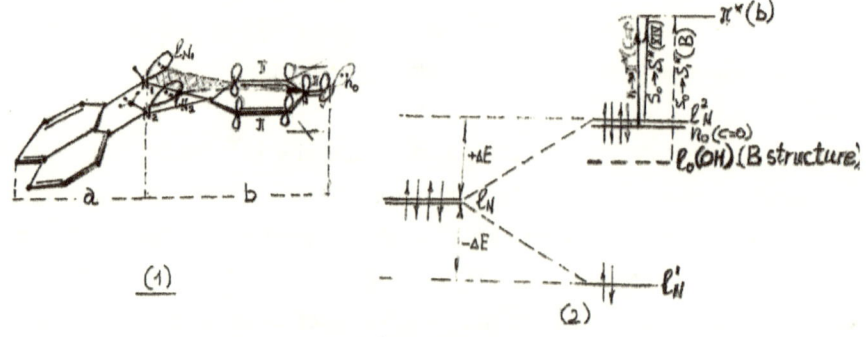

Scheme 5.19
The nature of $S_0 \to S_1^*$ transition of SPP structures XIV.
The qualitative structural (1) and energetic(2).

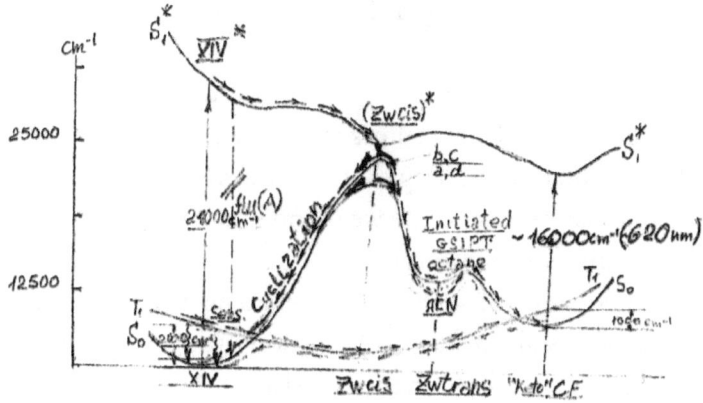

Scheme 5.20
The semiqualitative energetic scheme of the photochromic reaction for the SPP structures (XIV in the nonpolar solvents (see also the scheme 5.18$_a$, tables 5.13, 5.14).

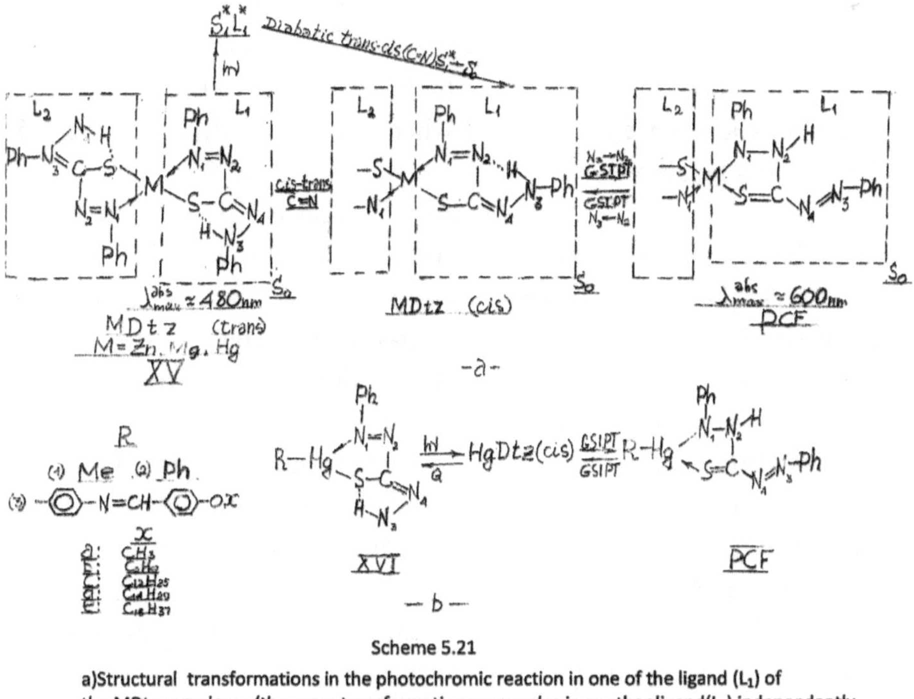

Scheme 5.21
a) Structural transformations in the photochromic reaction in one of the ligand (L_1) of the MDtz complexes (the same transformations occur also in another ligand(L_2) independently.
b) The photochromism of single ligand complex.

The colored Form can be also generated as a result of sensitization by the triplet-triplet energy transfer from the triplet donors $(E(T)=18000-19000 cm^{-1})$ with the marked quantum yield $(F(ST) \approx 0$.

12) in the oxygenless solutions that falls sharply in the presence of the molecular Oxygen (F(ST)≈0. 07).

At the same time the quenching of the photochromic reaction by even very low energy triplet acceptors is not observed

Thus there is the reactive Triplet state with the very low energy and the very low population due to the weak intersystem crossing along the path-way of the direct photoreaction in the S_1^* state.

The estimated value of such a triplet state energy does not exceed $\Delta E(So-T) \approx 2000 cm^{-1}$ above the So PES along the path-way from the initial structure (XIV) towards final one (KCF) (Sch. 5. 20).

That low-energy triplet state can obviously play the essential role in the reverse dark reaction(see below).

The reverse photoreaction of the bleaching upon irradiation in the region of the absorption band of the KCF ($\lambda \approx 620-630 nm$) is not observed whereas the dark reverse reaction of discoloration in the ground state runs slowly (Tab. 5. 14)with the rate which clearly drops(the characteristic time $\tau(i)$ increases) when the potential barrier (Sch. 5. 20) is being heightened.

The height of the latter ($\Delta Eo \approx 15-20 Kcal/mol$)is determined by both the steric factors with the increase of the R substituent size (the structures XIVa-c) and also by the electronic ones (the delocalization of the n-electrons along the phenyl ring π-system in the structure XIVd). The steric factors connected with the medium rigidity are manifested by the rate drop ($\tau(i)$ increase) in the polystyrene matrix.

The influence of the electron factors(the stabilization of the polar transition Zw cis state followed by recylization of the initial spirostructure(Sch. 5. 20))is displayed as the rate rise ($\tau(i)$ fall) in the polar solvent (ACN).

The observed variation of the kinetic parameters in the ground state and those of the photoproduct generation (see above) are correlated with the change of the height of just the same potential barrier in the ground state.

At the same time the back reaction rate increases sharply (characteristic time falls) in the solvents saturated by Oxygen (for instance $\tau=12400$ vs 1430s for XIVd in Octane) and vice versa.

This dependence can be explained by the participation of the very low triplet state(see above and Sch. 5. 20)which population is caused by the Spin-Orbital interaction controlled by the interaction with molecular Oxygen. With the growth of the triplet state population the contribution of the low-barrier reverse reaction along the triplet PES can be increased stipulating the marked rise of the reaction rate.

Thus the efficient photocoloration of the SPS solvents is accompanied by the unique opportunity of the dark bleaching rate change with variation of molecular Oxygen concentration in the solvents, and SPS compounds in various media represent the new photochromes of the T-type.

5. 2. 2. 3. Photochromic Metal-Dithizonate complexes.

The photochromic behavior of the mercury (II) dithizonate complexes(Hg Dtz)(Sch. 5. 21a, Structure XV) in the solvents was independently reported first in the several works about sixty years ago /52-54/.

According to the observations their benzene or chloroform solutions change on irradiation with visible light their normal orange-yellow color (initial form) to an intensive royal-blue that (photocolored form, PCF). The PCF returns slowly (with time about several minutes)in the dark to the initial one, and this process can be multiply repeated(see also /23/).

Table 5.15
Visible absorption bands [1/] for the initial and photocolored forms of HgDtzs (see Sch. 5.21)

Complex XVI((3)	Initial structures		Photocolor.form	
	λ^{max}_{abs} (nm)	ε $M^{-1}L\,cm^{-1}$	λ^{max}_{abs} (nm)	$\tau_{1/2}$ (min) 2/
a	485	35500	596	3
b	484	39500	596	3
c	487	44000	597	5
d	486	38500	597	5
e	487	42500	597	6

1/In $CHCL_3$ at ambient temperature.2/Half-life of the photocolored structure.

Scheme 5.22

The examples of aromatic o-methoxy carbonyl structures and their probably photochromic transformations involving CH-->O ESIPT with formation of colored structures.

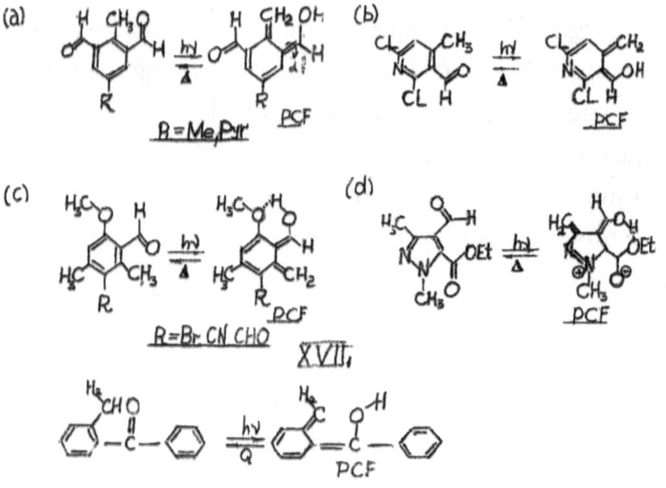

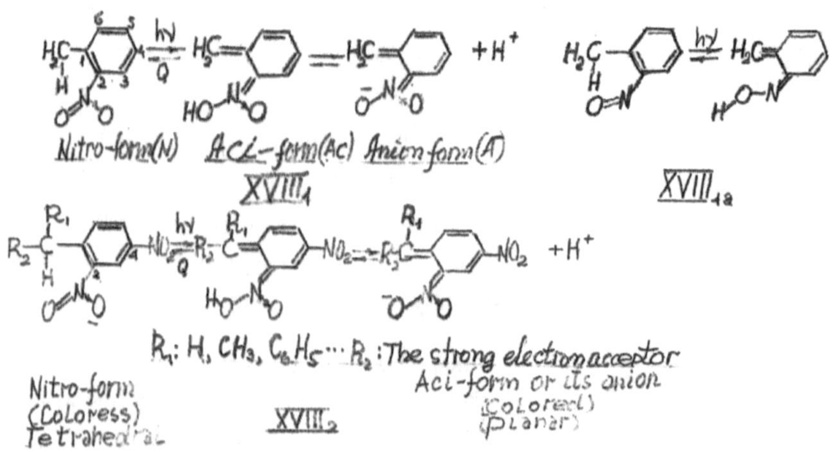

Scheme 5.23

The examples of the photochromic Alkyl substituted Nitro derivatives (XVIII_n) with the H transfer from atom C to atom O of 2-Nitro group.

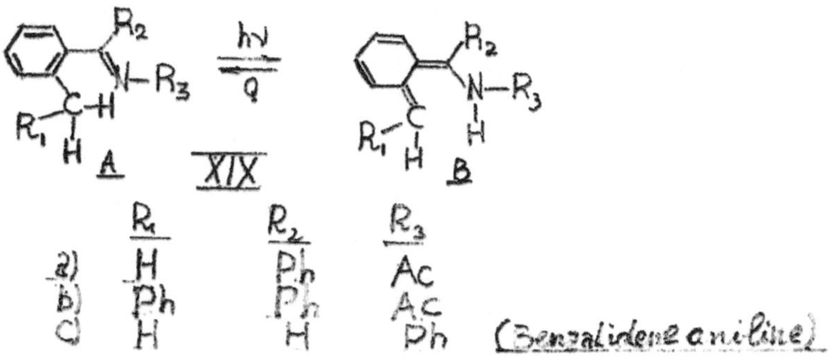

Scheme 5.24
Photochromism of o-Alkyl aromatic imines with the H-transfer from atom C to atom N of imine group.

However only after fifteen years it had been established that the similar behavior is typical also of Dtz of another heavy metals (MDtz, were M=Mg, Zn, see Sch. 5. 21a) in the solvents but unlike the mercury(II)complex (where half-lifetime, τ(1/2), of PCF is the order of 1min) the rates of the thermal back reactions are too fast for the conventional measurements under steady-state irradiation /55/.

Thus the central metal atom determines the stability of the PCF structure, and in some cases also can influence the photochromic band intensity. The analogous photochromic transformations have been observed also in the crystal for HgDtz (XV)/55/.

The kinetic and the spectral IR studies /56/ have shown that the photochromic reaction can occur in the both ligands(L_1 and L_2) independently, and proposed reaction mechanism involves the excitation of one of the ligands, the H-bond(N-H...S) rapture with obviously the diabatic trans-cis isomerization around C=N bond, and the H-transfer from N^3 atom to N^2 one in the ground state with the formation of the PCF structure with the long-wavelength absorption band (λmax≈600nm)(Sch. 5. 21).

The ligands' independence is insured by the tetrahedral structure of the chelate knot in the MDtz complexes with the nontransition metals where sp^3 hybridization of the metal orbital is realized.

One can believe however that photocromic behavior of the MDtz complexes with the transition metals (e. g. Cu, Ni) would be modified by the ligands' interactions due to the planar structure of the chelate knot, and in our view, the special investigation is deserved for this problem.

It has been shown more later/57, 58/, and confirmed by the X-ray structural analysis/59/that the the analogous photochromic reaction occurs in the HgDtz XVI_1, $XVI2(R=Me, Ph$ correspondingly) with only single Dtz ligand (Sch. 5. 21b).

The characterization of such complexes as photochromic compounds has given an opportunity for the synthesis of Organomercury (II)DTz with modification of the photochromic properties.

The long-chain azomethine derivatives of Hg DTz XVI3 (a-e) (Sch. 5. 21b)have been investigated comparatively recently /60/.

All complexes are T-photochromes (yellow ⇔blue, λmax, col ≈596nm, chloroform) under irradiation by the visible light, with $\tau(1/2)$ of the "blue" form increased strongly as compared with that of Hgbis DTz (Tab. 5. 15).

The positions of the absorption bands of the both the initial and photocolored structures don't depend on the Alkil parasubstituent's length(x).

At the same time the absorption coefficient (ε)of the initial band increases a little with the alkil radical length, and are virtually half ($\varepsilon \approx 35500 M\ l^{-1}cm^{-1}$)of that found for the biscomplex Hg Dtz(XVI, M=Hg, $\varepsilon = 70500\ mol^{-1}L\ cm^{-1}$) where there are two identical independently absorbing chromophores (see above).

The estimated value ε max for the colored form is about $39000 mol^{-1}L\ cm^{-1}$ reported for the colored form of the corresponding biscomplex HgDTz(XVI, M=Hg)/59/.

The qualitative estimation indicates that half-lifetime ($\tau 1/2$)increases with the increasing of the chain length of the alcoxy substituent, however the sensitivity of the molecular system studied to the smallest traces of water or other contaminants accounts for the poor reproducibility observed for the thermal return rates.

Unfortunately unlike bisHgDtz complexes XV(see above)the photochromism of HgXVI is not displayed sufficiently to produce the solvent and crystalline materials, and all compounds deteriorate with the prolonged exposure by initiating light.

In our view the structures on the base of MDtz complexes of XVI type (with different metals M) with R=Anils (SA or another imine of o-Hydroxyaldehyde) could provide the superposition and useful mutual modification of the photochromic properties of the both moieties to obtain the novel prospective photochromic compounds.

5. 3. The Proton or Hydrogen transfer from Carbon atom to Heteroatoms.

5. 3. 1. The H-transfer to Oxygen of Carbonyl group.

The clear classification of the photochromic o-Alkyl aromatic carbonyl compounds XVIII and desciption of the data of their photochromic properties in the crystalline state have been carried out in the reviews /7, 23/(Sch. 5. 22).

The first observation of the crystalline photochromism of the o-alkyl carbonyl compounds has been reported for alkylbenzophenon derivatives (XVII) in 1969/61-63/ and 1970 (in the crystal-line state)/64/.

After twenty years the crystalline photochromism has been found out of some alkylisophthalaldehydes (XVII_1a)/65/.

The colorless crystals (in the dark) turn to red upon exposure to sunlight for about 3-5s and become completely colorless in about for 5 min after storage in the dark.

According to the results of the X-ray crystallographic study of the colorless form there are no any significantly short intermolecular O...H distances in the crystalline packing.

At the same time the intramolecular distances between carbonyl oxygen and ortho-methyl hydrogen atoms are 2. 54 and 2. 58 Å which are considerably shorter than sum of the Van der Waals radii of the Oxygen and Hydrogen atoms (2. 72Å).

Thus the geometrical arrangement of the reaction site including its almost an ideal planarity derived from the X-ray crystallographic studies meets the requirements for a Norrish type II reaction involving the intramolecular y-hydrogen abstraction process followed by the photoenolization with formation of the colored structure.

The next step can involve, from our point of view, the cis-trans isomerization about C=C double bond (angle α) with the generation of one more colored structure.

However the additional spectral and kinetic information is necessary to confirm the existence of such a colored structure.

The photochromism of 2, 6-dichloro-4methyl-3pyrinecarboxaldehyde(DCMPA)XVII_1b in the crys-tal has been observed in 2000 /66/.

The colorless crystals of DCMPA turn in deep orangered about 3-5s upon exposure to the light and return to the initial colorless form in the dark at room temperature for about 40min.

The geometric parameters of the colorless structure obtained by the X-ray crystallographic studies, meet the conditions for the y-Hydrogen abstraction process with formation of the colored Enol structure(Sch. 5. 22b).

Unlike XIIa the possible cis –trans isomerization around C=C double bond of XVIIb (PCF) is prevented by the steric interactions with the bulky ortho Cl-atom. The prolonged irradiation in air of the solid PCF product results in the destruction of the crystal.

The idea of the stabilization of the photo-Enol structure through the Intramolecular O-H...O hydrogen bond has been realized by the synthesis and analysis of the photochromism of the crystalline o-anisaldehydes XVII_1c /67/.

These compounds are changed to a brick-red color upon the brief UV irradiation, and the red form which is attributed to (E)-xylyenol is remarkably persistent (unlike XVIIa, b) for several hours(c. a. 10h in the case of R=CN) owing to stabilization of the PCF structure by the O-H...O hydrogen bond. In contrast, p-anisaldehydes undergo irreversible cycli-zation.

At last it has been reported recently/68/about ethyl 4-formyl-1, 3-dimethylpyrazole-5-carboxylate (XVIId) as a new class of the T-photochromic crystals.

The coloration from colorless to red upon 360nm light has been observed in the crystalline state at room temperature. The PCF structure is highly stable to irradiation of the visible light but can return to the initial form either by melting or dissolving into different solvents. It is considered that the primary step of the photo-chromic reaction might be abstraction of the hydrogen atom on the C-3 methyl group by the carbonyl oxygen atom of the formyl group(Sch. 5. 22d).

Thus the described results concerning the photochromism of carbonyl compounds XVII_1a-d are explained by hydrogen abstraction from C-atom and its addition to oxygen atom of C=O group.

However it has been concluded indirectly on the base of the structure of the initial form in the crystalline packing only. While the evidence of the intramolecular mechanism of the photochro-mic reaction has to be based on its occurrences also in the amorphous solid matrixes or in the liquid solvents.

5. 3. 2. The H-transfer to Oxygen of Nitro-group.

The generation of the short-lived colored forms of ortho(2)-nitrotoluene XVIII_1 has been observed first (1960, 1963) when its aqueous solutions were exposed to UV light with use of the flash photolysis /69, 70/(see also /23/), and it has been shown that the spectral and kinetic characteristics of both XVIII_1 and corresponding 2, 4 Nitrosubstituted structure were identical.

Therefore it has been proposed that the Aci structure (Ac) is formed (Sch. 5. 23) with participation only the 2-nitrogroup by the abstraction of H-atom from CH3 group and its addition to Oxygen of the 2-NO2 group in the Nitro-form XVIII_1(N).

It can be supposed that just this AC-structure is short-lived colored species in the aqueous solutions at low pH. At high pH the Anion structure (A)represents the additional colored species in the solvent.

Thus anion is one of the photochemically produced colored species /70-72/. Although the direct evidences for such a mechanism of the photochromic reaction have not been given, the participation of the

Anion structure (A) can be shown by running the photolysis in D2O and posterior observation of deuterium into Methyl group/73/.

The colored Ac-form can be formed also by the photolysis of the N-structure in the solution at low temperature, and further photolysis can lead to another product/74/.

This photochromic mechanism is common also for dinitro(2, 4) phenyl methans XVIII /72, 77/ in which Hydrogen is transfered from methylene Carbon to Oxygen of the ortho (2) nitro group producing a colored aci-quinoiud structure in equilibrium with its anion. (Sch. 5. 23).

The electrophylic substituent R has to promote the increase of the ionizing ability of the central C-H bond without interfering with light absorption of the dinitro phenyl moiety entering into conjugation with the π-electron system of the quinoid structure of aci-form.

Thus the photochromism with H-transfer from Carbon atom to Oxygen one of the Nitro group unlike that to Oxygen of the Carbonyl group involves the formation side by side with the colored aci-quinoid structure also anion one.

It is necessary to notice that the latter has to be included in the close ionic pair with cation of the solvent molecule ((e. g. H2O), involving H⁺, to provide the high rate of the reverse dark reaction.

An analogous Hydrogen abstraction of o-nitrosotoluene (XVIII_1a, Sch. 5. 23) with formation of the colored form under irradiation has been also observed/75/.

5. 3. 3 The ESIHT from Carbon atom to Nitrogen one (O-Alkyl Aromatic Imines).

The UV irradiation of N-acetyl derivatives (R =Alk) of o-methyl (R_1=H) (XIXa) and R2 =Ph (XIXb) (Sch. 5. 24) at 77K in unhydrous propan-2-ol-methanol (1:1) glass results in appearance of a new long-wavelength absorption band with the two vibration maxima typical for the quinoid (keto) azomethine molecular structure(see Ch. 2 and 3) which is strongly shifted towards red($\Delta\lambda\approx$30nm) from 405 and 430nm to 430 and 480nm with the replacement of R =Me by Ph for XIXa and XIXb correspondingly /78/.

The spectral location of the absorption band and its strong red shift with the above structural changes (XIXa->XIXb) correspond well to the idea of the quinoid form of the colored species like anils and o-alkylbenzophenone /61-63/(see Sec. 5. 4. 1).

Like SA the colored form is stable at low temperatures and reverts to the initial structure at the room temperature in the liquid solvents and the photochromism is not observed on steady-state irradiation.

At the same time the similar quinoid transient species at room temperature in heptanes solu-tions on flash photolysois /23/and stable species in the glassy IPP solvent at 77K on steady-state irradiation /8/ have been observed for o-methylbezylidene aniline (XIXc). However no similar photochromism is observed for unsubstituted and o-methoxy benzylideneanilines/8/.

Thus the photochromism of imines XIXa-c is caused most probably by the ESIHT from Carbon atom of o-alkyl group to Nitrogen atom of the C=N group with the generation of the quinoid colored form(Sch. 5. 24).

5. 4. The double-step ESIPT(ESIHT) from C atom to both O and N ones. The derivatives of 2-(2', 4'-Dinitrobenzyl) pyridine (DNBP).

Historical introduction.

The derivatives of 2-(2', 4'-Dinitrobenzyl)pyridines (DNBPs) (Structures XX Sch. 5. 25) attract the special attention from both the scientific and the practical points of view. (see/79/and references below).

When irradiated with near ultraviolet light ($\lambda \approx 200\text{-}300$nm)the crystals and solvents of DNBPs undergo the characteristic intensive reversible changes of the color from pale-yellow(λabs, max≈ 254nm) to blue-purple(λabs, max≈ 600nm)which can persist in dark at room temperature during period of 4-5 hours in crystals and strongly solvent dependent (T-photochromism).

There is also more short-wavelength, short-lived photocolored forms (λabs, max≈435nm) which can be observed mainly by the impulse irradiation (L-photochromism).

Although the photochromic structures XX(1) is known from as early as 1925 /79/, the study of such structures has been started only in the early sixties /80-87/, continued in 70-80th /88-102/, and now are in progress.

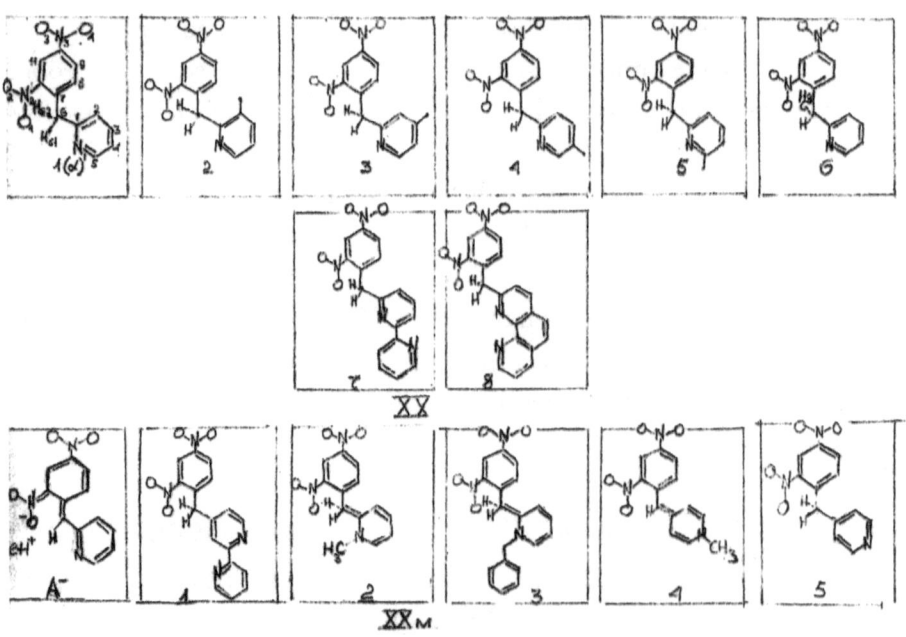

Scheme 5.25

ESIPT Photochromism

Table 5.16
The structural parameters of the compounds XX.
(a) Initial structure (CH$_2$ form).

Struct.		Method	Pyrid.—phenyl (deg)	o-Nitro Phenyl (deg)	p-Nitro Phenyl (deg)	N$_1$-H$_{61}$ (Å)	N$_1$-H$_{62}$ (Å)	O$_1$–H$_{61}$ (Å)	Ref.	Notice[1]
1		X-ray	65.1	31.7	12.30	3.02	3.06	2.35	/103/ /104/	+
		X-ray	66.8	-----	-----	3.03	---	2.37	/87/	+
		X-ray	-----	30.4	40.2[2]	-----	-----	2.28	/108/	
		Cal PM$_3$	87.6	-----	119.9[3]	2.81	-----	2.40	/107/	+
2	A	(X-ray, NMR)	56.1	40.8	14.2	3.12	3.01	2.47	/109/ /111/	+ <310K
	B		68.3	29.7	8.2	3.13	2.95	2.38	/109/ /111/	-- T.>320K
3		X-ray	56.9	41.7	15.3	2.95	3.12	2.35	/110/ /104/	+
4		X-ray	90.7	161.3	167.5	3.03	3.04	2.40	/104/	+
5		X-ray	55.0	40.1	12.3	2.96	3.09	2.43	/104/	+
6		X-ray	~115	~36-44	63-71	~3.2	----	2.47	/110/ /104/	+ The 2non equiv.mol. in same unit sel
7	α	X-ray	~57-58	36-37	18-21 106.8[2]	2.90	>3 3.2	~2.35	/103/ /112/	+ See also /114/
	β	X-ray	65.1	23.3	15.5	>4.00[4]	3.19	~2.4	/103/ /112/	--
			65.1	21.8	25.5[a]	>4.00[4]	4.00[4]	2.4	/112/	--
	γ	X-ray	51.1	35.5	113.9[3]	2.98	3.05[4]	2.47	/112/ /113/	+
8		Calcul.	86.8	----	111.3[3]	3.02	---	2.34	/107/	+
		PM$_3$	81.3	----	112.1[3]	3.36	---	3.78	/107/	+

1/ +,- Photochromic and nonphotochromic (The room t-re) crystals respectively. 2/The angle C$_6$-H$_{61}$-N$_1$ 3/The angle C$_1$—C$_6$---C$_7$. 4/The distance N$_1$-O$_1$.

(b) Parameters of the Photocolored structure (calculated by PM3-MRD-CI method).

Structure	OH form				NH form				Ref.
	Pyr-Ph (degree)	C$_1$-C$_6$-C$_7$ (degree)	O$_1$-H$_1$ (Å)	N$_1$-H$_1$ (Å)	Pyr-Ph (degree)	C$_1$-C$_6$-C$_7$ (degree)	O$_1$-H$_1$ (Å)	N$_1$—H$_1$ (Å)	
1	2.6	130.2	0.978	1.730	78.8	126.5	2.645	0.994	/107/
7	4.6	130.4	0.980	1.728	87.3	125.5	3.016	0.995	
8	5.1	130.8	0.982	1.743	129.9	126.4	4.820	0.996	

Table 5.17

Intermolecular structural parameters of the crystal molecular systems XX

Parameters→ Structures XX ↓		The angles(degree) and the distances(Å)						Refer.	1) Notices
		Ph-Ph′ Å	Ph-Ph′ degree	Ph-Pyr′ Å	Ph-Pyr′ degree	Pyr-Pyr′ Å	Pyr-Pyr′ degree		
1		4.76	15.9	5.89	65.15	>6	--	/103/	+
3 and 6		≈4.3 2)	---	3.7 3)	3.9 3)	4.31	0	/110/	+
7	α,γ	5.16	~0.05	>6	---	3.8	13-15	/114/ /112/	+
7	β	5.62	0	3.7-3.8	5.6-7.4	5.76	~6.2	/114/ /103/	-
Parametes→ Structures ↓		The distances(Å)						/109/	Decay of NH str.
		O_1-O_1'	N_1–O_3'	H_{Me}-O_2'	O_3'-H_2	O_1'-H_1	O_1'-H_2		
2		3.48	3.7	2.5	2.6	2.8	>4		293K slow
2		>4.0	3.5	2.5	2.9	>4	2.8		343K fast

1)+ and – are photochromic and nonphotochromic crystals respectively.2)The data for the structure 6. 3)The data for the structure XXM.4)The strokes marked with the relevant atoms of the adjacent molecules in the crystalline packing.

Table 5.18[1/]

Structural parameters of the transient states (calculated by PM3 /107/ [2/]

Transient structure → / Struct.param.↓		OH→CH_2 (OC)	OH→NH (ON)	NH→CH_2 (NC)
Distance (Å)	C_6-H_{61}	1.51	2.66	1.65
	N_1-H_{61}	3.24	1.22	1.28
	O_1-H_{61}	1.11	1.22	1.81
Angle (deg)	C_1-C_6-C_7	121.3	134.0	121.3
	Ph-∠-Pyr	-147.4	1.5	134.5

1/See scheme 5.26b and compare with the corresponding data of the table 5.17.
2/The values are rounded to the nearest hundredth (Å)and the tenth (degree).

The scheme 5.26 represents the courses of the reverse reactions of discoloration ,supplements the data of the tables 5.16 and 518 and visually pictures both the base (CH_2,OH,NH) and the transient (OC NC,NH) structures.

Scheme 5.26

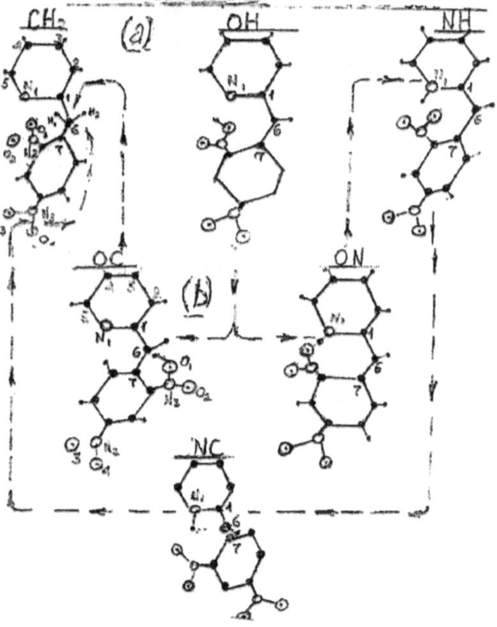

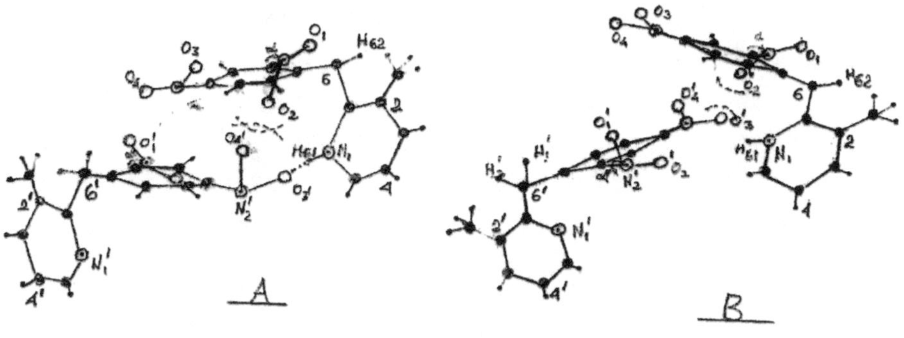

ESIPT Photochromism

Scheme 5.27

The suggested NH structure and location of the neighboring molecules XX(2) in the phases A and B. On the basis of the data /111/(fig1) with taking into account experimental/108,120-122/ and calculated /106,107/ findings. See also tables 5.16a,b ; 5.17;and schemes 5.26a,5.32c.

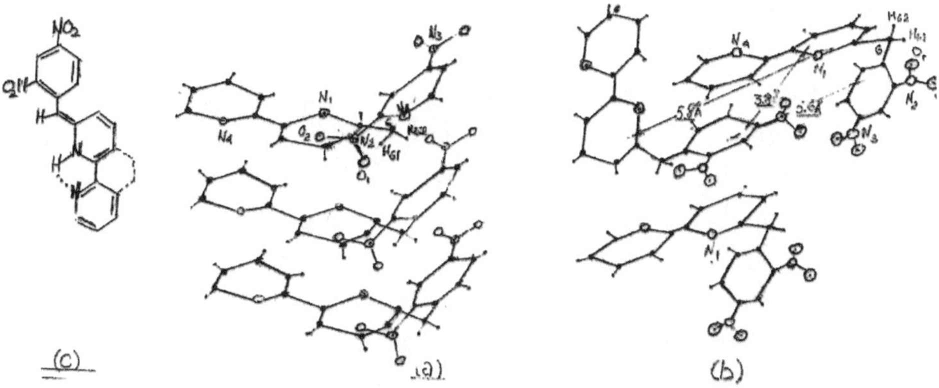

Scheme 5.28

The molecular and crystalline structures of three polymorphs of the compound XX(7): α and γ- two photochromic forms differing a little (a) and form β—the nonphotohcromic one (b). (see also tables 5.17;5.18) and/103,111,112/.
The (NH)structure with the Intramolecular H bond (C).

It has been shown that observed photochromic transformations can be described by the general qualitative structural scheme /100/ (Sch. 5. 30) as a result of the photoinduced or thermal activated proton transfer (ESIPT or GSIPT respectively) CHOH, CH⇔NH, OH⇔NH between C, O, and N atoms of the CH_2, NO_2, and N-pyridine groups correspondingly in the structures XX.

Although this scheme is widely practiced it however does not reflect completely the peculiarities of the structural transformations and gives no idea of their energetics.

In this section the up-to-day findings and ideas of the photochromic transformations in the compounds XX have been presented involving the detailed structural-energetic scheme of the photochomic and accompanied processes in the solutions and crystals.

5. 4. 1. Structure and electronic spectra of the initial form in solvents and crystals.

The very similar findings of XX(1) (CH_2)molecular structure by both the X-ray structural analysis earlier /87/, much later /103, 104/, NMR study /105/, and by the quantum-chemical calculations for the gas state and the crystal /106, 107/are evidence of the strong acoplanarity of the pyridine and nitro-phenyl rings (angle>65°) and also the nitro-group and the phenyl ring (angle>30°)(Tab. 5. 16) which can be caused by the considerable contribution of the sp³ hybridization of the C_6 orbitals and the steric interactions respectively.

The crystalline environment of the molecule XX(1) does not affect greatly the relative energies of the CH_2 and other isomers, and the packing of the molecules in the crystals is loose enough, so a strong effect of the molecules on each other is not to be expected /106/.

The data of the influence of the substituents in the structure on the structure of the reaction site and the arrangement of the of the crystalline packing obtained by various scientific groups are produced in the tables (Tab. 5. 16, 5. 17 with corresponding references)and the schemes for the special cases(Sch. 5. 27, 5. 28).

In all the structures the pyridine-benzene interplanar angles vary somewhat along the series XX(1-6), and plane of the nitro-group deviates from the plane of the benzene ring so that the o-nitro-group and N_1 atom of the pyridine moiety are positioned on the same side of molecule.

However in spite of significant variation of the angles between the pyridine and phenyl rings (from 60 to115°) with insertion of the CH

$_3$ group in the various positions of the pyridine ring (XX(3-6)), the oxygen atom of the nitro-group and nitrogen atom of the pyridine ring are in the close contact to one of the benzylic hydrogens, i. e. interact with the same (H_{61}) benzylic hydrogen, and distances between H_{61} and O_1 and N_1 atoms change a little((2. 35-2. 40 Å) and (2. 95-3. 00Å)) respectively.

A great intermolecular distances between the adjacent molecules (>4. 3Å) and their unparallel mutual disposition in the crystalline packing of the compounds XX(1, 3-6) (Tab. 5. 17) prevent the strong intermolecular interactions and create the considerable free space providing the feasibility of the conformational transformations of the separate molecules in the crystalline packing.

At the same time the insertion of the 2-Me group (XX(2)) promotes the origin of the two different forms of the crystalline packing (polymorphism)(Sch. 5. 27) /109-111/differing strongly in the distance between the atoms of the reaction sites of the adjacent molecules (Tab. 5. 17) with almost the same intramolecular parameters (Tab. 5. 16).

The addition of the second pyridine ring (XX(7))results in the formation of the three polymorphous structures (Sch. 5. 27)/103, 112, 114/. In this case the two very similar but not identical forms (α, y)differ strongly from the third (β)one in the intramolecular parameters(Tab. 5. 16).

The o-nitro group and nitrogen N_1 are positioned on the same and opposite sides of the methylene bridge in the α (or γ) and β forms respectively.

Corresponding angles (C_6-H_{61}-N_1) are >100°and 25° for α (y) and β forms respectively.

In the latter the o-nitro group is remote from both the pyridine ring (N_1...O_1>4Å) and the H_{61} atom of the methylene bridge (O_1...H_{61} >3. 5Å) while N_1...O_1=3. 2Å and O_1...H_{61}=2. 3Å for α (or y) forms.

The crystal structures of the β form is also significantly different from those of α (or y) one (Sch. 5. 28). The distances and angles between the phenyl and pyridine rings of the adjacent molecules are >6Å and 3. 8Å for α (or y)and β structures respectively, and they almost parallel.

So there is partially tight stacking between the pyridine and phenyl rings of the adjacent molecules in the β form.

The difference in the photochromic properties of such polymorphic forms of the initial molecular structures XX(2) and XX(7) is manifested very distinctly (below, Sec. 5. 55, 56).

The very broad long-wavelength absorption band of the initial CH_2 structure(λmax, abs≈250nm)/103, 105/in the solvents, poly and single-crystals stretches to about λ≈400nm securing the excitation of the So--S_1^* transition under various conditions (Tab. 5. 19).

One can suppose this band being superposition of the electronic transitions is located mainly on the different acoplanar moieties with the pyridine and nitro-phenyl rings, and the most long-wavelength region about 330nm is caused by the CT-transition from the CH_2 group towards the two rings with N atom and the two nitro-substituents.

The CT transition is secured by the considerable interactions of the σ(C-H), π(rings), n(N), σ(N-O), n(O) orbitals due to the acoplanarity of the corresponding molecular moieties.

The estimation of the relative positions of the ππ*(CT) and nπ* states on the base of the calculated and experimental data for pyridine and nitro-phenyl substituted structures/107, 118, 119/shows that S(nπ*), T(nπ*) states localized on the N(pyridine) and NO_2 groups are positioned above and under S^*_1(ππ*(CT)) state correspondingly (Sch. 5. 32, inser. CH_2).

The spectral position of the long-wavelength So--S_1*transition is weakly dependent on the molecular structure and medium (solvent nature or crystal structure) but shifted markedly towards lower energy ($\Delta\lambda$≈40nm) with the extension of the π-ring electronic system owing to the addition of the second pyridine ring.

5. 4. 2. The absorption and the structures of the photocolored form(PCF) in solvents.

The nature of the ESIPT (or ESIHT) in the molecules XX is closely bound up with the structure of the initial (CH 2)(λmax, abs >250-300nm)(see Sec. 5. 51)and final(proton transferred, PT, photocolored transient forms which can be characterized by bands with λmax, abs≈550-580nm (long-lived transient I), λmax, abs≈370-410nm(very short-lived transient II), and λmax, abs≈480-490nm(short-lived transient III).

On the base of the comparative results obtained for the compound XX and the reference structures in the sixties and the early seventies by UV-VIS absorption measurements /72, 76, 82, 83/, the IR spectra of the irradiated XX(1)/81/, and the semi-empirical quantum-chemical calculations/123-125/it was difficult to understand unambiguously which a structure–enamine, "NH" /76/ or aci-nitro "OH" –anionic (O)/123-125/ can be attributed to the "blue" transient I.

However the investigations which have been carried out later have led to the clear interpretation of the I-III transient structures.

For the interpretation of the transient I-III in the solvents the combination of the spectral and structural methods with the quantum-chemical calculations have been used with utilization of the model structure XXM(Sch. 5. 25).

5. 4. 2. 1. The long-lived transient (I).

A. The UV-VIZ absorption spectra.

The transient I is not detected among the photocolored structures of the reference compounds without the pyridine ring (tetranitrophenylmethanes)by the nanosecond flesh photolysis in the nonpolar solvents under the similar conditions /89, 91/.

At the same time the transient absorption band of XX(I) (560nm) arising by the flashphotolysis/94/or steady-state irradiation in ethanol at 173K/105/ and that of the reference compound XX(M2) are very similar and insensitive to the solvent polarity.

At the same time the absorption band of the transient I is quite weak in the reference structure XX(M5)which is favorable to the formation of NH-structure.

Thus the date presented above are strongly favorable to the assignment of the transient I to the low-polar NH quinoid (not Zwitterion) structure but not to the OH(acinitro)one.

A. The IR-spectra.

The IR-spectra of XX(1) have been recorded in NACL and KBr pellets after irradiation (t=4min) by the filtered light (250-380nm) of the 150 w xenon lamp of the samples with the very high degree of photoconversion into the colored form (30-40%)/105/ unlike the previous studies /81/which results were very doubtful due to the very low conversion to the PCF (1-4%).

Additional support comes from the comparison with the IR spectra of the reference structure XX(M2)/105/.

The most important results as following:

(i). The appearance of the line at 3338cm^{-1} after irradiation which is assigned to the NH form;

(ii). The intensity drop of the band at $\approx$1567cm^{-1} after irr-adiation which is not present in the NH form and also in the model structure XX(M2M);

(iii). There are the two common characteristic bands of the irradiated XX(1) structure and of the reference XX(M2) one which are absent of the non-irradiated XX(1) at 1290 and 1635cm^{-1}. The latter is attributed to the NH-structure with the quaternary N in the Zwitterion structure already proposed earlier /86/, and corresponds to the band at 1640cm^{-1} of the reference structure XX(M2);

(iv). The isotope shift of the IR band from 3388cm^{-1} in XX(1) to 2513cm^{-1} in d XX(1) has a magnitude expected for the variation from N-H to N-D.

Thus according to the data of the IR spectroscopy the "blue" colored form I(λmax, abs$\approx$560nm) is attributed to the NH form which is represented by superposition of the two limited configurations, Quinoid and Zwitterion, with the small contribution of the latter (see also /93, 124/) that gives an explanation of the relative insensitivity of the absorption band (λmax, abs$\approx$560nm) to the solvent polarity.

C. The Resonance Time Resolved Raman Spectroscopy (RRS).

The RRS spectra are measured in ACN solvent at room temperature under pumping($\approx$308nm) and probing into the band with λmax, abs$\approx$560nm (e. g. λ=485nm). The two bands at 1303cm^{-1} and 1635cm^{-1} observed in RRS spectra were considered to be characteristic of the transient I structure.

The 1635 cm^{-1} band (see above, the data of the IR spectroscopy) was assigned to the C=C stretching of a quinoid of the azamerocyanine form /97/.

A comparison with the RRS of the irradiated XX(1) and XX(M5) in methanol and with the corresponding N-methylimines as well as with the molecules having respectively the structure of the quinoid form and of the aci-nitroform /99/ have led to also to the conclusion about of "NH" nature of the transient I(550nm).

It has been shown /115/ that RRS spectra of the transient strongly resemble that of the model molecule XX(M2) which has quinoid NH structure.

Thus the data of the RRS spectroscopy are evidence of the "NH" structure of the long-lived transient.

D. The PMR spectra.

No ^{1}HNMR spectra of the PCF of XX(1) in the solution could be obtained because of the too short time for the recording of the spectra.

However the assignment of the PCF of the irradiated phenantroline derivative XX(8) is possible owing to the long lifetime (about several hours) and comparatively high conversion (about 8%) in toluene d at 213K.

Such a PCF structure with the absorption band (λmax, abs≈500nm) is the same that of XX(1) as well as of the model compound XX(M3)/105/.

The singlet at 6.6 ppm observed in the photoproduct of XX8 can be assigned to the H 6 2 proton of the enamine double bond appearing also at 6.2ppm of the reference structure XX(M3). Moreover the analysis of the NMR spectra taking into account the proton-proton interactions (2DNOESY ^{1}HNMR spectra) indicates that the cis configuration is dominant in XX(M3). One can also be assumed the same NH structure for another XX(1-6) compounds.

E. The quantum-chemical calculations.

The calculations have been carried out by the semiempirical PM3 –MRD-CI (Energy states) and PM3-SCF(Structures) methods. The location of the absorption band (S_o—S_1^*)(λabs, max≈534nm) for the calculated NH structure /107/ and the experimental data of the absorption band positions for the transient I in various solutions (λmax, abs≈530-580nm)/89, 91, 105, 115, 116/are very similar (Tab. 5. 19). Therefore the band can be attributed to the NH structure.

The distinctive features of the NH form are the very short N-H distance and acoplanar structure similar to the initial (CH 2) one (Tab. 5. 16 b, Sch. 5. 26a).

The structural characteristics depend insignificantly on the methyl substituents in any position of the pyridine ring and the bridge(structures XX(1-6)) but the Pyridil –Phenyl angle and especially the O-H distance depend markedly on the nature of the pyridine moiety (structure XX(7, 8), Tab. 5. 16A).

At the same time according to the data of the calculations the contribution of the Zw-ion structure in the NH form is very small that contradicts to the findings of the vibration spectra but corresponds to the data of the UV-Vis spectroscopy(see above).

Thus the all above described results point clearly to the NH structure as the long-lived "blue"
transient I.

5. 4. 2. 2. The very short-lived transient (II).

Interpretation of the transient II nature is more difficult due to its very short lifetime and the strong overlap of the II(λmax, abs $\approx$370nm) and III(λmax, abs$\approx$490nm)absorption bands.

Therefore although the RRS spectrum of the transient II was measured by probing at 390nm (solvent ACN)which is close to the absorption band peak($\lambda\approx$370nm)of the transient II/115/, its identity is not clarified by the Raman spectra.

However the NOH (acid)form (Sch. 5. 26) is the only transient precursor of A$^-$ (Anion) structure (Sch. 5. 32b), and therefore the transient II must be assigned to the acinitro acid isomer (OH)/115/.

Really the decay and the rise times of II and III correspondingly are almost identical. In addition the bands of II and III are very intensive, and I one is very weak at the early stage of the photochromic reaction for the reference molecule XX(M5) which structure is very favorable to the formation of the OH form as compared with the NH one/115/.

The assignment of the transient II to the OH structure can be based also on its sensitivity to the polarity and pH of the solvent as a result of the OH$\Leftrightarrow$A$^-$ equilibrium by the analogy with the photochromism of 2, 4-dinitrotoluene /72/.

The position of the long-wavelength absorption band of the OH structure calculated by the semiempirical method (λ=454nm)/107/and that of the transient II in the solvents (λmax, abs$\approx$370-450nm)/89, 91, 105, 115, 116) (Tab. 5. 19)are sufficiently close.

The typical structural characteristics of the OH-form(Tab. 5. 16b, Sch. 5. 26)are a very short distance O-H, not a very long N-H distance, and almost planar structure (Phenyl-Pyridine ring angle $\leq$5°, i. e. almost complete sp^2 hybridization of the C6 orbitals)unlike the initial (CH2)and the colored (NH)forms.

The structural characteristics of the reaction site don't depend practically on the substituents and the nature of the pyridine moiety.

5. 4. 2. 3. The short-lived transient (III).

The transient III is closely bound up with the II one which is interpreted as the acinitro structure and represents a precursor of III under conditions which is favorable to the acid dissociation and to the shift of the OH⇔A⁻ equilibrium towards A⁻ structure.

The transient absorption III with the band of λmax, abs≈490nm is assigned to the deprotonated anion of XX(1) ((NO⁻) or(A⁻)) by comparison with the spectrum of the chemically produced anion /83/.

The RRS of the transient III which are calculated from the spectra obtained by excitation at 308 and probing at 465nm are identical to those of the acinitro anion (XX O⁻) of XX(1) produced by addition of the base/115/.

In the model structure XX(M5) which is favorable to the formation of the OH and A⁻ structures in contrast to XX(1), the absorption bands II($\lambda \approx 370nm$) and especially III($\lambda \approx 480nm$) are very intensive, and I($\lambda \approx 565nm$) is quite weak/115/.

At last the intensive absorption band of III ($\lambda \approx 490nm$) which arises along with the decrease of the bands I, II after irradiation of the dry toluene solution of XX(1) with cryptand (C 211) is caused by the formation of anion A⁻ by deprotonation of the OH structure in the cryptand cavity/116/.

Thus the totality of the above adduced findings shows the origin of the three photocolored structures in the solvents.

The two from them are generated as a result of the Intramolecular Proton Transfer and the third one is formed by the acidic dissociation (deprotonation) in the ground state especially under favorable conditions.

5. 4. 3. The kinetics of the generation, and the interplay of the colored forms in the solvents.

It has been supposed in the earlier experiments /79, 80/ and supported by a lot of the more recent ones (e. g. /76, 81, 83, 89, 95, 96, 98/) that the photochromic reaction of XX in the solutions is caused by the Intramolecular Proton Transfer from the methylene group (CH_2

)in which one of the three (OH, NH) and (A⁻) structures with the absorption band I, II, III respectively (see 5. 5. 2.) is produced.

However it had not been known exactly until nineties which was the primary step of the photoreaction, how many and in which sequence the products were been formed, and which a structure is responsible for the "blue "(λmax, abs$\approx$560nm) colored form(see 5. 5. 2).

The key experimental data for the understanding of the mechanism and kinetics of the photochromic reaction of XX(1) in the solvents have been obtained by the time-resolved absorption and the Resonance Time resolved Spectroscopy in the polar (ACN) and nonpolar (toluene) solvents with the utilization of the reference compounds and the proton acceptor species (cryptand C211 or TEA)/115, 116/(see Sch. 5. 32b).

The absorption band II ($\lambda\approx$370nm) increase follows exponential law with the very low bimolecular contribution /116/. Therefore the photocolored (OH, acinitro) structure is produced by the direct ESIPT (CH-->NH) solely in the S_1^* state of the initial (CH 2) form with the sufficiently high rate.

The rate grows strongly with increase of the solvent polarity (from $k_1'\approx 10^6 s^{-1}$ in toluene to $k_1'>10^9$ s^{-1} in ACN)(Tab. 5. 20)/115, 116/as a result of the stabilization of the excited (OH*) Zwitterionic structure generated by H$^{+\delta}$ (almost Proton but not Hydrogen transfer).

The transient absorption band III of anion (A⁻)(λmax, abs$\approx$480-490nm)arises and its intensity increases in the polar solvent (ACN) by the exponential law with τ(rise)$\approx\tau$ 4 $\approx$5 µs that corresponds to the (OH) structure decay time (τ(decay)$\leq$5µs)(Tab. 5. 20)/115/.

This band is not detected in the nonpolar solvent (dry toluene) without appreciable (c <10⁻³M) amounts of the proton acceptor (cryptand C211 or TEA)/105, 116, 126/.

The new transient absorption band (λmax, abs $\approx$490nm)appears with the increase of the cryptand concentration, and is clearly observed also with the equal concentrations of XX(1)(10⁻³M)and TEA.

In the equimolar (10⁻³M)solution of XX(1) and cryptand C211 in addition to the fast (τ<10⁻⁵ s) initial rise, a slower(τ(rise)$\approx$10⁻⁴ s) growing component is observed which parallels the (NH) decay(Tab. 5. 20)/116/.

The experiments described above, clearly demonstrate the generation of anion(A⁻) from the trans (OH) structure, especially promoted by the proton acceptors which can be also an intermediary in the formation of anion from (NH) structure.

Thus the photochromic reaction of the structures XX in the solvents results in not only the generation of the "blue" form (see below) but also in an equilibrium between aci-nitro structure (OH) and its anion(A⁻) which is being shifted towards the anion in the basic media similar to other ortho-nitro photochromic derivatives /76, 101/.

The increase of the transient absorption band I which is attributed to the NH("blue") photocolored structure can be describe in the polar solvent (ACN) by various exponential processes with the three different times(τ_1, $\tau 2$, $\tau 3$)(Tab. 5. 20)/115/.

(i). τ_1(rise)<10^{-7} s which is temperature independent and below of the lowest limit of the time resolved setup ("instant" process) can be attributed to the direct formation of the (NH) structure by the ESIPT.

(ii). 2 (rise)≈2×10^{-6} s (k 2 ≈5×10^5 s^{-1}) corresponds to the decay time of the transient band II attributed to the (OH) form (Tab. 5. 20).

At the same time for the reference structure XX(M5) in contrast to XX(1) the relative intensity of the bands is II(365)/I(570)=1. 2 (very high) at the early stage of the photochromic reaction and falls sharply (till zero) within 5μs so that the only I absorption band (λ≈570nm) remains.

Such a time dependence shows that the direct route via CH->N transfer is missing for XX(M5), and only the one path-way (CH2)->(OH)->(NH) occurs. The RRS spectra give the same results.

Thus for XX(1) in addition to the direct path-way (CH2)->(NH) the route (CH2)-k_1->(OH)-k2 (GSIPT)--> (NH) can occur with the high probability, and conversion (OH)-k2-(GSIPT)->(NH) takes place much faster then dissociation to the form(A⁻) (k 4)(k2<k_1, $\tau 2$ >τ_1, Tab. 5. 20)/115/.

(iii). At last $\tau_3 \approx 30\mu s$ ($k_3 \approx 3.3 \times 10^4 s^{-1}$) is very close to the decay time of the anion(A^-) structure, $\tau(dec) \approx 40\mu s$ ($k \approx 2.5 \times 10^4 s^{-1}$) and therefore can be attributed to the reaction $(A^-)-k_2->(NH)/115/(Tab. 5. 20)$.

In the nonpolar solvent (toluene) /116/ the rise of the transient absorption band I follows exponential law with the time significantly longer ($\tau(rise)=10^{-4}-10^{-5}$ s)(Tab. 5. 20) in comparison with τ (rise) in polar solvent and can be attributed to the production of the (NH) structure via (OH) one because the decay time (τ (OH), decay) of (OH) and the rise time (τ(NH), rise) of (NH) are very close under such conditions (Tab. 5. 20).

The observation that (NH) is generated and can be detected even at very high cage concentration (1×10^{-2} M)(Tab. 5. 20) confirms that (NH) doesn't proceed only via (OH) but in part also generated directly from the excited $(XX(1))(S_1^*)$ via short-lived excited intermediate, since the (OH) tautomer as a precursor of (NH) is efficiently quenched in such concentrations of the proton acceptor (C211).

In this case the rise is expected also to have a bimolecular contribution via proton acceptance by analogy with the polar solvent $(A^-)-k_3->(NH)$.

Thus in the nonpolar solvent especially with an addition of the proton acceptor, the (NH) form can also be generated via several parallel routes.

As it has been shown by the measurements of the UV-Vis and IR spectra /105/ the efficiency of the photocoloration under such conditions can research 30-40% for the compound XX(1) but falls strongly (till 8-13%) with the expansion of the π-electron system in the pyridine moiety (compound XX(8))(Tab. 5. 20).

Such a low photocoloration can be caused also by the decrease of the excitation transition probability with the extension of the conjugation by phenantroline group /126/.

5. 4. 4. The kinetics and structures' interplay of the discoloration reaction in solvents.

One can believe on the base of the kinetic findings of the previous section(Tab. 5. 20) in the polar solvent (ACN)the dark decay of the short-lived (OH) and (A$^-$)colored structures is accompanied by the corresponding formation of the long-lived (NH) structure, and their direct transformations to the initial colorless (CH$_2$) one occur with the very low rate(see sec. 5. 5. 6. 4.).

The conversion of the aci-nitro (OH) structure to the (NH) tautomer takes place much faster ($\tau<5\times10^{-6}$ s)than its dissociation to form aci-nitro anion (A$^-$)($\tau\approx5.5\times10^{-6}$ s), and the conversion of the aci-nitro form to the (NH) structure($\tau_3 \approx30\times10^{-6}$ s) is quite slow /115/.

The last stage of the dark bleaching reaction, the GSIPT NH->CH, is the rate-limiting step and proceeds with the very low rate which decreases strongly in going from the nonprotic solvent (ACN, $\tau\approx0.5$s) /233/ to the protic one (alcohol, $\tau\approx5$s)/116/(Tab. 5. 20).

Thus the strong sensitivity of the (NH) structure to the formation of the Intermolecular H-bond with the solvent leads to its stabilization in alcohol.

The absolute instability of the anion structure (A$^-$) and the sharp fall of the stability of the Zwitterionic (NH) structure in the nonpolar solvent (toluene) as compared with the polar one make them undetectable and shirt-lived ($\tau\approx10^{-2}$s)respectively/116/.

At the same time the stability(and lifetime) of the low polar quinoid ground state (OH) structure changes insignificantly in going from the polar solvent to the nonpolar one, and lifetime of the latter becomes smaller by one-two order of magnitude than the rise time of the(NH) structure under the same conditions(Tab. 5. 20).

Therefore the decay of the colored (OH) structure proceed mainly via the direct transformation (the GSIPT OH->CH) to the initial (CH$_2$)form. Thus in the nonpolar solvent at the room temperature the decays of the (NH) and (OH) colored forms can proceed by the two independent routes. Nevertheless the (NH) form under such conditions remains the most stable.

The addition of the proton acceptor cage (C211) or TEA in the nonpolar solvent changes sharply the photochromic behavior of XX(1)/116/, and results in the following changes (Tab. 5. 20):

(i). Disappearance of the(OH) tautomer at the time resolution of the experiment,
(ii). A shortening of the decay time of the (NH) tautomer with the increase of the cage concentration, and
(iii). (The appearance of the comparatively long-lived anion form (λabs, max$\approx$490nm, τ(decay)=10^{-5}-10^{-6} s).

It has been shown by the kinetic investigations/116/that the above findi0ngs are bound up with the following characteristic phenomena:

1. The PT takes place from both(OH)and(NH) tautomers to the cryptand.
2. The PT from (OH) to the cage is diffusion controlled and not reaction rate liming stage.
3. The PT from (NH) to the cage (decay of (NH))occurs with the rate that is proportion-al to the cage concentration with the bimolecular rate constant k$\approx$2. 5x10^7 $s^{-1}M^{-1}$which is smaller than the collision rate by more 2 orders of magnitude and therefore not diffusion limited.
4. The decay of the anion(A^-)is due to the PT from the protonated cage to anion which regenerates the ground state (CH_2) structure.

The accumulation of the (A^-) (λabs, max$\approx$490nm)is a result of the following processes:

While the rise of the anion follows the exponential law($\tau\approx10^{-6}$ s) its decay is the bimolecular process involving equal population of the anion and protonated C211 or TEA.

The estimated rate constant is about 2x10^5 $s^{-1}M^{-1}$ and smaller by more 4 order of magnitude than collision rate. Hence the anion decay rate is limited by the escape of the stabilized proton from the cage as a result of its structural changes.

Such structural changes are necessary to the cage deprotonation that is supported by the findings of the corresponding bimolecular constant for TEA which the lowest limit of 7×10^9 s^{-1} M^{-1} is higher by factor 3×10^4 than rate obtained for the cage.

Thus the adjustment of the reverse PT reaction (i. e. discoloration)rate and supplementary coloration (absorption band with λabs, max$\approx$490nm) can be conducted by addition of the proton acceptor to the solvent /116/.

The use of the sensitivity of the (NH)form stability to the formation of the H-bond (see also above) can be considered as an effective way to the modification of the photochromic properties.

In this connection 1, 10 phenantroline derivative XX(8) had been especially designed and synthesized with a purpose of stabilization of the(NH) structure /126/.

Really the lifetime of XX(8) (NH) tautomer at the room temperature in the toluene solution was found to be increased by a factor 5×10^3 compared to XX(1)(Tab. 5. 20).

The activation energy for the back GSIPT was found to grow from 2. 9 to 8. 2 kcal/mol in toluene solution due to stabiliza-tion of the(NH)structure by 5kcal/mol rather than increase of the transition state energy that is conformed also by the results of the quantum-chemical calculations(PM3)/107/.

This stabilization was attributed to the H-bonding between the transferred proton and the additional adja-cent pyridine nitrogen /105, 126/, and it is sufficient to make the PT state thermally accessible so that this compound is also thermochromic.

The identity of the thermo and photochromic structures has been supported by the UV-Vis absorption and NMR spectra in d toluene at 317, 323, and 343 K/105/.

Such a thermochromism in the liquid solutions that takes place due to the low rate of the GSIPT is an unusual phenomenon for the ESIPT photochromism (e. g. see photochromic anils).

5. 4. 5. Photochromism of the crystals.

5. 4. 5. 1. The structure and the absorption spectra of the colored forms.

Upon the steady-state or impulse radiation of the colorless crystal(single crystal, polycrystalline film)in the region of the absorption band of the initial form (λmax, abs$\approx$250nm)the two types of the new transient absorption bands appear in the visible region –the longlived band in the region of 550-620nm (I) and the shortlived one in the region of 430-480nm (II).

The semi-empirical (PM3)/107/ and Density Functional Theory (DFT)/106/ calculations of the colored (NH) structure give the results that the both initial and (CH_2)and (NH) forms have the essential nonplanar structures (Tab. 5. 16a, b, Sch. 5. 32c) due to the large contribution of the sp^3 hybridization at the methylene bridge by that the two rings are connected.

The first direct evidence of the (NH) isomer with the absorption band (λabs, max$\approx$600nm) (I)has been obtained by employing the two-photon impulse excitation technique (with λ=502nm) for preparation of the diffraction quality single crystal with the high concentration of the (NH) blueisomer (C$\approx$36%) and with suppressed concentration of the (OH) isomer (λabs, max$\approx$435nm) (C<0. 5%)/108/.

The data of the calculations (DFT and PM3)(Tab. 5. 16b) are corroborated with the experimental results completely (Sch. 5. 32c). The (NH)isomer as compared with (CH_2)one shows minor structural rearrangement, and only one dynamic proton H_{61} transferred over distance of nearly 3. 16Å (from C to N_1). As a result of the subsequent structural relaxation, the planar phenyl ring shifts within its plane together with p-nitro group, the relative orientation of N_2-O_1 to the phenyl ring and the rotation around C_6–N_2 bond are changed a little from (CH_2) to (NH) form. Rehybrydization in the 2-benzylpyridyl fragment, leading to the loss of coplanarity of the pyridine ring corresponds to enamine structure of (NH), and remaining proton H_{62} shifts a little (by about 0. 2Å).

Thus the blue (NH)isomer with the absorption band about $\lambda \approx 600$nm differs from the (CH$_2$) initial structure by the proton position and only a little by the positions of the skeletal atoms and shape of the molecule /108/.

The broad absorption band of the (NH)structure for all compounds XX in the crystal is composed of at least the two partially overlapping bands with the different relative intensities and the different polarization (Tab. 5. 19) /103/.

The bands decay homogenously with the identical decay rates indicating that all of the transitions composing this band originate from the same molecular structure. It could be supposed due to the orthogonalllity of the two bands' polarization that the corresponding transitions with the ICT are localized on the separated Nitrophenyl and Pyridine moieties positioned in the almost orthogonal planes.

The peculiarities of the band's behavior can be explained by its complex nature.

Really the marked effect of the methyl substituents (XX(1, 2, 5)) and of the second pyridine ring (XX(7)) on the shift of the λabs, max in the series of the structures XX (5<1<2<7) and on the relative intensity of the partially overlapped bands with the different polarization is observed (Tab. 5. 19).

Such an irregular dependence is hardly connected only with the intramolecular reasons, and can be caused by the above discussed complex nature of the "blue" band which position and shape can be determined by the both the influence of the extension of the π-electron system (XX(7)) and the mutual disposition of the almost orthogonal moieties under influence of the interaction between the adjacent molecules in the crystalline packing.

The very short lifetime and low concentration of the photocolored form with the absorption band II(λabs, max$\approx$430-480nm) don't allow to utilize the direct spectral and structural methods for determine its molecular structure in the crystal.

However the likeness of the spectral and kinetic characteristics of the short-lived form in the crystal and the solvents (see above and Tab. 5. 19) allows to attribute them to the same (OH) structure. At the same time the coincide of the structural and spectral data obtained by the experimental (X-ray analysis, UV-Vis spectroscopy) and the theoretical (PM3 and especially DFT) methods for the initial (CH_2) form /106/ allows to consider the latter as the reliable ones to determine of the (OH) structure.

The DFT /106/ and semiempirical PM3/247/methods yield the same results for the (OH) structure in the crystal which coincide practically with those of the free molecules /106/. According to the calculated findings the (OH) form, unlike (CH)and (NH) ones, is planar (the pyridine-phenyl angle $\alpha \approx 2.6°$ for the structure XX(1) (Tab. 5. 16b) and close to the keto-structure(Sch. 5. 3b) in which the phenyl ring is somewhat acoplanar with the partially double C:::N-bond /108/, and the dynamic proton H_{61} is positioned close enough to the pyridine Nitrogen N_1. The planarity is slightly broken with the π-system expansion in the pyridine moiety (XX (7, 8)), $\alpha \approx 5°$).

The absorption band of the (OH) form in the crystal, unlike (NH) one, includes the single electron transition in the region of 22000-23000 cm^{-1} that is reproduced well by the calculated results ($v \approx 22000 cm^{-1}$, PM3) which represent the electron transition energy for the free molecules(Tab. 5. 19).

That is the electronic intermolecular interactions for the (OH) structure in the cryst-alline packing are negligible at least for the molecules XX(1-5, 7). The small blue shift($\Delta\lambda \approx 10nm$) is connected with the π-system expansion by addition of the second pyridine ring (XX(7, 8), Tab. 5. 19).

The anion structure with the band ($\lambda \approx 480-490nm$) III is absent in the crystal due to its absolute instability under such conditions.

5. 4. 5. 2. The dynamics of photocoloration in the crystals.

The real reversible photocoloration(T-photochromism)in the crystalline compounds XX is bound up with the origin and decay of the long-lived "blue" (NH) structure ($\lambda abs, max \approx 600nm$).

The highest observed degree of the conversion into the (NH) form which has been determined by the absorption spectra of the thin polycrystalline film (0. 5μm) in the band with λabs, max≈ 600nm, as well as from the IR measurements, is at least 16% /105/.

The existence of the different temperature ranges for the generation of the colored forms is the characteristic feature of the crystalline compounds XX((Sch. 5. 31, Tab. 5. 21)/103, 117/.

a) The fast direct (NH) structure formation ($\tau < 10^{-8}$ s) is nearly temperature independent below 100-200K shows the marked deuterium effect($\Phi(H)/ \Phi(D) \approx 20/103/$)including obviously the tunneling ESIPT reaction.

However the drop of the (NH)concentration(optical density at 600nm) at very low temperature(T<30K) shows the very small barrier in the excited state (0. 023kcal/mol) when forming the (NH) structure from the(CH_2) one. Meanwhile the higher barrier (around 0. 4 kcal/mol)has been predicted for that process by the quantum-chemical calculations/107/.

b) At higher temperatures(T>200K)the thermally activated (with the relatively short time, $\tau \approx 10^{-3}$--10^{-2} s) process gradually increases the overall quantum yield (only /103/). This process may be attributed to the "direct" PT from the (CH_2)* excited form to the (NH)one with the barrier which be found to be 1-2. 6 kcal /mol.

c) The slow, temperature dependent, (NH) structure formation (increase of the optical density at λmax≈600nm)above T≈230K with τ≈8-10s at the room temperature does not occur immediately after impulse excitation (the findings adduced in /117)and parallels the decay of the (OH) form concentration(optical density at 435nm)/117/. It points to the existence of the potential barrier about $\Delta E \approx 19$kcal/mol on the path-way of the (NH) form generation from (OH) one in the ground state.

The fast temperature independent formation of the (OH) structure ($\tau < 10^{-7}$ s)below≈150K(CH--hv->OH) involves obviously the tunneling ESIPT reaction /103/results in the coloration with the ve-ry low intensity due to both the small concentration and low absorbance($\epsilon(OH)/\epsilon(NH)\approx 0.1$) of the product.

The slower thermally activated formation of the (OH) structure has complicated nature caused by existence of the only one precursor state (CO). The kinetic analysis shows that the main contribution in the temperature dependence is caused by the overbarrier process of the population of the (OH) structure from the precursor (CO) ((CO)->(OH)) ($\tau \approx 3 \times 10^{-4}$ s) at the room temperature, $\Delta E \approx 2.3$ kcal/mol)(Tab. 5.21).

The identical temperature dependent increase of the overall coloration efficiency(at T>100K) with almost the same activation energy ($\Delta E_{act} \approx 2.6$ kcal/mol, Tab. 5.21) as a result of the population of the colored (OH) form has been observed also in /103/ but attributed obviously insufficiently correctly to the direct thermally activated population of the (NH) form((CH 2-hv---->NH)).

The second process is the competing (CO) state deactivation to the initial structure over potential barrier in the ground state ($\tau \approx 3 \times 10^{-4}$ s at room temperature, $\Delta E \approx 0.08$ kcal/mol) (Tab. 5.21).

Meanwhile the population of the (CO) state which is caused by the ESIPT followed by the small structural deformations, occurs above the potential barrier in the excited S_1^* state ($\tau \approx 10^{-5}$ s at the room temperature).

That process can be distinguished by the treatment of the kinetic data in the temperature range $\Delta T=150-240K$ in which the accumulation of the only (OH) form takes place with the constant concentration of the (NH) form (see/117/).

The manifestation of the two last processes is bound up with the delay of the population of the metastable(OH) form after the ESIPT reaction (CH->OH) clearly indicate the existence of the (CO)-precursor state.

At last the colorless intermediate in the ground state (but it is not the triplet state!) can also be trapped during several minutes after the end of the excitation at the low temperature since just by increasing of the temperature a colored form can be generated (see also /85/).

The kinetic experimental findings of the substituent's influence the generation of the colored forms are very scars and show clearly only the existence of the temperature independent fast direct formation of the (NH) structure from the initial (CH_2) form of the all structures XX with the methyl substituent in the different position of the pyridine ring (Tab. 5. 21).

The variation of the coloration efficiency depending on the position of the methyl substituent $\Phi_1 \approx \Phi 3 > \Phi 2$ /103/ is caused apparently by the crystalline structure which can provide the deactivation of the excited state competing with the ESIPT.

However the addition of the second pyridine ring(XX(7)) can promote the cardinal change of the photochromic properties of the crystals resulting in the existence both the photochromic (XX(7α) and photoinert (XX(7β)) crystalline structures(Sch. 5. 28) and provides the marked modification of the kinetic characteristics of the photochromic reac-tion in the photochromic crystals XX(7 α).

The detailed discussion of the problem concerning the existence of the photoinert crystalline structure XX(7 β) will be carried out in the next section.

In the photochromic crystal XX(7 α) just as in XX(1) the high rates ($\tau \approx 10^{-4}$s) of the (NH) and (OH) structures' generation are temperature independent and undergo the deuterium effect.

However unlike XX(1), the activation energy of the (OH) colored structure formation determined by the temperature dependence of the overall coloration efficiency (T>200K)(see above) (Tab. 5. 21)falls drastically (from $\Delta E_1 \approx 2.6$ kcal /mol to $\Delta E_\gamma \approx 0.95$ kcal/mol), and the overall coloration efficiency increases correspondingly as compared with the compound XX(1)($\Phi_\gamma/\Phi_1 \approx 8$). Such a dependence can be caused by stabilization in XX (7_α) of the almost planar (OH)structure comparing with a nonplanar transient structure (arising from the ESIPT (CH)--------->(OH)) in the molecule with the extended π-conjugated electron system.

That conclusion is opposite, however, to the supposition of dependence of the (OH) stability on the XX(1)-->XX(7_α) structure change/103/.

Table 5.19

The absorption data of the photochromic structures XX (see Sch.5.29)

Media, Method	Longwavelength absorption bands (λ_{abs}^{max} nm) of the structures				Reference	Notes
	Initial CH_2	Quinoid NH	Aci-nitro OH	Anion A		
Polycrystalline layer Steady state and time res. spectr.	250	XX(1)567,592 XX(2)571,621 XX(5)558,608 XX(7)583,642	440 430 433 450	— — — —	/103/	----
Single crystal, α(photochromic) form of the structure XX(7) Steady state and time resolved Spectroscopy.	250	0° 550 I_1 ---- ≈1.0 580 I_2 90° 550 I_1 ---- ≈ 0.8 600 I_2 0 580 I_1 ----≈0.7 620 I_2 90° 580 I_1 ----≈1.4 620 I_2	470	----	/103/ /114/	The data of the absorption with the crossed polarizators are evidence of the two different electron transitions of NH form and possib. of two NH forms
Liquid ACN. Time-resolved abs.,Resonance Raman spectr.	< 308 ex	565	370	490	/115/	For initial form the excitation wavelength is presented only.
Liquid polar. and nonpolar solvs. Nanosec.abs. spectr	—	510-580	390-410	—	/89/	OH and NH - the short and long - time forms resp.
Single cryst.T≤300K Time res.ns abs.spe.	~400 ex	600	435	—	/100/	For initial form Only excit. λ is presnted
Nonpolar solv. ns abs. spectr.	—	550	—	—	/89/ /91/	Longlived form
Metanol Room t-re Steady-state spectr.	253 267	580	~430	~480	/105/	For str XX(8) I_{OH+O}/I_{NH}=0.33
Dry toluene Trans abs.ms,μs	355 ex	546sh 588	<400	490	/116/	Anion appear in In pres.of (C 211)
Polycryst.film Trans a.steady st ab.	355ex	590	434	----	/117/	----
Single cryst. Steady- state abs.sp	~360ex	611	----	----	/109/	For struct.XX(2)
Metanol 173 K Steady- state spectr.	253	~560	360 --370	----	/105/	$I_{OH}^{abs}/I_{NH}^{abs}$=0.2
Quantum S_0-S_1^* nm Chemical S_0 -T_1 nm Calculat. PM3 S_1^*-T_1 cm^{-1}	334 445 7520	534 750 5100	454 800 3200	---- ---- ----	/107/	The satisfactory coincidence of exp. and calc.res. (comp.data for S-S trans.) allows to learn of transitions nature.

Table 5.20
The kinetic parameters for the structures XX in the solvents

Solvent, Excitation conditions	Photocolored structures							References	Notes
	OH		NO		NH				
	$\tau(s)(\lambda^{max}_{abs\,nm})$		$\tau(s)(\lambda^{max}_{abs\,nm})$		$\tau(s)$	$(\lambda^{max}_{abs\,nm})$			
	rise	decay	rise	decay	rise η %- reac.eff.	decay			
ACN, decay of NH in CH_3OH. Room temper. Laser 308 nm	$\leq 10^{-7}$ (370)	$<5\times10^{-6}$ (370) 1/	5.5×10^{-6} (490)	4×10^{-5} (490)	(1) $\sim 10^{-7}$ (2) $\sim 2\times10^{-6}$ (3) 3×10^{-5} 2/	≤ 0.5 a/ ≤ 5.0 a/ (565)	/115/ a//105/	-----
Toluene, room tem Abs.transient sp.	---	---	---	---	$\eta(1)=21\text{-}40$ $\eta(8)=8\text{-}13$	$\tau(8)/\tau(1)=$ $=5\times10^3$	/105/ /126/	-----
Toluene 0 $XX(1)10^{-3}M <10^{-3}$ +[C211]M : 10^{-3} 10^{-2}	10^{-6} not det --"-- --"--	5×10^{-5} not det. --"-- --"--	not det no quan 10^{-6} 10^{-6}	not det nquan. 4×10^{-5} 5×10^{-2}	$10^{-4}\text{--}10^{-5}$ 10^{-5} 10^{-6} 10^{-6}	10^{-2} 3×10^{-4} $\sim 10^{-4}$ $\sim 2\text{-}3\times10^{-5}$	/116/	Strong Incr. and Decr.stab. of NO,NH respectiv.

1/ τ of rise is below of the lowest limit of time resolved setup. 2/ The number of the structures in brackets

Table 5.21
The kinetic parameters for structure XX in the crystal (see also tables 5.17, 5.18)

Crystal Tape Excitation	Structure XX (n)	Photochromic structures							Ref.	Notes
		OH				NH				
		$\tau(s)$ 293 K 1/		ΔE kcal/mol		$\tau(s)$ 293 K		ΔE kcal/mol		
		rise	decay Temp.	rise	decay	rise	dec. Temp	rise	decay		
Single crystal 250 nm	(1)H_2 (1)D_2	--- --- ---	55×10^3 198K 209×10^3 207K	2.63 --- --- ---	16.4 (14.8)* 17.2 (17.0)*	--- $\Phi_{H\,1/}$ $-=20$ Φ_D	750 325K 3080 325K	(0.4)* --- --- ---	21.6 (39.5)* 23.2 (40.9)*	/103/ /247/	*Results of calculations in brackets
Single Crystal 250nm	(2) A B	--: ---	--- ---	--- ---	--- ---	--- ---	400 325K 270 325K	--- ---	26-27 18-19	/103/ /104/ /105/	The two polymorphic forms
	(3)	---	---	---	---	$<10^{-8}$	4450 312K	---	19.7	/104/	---
	(4)	---	---	---	---	$<10^{-8}$	545	---	15.9	/104/	---
	(5)	---	Pratically Stable <198 K	---	24.5	$<10^{-8}$	3290 312K 910 325K	---	22.8	/104/ /103/	---
	(6)					$<10^{-8}$	3640 312K		14.5 (18)*	/104/ /107/	One of polymorphs *Results of calcul.

It has been taken notice in the section 5.5.5.1. in all crystalline compounds XX, side by side with the photochromic reaction caused by the ESIPT, the open-chain reactions occur that may result in the low-yield latent (L) photochromism and photo-fatigue /112/.

This important fact has not been considered in the attempt of the structure-reactivity correlation/110/(see below).

5. 4. 5. 3. The date of the dark and photo-discoloration in the crystals.

For all crystalline compounds XX the (NH) form slow decays homogenously in dark pointing out the thermally activated reaction (NH)->(CH$_2$).

For the compound XX (1, 3-6) this process follows monoexponential law with the low time-rate constant τ(decay)$\approx$400-5000s at the temperature range 290-325K and with the high activation energy $\Delta E \approx$14-27 kcal/mol /103, 104, 109, 117/ (Tab. 5. 21).

Although all the molecules have the similar or identical internal coordinates (Tab. 5. 16) around reaction center, they have the quite different decay rates(e. g. compare 1 and 6, 7 or 4 and 7), and there is no correlation between the intramolecular parameters and the kinetic characteristics.

Moreover the molecular structure XX(2) in the crystals has a decay with the biexponential nature indicating the coexistence of the two different types of the (NH) tautomer crystalline structures with the two thermal decay processes having the different barriers and preexponential factors (Tab. 5. 21)/103, 104, 109/.

Thus the supramolecular structure (environment)and the intermolecular interactions in the crystalline packing play obviously a decisive role in the stability and decay characteristics of the structures XX. The supposed mechanisms of a such effect are discussed in the next section.

Unlike the SA molecules(Ch. 3), a significant isotope effect on the rate of the "blue" form thermal decay (τ(D)/τ(H)$\approx$4-5) is observed without any marked changes of the activation energy /103, 114/. Such an effect emphasizes that the PT unlike SA is involved in the rate determining step of the thermal (NH) decay.

The dark decay of the (OH) form is a spontaneous process including the two parallel path-ways:

(i). the formation of the (NH) structure (OH)-Q->(NH) (k(ON)), and
(ii). the return to the initial co-lorless (CH_2) form ((OH)->(CH_2))(k(OC))/103, 114, 117/(Tab. 5. 21), and it is supposed /117/ that k(ON)>>k(OC) in all temperature range where the thermal bleaching takes place.

In the reaction (i) the (OH) decay concomitant with the (NH) structure formation is observed by increase of the "blue" band intensity. This GSIPT NOH->NH occurs with a rate of a few second at the room temperature (see Tab. 5. 21 and Sch. 5. 31, T=225-300K)/117/. Within this temperature range the kinetics of the (OH) structure thermal bleaching involves the several reactions (see also below) but in the beginning of the temperature interval (T=200-250K) it is mainly the first order process with the activation energy ΔE=18-19kcal/mol (see also /100/)(Tab. 5. 21).

The time decay constant, τ(dec), varies strongly with the structure of XX. For instance at the temperature about T=200K, τ(dec)=5500s and 6. 4s for the compounds (1) and (7H) respectively and practically no decay for (5) under the same conditions.

The strong isotope effect for the (OH) structure decay (but less than for (NH) decay, e. g. $\tau(D)/\tau(H)$
≈ 4 for(1), (Tab. 5. 21)/103/ implies that the rate-limiting step of the (OH) structure decay is also dominated by a Proton translocation OH->NH and OH->CH_2.

As it follows from the data, the very similar structures (Tab. 5. 16a, b) exhibit the very different parameters (e. g. compare ΔE and τ(decay) for 1, 5, 7)(Tab. 5. 21).

Especially it is evidently for the remarkable stability of the (OH) form in the crystal XX(5) which is more stable than the (NH) form unlike all other photochromic crystal XX and other noncrystalline phase/100/ that can not be explained by the intramolecular effects.

Hence the all above phenomena can be realized only with the participation of the intermolecular interactions in the supramolecular arrangement of the photochromic molecules in the crystalline packing.

Thus the kinetics of the dark decay both the (NH)and (OH) photocolored forms in the crystals XX caused by the GSIPT, determined mainly by the intermolecular interactions in the crystal packing.

At high temperature (T>250K)(e. r. for (1), Sch. 5. 31/117/) the concentration of the (OH) form decreases not only due to the generation of the (NH) form ((OH)->(NH)) (i) and discoloration ((OH--->(CH$_2$) (ii) but also as a result of the formation of another structures not involved in the reversible process and responsible for the photofatigue. The lifetime of the (OH) form under such conditions estimated from the kinetic data with continuous irradiation to be $\tau \approx 65s$/117/.

The similar structures can also be formed due to the transformation of the (OH) and (NH) structures under the impulse UV irradiation.

Although the photochromic polymorphic crystal structures (7_α)and (7_γ) are similar(unlike nonphotochromic(7_β)) and completely bleachable in the dark ($\tau \approx 14$ and 12h at 298K correspondingly)however the prolonged(several minutes)UV irradiation brings for (7γ) (unlike(7α) only the partial discoloration with the permanent red coloration and the incomplete recovery of the(CH$_2$) form. The spectra show the significant decrease of the initial (CH$_2$)form absorption and the appearance of the strong absorption continuum in the range 200-800nm with a part of which is unbleachable.

The reversibility of the coloration in the photochromic crystals XX is realized not only by the dark back PT reaction but also by the reversible photoprocess that can involve the ESIPT(NH-hν->OH)/103/.

Upon the irradiation by the visible light in the spectral range of the "blue" NH band ($\lambda max \approx 600nm$) the observed weakening of this band is accompanied by various spectral change depending on the structure of the molecule:

    (i). the formation of the new absorption band ($\lambda max \approx 435nm$) with various intensity attributed to the (OH) form (compounds (1) and (5)), (ii). the lack of the marked increasing of the (OH) band intensity (crystal of (7α) and (3)), and (iii) the parallel but significantly less fall of the (OH) band intensity (crystal (2)).

5. 4. 6. The mechanism and the general structural-energetic scheme of the photoinduced processes.

This part is based on the data given in the previous sections 5. 4. 1-5. 4. 5.

The comparative consideration and the additional interpretation of the findings in the solvents and the crystals on the base of the common structural-energetic scheme are presented.

The attempt of the discussion on the base of the available data of the different and often contradicttory views at the mechanisms of the processes of various scientific groups has been carried out in accordance with the general scheme (Sch. 5. 32 a, b).

The main distinctions of the photochromic behavior of the molecules XX in the crystals and the noncrystal media have been discussed along with the consideration of the schemes.

In conclusion the comparative examination of the photochromic mechanisms of the molecules DNBPs(XX) and Anils (see Ch. II, III) has been conducted.

5. 4. 6. 1. About the nature and the intrinsic mechanism of the ESIPT and the GSIPT in the two-step PT reaction.

After the electronic excitation $S_o\text{-}h\nu\text{->}S_n^*\text{-->}S_1^*$ of the (CH 2) structure the effective radiationless deactivation of the $S_1^*(\pi\pi^*)$ state can occur involving the Internal Conversion (IC) and the InterSystem Crossing (ISC) via the low lying $n\pi^*$ states localized on the Nitrogroups and with the participation of the close (but high) $n\pi^*$ state localized on the N pyridine atom (the proximity effect) (see Sec. 5. 5. 1. and the Sch. 5. 32a with the insertion). Such a fast deactivation of the energy can proceed with the time rate constant $\tau(d) \approx 10^{-10}\text{-}10^{-9}$ s /118/.

Therefore the emission $S_1^*(\pi\pi^*, CT)$-hv>So($\tau < 10^{-9}$ s)(see e. g. /128/) of the initial (CH2) structure is suppressed completely, and the fluorescence of the (CH 2) structure is not observed.

However in confirmation of the key role of the radiationless process in the complete fluorescence quenching in the (CH2) structure the supplementary study of the emission properties of the model structure without the ESIPT reaction (e. g. double methylsubstituted DNBP) must be conducted.

Nevertheless the ESIPT reaction CH2-->HONO in the S_1^* excited state(see below)occurs in the solvents and the crystals competing successfully with both the radiationless and radiative deactivations of the S_1^* state.

Hence the ESIPT CH2->HONO has to have the time rate $\tau<10^{-1}$ s, and it is not a too slow process both in the solvent and in the crystals.

However in the nonphotochromic polymorphic crystal structure XX(7 β) (unlike (7α) and 7(y)/114/and XX(1-6)) where the dinitrophenyl ring is tightly sandwiched between the pyridine rings of the two adjacent molecules (Sch. 5. 27, Tab. 5. 17, Sec. 5. 4. 1) the strong electronic coupling between the π-systems of the excited chromophore and the neighboring molecules can take place.

According to /105, 114/ such a coupling can be responsible for the effective additional way of the electronic energy de-activation of the $S_1^*(\pi\pi^*, n\pi^*)$ state of the (CH 2) structure caused by the intermolecular charge transfer which suppresses the ESIPT reaction completely. But at the same time as it will be discussed later, such an "electronic" factor is not the only possible cause of the photoactivity quenching in the crystals (7β) and (2B).

Unfortunately there is apparently no single idea of the intimate mechanisms of the ESIPT which are probably can be different for the solvents and the crystals.

The ESIPT(and photochromism) in the crystal is not a surface or defect sites' effects because of the relatively high degree of the conversion into the colored form/103/(Sec. 5. 4. 5. 2.), and is bound up tightly with the bulk crystalline structure.

It is very important that none of other (unlike DNBPs) 2, 4-dinitro benzyl derivatives was reported to be photochromic at in the crystals/76/ and NO 2 group is obviously the necessary condition for the "blue" photochromism.

Thus /103/the break of the C-H bond in the structures XX is activated by the only both abstracting groups in the appropriate positions which are close to the same H 6_1 atom of the CH2 that. These groups can very strongly lower the activation barrier in the S_1^* state(to $\Delta E \approx 1.4$ kcal/mol) /107(the path-way 1, Sch, 5. 32).

As it will be shown below (Sec. 5. 5) any process underlying the mechanism of the (NH) structure formation and decay in the solvents and the crystals is the two-step PT in the excited and (or) ground states including CH$\Leftrightarrow$ HONO and HONO$\Leftrightarrow$NH PTs as the first and the second steps and vise versa respectively. Therefore the questions arise of the mechanism and the nature of each stage, and which of the stage is the really determining one.

The temperature independence, high rate of the ESIPT reaction, small barrier height, and the significant isotope effect of the direct pathway of the (NH)form generation /103/(Tab. 5. 21)allow to suppose the tunnel mechanism of any (or both) ESIPT fast reaction involved in the first or the second rate determining step of the (NH) formation: (CH 2)*-ESIPT->(CO)*->(OH)*->(ON)*-ESIPT->(NH)*(Sch. 5. 32a).

The contribution of the tunnel mechanism in each stage of the ESIPT can be estimated only as a result of the separate study of the (OH) formation mechanism.

Unfortunately such an information is absent, to the best of our knowledge, even in the special investigations/103/.

This mechanism can be discussed only for the crystal because of the lack of the information of the isotope effect in the solvents.

The GSIPT ((NH)->(CH), (OH->CH))is the main back reaction responsible for the dark bleaching in the solvent and crystals. The strong isotope effect on the dark bleaching is observed ($\tau(D)/\tau(H)$ =4. 1and 5. 6 for XX(1)and XX(7α) correspondingly).

Crystal structures were found to be independent on the isotope substitution(unlike the ratio of the polymorph formation of XX(7) and XX(2)), and it is evidence that the difference in the lifetimes of the thermal discoloration must be attributed only to the isotope effect.

Thus the back PT just as the direct ESIPT (see above) is the limiting step for the dark bleaching and the photocloration correspondingly. However the calculated values for the preexponential factors and the experimental ones of the activation energies for the H and D crystals are in the good agreement with the thermal activation mechanism in the GSIPT back reaction but not with the tunneling way as the direct ESIPT reaction (see above).

The activation energy of the (NH->(CH2)conversion varies from 20 to 27kcal/mol depending on the crystal structure and considerably higher than the kinetic barriers in the solutions/103/.

These barriers reflect the need of the high energy for the O-H covalent bond break and the C-H covalent one formation.

Such a high and broad barrier along the path-way of the PT reaction involving the rotation of the H-ONO group on the way of its approaching to the C6 –H62 one (Sec. 5. 4. 4) excludes the tunnel mechanism of the PT in the Ground state unlike in the S_1^* one.

Thus the limiting kinetic characteristic of the photocoloration and the dark bleaching are caused in the crystals completely by the long-time PT in the direct and reverse reactions, and the crystal structure XX appears to be first long-lived photochromic systems based on a pure proton transfer process.

However as it follows from the results of the quantum-chemical calculations/106/ the proton transfer itself in the gaze phase even in the ground state occurs in the time scale of roughly 100fs.

Obviously such a fast process can't be involved in the rate-determining step of the discoloration in the crystals and solvents. However it may be supposed that either the fast PT is typical only for the free molecule or the real PTrate is overestimated very much by the calculations.

5. 4. 6. 2. The generation of the (OH) and accompanying colored forms.

The structure of the (OH) and initial (CH$_2$) forms change a little with the crystallization. Therefore they differ strongly from the each other both in the solvents and the crystals by the phenyl-pyridine angles in accordance with the different contributions of the sp^3-hybrydization of the C-orbitals in the methylene bridge for the (CH$_2$)form(very acoplanar) and (OH) one (almost flat)(see Tab. 5. 16a, b and Sch. 5. 26a).

Thus the generation of the ground state(OH) colored form could need the marked activation energy as a result of the considerable structural transformations.

However the potential barrier is strongly lowered either by the stabilization of the flattened Zwitterionic (OH)*structure in the S$_1$*excited state or as a result of the direct diabatic process (see below).

The generation of the flat (OH) structure in the So state can occur as a result of the ESIPT (CH->

HONO)by the two routs /11/(Sch. 5. 32a, b, Tab. 5. 20, 5. 21) when approaching of the o-NO2 group towards the C 6H 6 $_1$ one over a low potential barrier for formation of the interaction C6H 61.......O$_1$NO.

The first, fast path-way with the temperature independent rate ($\tau \approx 10^{-6}$-10^{-9} s)can take place after the ESIPT (CH 2)*->(CO)*->(OH) from the acoplanar transient excited (CO)*structure by either the adiabatic process (CO)*->(OH)* ~~>(OH) in the S$_1$* state or by the diabatic reaction (CO)*->(OH) via bifurcation point "o" in competing with the adiabatic (NH)* structure formation.

The diabatic reaction is much more likely for the several reasons:

(i). The typical ESIPT fluorescence(with the Anomalous Stokes Shift-ASSflu)((OH)*-hv->(OH)) is absent.

(ii). The deactivation of the $S_1^*(OH)^*$ state via $ISC(S_1^* \to T_1 \to S_o)$ without low lying $n\pi^*$ states (in the (OH) structure $E(n\pi^*) \gg E(\pi\pi^*)$) is too slow process.

(iii). The energy gap $\Delta E = S_1^* - S_o$ in the vicinity of the (OH) structure is sufficiently narrow.

The diabatic path-way proceeds both in the crystal and the solvent and in the latter the rate of the (OH) generation increases strongly due to the stabilization of the (OH) Zwitterion structure especially in the polar media (Tab. 5. 20).

The second, slow($\tau = 10^{-4}$ at the room temperature), temperature independent route (CH)->(CO)*->(CO)->(OH) goes via precursor, ground state structure (CO) (Sch. 5. 32a, 5. 26b, Tab. 5. 18, 5. 21) which is assigned to a short-lived non-planar (OH) tautomer, not absorbing in the visible spectral range, and forming as a result of the radiationless deactivation of the (CO)* structure which arises from the ESIPT (see above).

The existence of such a precursor is proved convincingly by the kinetic data (Sec. 5. 5. 5. 2. and 5. 5. 6. 3.). The fast decay of the precursor (CO) proceeds not only to the (OH) colored form above the energetic barrier $\Delta E \approx 2.3$ kcal/mol /117/ or 1.8 kcal/mol /100/ but also over the very low potential barrier $\Delta E \approx 0.8$ kcal/mol ((CO)-Q->(CH2)) to the initial structure.

Although the existence of such a transient structure in the ground state has been predicted by the quantum-chemical calculations for the gas phase /107/, however in the solvents this form is not manifested in any way owing obviously to its extremely low stability at the expense of the fast transformation to the initial (CH2) structure.

Thus in the solvents the (OH) colored form is generated by the first (fast) path-way only.

The additional transformation of the photochromic forms is the result of the generation of the anion (A⁻) (the absorption band with $\lambda max \approx 490$ nm) by the exponential law ($\tau(rise) \approx 5\mu s$) from the (OH) structure ($\tau(decay) < 5\mu s$) in the polar solvents and other media favorable for the proton dissociation of the (OH) structure(see Sec. 5. 5. 3.)in the composition similar to the "closed ion pair" (A⁻...H+solv)(CIP).

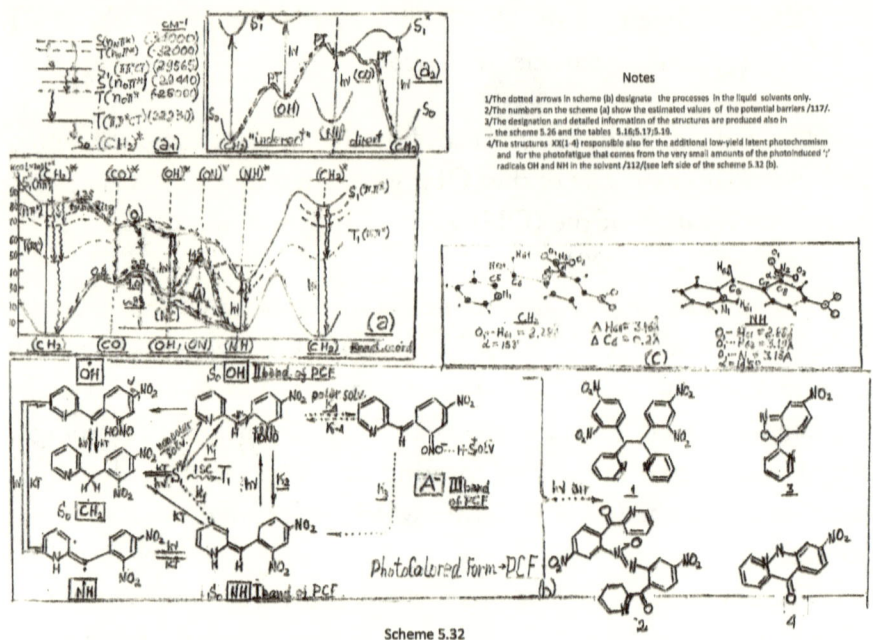

Scheme 5.32

a) The general energetic scheme (with the position of the lowest excited states (a₁)) and pthotobleaching reaction (scheme a₂)).
b) Corresponding structural scheme with the structures of irreversible reaction (1-4).
c)The structures of the initial (CH₂) and photocolored (NH)forms in the crystal.
The schemes correspond data presented in the text on the base of the findings of athors(see text).

The photoinduced processes of the compounds XX in the solvents and the crystals

a) The general energetic scheme (with the position of the lowest excited states (a_1)) and pthotobleaching reaction (scheme a_2)).
b) Corresponding structural scheme with the structures of irreversible reaction (1-4).
c) The structures of the initial (CH_2) and photocolored (NH) forms in the crystal.

The schemes correspond data presented in the text on the base of the findings of athors(see text).

Within such an ion pair the back reaction SolvH+-->(HO) occurs and an equilibrium (OH)⇔(A⁻) is established which is usually displaced towards (A⁻)form strongly in the basic media.

Under such conditions the CIPT can be also an intermediate in the generation of the (NH) structure in the dark reaction (se also Sec. 5. 3).

Thus unlike the crystals the photochromism of the compound XX in the polar solvents and the base media can be caused in addition by the anion (A⁻)structure.

5. 4. 6. 3. The generation of the (NH) colored form.

The (NH) form is the principal one providing the blue color of the photochromic product in the solvents and the crystals (absorption band with $\lambda max \approx 600nm$).

Its structure differs from the initial (CH2)one(both in the crystals and the solvents)mainly by the position of the proton (N-H or C-H correspondingly) but not by the rings' interplane angles.

And the essentially nonplanar structures are realized in the both structures due to the great contribution of the tetrahedral(sp³) hybridization in the C-orbitals of the methylene bridge/107, 108/ (Sch. 5. 26a, 5. 32c, Tab. 5. 16a, b)(See also Sec. 5. 4. 1., 5. 4. 2., 5. 4. 5. 1.).

Such a structural similarity of the initial (CH 2) and colored (NH) forms is bound up with the peculiarities of the mechanism of the (NH) structure generation especially in the crystals(see below).

The kinetic measurements reveal the occurrence of the four routs in the (NH) formation process /103, 114, 117, 127/(Sch. 5. 32a)(See also Sec. 5. 4. 5. 2.).

- (i). The "direct ", fast, nearly temperature independent adiabatic isomerization with the ESIPT (CH->NH)*: (CH2)*-ESIPT->(CO)*->(OH)*-E S--[(ON)*]--I P T->(NH)*~~>(NH) with the tunnel mechanism of the ESIPT (see below).
- (ii). The relatively slow "direct" thermally activated process(CH2)->(NH)via the ground state transient (CO) structure without formation of the (OH) ground state intermediate

structure is observed clearly in the crystals of XX(1) and (7) by the kinetic study (only in /103/) (the path-way (CH2)*-ESIPT->(CO)*-~>(CO)-GSIPT->(NC)->(NH).

One can be supposed that the nonplanar intermediate(CO) as a result of the ESIPT CH->HONO which is common for the planar colored (OH) and the nonplanar colored (NH) structures is kineticly destabilized completely in the solvents(and in the gas phase).

Therefore the route(ii) is realized in the crystals only via the potential barrier about 1-2kcal/mol competing with the formation of the (OH) colored structure.

> (iii). The competing, slow, thermally activated indirect route involves the ESIPT (CH->HO)* with the formation of the excited (CO)* transient structure, the (CO) ground state transient (only in the crystals)and the generation of the (NH) colored structure via the(OH) colored one above the potential barrier $\Delta E \approx 19$-20kcal/mol: ((CH)*-ESIPT->(CO)*-~>(CO)(cryst)->(OH)----20kcal/mol----->(ON)--GSIPT->(NH)) by the GSIPT reaction via the transient (ON)(Sch. 5. 32a, b, 5. 26a, Tab. 5. 18). The barrier is caused by the significant structural transformations of the flattened (OH) structure into the nonplanar (NH) one(rehybridization sp^2->sp^3 in the methylene bridge, and the PT reaction can obviously proceed without the tunneling.

Such a scheme for (i)-(iii) is agreed with the kinetic data of the generation of the colored forms (NH) and (OH) (Sec. 5. 5. 5. 2.) including the findings of the isotope effect.

The path-way(iii) occurs in the crystals, the polar and nonpolar solvents, and in the latters it is the single route.

This reaction proceeds in the polar solvents much faster than dissociation of the (OH) form to anion (A⁻) (Tab. 5. 20)(see above)/115/.

The reaction (OH)->(NH) rate, τ ((NH), rise)decreases strongly in the crystals in comparison with the solvents (compare the data of the Tab. 5. 20 and 5. 21) due to the stabilization (by $\Delta E \approx 20 kcal/mol$) of the nonplanar (NH) structure in the solvents (similar to the free molecule /107/)because of the unimpeded rotation of the dinitro-phenyl group with the rehybridization C_6 sp²->sp³ orbitals unlike the crystals /105, 107/.

> (iv). The generation of the (NH) structure with the participation of the ion pair entity (like the tight ion pair) ((A-)...(H+) Solv.-->NH(A-)(H+)+ Solv.)) with $\tau \approx 30 \mu s$ at the room temperature (see Sch. 5. 32b, path-way k_3 ((A⁻)->(NH) (Sch. 5. 32a)).

In this case the proton is transferred to N-atom of the pyridine ring (but not to the NO_2 group) via the formation of the (A-)...(H+)basic media ion pair by the electrostatic forces, and the basic media acts as a mediator (or a provider of the proton)on the route (CH_2)->(OH)—media->(A-)(H+)media->(NH) + media. Thus such a route can occur in the basic media (including the protic solvents).

The nearly temperature independent "direct" route (i) needs an additional discussion.

According to the experimental data this route involving the ESIPT has to occur with the comparatively high rate, practically without the potential barrier along the S_1^* excited state PES.

The (NH)and initial(CH_2)forms in the photochromic molecules XX have almost the same nonplanar conformations which differ one from another only by the proton position (Sch. 5. 32c)/108/.

At the same time in all photochromic compound XX(1-8)(Sch. 5. 25, Tab. 5. 16)the initial structure is unfavorable for the CH->NH ESIPT since the distance between the pyridine Nitrogen atom and the methylene proton H_{61} (N_1--H_{61})is too large (>3Å), and the directions of the N_1 lone pair and C_6-H_{61} bond are strongly different (C_6-H_{61}.... N_1 angle $\approx 40°$ (see also Sch. 5. 32c, 526a).

Thus one should expect the direct route (CH)*->(NH)* (right part of the Sch. 5. 32a) has to involve the too high potential barrier (in the So state)even with the diabatic process (Sch. 5. 32d) and does not meet the experimental findings.

In addition the break of the C-H bond cannot be activated by only pyridine Nitrogen atom and needs the participation at least one more abstracting group (Sec. 5. 4. 6. 1.).

At the same time there are the convincing evidences of the key role of the o-nitro group in the fast, temperature independent ("direct") route (CH->NH)/108, 112, 113, 117, 127/.

The (NH)form is not generated in the solvents and the crystals of the structures similar to XX ones but without the NO_2 phenyl groups while all compound XX with the benzyl nitro-groups exibits the photochromism with the "blue" NH form in the solvents.

However unlike the solutions both favorable and unfavorable molecular structures for the generation of the (NH)form, it can be realized in the crystals under influence of the intermolecular interactions in the crystalline lattice.

Therefore the "blue" photochromism can be exhibited only in the crystals of XX with the molecules which meet the necessary structural conditions.

Such conditions are revealed when studying the photochromism of the separate crystalline compounds XX.

(i). For the molecule XX(7) the three crystalline polymorph forms α, y, and β are realized /103, 114/(Sch. 5. 27).

The first two are photochromic structures and the latter is photoinert that. Such a difference in the photochromic activity cannot be explained by the difference between the values of the free volumes in the crystalline packing (like SA) and of the crystal density since these values are almost the same of all three crystals.

The same short distances $O_1...H_{61}C_6$ (d=2. 3-2. 4Å), and the favorable orientation for the formation of the $C_6-H_{61}...O_1H$-bond of the C_6-H_{61} bond relatively to the lone pair of O_1 atom($C_6-H_{61}-O_1$ angle is not large)are also typical for the initial (CH_2)forms of the all (photochromic and photoinert)structures (including 7 $α, _β,$ y) in the solvents and the crystals(Tab. 5. 16).

At the same time the distances $O_1...N_1$ or $H_{6\,1}...N_1$ of the 7 β inactive polymorph structure are much longer than of 7 α, y and other photochromic structures (1-6)(≥4Å, ≈3Å correspondingly).

It is evident that is the just the distance between the O_1 atom of the o-nitro-group and pyridine N-atom is the key factor which stimulates the photochromic activity.

Thus there is the distance limit (O...N≤4Å) above which the ONO_1H_{61} group can't reach the pyridine N_1-atom for the transmission of the proton from the O_1 atom to N_1 one with the formation of the covalent bond N-H.

(ii). The quantitative prediction /100/of the ability to the photochromic activity in the dependence of a small change in the molecular structure in the crystals of the molecules XX(1-8) can be conducted on the basis of the estimation of the intramolecular cavity around the reactive fragment N_1-C_5-C_6(H_{12})-C_7-C_8-NO_2 including the all atomic groups participating in the PT CH_2->NH with the rotation of the o-NO_2 group for the delivery of the Proton by the C_6H_{61}-PT->ONO_1H_{61}-PT-->N_1H_{61} double proton transfer without the marked structural transformations (i. e. the ring flipping) which results in the formation of the (NH) structure without the steric hindrances.

Thus the shape and the size of the cavity serve as a criterion for the generation of the (NH)structure. The calculated optimum cavity volume for the realization of the reaction CH->OH->NH is Vcav>20 Å³ (the structures 1-7 α y) while at the Vcav≤17Å (the structure 7 β, 8) the reaction in the crystals does not occur.

Within the limits of the structural conditions (i), (ii), the small rotations and (or) torsion of the o-nitro-group about the C-N bond accompanying the PT reaction, provide the capture and delivery of the H 6_1 proton from C 6 atom to the N_1 one.

This process proceeds with the high rates which exceed much more the intrinsic tunnel two –step PT rate CH->OH->NH ensuring the observed isotope effect (Tab. 5. 21). However the minute changes into the relative orientation and distances of the three (CH 2, NO2, Npyr) groups may induce an increase in the height and (or) the shape of the PT barriers affecting the efficiency of the (NH) generation (compare the compounds 7 α and 1, (Tab. 5. 21)) and even inhibit it.

In contrast, in the free molecule in some solvents (in dependence on its nature) the relative position of the rings is controlled by the hybridization at the methylene connecting bridge, and more stable and flattened (OH)-like transient structure can been formed in the S_1 that results in the marked fall of (NH) generation rate in comparison with the crystals (Tab. 5. 20, 5. 21).

Thus the fast "direct"(NH) generation as a result of the participation of the nitro-group with so called the "nitro-assisted" mechanism. In other words the NO2 group is relatively close to the transfered proton and the energy of the transition state along the path –way (CH 2)*–> (NH)*reaction on the S_1* PES is much more closer to the energy of the (NH)* structure than the (CH2)*one and the (NH)*state is easily reached /107/.

According to the scheme (Sch. 5. 32a) the (CH2)*->(CO)* reaction is beginning from the small rotation (torsion) of the o-NO2 group with a very low barrier towards the position of the interaction between the O_1 lone pair and the proton H 6_1. As a result the proton approaches the nitro-group rather by a tunneling through a small barrier(~ 1. 4kcal/mol) but is not coming close enough to form a true covalent bond O-H in the short-lived precursor. The (CO)* is adiabatically forming, very unstable, nonplanar transient structure on the S_1*PES with the completely broken C6 –H 6_1 bond.

The existence of such an (OH)-like (but different from (OH)*structure) short lived transient precursor species in the excited state is demonstrated also by the efficient production of the (NH) form even during low concentration of the (OH) form in the ground state.

According to the calculations /107/ there is also the very low stable transient (ON)* structure in which the proton becomes only weakly bound with the pyridine N_1 atom.

Such a structure is formed as a result of avoiding of a vicinity of the (OH)*structure via the bi(or three)furcation point "o" in which the competition with the generation of the(OH)* (adiabatically) or (OH) (diabatically) structures occurs. The stable excited (NH)* structureis formed by the barrierless process.

Thus the "direct" two-step ESIPT path-way along the S_1*PES can be represented by the adiabatic transformation (CH_2)*--ESIPT->(CO)*-ESI-(ON)*PT-->(NH)* where proton migrates from the carbon atom towards the o-Nitro-group but changes the direction (in the (NO)*state and forms the covalent bond with the pyridine nitrogen atom.

The generation of the "blue "(NH) structure in the ground state proceeds from the (NH)*state by the fast radiationless deactivation (NH)*~~>(NH) without the ASS emission (NH)*-hv->(NH) or by the direct diabatic process (ON)*~~>(NH) with some change of the molecular geometry.

As it appears from the aforecited discussion, the generation of the (NH) structure in any case-in both the indirect and the "direct" pathways, is caused by the two-step PT mechanism differing by the nature of the transient precursors which are relatively stable ground state planar (OH) structure for the indirect case and the utmost low stable (OH)*, (ON)* excited state structures for the direct case.

Thus the realization of the interaction with o-NO group in the S_1* state with the ESIPT (or almost the ESIPT)(CH)*->([H]ONO)*is the necessary primary step in the generation of any (OH), (A⁻), and (NH) colored structure.

5. 4. 6. 4. The reverse ground state bleaching reactions

The slow bleaching reaction in the solvents and the crystals is the distinctive feature of the photochromic compounds XX (Sec. 5. 5. 4. and 5. 5. 5. 3.).

The low rate of the reverse ground state reaction is determined by the three principle factors that affect the high broad barrier along the PT (NH->CH) involved in the limiting stage of the dark bleaching reaction.

(i). The structural transformations that provide back proton transfer.
(ii). The relative stability of the colored (NH) structure.
(iii). The intrinsic mechanism of the reverse PT in the ground state.

(i). The reverse process involves the two thermally activated reactions with the high extended barrier along the (NH)->(CH$_2$) route /103/that is explained by the peculiarity of the (NH) and (CH$_2$) structures. The colored (NH) and the colorless (CH$_2$) forms are structurally similar differing only by the proton location/108/(Sec. 5. 4. 1., 5. 4. 4.).

The too long distance between NO$_2$ and CH$_{1,2}$ groups and the very small difference ($\approx$0. 3Å)between this distances for various structures XX contradict the both the realization of the PT and the large variations in the observed barriers and lifetimes (Tab. 5. 21)with the structure variations.

However there is large void space around the NO$_2$ group, and one of the Oxygen atoms has no hindrances for the simultaneous shift backward when the aromatic ring remains nearly unaltered and the proton is free to undergo to reverse series of the events that secures the thermal reversibility PT (NH<->OH and OH<->CH).

Thus both the indirect ((NH)-> (OH)->(CH$_2$))and the "direct" ((NH)->(CO)->(CH$_2$)(Sch. 5. 32a) can proceed only with the participation of the o-NO$_2$ group which acts as a chaperone and escorts proton along its path.

The comparison of the results of the calculations for the transient structures by the semiempirical (PM3)method that is bound up with the definite parameterization and by Density Functional Theory (DFT) method that yields qualitative results for any geometry /106, 107/ shows that only the second method gives the correct data when the unusual functional groups(ONOH) are involved, and these results are an evidence of the unusual geometrical transformations bound up with (ONOH) group when (NH)->(CH2)reaction is occurring.

Really the ability of the NO2 group to assist in the GSIPT even along the "direct" route is manifested by a marked differences in the dependence of the activation energy on the interplanar angle between o-NO2 group and the phenyl ring in the different polymorph phases A and B of the crystal XX(2).

Thus in the photochromic structures XX the different molecular conformations (like 2 phases A and B) exhibit the different behavior as a result of the restriction of the o-NO2 group rotation (or libration)by the intermolecular interactions in the different crystalline lattice /104, 111/ (Sch. 5. 27, Tab. 5. 21).

It is obvious the rotation (or libration) of the NO 2 group is the principal structural transformation that provides the back proton transfer.

(ii). The energies of the destabilization of the (NH) structure relatively to the initial (CH2)one are about 10kcal/mol in the gas phase and about 30kcal/mol in the crystal for the compound XX(1), and vary a little with the change of the molecular structure (e. g. 7, 8)/107/.

The energy of the transition state along the route (NH)->(NC)->(CH 2)is about 20 kcal/mol for the structure XX(1) and depends on the restriction of the o-OH rotation in the different structures.

The kinetic study of the biexponential decay, the PM3 calculations, and the solid state NMR studies of the model crystalline compound XX(2) (Tab. 5. 21, 5. 16, 5. 17, Sch. 5. 27)/104, 109, 111/ and also /103/ show the existence of the same (NH) tautomer which embedded in the two phases A and B (in the equilibrium A<->B)related by a no-classical phase transition with the long-lived (LL)(NH) state ($\Delta E \approx 27. 2$ kcal/mol)and the short-lived (SL) (NH)state ($\Delta E \approx 18. 9$ kcal/mol).

The LL and SL are primarily populated at T>315K and T<315K correspondingly with the coexisting at various ratios over wide temperature range. The difference in the activation energy is explained by the stabilization of the NH form ($\Delta E \approx 1.49$ kcal/mol, PM3) that is achieved by the formation of the intermolecular H-bond (N_1-H...O'3) as a result of some decrease of the intermolecular distance N_1...O3' and the favorable directions of the O3' lone pair and H-N bond which are fitted in the low-temperature phase (A) by the orientation and interaction of about 16 molecules in the crystal packing (Sch. 5. 27).

Thus the above results indicate that even small changes in the supramolecular structure and the solvate shell of the reaction site can result in the large variation in activation energy andreaction rate.

Table 5.22
The main differences
in the properties between the liquid solvent and the crystalline photocromic systems on the base of DNBPS(XX)

Solvent	Crystal
The (CH$_2$) structure	
All identical molecules of the same compound have the same initial structures (CH$_2$), and the all systems photocromic.	There is the difference in the molecular structures between the (CH)initial forms of molecules, Promoted by the intermolecular interactions in the polymorphic crystals of some compounds which are not all photochromic or possess the different routes in the dark bleaching reaction(7β,2)
In nonpolar solvents λmax=334nm	The short wavelength shift, λmax=250nm
The (NH) structure	
The nonplanar quinoloid-zwitterions structure with the very short NH–bond. With high contribution of the SP^3 hybridization of C$_6$	More planar and less stabilized with much more hynoid character due to Impending of benzoring rotation and less contribution of SP^3 (C$_6$) hybridization.
The (OH) structure	
The same almost planar quinoid –like structure with the very short O-H bond.	
The long-wavelength absorption bands of the colored structures, λmax.	
(NH) 550-580 nm (OH) 370-450 nm (Polar solvents A$^-$) 480-490nm	(NH) 600nm with two partial overlapping bands of different polarization. (OH) 430-480nm (A$^-$) Absent
The generation of the (OH) structure	
The single rout with the ESIPT CH→O, fast, with the exponential law, temperature independent, with rate growing with solvent polarity	The two pathways: i: Fast, temperature independent ESIPT, CH→O, ii: Thermally activated, slower, via the So state precursor, ESIPT+ structural transformation
Solvent	Crystal
The generation of the (NH) structure	
The two routs both in polar and nonpolar solvents. i:The fast(but slower,<10^{-1} s than in crystal) ESIPT, temperature independent and controlled particularly by the rehybridization sp$_2^2$ –> sp^3 .ii: The slow, temperature dependent, via OH structure with the small potential barrier.III: In the polar solvent only the extra way from(A$^-$)H$^+$ (close ion pair)hto NH structure	The three routes. i:The "direct "fast(10^{-3} s),temperature independent, isotope effect (possib. tunneling, nitro assisted, without rehybridization. ii: The direct ,thermally activated(200K),above low barrier, via ground state transient structure. iii: The slow, thermally activated(above230K),via (OH)structure above barrier about 19 kcal/mole.
The dark decay of the colored structures (bleaching)	
The polar solvents The all bleaching reactions including decay of (A)proceed mainly via the (NH) structure which decays very slow to the initial form (CH$_2$)directly by the GSIPT assisted by the O—NO$_2$ group (NH->C) over high barrier. The nonpolar solvents The two separate routs occur: i. direct GSIPT reaction NH->C via the transient nonplanar (OH) structure controlled by the rehybridization of C$_6$ orbitales sp^3->sp^2->sp^3 over high potential barrier(E>20kcal/mol).	The (NH) structure decay proceeds above the potential barrier about 20kcal/mol .The GSIPT NH->C is the limiting stage of the the dark bleaching reaction assisted by the ortho-NO$_2$ group. The (OH)structure decay proceeds via two parallel path-ways: i. via the formation of the (NH) structure (GSIPT OH->N). ii. The direct return to the initial structure(CH$_2$)by the GSIPT OH->C, which is the limiting stage of the bleaching reaction. At the room temperature the time of the dark bleaching is 750—1400 s. and more in dependence on the structure.

All the data presented in the table are taken from the results of the authors of corresponding works (see references).

The table 5.23
The ESJPT photochromic types (groups), molecular systems, and structures*⁾

Photochrom types or groups	The ESIPT molecular structures and systems									Notes
L Latent	The prospective structures that are not colored in the liquid molecular systems									The atomic centers of ESIPT and GSIPT are shown in the 2th rows. The structure IV is not photochromic.
	IV	V	VI$_{1,2}$	IX$_{a,1-3}$	IX$_{b,1-3}$	IX$_{C1}$	XI$_{1a}$	XIII$_{1,2}$		
	O⇔N	O⇔N	O⇔N	O⇔N	O⇔N	O⇔N	O⇔O	N⇔N		
LT Latent-True	Molecular photochromic systems (Solid solvents and crystals)									The symbols Sv,Cr are rigid solv.and crys.where T-photochromism is displayed. Structures IX$_{c,1}$ and (XIII $_1$) are the struct. modific. of IX$_{C,1}$ and XIII$_{1,2}$ that display or can display T - photochromism respectively.
	I	II	III	VII	IX$_{c,2,3}$	X$_1$	XV II$_1$$^{a-d}$	XIX^{a-c}	(XIII$_3$)	SA
	O⇔N	O⇔N	O⇔N	O⇔N	O⇔N	O⇔N	C⇔O	C⇔N	N⇔N	O⇔N
	Sv	Sv	Sv	Cr	Cr	Sv	Cr	S v	?	Cr,Sv
T True	Structures that are photochromic in molecular systems with liquid solvents at ambient conditions.									The structure XI$_2$ is the structural modifcation of XI $_{1a}$that leads to the sharp change of the photochromic properties (L→T).
	VIII	X$_2$	XI$_2$	XII	XIV^{a-d}	XV	XVI^{a-d}	XVIII$_{1,2}$	XX$_{1-8}$	
	O⇔N	O⇔O	O⇔O	N⇔N	N ⇔N	N⇔N	N⇔N	C⇔O	C⇔O⇔N	

*) For numeration of the structures see text.
Explanation of the table 5.23 (see also the Notes of the table)
Photochromism, by the sense of this word, is the appearance taking place as a result of irradiation (photo),displays in the change of the coloration i.e. changes in the visible part of spectra that exist a sufficiently long (but limited) time in order to be seen by a naked eye(chromism) .The reverse, dark reaction rate can be a very little and can be equal zero. For the last case it is necessary to use irradiation in the spectral area of the colored form. And if the coloration is not disappear, this molecular system does¡'t show photochromism in view of irreversibility of reaction (by definition).
Thus the definition includes the observation of the appearance by the naked eye ,that it was not get earlier although it is illogically,taking also into account that almost all photochromes have been found out visually. If the molecule under study shows a change of coloration under steady-state irradiation in the liquid nonpolar solvent where it doesn't almost interact with environment, such a molecular system is attributed to the True (T) phototochroms.If the liquid(e.g. solvent) molecular system including molecule under study, shows fast change of visible spectra only under impulse irradiation is attributed to the latent (L)photochroms. Only those from structures that gain an ability to change their color under steady- state irradiation(trueT-photochromes with inclusion in another (more often –crystals, solid solvent, or polymer matrix)can be attributed to Latent-True(LT-photochomes). Such a classification can include almost all photochromes.

In addition the neighboring molecules in the crystalline packing of the compounds XX could be involved in the complex mechanism of the PT due to the proximity of the proton accepting groups, and therefore the PT rate is dependent not only on the distance N-C but also on the relative orientation and position of the electronegative atoms of the neighboring molecules that can play the important role in the stability of the (NH) form.

Really with the lack of the crystalline intermolecular interactions (e. g. in the solvents) the rates of the dark reaction increase by the several order of magnitude (compare the data of the tables 5. 20 and 5. 21). However according to the calculations/107/ for the gas phase, the absent of the crystalline environment mast secure the stabilization of the nonplanar (NH) structure and the corresponding decrease of the bleaching reaction rate in some contradiction with the experimental findings in the solvents.

The effect of the (NH) structure stabilization can be achieved not only by the intermolecular interactions but also by the formation of the Intramolecular H-bond ($N_1H...N$)(Sch. 5. 27c) in the solutions and the crystals of the compounds XX(7), XX(8) with the energy of the stabilization about 4 and 8 kcal /mol relatively compound XX(1) correspondingly which results in the sharp decrease of the bleaching rate (Tab. 5. 20)/105, 107, 126/. It follows from the above that the both inter and intramolecular stabilizations of the (NH) play the important role in the kinetics of the bleaching reaction.

According to the schemes (Sch. 5. 32a, b)and the data of the calculations /106, 107/ the reverse "direct "(NH)->(CH2)reaction along the route (NH)-GSIPT->(ON)-GSIPT->(CH 2)begins from the rotation of the o-NO2 group that is much slower than in the S_1^* state, and of the approaching of that group up to the distance of the Hydrogen bond with the Nitrogen of the pyridine ring.

In the transition state (NC) the short-lived quasi-hydrogen bond NO...H--N is formed along the limited proton migration and together with the lengthening of the N-H bond (Sch. 5. 26b, Tab. 5. 18). With the approaching of the transition state (NC) the O-H distance becomes much shorter ($\approx$1. 8Å).

After (NC) state the proton changes direction of its motion and does not return to the pyridine nitrogen but migrates to the carbon atom to the distance about 1. 1Å forming C-H bond. The NO2...H...N bond is being broken, and the O-H distance becomes more longer again (>3. 0Å). In this way the o-nitrogen group assists the "direct" thermal (NH->(CH) back proton transfer.

The indirect(NH)->(OH)->(CH2) back reaction begins also with the rotation of the o-NO2 group, but instead of the transient (NC) structure, the relatively stable (OH) form is generated above the barrier about 20 kcal/mol. It is followed by the transformation into the uncolored(CH2) structure (NH)-GSIPT->(OH)-GSIPT->(CH2).

The slow thermal isomerization of the (OH) into the (NH)structure with the PT OH->NH sproceeding in the solvents and in the crystals over barrier about 19 kcal/mol is one of the essential path-ways of the (OH) form decay. The sufficiently fast direct decay of the (OH)colored structure (OH)->(CH 2)via the transient (CO) and above the small barrier (0. 8kcal/mol)proceeds in the nonpolar solvents especially and in the crystals stimulated mainly by the intermolecular interactions in the crystalline packing.

In the polar protic solvents the significant portion of the (OH) form turns into the anion (A⁻)in the equilibrium (OH)<-> (A⁻), and the latter decays into the (NH) form (route A⁻->(NH))(Sch. 5. 32a, b). The subsequent reaction of the discoloration follows via the (NH) structure (see above).

The photobleaching reaction under excitation of the (NH) can proceed by the several routes in dependence on the molecular and (or) crystalline structures that is displayed as a different correlations (see sec. 5. 5. 5. 3) /103/between the decay of the (NH)form and the change of the population of the (OH)one (the cases i-iii in 5. 5. 5. 3.).

According to the supposition /103/the marked difference between the change of the relative population of these structures can be caused by the competition between the "direct" (NH)-hv-->(CH2) and the indirect (NH)-hv->(OH)-Q-->(CH2)routes. However the both "direct" and indirect path-ways can proceed due to the structural and energetic requirements only with the participation of the o-nitro group (see above)as a result of the diabatic reaction when forming the transient short-lived (CO) structure("direct" route)or the relatively stable (OH) colored structure (indirect path-way).

Thus the competition of the both routes in the photo-bleaching reaction depends on the fine structure of the reaction site involving the

CH2, o-NO2, and NH-ring groups in the crystalline packing or in the solvent shell.

It is very important the ability of the remarkable irreversible transformations can be displayed in some molecule XX in the crystal and the solvents under prolonged UV irradiation (e. g. XX(7y) unlike 7α)(see 5. 4. 5. 3.).

They can bound up with some details of the molecular structure (e. g. the great value of the interplanar angle ONO-phenyl ring for the y-structure unlike the α one (Tab. 5. 16).

The photoinduced formation of a small amount of the radical structures is followed by the generation of the stable side products (Sch. 5. 32b)/112, 113/.

Such reactions result in the undesirable phenomena of the "photofatigue" in the observed reversibility of the "blue" photochromism and the low-yield latent (L) "red" photochromism of the unknown nature.

The processes have to take into account when considering the kinetics of the photochromic reactions.

Thus the variation of the packing can result in both the change of the photochromic properties and the stimulation of the irreversible reaction reducing photochromic activity.

5. 4. 6. 5. The differences between the photochromic processes in the solvents and the crystals.

It follows from the above produced findings that the photochromism of DNBPs (XX) is displayed effectively both in the solvent and the crystals.

At the same time there are the marked differences in the mechanism and the kinetics of the generation and the bleaching of the colored forms between the liquid media and the crystals which have brought together in the table (Tab. 5. 22) (N1-8 numeration of the first column).

The note should be taken that the structures of the molecules in the crystals and the solvents can be different markedly under influence of the intermolecular interactions in the crystal packing.

Therefore the findings obtained in the solvents cannot be directly used in the general case for the prognostication and understanding of the mechanism of the photochromic transformations in the crystals and vice versa.

Really the quiet disparate photochromic properties in the solvents and the crystals are revealed by the compounds XX(7), XX(2)(Tab. 5. 22 N1 and Sec. above), and the difference in the longwavelength transition energies of the initial structure for the solvents and the crystals indicates the different nature of the excitation and of the deactivation of the initial (CH2) structure (N2).

The difference in the structures and in the stabilization of the "blue" (NH)form between the solvents and the crystals(N3) unlike (OH) one (N4)results in the marked distinction in the positions of the long-wavelength transition(N5)and the kinetics of the bleaching reaction.

The difference between the mechanism of the colored form((NH), (OH)) generation in the solvents and the crystals depends on the nature of the solvent and bounds up with the distinctions and the number of the reaction routes and with the stability of the transient structures(N6, 7).

In all cases however the direct (CH2)->(NH) reaction is faster in the crystal than in the solvents and assisted by the o-NO2 group being the rate-limiting stage in the reaction of the photocoloration in the solvents and the crystals.

The peculiarity of the polar protic solvents is the generation of the anion colored structure which is caused by the interaction with the base media(N5).

The dark bleaching reactions in the both solvents and crystals are the thermal activated ones.

The GSIPT reaction is the rate limiting stage, and the direct GSIPT NH->CH2 is assisted by the NO2 group. However in the solvents unlike the crystals, the reverse (NH)->(CH2)reaction is strongly controlled by the rehybridization of the C6 atom's orbital, and the bleaching rates differ strongly especially by those of the "blue" NH form which about three times as less in the crystals than in the solvents (N8).

Thus the parallel study of the photochromism in the solvents and the crystals provides the valuable information about the mechanisms of the photochromic reactions of the DNBP molecules.

Conclusion and the general discussion

The six types of the ESIPT and the GSIPT –(O⇔N), (O⇔O), (N⇔N), (C⇔O), (C⇔N) and the double step (C⇔O⇔N) involve about twenty different structural groups with the several molecular structures in each group.

According to the conception considered in the introduction, the compounds in the groups can be distributed between the three main types of the photochromic structures and systems (Tab. 5. 23).

1) The "True" photochromic structures (T-photochromes) display the visible photocoloration under the steady-state irradiation in the liquid media(including the solvents), and as a rule, in the rigid media (including the crystals with some exclusions).

The high stability of the photocolored (PC) structure in such molecules is determined mainly by the high intramolecular potential barrier ($\Delta E > 10\text{-}15$ kcal/mol) on the ground state PES along the path –way from the relatively stable colored structure to the initial one that can be connected with the structural transformations before the GSIPT or with the mechanism of the GSIPT itself (e. g. structures XX).

The intramolecular stabilization of the PCF structure without the marked participation of the media results in the accumulation of the PCF in the photo-steady-state and the visualization of the coloration under the common conditions.

However the spectral pattern and the efficiency of the absorption and lifetime of the PC form can be depended strongly on the media nature.

The investigation of the T-type molecules have been conducted in general by the adequate comparative methods including the steady-state, low time-resolved spectroscopy of the solutions and the crystals with the utilization of the quantum–chemical computations.

In some cases (e. g. the structures VIII, IX(2)) the modified mechanisms of the photochromic reactions corroborated with the experimental findings(unlike earlier supposed those)have been proposed in the present book.

The most detailed and convincing data of the high-time resolved spectroscopy have been obtained and discussed for the compounds XX(DNBPs).

It shell be noted that versalility of the investigations enabled the researches/ /to sort out and undersand the complex and unusual mechanism of photochromism of those.

These data are produced (in Tab. 5. 24) and reflect the peculiarities of the unique mechanism of the photochromism in which the ESIPT and GSIPT are the rate limiting stages in the direct and the reverse reactions correspondingly.

However in the most cases the intimate mechanism of the ESIPT and the following structural transformations responsible for the photochromism need the additional detailed studies.

2) The "latent-true" photochromic structures (LT-photochromes) display a relatively fast (invisible)specral changes (L-photochromism)in the liquid media under the impulse excitation and the visible change of the color(T-photochromism)in the rigid media including the crystals (in dependence on the crystal structure).

The low stability of the photocolored structure is insufficient for its visualization in the liquid media is determined by the low intramolecular potential barrier on the ground state PES ($\Delta E < 5$ kcal/mol)along the path-way to the initial structure from the main colored one that does not involve any significant transformations.

The L-photochromism is displayed by the molecules in the liquid solvents under the short-time irradiation. The nonspecific and the specific intermolecular interactions especially the steric those in the rigid media (including crystals) between the "photocolored "and the adjacent molecules result in the sharp stabilization of the colored structure. The T-photochromism is manifested by the same molecules under such conditions and the steady-state excitation.

The study of the majority structures with the photochromism of the LT-type is conducted without the utilization of any kinetic methods in both the liquid and the rigid media which are especially necessary for the elucidation of the mechanisms of the fast structural transformations responsible for the L-photochromism. As a rule the clear understanding of the reaction mechanism is impeded by the lack of the data of the quantum-chemical studies. In this connection the modified variants of the reaction mechanism have been proposed in some cases on the base of the produced findings (e. g. for the structure VII) in the chapter five.

The most detail studies among the LT-type photochromic molecules have been carried out for the structures X(1), XVIIa-d, and XIX. Nevertheless the level of the experimental and the quantum-chemical methods used for the study even these LT-type structures is more lower than that for anils which belong to the same LT-type.

Really the photochromic anils (SAs) considered in detail in the previous chapters which have acquired historically their specific name ("photochromic") after the visual discovery their photochromic properties in the crystals only in spite of the early numerous unsuccessful attempts to find out any signs of the photochromic changes in the solvents and the structural photochromic changes in the crystals (see Ch. 1) are the classical example of such LT photochromic structures.

On other hand the data obtained in the numerous studies of DNBP(XX)(T-type photochromism) yield to the findings of Anils (SA) only in the variety of the structures studied. Therefore their detailed comparative description can be carried out. (Tab. 5. 24).

According to the data of the table the fundamental difference between the mechanisms and the kinetics of the direct and the reverse reactions involving the ESIPT and the GSIPT results in the sharp distinction in the properties of the photochromic systems of the different (T and LT) photochromic types.

3) The "latent" (L) photochromism is characterized by a very short lifetime of the invisible "colored" form even in the rigid media at low temperature and, as a rule, with the efficient ASS fluorescence typical for the ESIPT.

Such a short-lived transient "photocolored" form attributed to the proton transferred structure(see Ch. 3) is the typical one for the great amount of the molecular structures with the ESIPT /1/. The insignificant post-ESIPT skeletal transformations due to the inflexible molecular structure (see e. g. /2/), and (or) a too low stable proton transferred one result in an inability of the "colored" form to its stabilization under even the most favorable conditions.

In other words the very low intramolecular potential barrier of the reverse reaction almost without the structural transformations from the instable "colored" structure cannot beheightened somehow by the intermolecular interactions in the rigid media (including crystals).

Thus the L-photochromism can only be manifested by such structures under any conditions.

The slight modification of the molecular structure can be an only way for the stabilization of the "colored" form.

In this case there is a good chance for the change of the L-type structures for the LT-type and even T-one of the compounds with the weakly flexible structures allowing the rotation (or twist) of the aromatic ring around the C-C single bond or the change of the mutual positions of the ESIPT reaction centers.

The later situation is realized in Hydroxychalcones with the modification of the structure XI (1a)(L-type) to the XI(2) one (T-type).

In the former case, e. g. for Bipyridines (see /3/), Triazines (VI), Azoles (IX) and Arylethylenes (XIII') the stabilization of the "colored" structure can be provided by both the electronic and the steric factors.

The typical example of such a modification is the Azolestructures(IXa(13), IX1b(13), IXc(1, 2)) (Sch. 5. 11_1) which are inflexible analogs of SAs structures of the L-type photochromes with the short-lived "photocolored" form.

In all these structures the rotation (twist) around C-C single bond is possible like SAs (Sch. 5. 11_1).

However the shortlived "photocolored" form like SA appears in the solvents(L-photochromism) only with the insertion of the S-heteroatom (IXc_1), and the inclusion of the bulky substitution in the phenyl ring leads to the both the L-photochromism in the solvents and to the stabilization of the of the twisted "colored "structure similar SA s and hence to the T-photochromism of the structures IX(2, 3) in the crystals(IX(2, 3) of the LT-type)).

The classification in the groups by the ESIPT nature bears the formal character convenient only for the consecutive consideration of the structures with the different H-bond.

Moreover the proton transfer is not a limiting step of the direct or reverse photochromic reactions(excluding the structures XX).

Therefore it cannot be expected for any correlation between these two kinds of the classifications (Tab. 5. 23).

The above data about the ESIPT photochromic molecular systems open the ones with both the known and the novel photochromic structures on the base of the ESIPT(ESIHT) and the GSIPT.

References

1. S. Formosinho, and L. Arnaut, Photochem. Photobiol. A. Chem., 75, 21(1993)(Review).
2. C. Kim, and T. Joo, Phys. Chem. Chem. Phys. 11, 10266 (2009).
3. K. Tokumura, O. Oyama, H. Mukaihata, and M. Itoh, J. Phys. Chem. A 101(8), 1419(1997).
4. R. Nakagaki, T. Kobayashi, and S. Nagakara, Bull. Chem. Soc. Jpn., 511(6)1611(1978).
5. P. Barbara, L. Brus, and P. Rentzepis, J. Am. Chem. Soc. 102(17)5631(1980).
6. M. Taneda, Y. Kodama, Y. Eda, H. Koyama, and T. Kawato, Chem. Lett. 36(12), 1410(2007).
7. K. Animoto, and T. Kawato, Photochem. Photobiol. C. Photochem. Rew. 6, 207(2005).
8. M. Knyazhansky, and A. Metelitsa, "Photoinduced Processes of Azomethines' molecules and their structural analogs". Publ. house of Rostov State Univer., Rostov-on-Don(1992) (in Russian).
9. M. Knyazhansky, A. Metelitsa, M. Kletskii, A. Millov, and S. Bezugliy, J. Mol. Struct. 526, 65(2000).
10. A. Grabowska, K. Kownacki, and L. Kaczmarek, Acta Phys Polon. A 88, 1081(1995).
11. The chemistry of Hydrozones, Yu. Kittaev, Ad., Publ. House "Nauka" (1977)(in Russian).

12. H. Shizuka, K. Matsui, Y. Hirata, and I. Tanaka, J. Phys. Chem., 80, 2070 (1976).
13. H. Shizuka, K. Matsui, Y. Hirata, and I. Tanaka, J. Phys. Chem. 81, 2243 (1977).
14. H. Shizuka, M. Machii, Y. Higaki, M. Tanaka, and I. Tanaka, J. Phys. Chem. 89, 320 (1985).
15. M. Moriyama, Y. Kawakami, S. Tobita, and H. Shizuka, Chem. Phys. 231, 205(1998).
16. M. Moriyama, M. Kosuge, S. Tobita, and H. Shizuka, Chem. Phys. 253, 91(2000).
17. W. Al-Soufi, K. Grellmann, and B. Nickel, Chem. Phys. Lett. 174, 609(1990).
18. J. Stephan, and K. Grellmann, J. Phys. Chem. 99, 10066 (1995).
19. C. Potter, R. Brown, F. Volmer, and W. Rettig, J. Chem. Soc. Faraday Trans. 90, 59(1994).
20. D. le Gourrierec, V. Kharlanov, R. Brown, and W. Rettig, J. Photchem. Photobiol. A. Chem., 117, 209 (1998).
21. M. Kasha, J. Chem. Soc. Faraday Trans. 2 (82)2379(1986).
22. J. Waluk, Polish. J. Chem. 82, 947(2008).
23. E. Hadjoudis, in "Photochromism, Molecules and Systems", H. Durr and H. Bouas-Laurent Eds., Elsevier, Amsterdam(1990).
24. R. Ellam, P. East, A. Kelly, R. Khan, J. Lee, and D. Lindsey, Chem. Ind. 74(1974).
25. S. Liu, X. Chen, Y. Xu, X. You, W. Chen, and Z. Arifin, Acta Cryst. C54, 1919(1998).
26. T. Arai, and M. Ikegami, Chem. Letters, 965(1999).
27. M. Barbatti, A. Aquino, H. Lischka, Ch. Schriver, S. Lochbrunner, and E. Riedle, Phys. Chem. Chem. Phys. 11, 1406(2009).
28. P. Stenson, Acta Chem. Scandinavica, 24, 3729(1970).
29. M. Taneda, R. Koyama, and T. Kawato, Chem. Lett. 36, 354(2007).

30. H. Harada, Y. Uekusa, and Y. Ohashi, J. Am. Chem. Soc. 121, 5809 (1999).
31. G. Spence, E. Taylor, and O. Buchardt, Chem. Rews. 70(2), 231(1970).
32. A. Metelitsa, O. Lyashik, N. Volbushko, I. Andreeva, M. Knyazhansky, E. Medyantseva, and V. Min-kin, J. Org. Khim. 22(7)1446(1986) (in Russian).
33. A. Metelitsa, M. Knyazhansky, V. Volbushko, B. Nedzvetsky, and V. Minkin, Khim. Heterocykl. Soed. (10), 1373(1990) (in Russian).
34. Y. Norikane, H. Itoh, and T. Arai, J. Phys, Chem. A, 106, 2766 (2002).
35. T. Teshima, M. Takeishi, and T. Arai, New J. Chem. 33, 1393(2009).
36. R. Matsushima, S. Fujimoto, and K. Tokumura, Bull. Chem. Soc. Jpn. 74, 827(2001).
37. H. Horiuchi, H. Shirase, T. Okutsu, R. Matsushima, and H. Hiratsuka, Chem. Lett., 29(2), 96(2000).
38. M. Knyazhansky, N. Makarova, A. Metelitsa, A. Shiff, V. Pichko, and E. Ivakhnenko, Mol. Cryst. Liq. Cryst. 298, 115(1997).
39. T. Ivakhnenko, N. Makarova, E. Ivakhnenko, V. Minkin, and M. Knyazhansky, Int. J. Photoenergy 1, 1(1999).
40. T. Ivakhnenko, N. Makarova, E. Ivakhnenko, I. Chuev, S. Aldoshin, V. Minkin, and M. Knyazhansky, Proceed. Russian Acad. Science 49(12), 1981(2000).
41. T. Arai, M. Moriyama, and K. Tokumaru, J. Am. Chem. Soc. 116, (7), 3171(1994).
42. T. Arai, M. Obi, T. Ivassaki, K. Tokumaru, and F. Lewis, J. Photochem. Photobiol. A. Chem. 96, 65 (1996).
43. M. Obi, H. Sakuragi, and T. Arai, Chem. Lett. 169(1998).
44. V. Komissarov, V. Kharlanov, L. Ukhin, and V. Minkin, Dokl. ANSSSR, 301, 902(1988)(in Russian).

45. S. Aldoshin, M. Novozhilova, L. Atovmyan, V. Komissarov, V. Kharlanov, L. Ukhin, and V. Minkin, Russian Chem. Bull. 1121(1991).
46. V. Komissarov, V. Kharlanov, L. Ukhin, Z. Morkovnik, V. Minkin, and M. Knyazhansky, J. Org. Khim. 26, 1106 (1990)(in Russian).
47. E. Postupnaya, V. Komissarov, and V. Kharlanov, J. Org. Khim. 29, 1915(1993)(in Russian). ss
48. V. Komissarov, E. Gruzdeva, V. Kharlanov, L. Olekhnovich, G. Borodkin, V. Khrustalev, S. Lindeman, Yu. Stryukov, V. Kogan, and V. Minkin, Russian Chem. Bull. 46, 1924(1997).
49. V. Minkin, V. Komissarov, and V. Kharlanov, in "Organic Photochromic and Thermochromic Compounds "Eds.:J. Crano, and R. Guglielmetti, v1, Ch. 8, Plenum Press, N. Y. (1998).
50. V. Kharlanov, J. Photochem. Photobiol. A. Chem. 122, 191(1999).
51. S. Bezugliy, M. Kletskii, and M. Knyazhansky, Unpublished results (2006).
52. J. Reth, and K. Gerisma, Recl. Trav. Chim. Pay-Bas 64, 41(1945).
53. H. Irving, G. Andrew, and E. Risdon, J. Chem. Soc. 541(1949).
54. J. L. A. Webb, I. Bhata, A. Corwin, and A. Sharp, J. Am. Chem. Soc. 72, 91(1950).
55. L. Meriwether, E. Breither, and C. Sloan, J. Am. Chem. Soc. 87, 4441(1965).
56. L. Meriwether, E. Breither, and N. Colthup, J. Am. Chem. Soc. 87, 4448(1965).
57. A. Hutton, and H. Irving, J. Chem. Soc. Chem. Comm. 1113(1979).
58. A. Hutton, and H. Irving, J. Chem. Soc. Dalton Trans. 2299(1982).
59. A. Hutton, H. Irving, L. Massimbent, and G. Gafner, Acta Crystallogr. Sec. B36, 2064(1980).

60. N. Comhout, and A. Hutton, Appl. Organometal. Chem. 14, 66(2000).
61. T. Okada, M. Kawanisi, H. Nozaki, M. Toshima, and H. Hizai, Tetrahedron Lett. 927 (1969).
62. A. Padwa, W. Bergmark, and D. Pashyan, J. Am. Chem. Soc. 91, 2653(1969).
63. B. Fraser-Reid, McLean, and E. Ushenvood, Canad. J. Chem. 47(1969).
64. P. Core, J. Hoskins, K. Lott, and K. Waters, Photochem. Photobiol, 11, 551(1970).
65. V. Kumar, and K. Venkatesan, J. Chem. Soc. Perkin Trans. 2, 289(19991).
66. T. Sakar, S. Chosh, J. Moodthy, J-M. Fang, S. Nandy, N. Sathyamurthy, and D. Chakraborty, Tetra-hedron lett., 4141, 6909(2000).
67. J. Murthy, P. Mal, R. Natarajan, and P. Venugopalan, Org. Lett. 10, 1579(20001).
68. Y. Yokoyama, Y. Karimo to, Y. Saito, M. Katsurada, T. Okada, Y. Osano, C. Sasaki, H. Tukada, M. Ada-chi, S. Nakamura, T. Murayama, T. Harazona, and T. Kodairo, Chem. Lett. 33, 106(2004).
69. G. Wettermark, Nature 194, 677(1962).
70. G. Wettermark, and R. Ricci, J. Chem. Phys. 39, 1218(1963).
71. G. Wettermark, J. Phys. Chem. 66, 2560(1962).
72. G. Wettermark, E. Black, and L. Dogliotty, Photochem. Photobiol. 4, 229(1965).
73. H. Morrison, and B. Migdalof, J. Org. Chem.. 30, 3396(1965).
74. A. Sergeev, R. Nurmukhametov, and R. Barov, Khim. Fiz. 8, 1096(1982)(in Russian).
75. E. Hadjoudis, A. Tsoka, and G. Wettermark, J. Photochem. 8, 233(1978).
76. J. Margerum, L. Miller, E. Saito, M. Brown, H. Mosher, and R. Hardwick, J. Phys. Chem. 66, 2434(1962).
77. G. Wettermark, J. Am. Chem. Soc. 84, 3658(1962).

78. N. Toshima, M. Saeki, and B. Hirai, Chem. Comm. 1424(1971)
79. A. Chichibabin, B. Kuindzhi, and S. Benewolenskaja, Ber. Dtsh. Chem. Ges. 28, 1580(1925).
80. W. Clark, and G. Lothian, Trans. Far. Soc. 54, 1790(1958)
81. R. Hardwick, H. Mosher, and B. Passailaigue, Trans. Farad. Soc. 56, 44 (1960).
82. H. Mosher, C. Sower, and R. Hardwick, J. Chem. Phys. 32, 1888(1960).
83. H. Mosher, E. Hardwick, and D. Ben-Hur, J. Chem. Phys. 37, 904(1962).
84. J. Sousa, and J. Weinstein, J. Org. Chem. 27, 3155(1962).
85. J. Sousa, and J. Leconti, Science 146, 397(1964).
86. H. Hiraoka, and R. Hardwick, Bull. Chem. Soc. Jpn. 39, 39, 380 (1960).
87. K. Seff, and K. Trueblood, Acta Cryst. B24, 1406(1968).
88. D. Ben-Nur, and R. Hardwick, J. Chem. Phys. 51, 2240 (1972).
89. E. Klemm, D. Klemm, A. Graness, and J. Kleinschmidt, Chem. Phys. Lett. 55, 113(1978).
90. E. Klemm, D. Klemm, A. Graness, and J. Kleinschimdt, Chem. Phys. Lett. 55, 503(1978).
91. D. Klemm, and D. Klemm, J. Prakt. Chem. 320, 551(1978).
92. D. Klemm, E. Klemm, A. Graness, and J. Kleinschmidt, Z. Phys. Chem. Leipzig 260, 555(1979).
93. D. Klemm, and E. Klemm, J. Prakt. Chem. 321, 404(1979).
94. E. Klemm, and D. Klemm, J. Prakt. Chem. J. Prakt. Chem. 321, 407(1979).
95. E. Klemm, D. Klemm, A. Graness, and J. Kleinschmidt, J. Prakt. Chem. 332, 348(1980).
96. A. Graness, E. Neumann, J. Klenschmidt, W. Trieble, D. Klemm, and E. Klemm, Adv. Mol. Relax. Process. 22, 89(1982).
97. K. Yokoyama, and T. Kobayashi, Chem. Phys. Lett. 85, 175(1982).

98. H. Takahashi, S. Hurikawa, S. Suzuki, Y. Torii, and H. Isaka, J. Mol. Structure, 146, 91(1986).
99. H. Takahashi, T. Kobayashi, N. Kaneko, I. Igarashi, and T. Yagi, J. Raman Spectrosc. 12, 125(1982).
100. H. Sixl and R. Warta, Chem. Phys. 94, 147(1985).
101. S. Atherton, and B. Craig, Chem. Phys. Lett. 127(1), 7(1986).
102. S. Chattopadhyay, and B. Craig, J. Phys. Chem. 91, 323(1987).
103. M. Scherl, D. Haarer, J. Fisher, A. Decian, J-M. Lahn, and Y. Eichen, J. Phys. Chem. 100, 16175(1996).
104. S. Knatib, S. Tal, O. Godsi, U. Peskin, and Y. Eichen, Tetrahedron 56, 6753(2000).
105. A. Corval, K. Kuldoba, Y. Eichen, Z. Pikramenou, J. Lehn, and Tromsdorff, J. Phys. Chem. 100, 19315 (1996).
106. I. Frank, D. Marx, and M. Parrinello, Phys. Chem. A, 103, 7341(1999).
107. I. Frank, S. Grimme, and S. Peyerimhoff, J. Phys. Chem. 100, 16187(1996).
108. P. Naumov, A. Sekine, H. Uekusa, and Y. Ohashi, J. Am. Chem. Soc. 124, 8540(2002).
109. Y. Eichen, M. Botoshansky, U. Peskin, M. Sherl, D. Haarer, and S. Knatib, J. AmChem. Soc. 119, 7167 (1997).
110. S. Knatib, M. Botoshansky, and Y. Eichen, Acta Cryst. Struct. Sci. B53, 306(1997).
111. A. Schmidt, S. Kababya, M. Appel, S. Knatib, M. Botoshansky, and Y. Eichen, J. Am. Chem. Soc. 121, 11291(1999).
112. P. Naumov, and Y. Ohashi, Acta Cryst. 360, 343(2004).
113. P. Naumov, and Y. Ohashi, J. Phys. Org. Chem. 17, 865 (2004).
114. Y. Eichen, J-M. Lehn, M. Scherl, D. Haarer, J. Fisher, A. Decian, A. Corval, and H. Tromsdorff, Angew. Chem. Int. Ed. Engl. 34, (22)2530(1995).
115. H. Kimura, K. Sakata, and H. Takahashi, J. Molec. Struct. 1-4, 293(1993).
116. K. Kuldoba, A. Corval, H. Tromsdorff, and J. Lehn, J. Phys. Chem. A, 101, 6850(1997).

117. O. Ziane, R. Casalegno, and A. Corval, Chem. Phys. 250, 199 (1999).
118. R. Nurmukhametov, "The absorption and luminescence of the aromatic molecules". Moskow, Publish house "Khimia" (1971)(in Russian).
119. S. McGlynn, T. Azumi, and M. Kinoshita "Molecular Spectroscopy of the Triplet State", Prentic. Inc. (1969).
120. G. Desiraju, Angew,. Chem. Int. Engl., 34, 2311(1995).
121. G. Desiraju, Acc. Chem. Res. 29, 441, (1996).
122. P. Langley, J. Hilliger, R. Thaimattam, and G. Desiraju, New J. Chem. 22, 1307(1998).
123. B. Tinland, and C. Decoret, Tetrahedron Lett. 27, 2467(1971).
124. V. Danilova, and G. Turovets, Izv. Vysch. Ucheb. Zaved. Fiz. 13, 141(1971)(inRussian).
125. G. Turovets, and V. Danilova, Izv. Visch. Ucheb. Zaved. Fiz. 15, 68(1972)(in Russian).
126. Y. Eichen,, J-M. Lehn, M. Scherl, D. Haarer, R. Casalegno, A. Corval, K. Kuldoba, and H. Trommsdorf, J. Chem. Soc. Chem. Comm. 713(1995).
127. R. Casalegno, A. Corval, K. Kuldoba, O. Ziane, and H. Trommsdorff, J. Lumm. 72-74, 78(1997).
128. N. Turro, "Modern Molecular Photochemistry", Univ. Science Books (1991). Dcddd

Chapter 6

The General Information About Possible Application Of The Esipt Photochromism.

Introduction and the principle requirements for the photochromic systems.

The commercial applications of the photochromic systems are possible in form of the various material, including both the crystals(polycrystalline films), polymer matrices(films and bulky species), and the liquid media (e. g. solutions).

The extensive investigations of the various applications of the photochromic materials based on the organic photochromic systems, including the ESIPT photochromism, have been started in the beginning of the seventies of the last century(see /1-39/in the chronological succession, and continue now actively in particularly in connection with the development of the technology of the nanoscopic materials (e. g. /40/).

The most widespread and effective applications of the photochromic systems for the molecular switches and the information storage on the molecular level in the electronic devices with the ultrahigh memory densities are based on the utilization of the crystal and polymeric (including nanoscopic) materials.

The application of the photochromic material in form of the liquid crystals and the Langmuier-Blodjet (LB)films can also be prospective for the electronic devices in which the optical and elec-trical parameters (electro-optical characteristics) change with the photochromic transformation.

The utilization of the polymeric materials is prospective also in the light filters, photochromic lenses and protective glasses.

The liquid media can be used for the actinometrical measurements, liquid light filters, for the analytic applications, and as structural marks in the molecular biology and the physico-chemical studies.

All applications of the photochromic systems are caused in principle by utilization of the interconvertion between the two forms A and B (bistability). At least one of which(B) is metastable colored(with the absorption band in the visible region of the spectra) structure generated by light and decays thermally (in the dark) or (and) under light irradiation.

The bistability is the necessary condition for the applications but the "true" (T) photochromism i. e. "colorability" as a result of the heightened stability of the colored structure(see ch. 5 conclusion)is often not utilized directly.

However as it follows from the results of investigations, side by side with bistability there is the series of other, often contradictory, requirements for the photochromic systems to be suitable for various practical devices (see e. g. /20/).

i. The both form should be clearly differ and readily detectable by the spectral or by some kind of other methods.
ii. The very effective photoinduced conversion A-hv->B and as low as possible efficiency of the thermal conversion A-Δ->B.
iii. There are several possibilities of the back bleaching reactions in dependence of on application type.

(a) The very slow (from hours to years) back thermal conversion B-Δ->A with combination of the high efficiency photoinduced back transformation B-hv->A("True" (T) photochrochromism) middle thermal decay time.

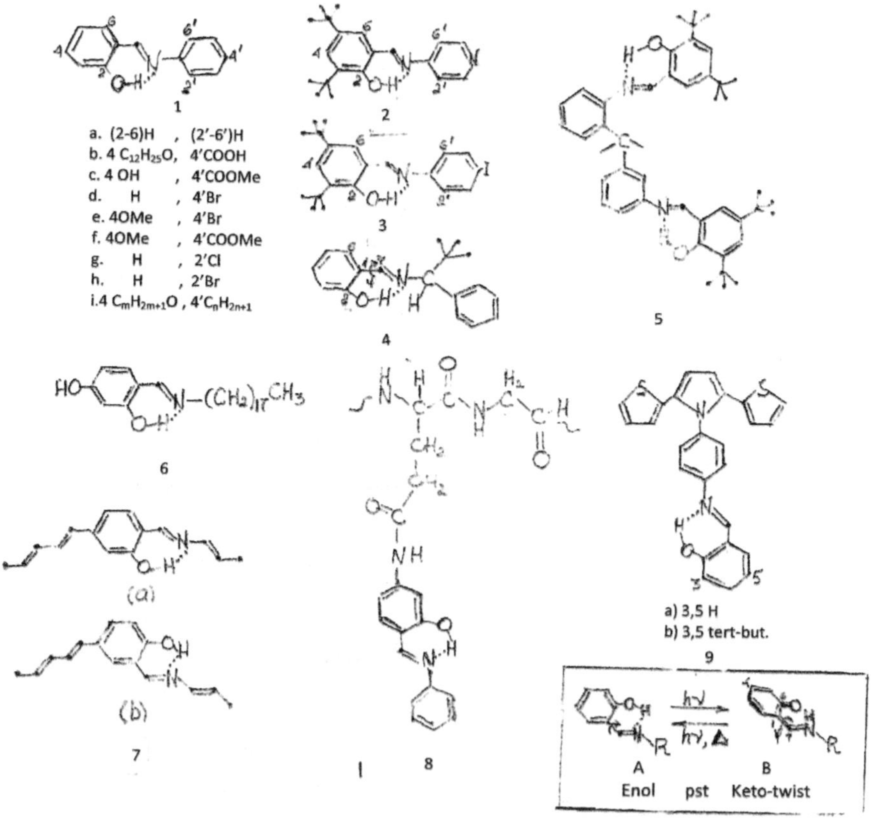

ESIPT Photochromism

Scheme 6.1

The compounds on a base on the Photochromic anils prospective for the applications. The Photochromic structural transformations (pst) (A ⇔ B) typical for anil molecules. (ESIPT and Twist) are shown in the frame.

Compound 1 substituents:
a. (2-6)H, (2'-6')H
b. 4 $C_{12}H_{25}O$, 4'COOH
c. 4 OH, 4'COOMe
d. H, 4'Br
e. 4OMe, 4'Br
f. 4OMe, 4'COOMe
g. H, 2'Cl
h. H, 2'Br
i. 4 $C_mH_{2m+1}O$, 4'C_nH_{2n+1}

Compound 9:
a) 3,5 H
b) 3,5 tert-but.

A Enol — pst — B Keto-twist

Table 6.1

The SHG switching characteristics of Photochromic anils (polycrystalline film, 1907nm fundamental beam) /43, 45, 49/.

General properties	Characteristics		Structure I 1-3			
			1c	1d	2	3
Structure	Crystal space group		P_{ca}	$P_{ca}2_1$	$P3_2$	----
SHG	SHG intens./Urea		10	2	3	Weak
	SHG contrast, %		10	10	30	50
Photochromism	UV-Irrad.	Wave length λ (nm)	405	365	365	200-400
		Irrad. Time (min)	10	1	5	5
	Dark decay time (s) Slow component		3.7	33	5.5x10^7	208

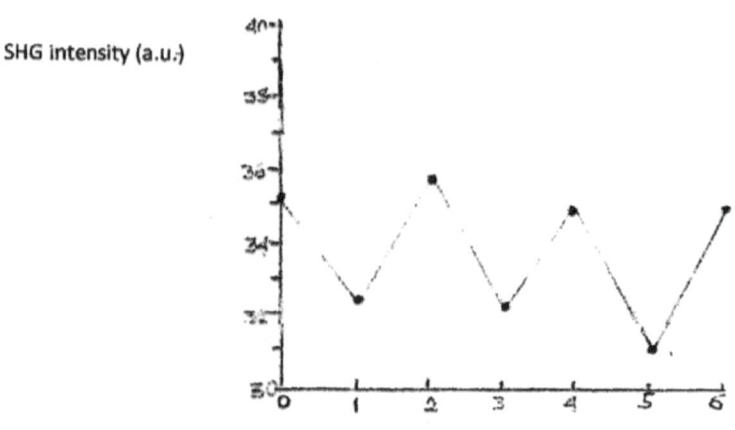

Scheme 6.2
Illustration of the SHG photoswitching at 1907nm.
On x axis odd and even numbers correspond to after UV and visible irradiation respectively.

- (b) The middle thermal decay time (several minutes-several hours) of the colored form (B-->Δ----->A) with the combination of high efficiency of the photoinduced back transformation (B--->hv->A (also T-photochromism).
- (c) The fast, spectral detectable only(invisible) thermal conversion B-Δ->A without participation of photobleaching ("Latent" (L) photochromism).

iv. As short as possible response time should be reached with use of the photoreaction mechanism involving the step starting the fast switching cycle.

v. A non-destructive read-out procedure should be available. The read-out method should not interfere with or erase the written data.

vi. High quantum yields of the photocoloration (A-hv->B) and the photobleaching (B-hv->A) (in the case iiia, b) should be achieved for the efficiency of the switching processes to avoid a long irradiation times.

vii. The photochromic material should be fatigue resistant i. e. the photo and thermal reactions A<=>B in the photochromic molecules interacting with the substances of the material should not have as far as possible some side path-ways leading to the stable side phototoproducts.

viii. The retention and useful modification of the switching kinetics and also spectral and other properties inculcated in the composition of the photochromic material should be reached.

The each of the above demands can be typical for only certain kinds of the applications but cannot be necessary or even can be bad for the other ones.

In spite of the widespread applications of the various organic photochromes there are no up to present to the best of our knowledge some generalized data(excluding the old reviews /7/, /23/) of the ESIPT photochromism application.

Therefore some examples of various especially recent applications will be discussed below on the base of the requirements i-viii.

6. 1. The materials for switching and information storage.

The earliest proposals for anils (I(1a, b)(Sch. 6. 1)as the possible candidates for the optical memory and switching devices have been made in the end of eighties/41/due to the strict fulfillment of the demands i(difference between absorption bands of the initial and colored forms $\Delta\lambda \approx 150\text{-}200 nm$), iv (the primary step, ESIPT, occurs in the subfemtosecond time scale (see Ch. 3)and especially vii(the enormous amounts (10^5) of the repeating photocoloration⇔photobleaching cycles /42/.

However the major problem for the information storage with such structures is noncompliant to the requirement iiia (the colored form fades in the crystal for time from the several minutes to the several hours).

Thus such ESIPT photochromic systems are notsuitable for the devices of the information storage at least under usual conditions, and need the special modifications for this purpose.

Meanwhile these compounds in form of the polycrystallic or plastic materials answer the requirement iiib in the wide temperature interval (-20°-60°C)(see Ch. 3).

At the same time it has revealed that the crystalline anile structures (I(1)) can exhibit the molecular second order nonlinear optical (NLO) properties manifested, in particular, in the Second Harmonic Generation (SHG)(/41, 43, 44/. Such properties are typical only for the anil molecules crystallized in the noncentrosymmetrical space groups (see e. g. Tab. 6. 1).

This is not surface effect which depends on the penetration depth of the excitation light/45/ ("3d" bulk material).

Table 6.2
Photochromic characteristics of the compound I_4 (SMB) in the crystal at ambient temperature.

Charac- Structeristics tures ↓	Spectroscopy		Kinetics		1)	Dipole moment(μ) & angles [2]		
	Absorp [3] λ^{max}_{abs} nm	Fluoresc. λ^{max}_{flu} nm	τ_{rise} (s)	τ_{decay} (s)	Heat of form Kcal/mol	Magnitude μ (D)	Angles (degrees)	
							α	β
Initial OH (E)	317	500	---	---	216	2.36	0	9.5
Photocolor. NH (K)	372	565 574	5	~300	25.3	4.65	68	39.8

1)Irradiation for τ_{rise} with λ=438 nm, τ_{decay} in the dark.2) See scheme 6.3 (below).3)CHCl$_3$,c=5x10^{-5}M.

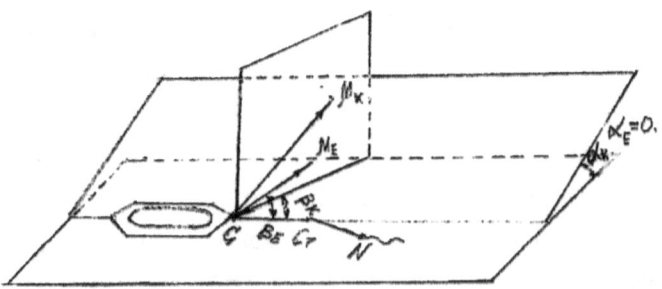

Scheme 6.3
The change of the dipole moment of the structure I_4 with transformation from the Enole (μ_E) structure to the colored keto (μ_K) structure/52/.

The NLO response is enhanced strongly of the molecular structures with the two "push-pull" substituents(e. g. OH and Ester groups at the opposite ends of the anil structure (Str. I(1c)Sch. 6. 1).

Really the findings of the detailed investigation /46/ of the two compounds, I(1c) and I(2), by experimental(UV-VIS absorption spectroscopy, Hyper-Rayleigh scattering method, and Electronic Field Induced Second Harmonic Generation technique), and theoretical(PM2, Density Functional

Theory(DFT) and time Dependent DFT) methods indicate the principal role of the first hyperpo-larizbility (β) in the NLO responses. Indeed, the ratio between values of the two Enol forms of the molecules I(1c) with "push-pull" substituents) and I(2) is equal to $\beta(I(1c))/\beta(I(2))=2.40$ that is in the good qualitative agreement with measured ratio $\beta(I(1c))/\beta I(2)=1.60$ and the SHG intensity increases accordingly for the "push-pull" molecular structure as compared with another anil structures without "push-pull" substituents (I(1d), I(2), I(3) and other) (see e. g. /47/) (Tab. 6. 1).

At the same time the values ratio for Enol (E) and NH-keto (K) structures calculated for the molecules I(1c) and I(2) is equal to 2-3 (both in the gas phase and the various solvents) demonstrating that the Enol structures have to give much more large NLO responses than the Keto ones in the different media that may be caused by the conjugation break between the aromatic rings in the acoplanar NH structure as a result of the ESIPT and "keto"-ring twist.

Thus photochromic(LT) anils can be act as NLO switches in the crystalline state also. Therefore some earlier /43/ and especially recently /45, 48, 49/ the compounds of the type(I) that are photochromic in the crystals and display the NLO properties in the polycrystalline films, and powders have been studied as SHG modulators of the NLO response.

Such a modulation can be achieved by utilization of the reversible photoswitching as a result of the photochromism of the I(1) structures with sufficently fast dark and especially photoreverse reactions (Sch. 6. 2).

More than twenty compounds of the I(1, 2, 3) type have been synthesized and studied but only eight ones both photochromic in the crystal and SHG active. Although the bulk substituent method (tert-butyl groups) has been used to obtain the photochromic crystal(see Ch. 3) only half of such structures can be utilized for the SHG switching.

The three main structural factors have been taken into account to obtain and study of the target crystalline photochromic anils.

(i). All crystalline compounds have to crystallized into the noncentricsymmetrical crystalline packing,
(ii). Realization of the "push-pull" molecular electronic structure with variation of the electron-donating and electron-withdrawing groups on the opposite ends of the conjugated system which provides the charge transfer into and from the central azomethine bridge group with the O-H...N Hydrogen bond and
(iii). The variation of the colored form stability (bleaching rate)with the insertion of the different substituent (including the bulk groups)for the SHG contrast control (modulation degree). The four compounds have been chosen for the detailed investigation of the SHG switching (Tab. 6. 1) /45, 49/.

Although in the liquid solutions the colored form disappears in the millisecond time scale (see Ch. 3)and can be detected only by the transient spectroscopy method, in the crystals the slow component of the colored form lifetime varies from the seconds to the ~1. 5 years at the ambient temperature. The SHG measurements have been made at fundamental wavelength 1907nm in order to avoid the reabsorption of the second harmonic beam by enol and/or colored isomers (unlike/43/ where fundamental wavelength $\lambda=1064$nm has been also used).

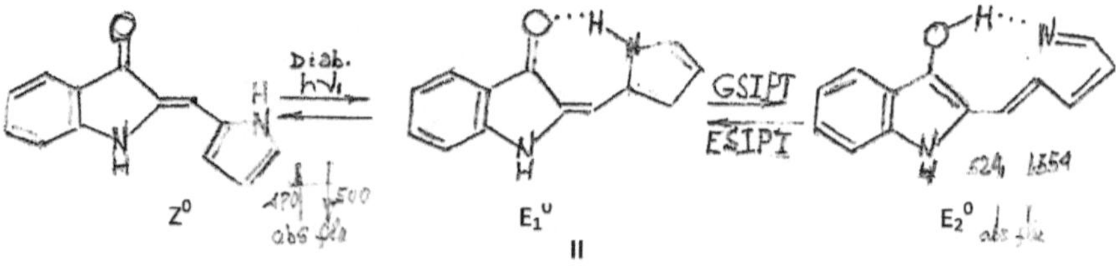

Scheme 6.4
The suggested structural scheme (with the additional structure E_2^0) of the Photochromic reaction for the possible storage and photoerasing of an optical information(see table 6.3 and also the section 5.2.5).There is no a dark bleaching.

Table 6.3
The parameters of the structure II concerning a recording of the optical information.

Structure	Spectral character.		Recording (Z^0----hv-->E^0)			Reading out		Erasing (E^0----hv--.>Z^0)		
	λ^{max}_{abs} (nm)	λ^{max}_{flu} (nm)	λ_{flu} (nm)	τ (s)	Efficiency	λ_{abs} (nm)	λ_{flu} (nm)	$\lambda_{abs,eras.}$ (nm)	τ (s)	Efficiency
Z^0	470 *)	500	366	420	0.3	0.3	----	----	----	----
E_1 or E_2	524	554	----	----	----	524	554	546	~100	0.03

*)The longwave lengfth absorption band.

Scheme 6.5
The simplified structural scheme of the photoreversible recording of optical information on the base of the photochromic molecule DNBP (see 5.3).

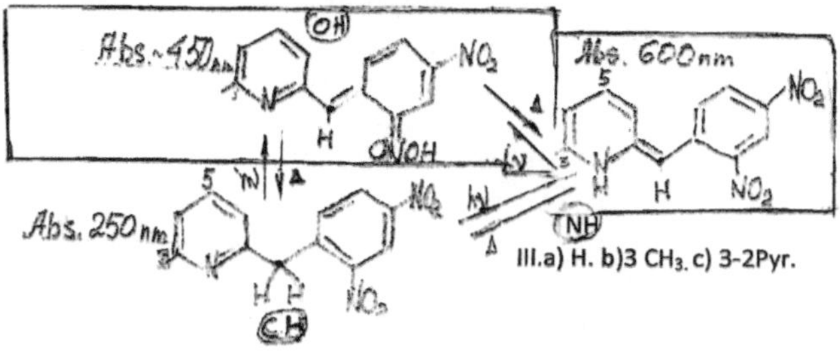

III.a) H. b)3 CH_3. c) 3-2Pyr.

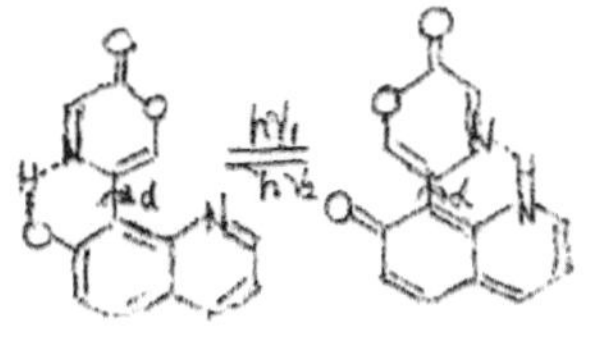

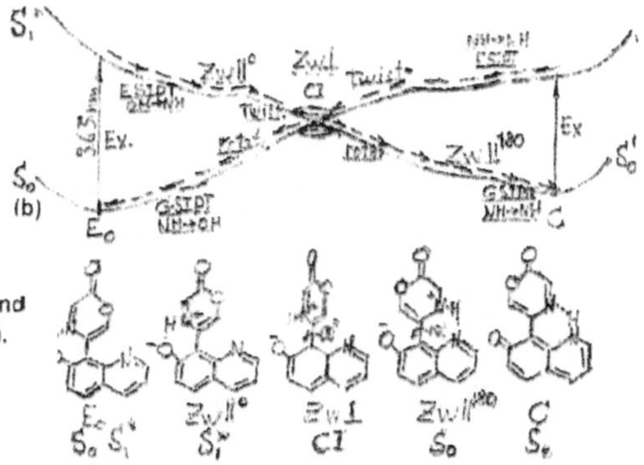

IV (a) Scheme 6.6
The hypothetic structure (7HXQ,see text)(a) and its application for switching and recording (b).

The SHG intensity was measured at irradiation time (1-10min) when the steady-state point was reached by the UV irradiation (by λ=200-400nm) and was compared with the initial value to deduce the SHG contrast that changes within the limits 10-50 %.

The reversibility was achieved not only by the dark reverse reaction but mainly with the visible irradiation (with λ=500, 532nm), and at least 40 cycles were observed (see also/42/).

The dependence of the SHG intensity(vs urea)on the molecular structure of the I(1-3) compounds is caused by the positions of the electron donor and acceptor groups involving in the conjugated π-electron system which may yield the optimal hyperpolarizabilities( ) providing the NLO properties /45, 50/.

The efficiency of the SHG modulations (SHG contrast) increases in the series I(1a, 1d, 2, 3) with rise of degree of the OH(E)->NH(K) conversion (i. e. with the increase of the colored form(K) time de-cay)(Tab. 6. 1).

The obvious advantages in the performance of the SHG switches on the base of the photochromic anils include the ultrafast generation of the colored form($\tau \approx$100-200ps)(Ch. 3) and an unique fatigue resistance/42/.

Such switches must be optimized by stabilization of the NH-form in the structures with "push-pull" substituents, better conversion in the solid state, and by optimizing of penetration depth of the excitation light beam in the bulk of material.

Such materials can be used as components in the optical signal processing and as optical memory devices /45/.

Irrespective of the SHG modulation some photochromic crystalline anils can be considered as perspective compounds for material of the recording and storage of the optical information.

The crystalline photochromic compound I(4) (N-salicylidenemethylbenzylaniline (SMB) meets almost the all demands for the devices of the molecular memory and switches(i-vii)/51/(see also Tab. 6. 2).

The fast response, high efficiency of generation, and clear spectral (including fluorescence) distinction of the colored form along with the excellent fatigue resistance are the very attractive properties for the application of I(4) in the material of switching and the information record.

Moreover unlike another compounds I, both the initial and colored forms of I(4) are fluorescent in the crystal that is very convenient for the fast recording of the weak signals of the optical information.

Unfortunately the marked difference in the energy of the stabilization of the Enol and Keto forms (Tab. 6. 2) and sizeable free space in the crystalline packing can't provide the sufficiently long duration for the information storage (demand iiib) in the crystal, and the peculiarities of the molecular and crystalline structures of I(4) cannot meet the demands of the SHG modulation(see above).

Thus the search of the rigid media for the photochromic systems including the compound I(4) that can be used for the information storage is the topical problem. This problem can be connected with the difference of the molecular and electronic structures of the initial and colored forms.

According to the data of the quantum-chemical calculations /52/ the structural transformations of the Enol form to the colored structure (see Ch. 3) are accompanied with the drastic-cally changes both the magnitude and the orientations of the electric dipoles moments (Tab. 6. 2, Sch. 6. 3).

In accordance with /52/the utilization of the photochromic monolayer of SMB disposed on a silicon surface can allow to write down any information with light due to the not only change of the reflection spectra but also due to the reorientation of the "colored" molecules on the surface as a result of the dipole-dipole interaction change.

Such a monolayer on the Si surface can also play role of the rigid media that can restrict the reverse dark reaction and stabilize strongly the photocolored form to obtain the material for the long-time information storage. It would be interesting to prolong the investigations in that direction.

The photochromic compound II(Sch. 6. 4)(see also Sec. 5. 2. 5. Str. viii) can also be considered as an example of the probable application. The photochromism of the compound II is observed in the liquid solvents as a considerable color change under irradiation, and can be considered as a "negative "photochromism (color shift towards "blue "side)/53/(see Sec. 5. 25). Under irradiation by the visible and UV light the relative intensities of the absorption bands are change in dependence of the irradiation wavelength.

The two or three structures ($E_1°$, $E°$, $Z°$)take part in the steady-state equilibrium (Sch. 6. 4). The equilibrium shifts almost completely towards $E°$ (λ=524nm) under irradiation by λ=366nm and backwards $Z°$(λ=470nm)under irradiation by λ=546nm.

The readout of information can be done on the λabs=524nm or by the fluorescence band with λflu=554nm. The considerable dark stability of the E° and Z° structures with the relatively low efficiency of the photofatigue(~1% per cycle)in the liquid benzene can be good precondition for the utilization of these compounds in the solid or liquid materials for the switching and the storage of information.

From the results of the detailed studies produced above (Sec. 5. 3, str. XX)the crystal compound III (Sch. 6. 5)have the strongly pronounced photochromic properties with color change from pale –yellow to blue-purple that allows to record and read out the optical information with light in the range of the absorption bands with λmax=250nm and 600nm correspondingly. The efficient photo-bleaching i. e. photo-erasing of the information can be done by irradiation in the range of λ=600nm (requir. iiib). In this connection the nondestructive readout of the information can be done with the IR light in the vibration band ν≈3338cm^{-1} typical for the NH structure (recuir. v).

The colored structure generation (absorption band λmax≈600nm) i. e. recording of the information is underline by the fast EIPT reaction (τ≈10^{-8}-10^{-1}⁰s)(requirement iv).

The evident advantage of the crystal compound III in terms of the optical information storage is the very high colored form stability that strongly depends on the temperature and the molecular structure, and the colored form lifetime (τ)can achieve 14h(298K) and 150h(200K) for compounds IIIa and IIIc correspondingly, and practically no decay observed for IIIb(298K) (require. iiia).

However the essential shortcoming of these crystalline compounds is occurrence side by side with the photochromic reaction caused by ESIPT, the open-shell reactions that can result in the law-yield formation of the stable structures causing the "photo-fatigue" in the cyclic process that efficiency depends strongly on the molecular structure and the medium nature (see Ch. 5 and the reference there).

Thus for the effective utilization of the storage and switching materials on the base of the com-pound III the special investigations are needed to obtain the molecular structures without side Irreversible reactions and with the optimal properties.

A novel ultrafast recording and thermostable molecular photoswitching photochromic structures on the base of the ESIPT reaction has been theoretically designed and characterized recently in terms of the Potential Energy Surfaces (PES) of the ground So and excited S_1^* states /54/.

The molecular structure of 7-hydroxy-(8oxazine-2-one)-quinoline(7HXQ)(IV)(Sch. 6. 6a) is suggested as an example of such compounds.

According to the Sch. 6. 6b, the UV excitation with $\lambda \approx 363$ nm results (via the high polar Enol excited state, $\mu(e)=11D$) in the fast ESIPT adiabatic reaction OH->NH on the S_1^* PES with formation of the Zwitterionic planar NH structure (ZwII°, $\alpha = 0°$) followed by the twisting (0°-->90°) to PES region of the Conical Intersection (CI)(with twisted Zw I structure ($\alpha = 90°$) with ICT, $\mu(e)=12D$.).

Thus unlike the most ESIPT reactions (including anils) in the case of IV there is no almost planar NH structure on the S_1^* PES responsible for the ASS fluorescence and the process has to proceed high rate ($\tau \sim 10^{-10}$s).

The fast generation of the colored polar ($\mu(g)=7.13D$) structure C in the So state proceeds via the funnel(S_1^*->So), instable planar Zw structure(Zw II) and the Ground State Proton Transfer(GSIPT) with the formation of the NH...N bond.

Thus the fast ($\tau \approx 10^{-8}$ s) recording of the optical information can be realized by light with $\lambda=360$nm. The colored Keto-structure (λabs, max= 454nm) is kinetically stable and can exist practically infinitely long time due to the high potential barrier for twisting (0°->90°)($\Delta E > 3$ev) on the pathway C->Eo of the reverse reaction in the So state.

However the reverse barrierless photoreaction under excitation So->$S_1^*(\pi\pi^*)$ in the region of the colored structure absorption band (λmax$\approx$450nm) goes on the $S_1^*(\pi\pi^*)$ PES to the area of the conical intersection(CI) by analogy with the direct photoreaction, and then to the global minimum of the initial (Eo) structure in the So state.

The conformers Eo and C can easily be distinguished not only by their UV-Visible absorption bands ($\Delta\lambda$max= 90nm) but also by their IR vibration spectra with the C=O stretching bands of the C structure.

Therefore the nondestructive reading-out can be done with the IR light in the range 1700-1800cm^{-1} corresponding such vibration bands in the C structure.

Thus according to the discussed above calculated data the molecular structure IV meets the pri-ncipal above mentioned requirements and is very prospective for the application in the devices for the ultrafast optical switching, recording and storage of the optical information at the mo-lecular level.

It is evident, for the realization of the application of the photochromic systems on the base of the structure IV, the detail experimental studies are necessary.

6. 2. Liquid crystalline properties, and photochromism.

The liquid crystals are well known as a dynamical anisotropic fluid material for the electro-optical applications and especially in the Liquid Crystal Display (LCD).

The main purpose of the studies described below is to obtain the novel molecular systems of the Schiff base possessing both the liquid crystalline and photochromic properties.

The anisotropic properties typical for the fluid nematic phase is usually manifested by the elongated fairly molecules having the molecular rigid central core with one or two flexible terminal alkyl or (and) alcoxy groups. The central core in this case has to have the SA skeleton with the ESIPT providing the photochromic properties. The wide groups of the compounds I(1i)(m=1-7, n=1-8) (see Sch. 6. 1) have been studied /55, 56/.

Seventy five percents of about sixty compounds show the photochromism in the crystals which can be observed with the naked eye (λabs, max $\approx$500nm), (T Photochromism)and almost all molecules (excluding the several most short ones)form the nematic liquid crystal phase with the transition temperatures θ=42-95°C.

However the visible photocoloration is observed only in a crystalline solid state phases but not in a liquid crystalline ones. According to

/57/ such a photochromic behavior can be expected in the liquid crystalline phases of the anisotropic nematic gel state in the mixture of any compounds I(1i) with the gelling agent.

Therefore the new liquid crystalline compound I(1i)(m=7, n=1), 4-methyl N(4nheptyloxysalicylidene) aniline (MHSA) which exhibits the visible photocoloration(T) in the solid crystal only has been studied in the mixture with the gelling agent, trans-(1R, 2R)(dodecanoylamino)cyclohexane (DAC). DAC as a gelling agent can provide preservation of the liquid crystalline properties of MHSA and form the three-dimensional network by an intermolecular H-bonding. These gel can be expected to have a great potential as a functional soft material combining the electro-optical and photochromic properties.

However the observed photocoloration (absorption difference at 440-550nm between UV irradiated and nonirradiated samples)falls sharply with the temperature increase within the solid crystal phase (from 20 till 60°C)and becomes zero in the liquid nematic phase jast the same as of LCMHSA.

The fading time does not depend on the DAC concentration (1-5%)in the solid crystal phase (~30 min at 20°C)but becomes markedly longer compared with that of the single liquid crystalline MHSA.

Thus one can hope that the shosen successfully gelling agent will be able to provide the steric inhibitions of the back bleaching reaction (see Ch. 3).

Neverlees the novel photochromic LC materials on the base of the single and gel compounds I(1c) are of interest for realization of the short-time changes (or modification)of the electro-optical parameters with a great advantage for the gel material.

6. 3. Photochromic Langmuir-Blodgett (LB) films.

The possibilities which have been open offered by the LB films can be connected with applications in the microelectronic with exploting their insolating. nonlinear optical, piroelectric, piezoelectric, semiconducted, sensing, and barrier properties (see e. g. /7, 58/.

The Photochromic LB films have to change their molecular structure and optical properties under irradiation by light.

The necessary structural peculiarity of amphiphilic azomethine molecules forming the LB films is usually the very long-length alkil chain connected with the molecular core.

Thus the amphiphilic azomethine molecule contains the longlength alkyl radical (Ph-Alk orNAlk).

In this case one can expect to obtain the T-Photochromic LB film with the o-hydroxyphenyl ring as a molecular core. Realy /59/the amphiphilic SA derivative with the long-length alkyl radical in the aniline moiety (Ia, Sch, 6. 1) forms the mulitylaer LB film that has a highly odered and densely packed structure which display typical for T-ESIPT photochromism photocoloration upon UV irradiation.

The comparatively low rate of the thermal back bleaching reaction is the same order of magnitude as that of crystalline SA ($k\sim 10^{-3}s^{-1}$ at room temperature) provided by the steric hindrances in the dence regular packing of the molecules in the LB film. The 9-monolayer LB film of the anil derivatives molecule with the long-length alkyl radical in the Imine-moiety (N-alkyl, I_6) has been obtained and investigated /60/.

The existence of the two types, Z and Y, of LB structures (Sch. 6. 8) in dependence on surface pressure has been revealed by the π-A isoterm (at 27°C, pH 6. 1) study of monolyer in pure water.

Table 6.4.
Identification of the molecule I_6 tautomeric structure in the LB film (see Scheme 6.7).

Structure	Assignation	Vibrational spectra (cm^{-1}) [4]			UV-VIS spectra (λnm)	
		LB film	Solvent or crystal [1]		LB film	Solv. or Cryst. [1]
		Before	Irradiation			
OH (E)	C=N	1612 IR	1615 IR 1620 IR 1621 Rm	CCl$_4$ Acet.—d$_3$ Cryst.	311	317-320
	Phenyl ryng deformation	1480 IR ~1550	1484 IR	Cryst. Hexen		
Zw NH	H C:::N +	1641 IR	1635 IR 1641 Rm	HFIP HFIP	380 440-600	390
		After	irradiation	[2]	(relative intensity)	
Color (C)	C:::O	1652 IR	~1650 IR		303-311(0.92) 380 (0.76) 440-650 (1.6) [3]	320 400 440-550 [3]

1)The data from the table 2.2(Ch.2) and ref.there.2)J.Lewis,C.Sandorfy,Can.J.Chem.60,1720(1982).3) Manifested clearly after .1h irradiation by 365 nm Hg. 4) IR,Rm –Infrared and Raman spectra respectively.

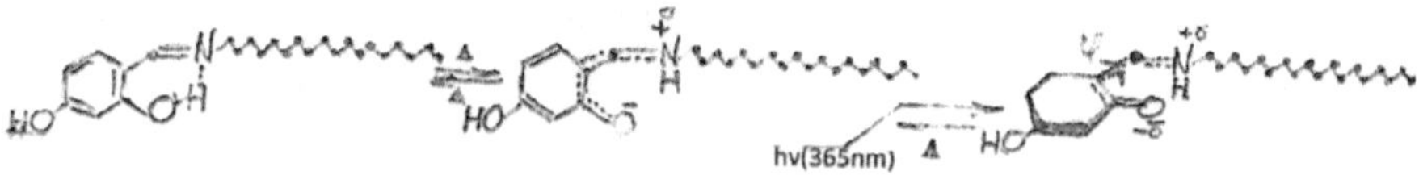

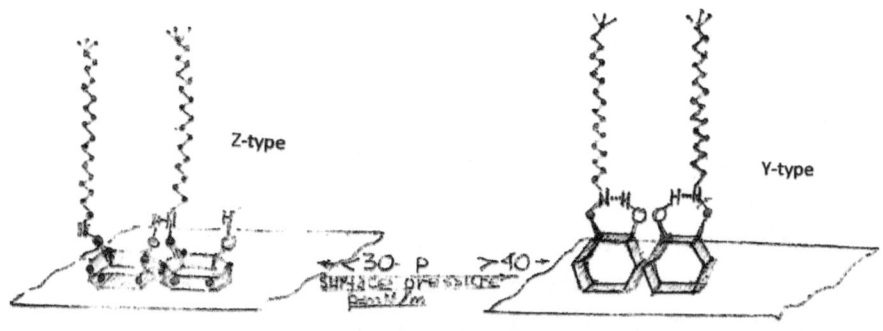

OH(E)　　　　　NH(Zw)　　　　　　　C(twist).
311 nm　　　　380-440nm　　　　　　440—650nm

Scheme 6.7
Compound I_6. Interpretation of the UV-Vis absorption spectra before and after irradiation including photochromism in the LB film(see Table 6.4).

Scheme 6.8.
Structure of the LB film for I_6 in dependence on the surface pressure $p_{mN/m}$. Photochromic Y (p>40) and Nonphotochromic (p<30)types.

At the low surface pressure (p<30mN/m) the aromatic rings lie "flat on" LB surface and intermolecular H-bond (O-H...N=C)takes place obviously (Z-type)between the different I(6) molecules with Enol structure.

At the high surface pressure (p>40mN/m)the "edge on "position with the intramolecular H-bond O-H...N is realized in the Enol structure (Y-type).

Thus the ESIPT followed by the Photochromic transformations can be occur only in the Y-LB type.

The following consideration of the structural aspects of the Photochromic reaction is based on the date discussed above {Ch. 2, 3) which have not been taken into account by the authots /60/.

According to those the equilibrium OH(Enol)⇔NH of the I(6) molecules(Alkilimines I(Ch2)in the Ground state especially in the polar solvents can be shifted strongly towards NH Zwitterion(Zw) structure(Sch. 6. 7).

The realization of such an equilibrium in the Y type LB film is supported by comparison of the spectral data for LB film /60/and those obtained by various authors for N-alk molecules(Tab. 6. 4 and Ch. 3).

Upon irradiation by UV light (e. r. $\lambda=365$nm) the ESIPT occurs followed by the generation of the colored form C after excitation of the photoactive E-structure (but not the NH one)in their Y-type LB film (E-hν->E*--ESIPT->C).

The dense packing of the molecules in LB film is not favorable for the large structural transformations like cis->trans isomerization about C=C bond in the structure K but can allow the twistinfg around that at an angle $\varphi<90°$.

Therefore the impeded formation of the deeply colored form C ($\lambda_{abs} \approx 500=650$nm)(Tab. 6. 4) is another important evidence of its twisted NH(but not trans) zwitterionic structure(see Ch. 3).

According to/60/ the formation of the colored structure has to induce a gradual local expantion of the LB film and partial rearrangement of the I(6) molecules.

In edition the equilibrium E(313nm)$\Leftrightarrow$NHZw(380nm) is shifted markedly towards E structure (the ratio NH/E)falls after irradiation (Tab. 6. 4). As a result of such a shift, the irradiation ($\lambda=365$nm)leads to the formation of the colored structures C (E--hν->C).

Thus the appreciable light induced change of the local optical, electrical, and mechanical parameters of the LB films occurs.

The inhibitions in the rigid and dense LB film provide a very low rate of the dark back bleaching reaction, so that no spectral changes were observed one our after the the film has been irradiated.

Unfortunetely there is no information of the photobleaching reaction and reversibility of the photoreaction.

Thus the application of the LB film on the base of the Photochromic anils are possible after realization of the necessary studies.

6. 4. The possible applications in the analytical, physical, and biological chemistry.

The ample opportunities fer the various applications in the chemistry and the biology can be provided by the sensivity of the spectral and photochromic properties of the molecules with the ESIPT to the environment.

It has been found recently (see Ch. 4 and /61/)that crystalline Photochromic molecule 3, 5 ditetrabutylsubstituted bis hydroxyazomethine (Sch. 6. 9)loses its T-Photochromic properties (photocoloration with the absorption band $\lambda=450-600$nm) in the medium of methanol vapor due to with the sharp decreasing of the room of the phototochromic structural transformations as a of the inclusion of the methanol molecules into the crystalline lattice.

Such crystalline photocromic anils can be utilized in the sensitive material for the detection of the methanol vapors. This phenomenon can be as a stimulus for the investigation of both the another vapors(guest molecules, e. g. CO_2, NH_3) and the crystalline LT-photochromes (host molecules, e. g. DNBP, Ch5, Sec5) for the construction of the of the new vapor sensitive material and devices.

The peculiarities of the localization o f the localization of the lowest electronic transition with the intramolecular charge transfer (ICT) in the specially sintesized of o-hydroxyazomethines containing an enolizable barbituric acid moiety (Sch. 6. 10 (a))can allow to utilize such structures as the novel UV/VIS probes for measuring of the polarity of the environment and the effect of the Hydrogen bonded complex formation.

These applications are provided by the clear spectral shifts and the strong intensity increase of the longwavelength absorption bands in the dependence of vthe solvent polatity and the H-bond complex activity respectively both with molecules of the solvent and the admisture ones.

The solvent spectral shifts of the absorption bands of the compounds 1a and 1b has been investigated in a set of 25 protic and aprotic solvents of the various polarity with use of the simplified Kamlet-Taft analysis to separate the individual solvation effect.

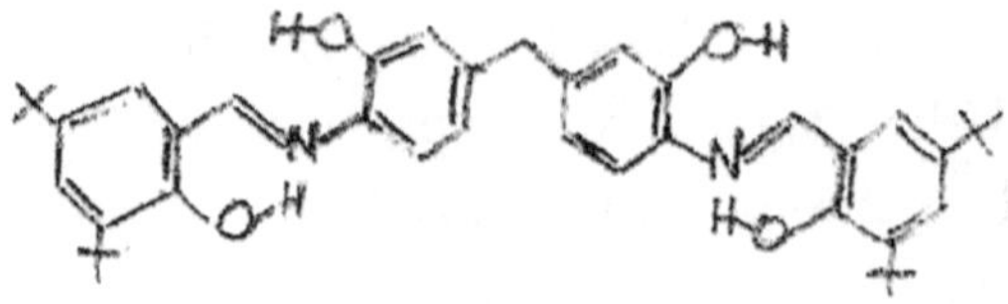

Scheme 6.9
Crystalline Photochromic compound suggested as methanol [(CH$_3$)OH] vapor indicator.

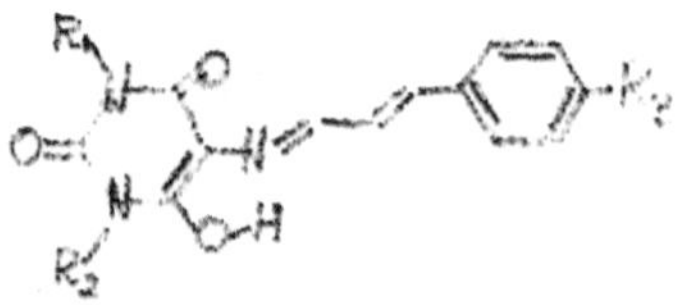

R$_1$=R$_2$=CH$_3$ 1a | 2a
R$_1$=H, R$_2$=n C$_4$H$_9$ 1b | 2b
     R$_3$ NO$_2$ | N(CH$_3$)$_2$

(a)   (b)

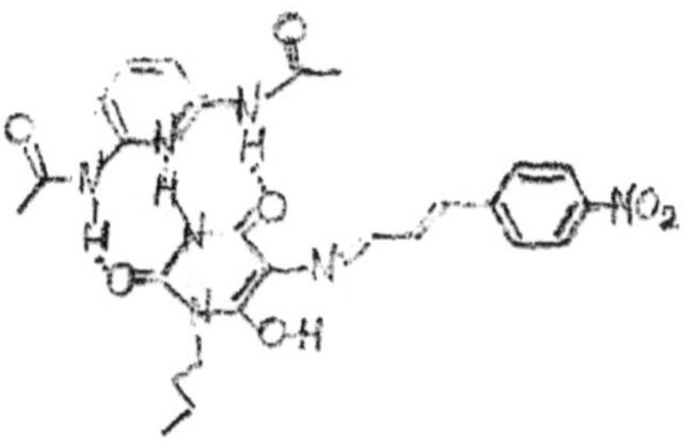

Scheme 6.10.
Compounds that can be suggested as an indicator of the solvent polarity and and the H-bond complex formation in the solvent.

(a) Novel solvatochromic compounds.
(b) Assumed structure of the complex between Schiff base 1b and DAC in DCM

The highest bathochromic and hypsochromic shifts are caused by the increase of the solvent polarity and the hydrogen binding activity respectively.

The maximum of the absorption band spectral shifts are 2160 and 2923 cm^{-1} for 1a and 1b respectively but these values for the compounds 2a and 2b are only about 730 cm^{-1}.

At the same time the intensity of the absorption band of 1b (λ max =463nm) increasing sevenfold with the heightening of the 2, 6-diacetamidopyrine (DAC) concentration from 0. 17 to 3. 34 mmol/L in dichlormethane due to H-bond complexing between 1b and DAC (Sch. 6. 10(1b)).

Thus according /62/only compound 1b unlike other studied molecules can be suggested as a sensitve indicator for both the solvent polarity and the H-bond complex effects at the same time.

At last it should be mentioned the ESIPT photochromes' applications for the Bioactive control.

Among such photochromic molecules the DNBP derivatives (see Ch. 5, Sec. 5. 5) are probably one of the most exploted phototrigger units in so named caged compounds for the controlled release of the bioactive molecules and the fluorescent dyes, and their use in the studies of the bioelecvtronic imaging and Patter Recognijtion /63/.

6. 5. Materials for photoprotection.

There are some areas of the human activity for which the problem of the protection from the light (including sunlight) of the both UV and Vis spectral ranges are topical.

For instance:((a) the UV light (including solar one)can damage the organic material (including photooxidation of polymers)with $\lambda \approx 290$-$320nm$(UV-B region); (b)overexpose of the UV light in the ranges of $\lambda \approx 320$-$400nm$(UV-A region)and also of UV-B region who causes skin cancer and can be responsible for a suppression of the human immune system, (c)Pulse light of the very high intensity of Lasers and explosion (including nuclear that) in the UV and Visible ranges which can cause the damage of the eye-sight.

In the case (a)for the protection against UV-A solar light the molecular screens of polymers are used.

The main requirements for the screen molecules are

(i). the high molecular extinction in the spectral range.

(ii). the strong overlapping of the UV absorption spectra with that ($\lambda \approx 300\text{-}400$ nm) of the solar actinic flux at the earth's surface (at 20° Zenith angle)-UV solar spectral function($f(\lambda)$).

(iii). the most possible photostability to the UV-irradiation.

As a result of the systematic studies /64/ of the organic compounds the molecules of salicylidene series (Sch. 6. 11) have been found to meet the requirements (i)-(iii) completely (Tab. 6. 5).

Really the high intensities ($\varepsilon = 1300\text{-}27000$ Lxmol^{-1}x cm^{-1}) and the spectral variability (λ^{max}abs$=340\text{-}380$) of the long-wavelength absorption bands and also the overlapping of the absorption bands in the region $\lambda \approx 300\text{-}340$ nm with $\varepsilon > 11000$ Lxmol^{-1}cm^{-1} are provided by

ICT nature of the lowest electronic transitions and regulated by including of the substituents/65/.

Table 6.5.
The spectral and photo-protection characteristics of SAs(see Sch.6.14 and text for designations)

No compound	λ^{max}_{abs} nm	$\varepsilon_{max} \times 10^{-4}$ (LxM^{-1}xcm^{-1})	SSPF IAS X10^5	SSPF (UV-A)	SSPF (UV-B)	SPF
1	338	1.32	54.1	47.7	6.3	57.0
2	341	1.50	64.9	58.2	6.7	68.4
3	349	1.99	96.5	89.8	6.7	101.7
4	384	2.72	167.7	163.4	4.3	176.8
5	351	1.57	92.2	87.4	4.9	97.3
6	344	1.82	131.6	119.0	12.6	138.8
Homosalate	----	---	4.02	1.67	2.35	4.24

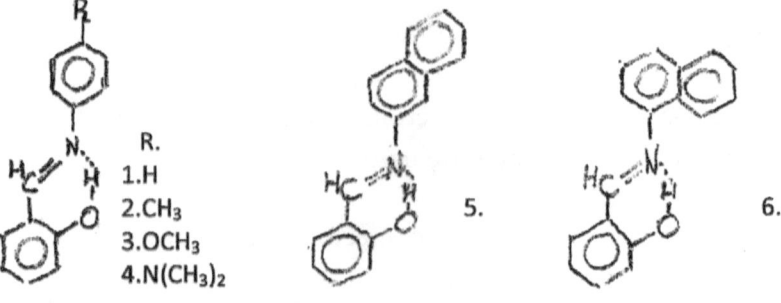

Scheme 6.11
Salicylidenes (SAs) compounds for the UV-radiation molecular screens.

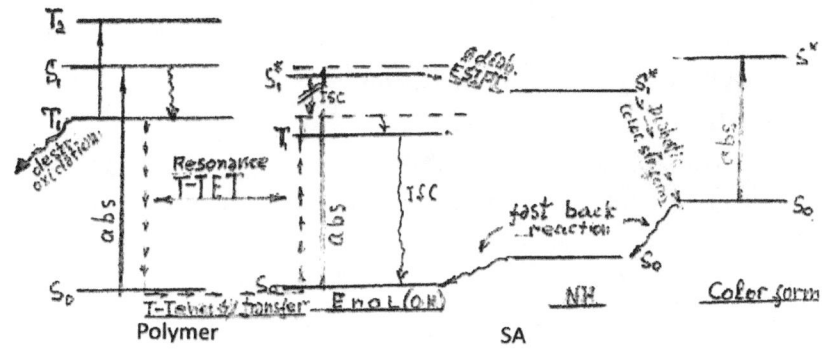

Scheme 6.12
SA for polymer photoprotection by T-T radiationless energy transfer (qualitative scheme).

According to /64/ the integral areas of Action Spectra (IAS) for every compound: IAS (comp) $\approx \int \varepsilon(\lambda) f(\lambda) d\lambda$ where $f(\lambda)$ is the UV solar spectral function simulated by means of polynomial $f(\lambda) \approx \sum a_n \lambda^n$ (usually n=1-4).

The Spectroscopic Scale Protection Factors /64/, (SSPF) total=10xIAS, (SSPF)A, (SSPF)B for the total, UV-A, UV-B spectral ranges respectively (see Tab. 6. 5).

The (SSPF) values meet the requirements (i), (ii) and exceed the (SSPF) values for the standard compound (homosalate) by one-two order of magnitude (Tab. 6. 3).

The third requirement is provided by the very high photostability typical for the compounds of SA series caused by the ultrafast processes of the electronic deactivation and the ultrafast reversible photoinduced structural transformations involving the ESIPT followed by the adiabatic processes, and the ground state structural changes with the back GSIPT reaction which suppress completely the possible competing irreversible reactions (see Ch. 3).

Thus the salicylidene derivatives can be used as the effective solar UV molecular screens in the polymeric films or UV-transparent solid polymeric matrices.

The polymer material interacting with UV-light can be undergone photodestruction or (and) photooxidation.

It has been shown theoretically and experimentally that the short lifetime and (or) low population of the triplet state Tst of the molecule-stabilizer are the essential condition for its use as a polymer photostabilizer.

Both these conditions are fulfilled for the molecular systems in which the nonradiative conversion processes are initiated by adiabatic structural transformations/66/.

With energy Tst less than that of Tpol of polymer (Tst<Tpol) such molecules can be a perfect triplet energy acceptor in the radiationless Tpol->Tst energy transfer.

Since ptotodestruction and especially photooxydation proceed mainly via the triplet state (Tpol), the deactivation of the latter by the way Tpol->Tst has to provide a photostabilization of polymer.

The molecules of the photochromic azomethines completely meet these conditions and can be perfect photostabilazers in accordance with the scheme 6. 12 (see also Ch. 3).

Thus photochromic anils dispersed into polymer can be both the UV molecular screens and photostabilizer.

On the other hand there are various urgent and very complex problems connected with the utilization of the photohromes in the eye protection from the extremely intensive visible shorttime repeated impulses(Lasers)and the separate flashes (i. e. explosions including nuclear ones).

The organic photochromes dispersed in the polymeric material can serve as positive optical filters having at the wavelength of the laser irradiation (λ_i) a high transmittance under the low Intensity (or in the dark)and and an extremely low transmittance for a high incident light intensity.

For the lasers in the visible range it becomes possible when the intensity of λi falls in the region of overlapping of the initial absorption band and the colored form one, and the energy of the laser light flow is limited by the latter.

Such intensity limiters have to meet also some demands: a sufficiently low limiting freshold corresponding to the incident light energy determining a sharp transmittance decrease, a fast response (e.

g. ps or faster), a broad bands as a response in visible range, and the large ratio of the transmittances at the low and highest possible incident ($>10^4$).

According to /67/ the photochromic compounds can only been used as optical limiters for lasers if the time constant for photobleaching in photochromes is slow compared with the pulse duration. However the most of them show the fast back thermal (μs) and photo(ps) bleaching reactions under condition used.

The kinetic studies conducted with mercury ditizonate complex(Sec5. 2. 2. 4) in the liquid solution show that the photochromic molecules can't be used under such conditions (in 10^{-9} s time scale/67/ as a optical limiter.

Thus for the search of the prospective optical limiters among the effective photochromes the study not only the new structures but also various media (including the solid material) is needed.

The photochromic structures similar anils cannot be used for the eye protection from laser irradiation because of the absence of overlapping of the initial and the colored forms' absorption bands.

At the same time the comparative low sensibility, the very fast response (~1. 5ps) and bleaching reaction(~ms)(see Ch. 3 and /68/ make such systems very prospective for the eye protection from the light of the extremely high intensity in the visible and UV range as flashes of the explosions including the nuclear one.

Such photochromic materials must response instantly and recover fast repeatedly.

6. 6. The prospective Photochromic materials.

6. 6. 1. Photochromic polymers (intrinsic photochromism).

The polymers that display their intrinsic photochromism (Photochromic polymers) can be excellent functional optical materials for the various devices starting from the glasses and ending by the electronic optical systems/69/.

It is evident, the copolymer with the photochromic structures in the lateral chain should probably be Photochromic. By the analogy to the photochromic polypeptids containing photoisomerisable azoaromatic chromophores the new polymer photochrome on the base of SA(IV, Sch.. 6. 13)has been proposed (see /70/)and refs. there). Unfortunately such a polymer has not been synthesized and studied up today.

Two examples of another types(V, Sch. 6. 14)of the polymer-like photochromic structures on the base of the ESIPT have been suggested in /71/.

The observed charge transfer (conductivity) in both cases is provided by the regular alteration of the single and double bonds both in the left and right hands of the polymer-like molecule.

Thus the switching from the "on" state to the "off" one may be controlled by the light and provide a modulation of conductivity.

The attempt to obtain the polymers combining both the electrochromic and photochromic properties has been undertaken/ 72/.

The electroactive polypyrroles and poly(2, 5-dithienyl pyrroles) have been obtained by polymerization of the monomers VI, VII (Sch. 6. 15) respectively bearing fully conjugated pendant salicylidenaniline group. Only monomers VIa and VIIb are photochromic in the crystal powder.

All polymer films are found to be electrochromic but no phtotochromic response with the naked eye is observed for the polymer in the thin film form.

Most probably this effect is bound up with high rate of the bleaching reaction due to unsufficient microrigiity of the polymeric matrix.

Thus for the realization of the dual photochromic-eletrochromic polymeric material based on the salicylideneaniline photochromes the elaboration of the polymers with the heightened visosity is needed.

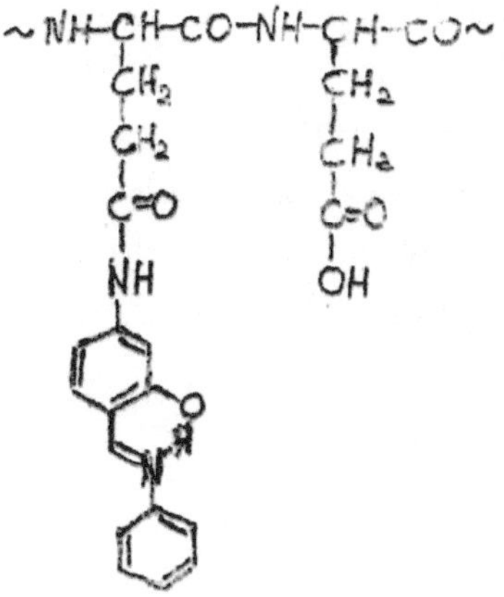

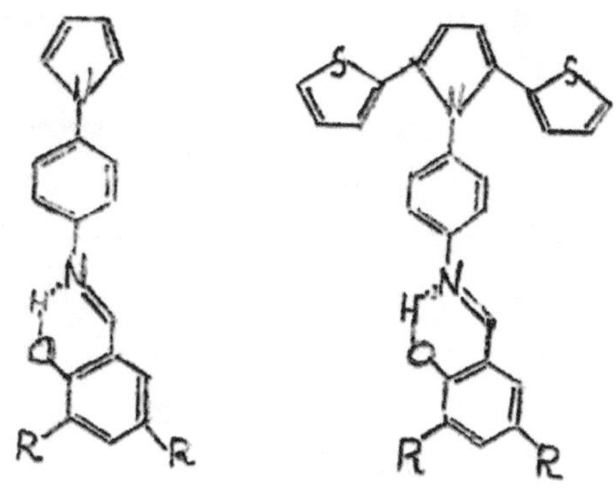

IV
Scheme 6.13
Proposed photochromic polymer on the base of SA

(a) R=H;
(b) R=tert-butyl

VI   **VII**
Scheme 6.15
Photohromic monomers with the fully conjugated pendant SA group.

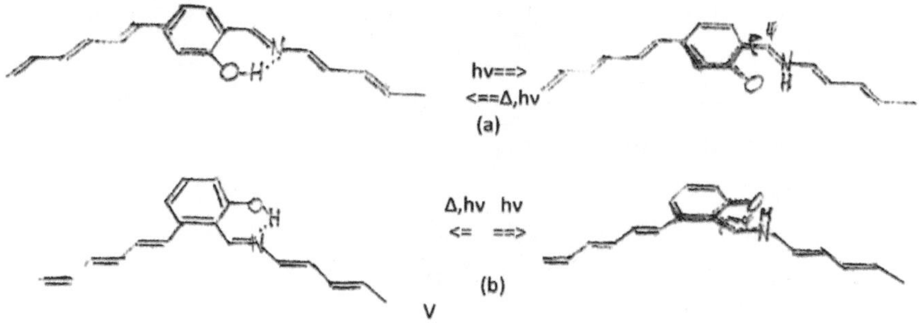

V
Scheme 6.14
Polymeric type switchers based on the Photochromic azomethines.

6. 6. 2 Possible amorphous molecular materials.

As it has been shown(see rew. /73/and ref. there) the amorphous glasses are exceptional candidates for the materials with various applications because of their good processability, transparency, and homogenous properties which are caused by the separated, weakly interacted molecules, unlike crystals.

In term of the photochromic molecules with the ESIPT (especially anils) it means the availability of the photochromism for any solid independently of the crystal structure(see Ch. 3-5).

Thus the amorphous molecular glasses and films based on the ESIPT photochromism provide whith the ample opportunity for the creation of the new Photochromic Amorphous Molecular Material (PAMM)with various applications.

There are several important guidelines for the molecular design of PAMM /73/.

(1) Molecules should posses nonplanar structure.
(2) The existence of the different nonplanar conformers (rotamers) is required.
(3) The incorporation of the bulky and heavy substituents and the enlargement of the molecular size provide the stability of the amorphous glasses and rise of the glass-transition temperature (Tag).
(4) The latter can be increased also by the introduction of the structurally rigid moieties.

According to /73, 74/ many of the compounds readily form amorphous glasses from polycrystals obtained by recrystalization from solutions.

The glasses have been obtained when the melt samples are cooled on standing in air or rapidly cooled with the liquid nitrogen.

Usually the amorphic glasses are characterized by well defined temperature Tg at which the amorphic unequilibrium state begins to decay due to a motion of the molecular groups resulting in the change in the gravity center of molecules and being accompanied by a specific heat change /73/.

The formation of the amorphous glassy state can be confirmed by polarizing microscopy, X-ray diffraction, differential scanning calorimetry, and Raman spectroscopy /74/.

The molecular structures satisfying the requirements (1)-(4) on the base of azobenzene and dithienyl chromophores which have been studied earlier extensively and known as fatigue–resistant thermally stable photochromic compounds have synthesized.

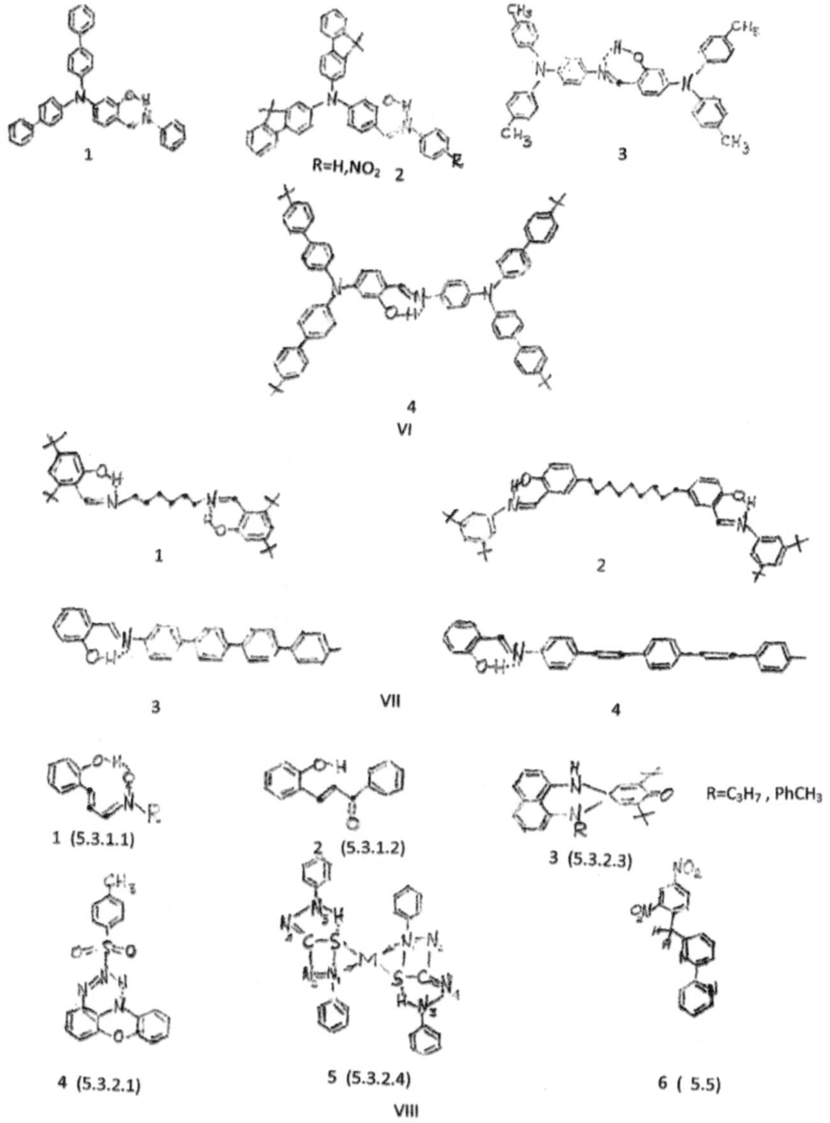

Scheme 6.16
The structures proposed as the photochromic amorphous molecular materials. The corresponding section of the Ch. 6 in the parentheses.

With use of the cooling methods the amorphic glasses have been produced and identified by above menthioned spectral-structural methods. All glasses show the photochromic properties typical for the corresponding benzene and dithienil molecular systems the new PAMM have been created with Ts from~60°C till~180°C.

Unfortunately for the best of our knowledge no any attempts to obtain PAMM on the base of The ESIPT photochromes have been undertaken in spite of simplicity of their synthesis and evident advantages in their photochromic properties (fast colored form generation and unprecentented fatigue resistance, and variety of the spectral-kinetic characteristics).

Therefore it seems to be advisable to propose the photochromic molecular structures which can meet the requirement (1)-(4).

The most prospective azomethine molecular structures VI(1-4)(Sch. 6. 16) differ formally from the corresponding azastructures produced in /73/ by the replacement of the N=N group by C=N one and by insertion of the o-OH group in the phenyl ring.

The bisazomethine structures VII(1, 2) and polyphenyl azomethine Photochromic structures VII(3, 4)(see Ch. 4 and /75/) meet main requirements (1)-(4) and also may be prospective candidates for the PAMM.

At last the various photocromic nonplanar molecules VIII(1-6) with the expanded π-electron Systems and the different mechanisms of the proton transfer discussed in the ch. 5(see 5. 2. 2. 1.,

5. 2. 1. 2., 5. 2. 2. 1., 5. 2. 2. 3., 5. 2. 2. 4., and 5. 4 for the compounds 1, 2, 3, 4, 5, 6 correspondingly) meet the main conditions (1)-(4) and can be pretended to the utilization for PAMM.

One be believed no significant difference between the glass transmission temperature (Tg) of the azomethine amorphic VI(1-4) and corresponding azastructures from /73/.

However one be expected the significant fall of the colored form lifetime for the molecule VI, VII that has to depend strongly on the microviscosity of the amorphic films. Meanwhile the discoloration of the amhorphic glass of the compounds VIII must be much slow process.

Thus the PAMM on the base of the ESIPT reaction can be distinguished by the great variations of the spectral and kinetic parameters which can provide with the information about the microstructure of the amorphic glasses. In connection with the structural similarity of the azomethine and azabenzene molecules the phenomenon of the Surface Relief Grating (SRG)formation deserves the brief discussion (see /73/). Formation of SRG was observed for azobenzene functionalized polymers.

Such a phenomenon arises interest from both the scientific and applied viewpoints especially for the for the erasable and rewritable holographic memory.

At the same time when amorphous films of azabenzen-based PAMM, e. g. the structural analogs of V(1) and VI(2)(R=H) were irradiated with the two linearly polarized 488nmAr+ laser beams, SRG formation also took place.

The SRG is restored after heating above Tg followed by irradiation with two writing bims at room temperature. A marked effect of the molecular structure on the SRG formation was observed for PAMM.

The analogous phenomenons are expected also for PAMM based on corresponding photochromic anils, however with the peculiaries caused by their photochromic properties.

By analogy with photochromic diarylethens(/73/)and refs. there) the bistable molecular memory with a nondestructive readout method can perhaps be realized also for PAMM based on photochromic anils.

Thus the production and the detail investigations of PAMM based on phochromic anils open the longterm prospective

6. 6. 3. Possible perspective nanoparticles.

6. 6. 3. 1. Information of the probable semi-manufactures for anil nanomaterials.

A new SALEN-based organogelator (SBO) with N, N'bis(salicylidene)ethylenediamine(SALEN) has been synthesized (Sch. 6. 17a) and studied by the Electron microscopic, Small-angle

X-ray diffraction, absorption and fluorescent spectroscopy methods with the semi-empirical quantumchemical calculations (AM-1) /76/.

The effective gelatinize organic solvents including cyclohexane, toluene, benzene, and some mixed solvents have deen used. The SBO moleculr in gel phase were self-assembled into 1-D nanotube of 25-100nm in width which further crosslinked to form 3-D networks with molecules packed into the lamellar structure with period of 3. 01nm.

The thermochromic, fluorescent, and photochromic properties are caused by the GSIPTand the ESIPT, i. e the OH ⇔ NH tautomeric equilibrium shifts in the Ground and Excited states in SALEN moiety of SBO with the GEL phase formation, temperature change, and irradiation. The principal findings /76/are summarized and produced in the Sch. 6. 17b and the Tab. 6. 6.

The thermochromism connected with the reversible change of the absorption spectra (appearance of the band with the $\lambda^{max} \approx 480nm$)with the temperature increase is caused by the two reasons(Sch. 6. 17b)Tab. 6. 6) as a result of:

(i). aggregation induced (AI)thermochromism (Sol(333K)≈Gel(293K)), i. e. formation of the gel phase with the strongly stabilized NH structure (as well as with the increase of concentration),

(ii). the OH⇔NH equilibrium shift towards the NH structure in the gel phase (gel(77K)≈gel(293K).

The monomer weak fluorescence ($\lambda^{max} 462_{nm,}$ $\lambda^{exc} 335_{nm}$)in solvent is caused by the formation of the fluorescent (NH)*state as a result of the ESIPT and quenched strongly by the generation of the shortlived colored NHtw (twisted or trans-keto)structure diabaticly or via S1*excited state.

Such a structure is very low stable and can be observed only by the method of the transient absorption spectroscopy (see Ch. 3).

At the same time the formation of the J-aggregates in the gel phase leads to the strong stabilization of the fluorescent (NH)*structure(Sch. 6. 17b)and restrict the twist about C:::C bond.

As a result the intensive Aggregation Indused ESIPT(OH*->NH*) Enhanced (AIEE)emission ($\lambda^{max} \approx 523$nm) can erise with excitation of 337nm (OH structure) or 480nm (NH structure with the quantum yield is approximatlyely 600 times more than that of the solution.

The gel state of SBO exhibits also a distinct so called negative photochromism caused by the photobleaching under irradiation in the spectral range of $\lambda = 365$nm. Such a phenomenon can be connected with the excitation in the intensive absorption band S0->S2* of the colored structure followed by the S2*->S1* conversion with the population of the S0 (OH) structure via the S1*->S0 emission and radiationless transitions (Sch. 6. 17b).

Thus the both color and emission switches could be achived by sol<->gel phase transformation.

Such a functional gel material with the effective fluorescence and photochromic properties can be prospective for the fabrication of the fluorescent photochromic nanoparticles (6. 7. 3. 2).

The attempt to obtain the colloid solution of BSP (Tab. 6. 8) with the particles of less than 100nm in size has been undertaken on the base of the mesoporous molecular sieves (MSM-41) /77/.

Unfortunately this attempt was not fully successful since only part of particles with the needed size formed the colloid solution that was used for the study of the absorption and fluorescent spectra.

The stabilization of the NH(keto) structure and the shift of the OH⇔NH tautomeric equilibrium in the BSP structure towards strongly fluorescent NH form with the participation of the intermolecular hydrogen bonds with the silanol group in mesoporous material lead in the beginning to the appearance of the absorption and intensive fluorescence (AIEE phenomenon) bands with $\lambda^{max} = 450$nm and 570nm correspondingly.

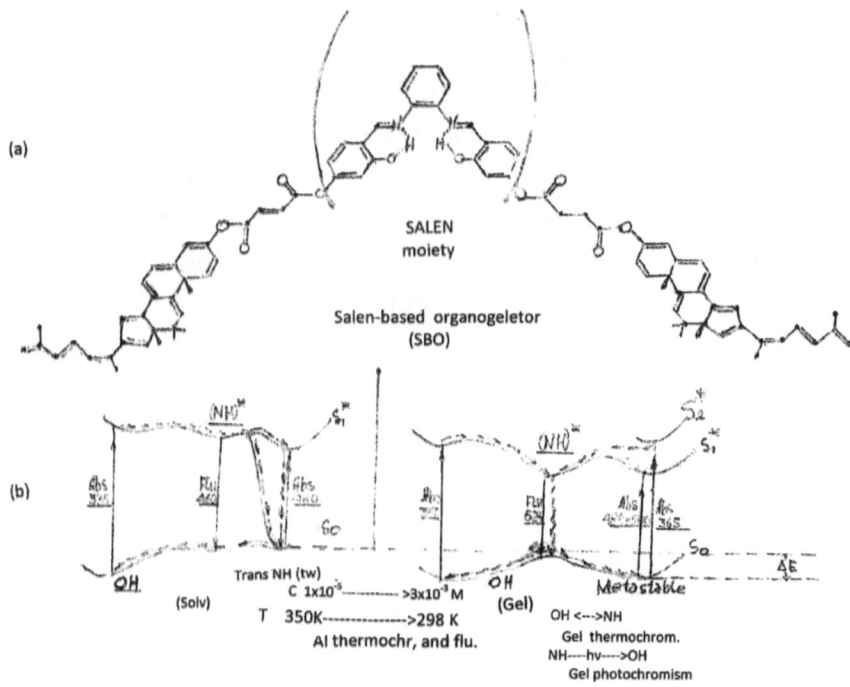

Scheme 6.17

(a) Molecular structure of Salen-based organogeletor.(b) Structure-energetic schemes of Thermo-and Photochromic processes in SBO

Table 6.6
Fluorescence ,Thermochromic and Photochromic characteristic of SBO in the cyclohexane (CH) solution.

State	Absorption,$\lambda^{max}_{abs,nm}$	Fluorescence,$\lambda^{max}_{flu,nm}$	Thermochr.$\lambda^{max}_{abs,nm}$	Photochr.$\lambda^{max}_{abs,nm}$
CH	337	462	337	---------
Gel	335,480	Δ ↑↓ 1) 462⇔523 2)	Δ ↓↑ 1) 335 ⇔ 480 2)	480----hv----->335 335------Δ------->480

1)Al and Al flu thermochromism. 2)Gel and Gel flu thermochromism.

After several hours, BSP in the colloid MCM-41 undergoes hydrolysis process(likeSA in the protic solvent with addition of water) with shift of absorption and fluorescent bands to the short wavelengths.

In the crystal powder of BSP encapsulated in mesoporous sieves the hydrolysis process is not observed. At the same time after irradiation with UV light (λ<400nm) persisting color changes occur with the intensive longwavelength bands (λ≈550-650nm) in the diffuse reflectance spectra and the very long lifetime of the colored form (twisted or planar trans-keto structures) without photobleaching.

The final colored form (with absorption bands $\lambda \approx 550\text{-}650nm$) is generated via precursor state (the absorption band with $\lambda^{max} \approx 450nm$) with the almost planar cis NH structure.

Thus in spite of significant differences in the methods of preparation and microstructure of the media involving the active molecules the active molecules, there are the common peculiarities – the shift of the tautomeric equilibrium towards the NH structures, strong enhancement of the fluorescence efficiency (AIEE phenomenon), and display of the photochromic properties which can be typical also for the generated nanostructures securing their applications.

Realy/78/the AIEEphenomenons is clear manifested in the nanoparticles of BSP molecules that have been prepared with the repricipation method in water.

The fluorescence intencity is increased beyond 60 times in the nanoparticles compared to that in solution and remanes high with the spontaneous change of the nanoparticles' shape from the spherical to rod-like one and to the belt-like aggregates.

A mechanism of the AIEE phenomenon can be connected with the stabilization of the fluorescent NH structure with the restriction of the structural transformations caused by the intermolecular, mainly sterical, interactions in the matrix forming nanostructures.

The results of the studies of the optical and electrochemical properties of the anil ligands with the gold nanoparticles have been reported recenrly /88/. The modification of the optical and fluorescent properties depends on the isomer structures and their attachment orientation with respect to the metal nanoparticle.

At the same time as it has been shown earlier(see ch. 3) the photochromism typical for anils is observed also for the complex compounds of the metals with anil as a result of the photoinduced shift of the E⇔K equilibrium in the ligands towards the K-structure, and the corresponding changes of the chelate knot structure.

Thus one can expect to obtain the photochromic nanomaterials on the base of the metal nanoparticles covered by the photochromic anil molecules.

Unfortunately the information of the Photochromic properties for the nanoparticles with BSP or anils is not reported.

6. 6. 3. 2 Photoswitchers on the base of the fluorescent photochromic nanoparticle.

The application of the photochromic nanomaterials is especially prospective as photoswitchable fluorescent nanoparticles both with (see also /79/)and without utilization of the energy transfer modification.

The general brief information including the structure, the methods of preparation of the fluorescent nanoparticles on the base of the organic (or inorganic) luminophores and the organic photochromes, the principle mechanisms of the formation and the switching of their emission color, and also their application in the special materials and devices has been taken from the data of /80/, the recent studies / 81, 85-87/and with taking into account the findings of /82-84/(Sch. 6. 18).

The various organic luminophores –Rodamine, Perilene, Diarilmethanes, Indoline spiropiranes, Nitrobezoxadiazolil, and organic photochromes –Indoline Spiropyranes, Spirooxazines, Diheteroarylmethane have been used in the material for photoswitchable fluorescent nanoparticles.

To the best of our knowledge the compounds with the ESIPT photochromism and fluorescence did not utilized up to date in such materials in spite of their advantages in many cases.

However the unpublished communication about possibility of such applications has been appeared only recently(private communication).

Therefore we pay some attention for this problem.

There are the several main methods of the preparation of the Photoswichable fluorescent nanoparticles:

i. The facile and quick reprecipitation method were the nanoparticles are prepared by injection of a the small amount of

concentrated stock solution of the organic compounds dissolved in the "good" solvent into excess the "poor" solvent.
ii. The utilization of the Quantum (Dots) as fluorescent nanoscopic high efficient emitters on the modified surface of the inorganic crystals with the photochromic organic switchers anchorted on the surface metal sites of the QDs.
iii. The Sol-Gel method used to synthesize hybrid organic-inorganic nanoparticles with covalent-incorporated organic phototochromic and fluorescent moieties in the host(commonly silica)polymeric network using hydrolysis and condensation reactions.
iv. The microemulsion polymerization method is used to obtain very small spherical nanophases(microemulsions) with diameters 10-100nm as a result of the polymerization into oil-in-water(o/w) or water-in-oil(w/o) droplets in the present of surfactant.

The functional components(luminophores and photochromes, in particular)can be integrated in such a polymeric nanoparticle using either covalent linkage or nonvalent doping.

v. The self-assembly method is used to construct highly organized self assembled structures from amphiphilic molecules involving functional (luminophore or photochrome)groups.

The most widespread methods are the varieties and modifications of the microemulsion polymerisation strategy allowing to prepare the spherical nanopartical of extremely different size of the very good monodispersity with the polydispersity index <10%.

According to /80/ the processes of the reversible fluorescent photoswitching in the photoswitchable fluorescent nanoparticles can be realized mainly with the mechanism of the four types (Sch. 6. 18).

The mechanisms of the first two types, I and II, occur in the nanoparticles including the photochromic molecules with the nonfluorescent initial structure. In such molecules the photocolored form generated as a result of the excitation of the initial structure, plays role of the Energy acceptor in the process of the Fluorescent Resonance Energy Transfer (FRET) from the Emitter (organic luminophore)-Donor of the Electronic Energy to the Acceptor of the later (photocolored form of the Switcher).

The necessary condition of the FRET is the considerable overlapping between the fluorescence and absorption bands of the Donor(Emitter) and the Acceptor (Switcher's Photocolored form)correspondingly.

Under such a condition the shortwavelength (usually "blue" or "green")emission of the Emitter is quenched by the competing process of the FRET(i. e. Singlet-Singlet resonance energy transfer by Forster).

The photo-(under irradiation in the spectral range of the colored form absorption band)or thermo-decays of the colored form result in the restoration of the shortwavelength emission, and the whole process is reversible.

In the case of the nonemitting photocolored form there is only the single-color fluorescent nanoparticle with the fluorescence intensity modulated by the Switcher Photochrome (mechanism of the Ist type).

In the case of the fluorescent photocolored form (usually longwavelength "red" fluorescence)it becomes the second Emitter, and the "green" (or "blue")fluorescence is changed by the "red" one, and the dual alternating color fluorescence photoreversible "green" ("blue")⇔ "red" mechanism of the type II occurs.

The more simple mechanisms of another two types(III, IV)are realized in the nanoparticles including both the photochrome and the luminophore (i. e. Switcher and Emitter)in the single molecule in which at least one of the forms (initial or photocolored) (type III) or both ones (type IV)are fluorescent.

The generation of the fluorescent (usually "red")colored form and reverse photo (or dark) process lead to the realization of the direct single color fluorescence on/off (type III) or dualternanting color fluorescence (type IV) mechanism with the nonfluorescent and fluorescent initial form correspondingly.

There is a very major and complicated problem associated with the photoearasing of the fluorescent form (and photoswitching)when exciting its emission.

This problem can be solved only partially by use of the considerably different, low and high, intensity for the excitation of the flurescence and photoswitching correspondingly.

Thus the lowest limit of the ratio (Φfl)/Φ(sw) (Φ(fl and Φ(sw)-the quantum yield of the fluorescence and the photoreaction of switching correspondingly)can be estimated for each of the structure for its display as Emitter in a fluorescent switchable nanoparticle. For instance in the case of II(FRET mechanism) it is concerned to the photochrome structure, and in the cases of III, IV it be concerned to the both ones.

Thus the finding produced above put the general information about the certain requirements to be answered by photochromes as a basis of the photoswitchable fluorescence nanoparticle.

Unfortunately to the best of our knowledge there is no information about nanomaterials based on the photochromic molecules with the ESIPT.

Meanwhile there are some peculiarities of the crystalline anils that can provide practically important advantage over another organic photochromes.

These peculiarities are caused by the ultrafast primary step (ESIPI) of the photoreaction and by a specific twisted structure of the photocolored form (PC) which decay time depends strongly on the ingtermolecular interactions with the adjacent molecules in the crystalline lattice. At the ambient temperature in the bulk crystal the PC is generated with the efficiency 10-30%, and its accumulation starts at ~200ns after beginning of the excitation and is completed at ~ 9µs.

The accumulation of the PC form(unlike thermochromic one) occurs mainly on the surface of the crystal owing to the surface shielding, and depend strongly on the intermolecular interactions that influence the back reaction rate.

Thus the photochromism of the crystalline anils is mainly a superfacial phenomenon. Therefore the efficiency of the photocoloration of the nanocrystalline structures must be a size-depended value which can be more high than bulk one and increased sharply with diminution of the nanoparticle size.

Scheme 6.18
The principal mechanisms underlying generation and switching of the emissions with different colors in the nanoparticles on the basis of the organic photochromes and luminophores. (A) Energetic schemes. (B) Visual pictures.

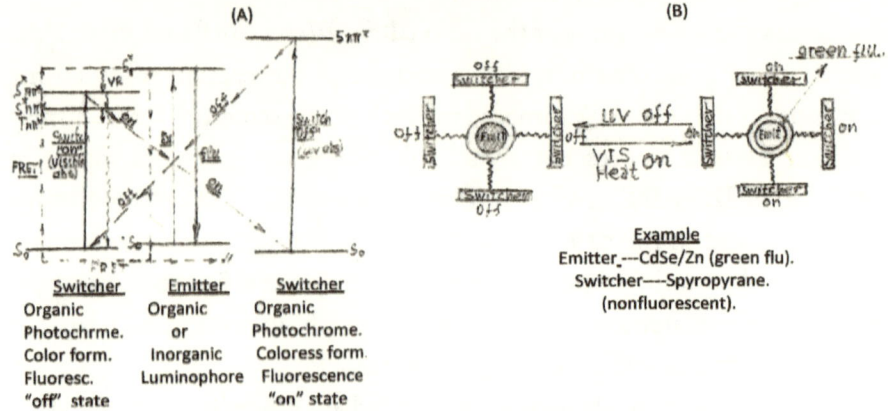

I. FRET single color flu photoreversible on/off mechanism. *)

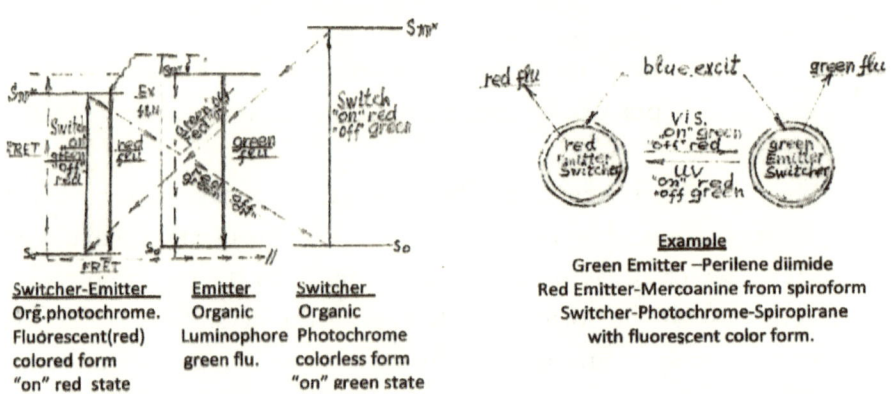

II. FRET dual alternative color flu photoreversible "on"green/"off"red mechanism. *)

*) FRET –Fluorescent Resonance Energy Transfer.

ESIPT Photochromism

Scheme 6.18 (Continuation)

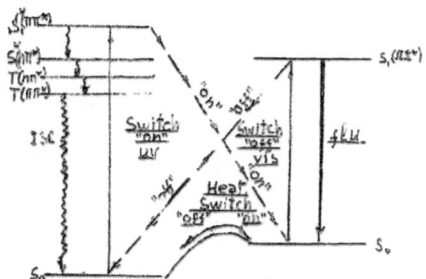

Switcher-Emitter (The same molecule)
Photochrome-Luminophore
Fluoresc."off"   Fluoresc."on"
state   state

Example
Switcher- Spiropyrane (Closed form)
Emitter-Spiropirane (Merocianine form)

III. Direct single color flu on/off reversible mechanism.

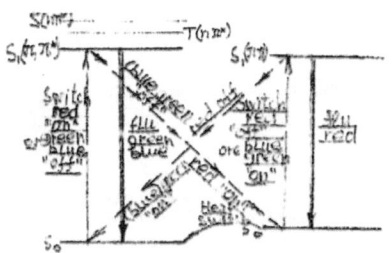
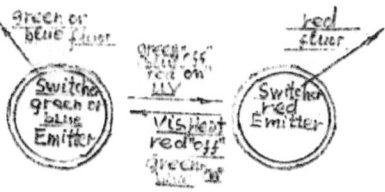

Switcher-Emitter (Two different mol.)
Photochrome-Luminophore
red flu "off" state   red flu "on" state
blue or   blue or
green flu "on" state   green flu "off" state

Example
Emitter green-Spiropirane (closed form)
or blue
Emitter red –Spiropyrane (open form)
Switcher-spiropyrane (closed ⇔ open)

IV. Direct dual alternative color flu reversible mechanism.

Table 6.8
The ESIPT based photochromes and luminophores as candidates for the photoswitchable fluorescent nanopartical (crystal, room temperature).

Luminophore (Emitter)				Photochrome (Switcher)				Dark decay time (room temperat.) τ_{decay} s
Structural formula	Substituents	λ^{max}_{flu} nm	λ^{max}_{ex} nm	Structural formula	Substituents	λ^{max}_{col} nm	λ^{max}_{ex} nm	
1E	R a) CH$_3$ b) OCH$_3$ c) N(CH$_3$)$_2$	530 570 600	340 325 365	1S	R a) H b) Cl c) Br	498 490 500	340 340 340	10700 37037 217400
2E	---	430 560	335 335 450	2S	---	450	335	~450
3E	---	640	460	3S	---	620	370	~700
4E	X a) O b) NH	500 470	320 365	4S	R CH$_3$ C$_2$H$_5$ C$_3$H$_7$ PhCH$_2$	626 626 633 615	415 420 420 420	6430 32100 33430 30370
5E	X a) O b) NH	460 550 580 470	365 365	5S	R H 3 Pyr	590-690 580-620	250 250	695 13200
6E BSP *)	---	570	335 450	6S **)	---	550-650	< 400	Very long life-time

*) Colloid solvent of BSP on the base of the MOM-41 with AIEE phenomenon.(see text,sec.6.7.3.1.). **)Crystal powder of BSP.

Thermochromism unlike photochomism is the typical bulk phenomenon caused by the interactions of the regularly positioned molecules in the crystal lattice (see Ch2).

The surface dislocations are unfavorable for the thermochromism caused by the GSIPT stimulated by the intermolecular dipole-dipole interactions inside of the separate crystalline domains. Meanwhile such dislocations can stipulate the photochromic transformations caused by the ESIPT.

Thus the preparathion of the crystalline nanoparticles can be considered as an effective method for the generation of the novel photochromic crystalline anils.

Such effective organic photochromes can be distinguished from another ones by such advantages as the high rate of the switching and the unique (practically 100%) fatigue resistance.

We suggest some possible double combinations of the ESIPT molecules as candidates for the Photoswitchable fluorescent nanoparticles (see also Ch 5 for LT photochromes).

Usually the Photochromic anils in both the Enol and Colored forms do not display a marked fluorescence.

Therefore such anils can be good candidates for the photo(thermo) switchable nanomaterial on the base of the single color FRET mechanism (I) only (Sch. 6. 18) together with corresponding organic luminophores based on the ESIPT reaction(1E-5E)(Tab. 6. 8) that provides the good overlapping of the ASS fluorescence and colored form absorption bands needed for realization of the effective FRET mechanism (Tab. 6. 8, 1s, 2s).

However the AIEE phenomenon revealed recently (Sec. 6. 7. 2)for BSP(2E)(Tab. 6. 8)can allow to use the direct single color fluorescence mechanism III(Sch. 6. 18, Tab. 6. 8, 6E).

Another examples of the possible candidates for the switchers in the fluorescent nanoparticles on the base of the ESIPT are given also in the Tab. 6. 8. Unlike anils the intensive absorption bands of the colored form of the compounds 3s-5s are diposed in the longer wavelength range and can be effective switchers suppressing the red fluorescence in the ESIPT process.

Such fluorescent switchable nanoparticles are especially useful for application in the "PULSAR" nanoscopy method/80/. The structures on the base of the DNP (3s)(Ch. 5, Sec. 5. 5)deserve a special attention owing to the very efficient p;hotochromic transformation with the intensive absorption band, the comparatively long time of the colored form, and sufficiently high fatigue resistance in the crystalline state.

Thus the example discussed above show the good perspectives in applications of the photochromic nanoscopic materials on the base of the ESIPT and GSIPT reactions.

Brief general discussion and concluson.

There are two main varieties of the problems concerning the studies on the applications of the ESIPT photochromism considered above.

Firstly, the real applications and materials which stand in need of improvement and (or) the expantion of potentialities, and Secondly, the supposed applications based on the analogy with the similar structures and with the organic photochromes of another nature.

The most important real applications are connected with the material of the switching and the

Information storage, especially with the modulation of the Second Harmonic Generation (SHG) as a NLO response of the crystalline anils.

The latter is caused by the unique coincidence of the specific crystal structures and photochromic properties of the crystalline anils but is not connected immediately with use of the changes of the color.

The important advantages of anils in the utilization for the molecular switchers, recording, and rewriting of the optical information are the fast response for the recording signal and the unique fatigue resistance.

At the same time the comparatively low efficiency of the generation along with the short lifetime of the colored form, that do not provide its sufficient accumulation in the ambient conditions, and the requirement of the non-destructing readout of the optical information along with the weak fluorescence of the both (initial and colored) structures impede utilization of anils as switchers and for recording of information, and result in need of the modification of the anil structures and search of another ESIPT structures and media for the novel Photochromic systems on the base of the ESIPT photochromes.

From such a viewpoint the some of the bichromophoric molecules on the base of anils in the crystals and the polymeric matrixes (Ch. 4)

and the structure III (DNBP) (Ch. 5 (XX)) and IV can be considered as the candidates for applications.

However although the first two structures have the very good photochromic parameters they need additional studies to improve they fatigue resistance and IV have been suggested for synthesis only theoretically by the quantumchemical calculation.

The another real applications of anil is connected with the material for the UV-photoprotection.

It is especially important that anils owing to the low position of the triplet state in the initial structure can be the excellent antidestructive additions to polymers. At the same time they play very important role as photoprotectors in the UV area of spectra and their photoprotection ability increases with the appearance of the very intensive So—>S2*transition in the NH-structure, forming with the ESIPT (OH->NH). For the elucidation of the optimal conditions for the utilizing of such unique properties of anils in polymers the detail investigation in each concrete case is necessary. It can be pointed also to the probable application of photochromic anils for protection from the short-time repeated flash light of the extremely high intensity.

In this case for the search of the prospective optical limiters among the effective photochromes the studies of the new structures (e. g. molecules of ditizonate-metal complexes) and the various media (including the solid materials) are needed.

Among such structures the self—protective photochromic polymers can be suggested for the application in various devices. Unfortunately the polymers with the photochromic SA moiety in the lateral chain analogous to azaaromatic ones have not been synthesized. It is the immediate point in this direction in spite of an unsuccessful attempt to obtain the photochromic polymer from the corresponding monomers.

In the region of the application of the materials for the modulation of the nonlinear optical, insolating, and semiconducting properties, the photochromic LB films of the amphophilic salyciliden derivatives are also investigated, and the immediate task of these studies is the modifycation of the of the photochromic properties especially of the dark reverse reactions rates and the efficiency of the photobleaching reactions.

The novel photochromic LC material on the base of the anil molecules with the long chains and those with the jelling agent are of interest for realization of the short-time photochromic changes and modulation of the LC electro-optical parameters with a great advantage for the gel material.

The utilization of the ESIPT photochromic molecules for the investigation of the physical and chemical properties of the solvent, and also for studies of the bioactive molecules and fluorescent dyes are the first steps of the use of the ESIPT photochromism in the physical and biological chemistry.

The very prospective investigations can be conducted in the region of the creation and the study the photochromic amorphic material (PAMM) by analogy with the similar structure of the azabenzene molecule.

In present the most urgent problem to create photochromic nanomaterials on the base of the ESIPT photochromic molecules. However the only initial attempts have been undertook to obtain such nanoparticles and there are no reliable findings about their photochromic properties.

Meanwhile the discussion of the probable applications of the ESIPT photochromes for the photoswitchable and fluorescent nanoparticles has been carried out on the base of the comparative findings of the "working "organic photochromes and luminophores of the different types, and the prospective ESIPT photochromes and luminophores have been proposed as the candidates for that.

For widening of the practical utilization of the ESIPT photochromes it is necessary the effective study of the irreversible Photo and Thermoreactions acompying the photochromic processes, especially in the compounds considering in the Ch 5, to identify the reasons and reducing of "fatigue".

Thus although real applications of the ESIPT photochromes are comparatively limited at present, there are the very wide prospects for their utilization owing to the favorable peculiarities of their properties, and the comparative easiness of the special molecular structure variations.

References

1. Photochromism, Technicues of Chem. VIII, ed. G. Brown, Wiley Interscience, NY (1971).
2. R. Ellam, P. East, A. Kelly, R. Khan, J. Lee, and D. Lindsey, Chem. Ind. 47 (1974).
3. H. M. N. H. Irving, Dithisone, Analitical Science. Monograth Ns, The Chemical Society, London(1977).
4. G. Castro, D. Haarer, M. Macfarlane, and H. Trommdorff, Frequency selective Optical Data Storage System, US Patent 4101976 (1976).
5. H. M. N. H. Irving, CRC Crit. Rev. Anal. Chem. 8, 321 (1980).
6. H. Rau, and E. Liddeke, J. Am. Chem. Soc. 104, 1616(1982).
7. R. Munn, Chem. in Britain. 518(1984).
8. J. Friedrich, and D. Haarer, Angew. Chem., 96, 96(1984).
9. J. Devis, and M. Thomas (San Diego State Univ.) Photochromic material study, Rome Air Development Center, Final Technical Report No Radc-TR-85-1777, Sept. (1985).
10. E. Hadjoudis, Vitoracis, and Moustakali-Mavridis, Tetrahedron, 43, 1345(1987).
11. Persistent spectral hole burning: Science and applications, Ed. W. Moerner, Springer, Berlin (1988).
12. M. Irie, and M. Mohri, J. Org. Chem. 53, 8903(1988).
13. S. Kawamura, T. Tsutsui, S. Saito, Y. Hurao, and K. Kina, J. Am. Chem. Soc. 110, (509)(1988).

14. Optical processing and computing, Ed. A. Arsenatt, Academic Press., NY. (1989).
15. Molecular Electronic Devices, Eds.:F. Carter, Siatcowsky, and H. Woltjen, North Holland (1989).
16. R. Birge, Ann. Rev. Phys, Chem., 41, 683 (1990).
17. E. Hadjoudis, K. Ishumira, U. Wild, and A. Renn, "Photochromism:Molecules and Systems" Studies in Organic Chemistry 40, H. Durr, H. Bouas-Laurent (Eds.), Elsevier, Amsterdam, 685, 903(1990).
18. Supramolecular Photochemistry, Eds. V. Balzani, F. Scandola, and E. Harwood, N. Y. (1991).
19. R. Ao, S. Jahen, L. Kummerl, R. Weiner, and D. Haarer, Jpn. J. Appl. Phys. 31, 693(1992).
20. B. Feringa, W. Tager, and D. De lange, Tetrahedron 49, 5525 (1993).
21. Ultrafast laser pulsus 2nd topics, W. Kaiser(Ed.), in Applied Physics, v. 60, Springer, Berlin (1993).
22. G. Tsivgoulits, and J-H. Lehn, Angew. Chem. Int. Ed. 341119(1995).
23. E. Hadjoudis, Photochrom. and Thermochrom. Anils, Molec. Engineer. 5, 301 (1995).
24. J. Corrie, and D. Trienthan, Methods in Enzymology, Mariott(Ed.), N. J., Ac ad. Press (1988)'
25. M. Irie, Chem. Rev. Special Issue on Photochromism. Memories and Switches100, 1683(2000).
26. Y. Yokoyama, ibid, 1717(2000).
27. A. Benniston, Chem. Soc. Rev. 33, 573 (2000).
28. S. Kawata, and Y. Kawata, Chem. Rev. 100, 1777(2000).
29. T. Norsten, and N. Branda, Adv. Mater. 13, 347(2001).
30. T, Norsten, and N. Branda, J. Am. Chem. Soc. 123, 1784 (2001).
31. M. Irrie, T. Fukaminato, T. Sasaki, N. Tamai, and T. Kawai, Nature 420759 (2002).
32. L. Giordano, T. Govin, M. Irrie, and E. Jares-Erijman, J. Am. Chem. Soc. 124, 7481(2002).

33. D. Tyson, C. Bignozzi, and F. Castellano, J. Am. Chem. Soc. 124, 7481(2002).
34. T. Fukaminato, T. Sasaki, T. Kawai, N. Tamai, and M. Irie, J. Am. Chem. Soc. 126, 14843(2004).
35. S.-J. Lim, B. K. An, S. D. Jung, M.-A. Chang, and S. Park, Angew. Chem., Intern. Ed. 43, 6346(2004).
36. Y. Ferri, M. Scoponi, C. Bignozzi, D. Tyson, F. Castellano, H. Doyle, and G. Redmond, Nanolett. 4, 835 (2004).
37. T. Golovkova, D. Kozlov, and D. Neckers, J. Org Chem. 70, 5545(2005).
38. F. Raymo, and M. Tomasulo, J. Phys. Chem. A109, 7343(2005).
39. M.-Q. Zhu, L. Y. Zhu, J. Hun, W. Wu, J. Hurst, andA. Li, J. Am, Chem. Soc. 128, 4303 (2006).
40. Z. Hu, Q. Zhang, M. Xue, Q. Sheng, and Y-g. Liu, J. Phys. Chem. Solids, 69, 206(2008).
41. H. Sixl, and D. Higelin, in Molecular electronic devices 2, F. Carter(Ed.), Dekker Inc. (1989).
42. R. Andes, and D. Manicovski, Appl. Opt. &, 1179(1968).
43. K. Nakatani, and J. Delaire, Chem. Material, 9, 9682(1968).
44. F. Poineau, K. Nakatani, and Delaire, Mol. Cryst. Liq. Cryst. 344, 29(2000).
45. M. Sliwa, S. Letard, I. Malfant, M. Nierlich, P. Lacroix, T. Asahi, H. Masuhara, P. Yu, and K. Nakatani, Chem. material, 17, 4727 (2005).
46. A. Phanquet, M. Guillaume, B. Champagne, l. Rougier, F. Mancois, V. Rodriguez, J. Pozzo, L. Ducasse, and F. Castet, J. Phys. Chem. C. 112, 5638 ((2008).
47. M. Guilaume, B. Champagne, N. Markova, V. Enchev, and F. Castet, J. Phys. Chem. A. 111, 9914 (2007).
48. M. Sliwa, K. Nakatani, T. Asahi, P. Lacroix, R. Pausu, and H. Masuhara, Chem, Phys. Lett., 437, 212 (2007).
49. M. Sliwa, A. Spabenberg, R. Metiver, S. Letard, K. Nakatani, and P. Yu, Res. Chem. Intemed. 34(2-3), 184(2008).

50. Y. Zhang, C.-Y. Zhao, W.-Fang, and X.-Z. You. Theor. Chem. Acc. 96(2), 129(1997).
51. J. Zhao, B. Zhao, J. Liu, T. Li, W. Xu, J. Chem. Research 416(2000).
52. M. Zgierskii, J. Chem. Phys. 115(18)8351(2001).
53. T. Arai, M. Ikegami, Chem. Letters, 965 (1999).
54. A. Sobolewski, Phys. Chem. Chem. Phys. 10, 1243 (2008).
55. S. Sakagami, T. Koga, and A. Takase, Liquid Crystals, 28, 1199(2001).
56. A. Takase, K. Nonaka, T. Koga, and S. Sakagami, liquid crystals, 29, 605 (2002)
57. K. Miyakawa, I. Ishuzu, T. Koga, S. Sakagami, and A. Takase, Liquid crystals 32, 373 (2005).
58. D. Mc Cullough,, and S. Regen, Chem. Comun. (24)2787 (2004)
59. S. Kawamura, t. Tsutsiu, S. Saito, N. Murao, and K. Kina, J. Am. Chem. Soc. 110509(1988).
60. S. Pang, and Y. Liang, Materials Chem. and Phys. 71, 103 (2001).
61. M. Taneda, h. Koyama, and Kawato, Chem. Letters 36(3)354(2007).
62. I. Bolz, C. May, and S. Spange, New J. Chem. 31, 1568(2007).
63. Methods in Enzimology, V. 291(Ch. "Caged compounds") N. Y. Academic Press, J. Abelson, G. Mariotti, and M. Simon (Eds) (1988).
64. R. Morales, G. Jara, and V. Vargas, J. Photochem. Photobiol. A. Chem. 119, 143(1998)
65. R. Morales, G. Jara, and V. Vargas, Spectroscopy Lett. 34(1)1(2001).
66. V. Plotnikov, and A. Efimov, Russian Chem. Rev. 59, 792(1990).
67. P. Fineyrou, F. Soyer, P. Le Barny, E. Ishow, M. Sliwa, and J. Delaire, Photochem. Photobiology Sci. 2, 195 (2003).

68. M. Sliwa, N. Mouton, C. Ruckebusch, L. Poisson, A. Idrissi, S. Aloise, L. Potier, J. Dubois, O. Poizat, and G. Buntinx, J. Photochem. Photobiol. Sci. 9, 661(2010).
69. G. Kumar, and D. Neckers, Chem. Rev. 89, 1915(1989).
70. E. Hadjoudis, Mol. Eng., 5, 301(1995).
71. T. Inabe, New J. Chem. 15, 129(1991).
72. B. Thompson, K. Abbod, J. Reynolds, K. Nakatani, and P. Andebert, New J. Chem. 29(9)1128(2005).
73. Y. Shrota, J. Mater. Chem. 15, 75(2005).
74. Y. Shirota, J. Master. Chem. 10, 1(2000).
75. M. Knyazhanskiy, and A. Metelitsa, "Photoinduced Processes in molecules of azomethine and their structural analogs" Rostov State Univ. Publ. House, Rostov-on-Don, (1992), (in Russian).
76. P. Chen, R. Lu, P. Xue, T. Xu, G. Chen, and Y. Zhao, Langmuir 25(15)8395(2009).
77. M. Ziolek, and I. Sobszak, J. Incl. Phenom. Macrocycl. Chem. 63, 211(2009).
78. S. Li, L. He, F. Xiong, Y. Li, and G. Yang, J. Phys. Chem. B. 108, 10887(2004).
79. F. Raymo, and M. Tomasulo, Chem. Soc. Rev. 34, 327 (2005).
80. Z. Tian, W. Wu, and A. Li, Phys. Chem. Chem. Phys. 102577(2009).
81. Z. Xu, Q. Zhang, M. Xue, Q. Sheng, and Y-g. Liu, J. Phys. Chem. Sol. 69, 206 (2008).
82. J. Cusido, E. Deniz, and F. Raymo, Eur. JOC N13, 2031(2009).
83. I. Yildiz, E. Deniz, and F. Raymo, Chem. Phys. Rev. 38, 1859(2009).
84. P. Liu, H. Cui, C-X. Wang, and G, Yang, Phys. Chem. Chem. Phys. 12, 3943(2010).
85. F. Benedetto, E. Mele, A. Camposeo, A. Athanassio, R. Cingolau, and D. Pisignano, Adv. Mater. 20(2), 314(2007).

86. J. Chen, F. Zeng, Sh. Wu, Q. Chen, and Z. Tong, Chem. A. Eur. J. 14(16)4851(2008).
87. J. Chen, F. Zeng, Sh. Wu, J. Su, and Z. Tong, Small 5(8)970 (2009).
88. J. Abad, M. Revenga-Parra, T. Garcia, M. Gamero, E. Lorenzo, and E. Pariente, Phys. Chem. Chem. Phys. 13, 5668(2011).

www.ingramcontent.com/pod-product-compliance
Lightning Source LLC
Chambersburg PA
CBHW020718180526
45163CB00001B/26